First-Principles Calculations for Cathode, Electrolyte and Anode Battery Materials

First-Principles Calculations for Cathode, Electrolyte and Anode Battery Materials

Edited by

Thi Dieu Hien Nguyen

Department of Physics, National Cheng Kung University, Tainan, Taiwan

Shih-Yang Lin

Department of Physics, National Chung Cheng University, Chiayi, Taiwan

Hsien-Ching Chung

Super Double Power Technology Co., Ltd., Changhua, Taiwan

Ngoc Thanh Thuy Tran

Hierarchical Green-Energy Materials (Hi-GEM) Research Center, National Cheng Kung University, Tainan, Taiwan

Ming-Fa Lin

Department of Physics/Quantum Topology Center (QTC)/Hi-GEM, National Cheng Kung University, Tainan, Taiwan

IOP Publishing, Bristol, UK

ISBN 978-0-7503-4685-6 (ebook)
ISBN 978-0-7503-4683-2 (print)
ISBN 978-0-7503-4686-3 (myPrint)
ISBN 978-0-7503-4684-9 (mobi)

DOI 10.1088/978-0-7503-4685-6

Version: 20211201

IOP ebooks

British Library Cataloguing-in-Publication Data: A catalogue record for this book is available from the British Library.

Published by IOP Publishing, wholly owned by The Institute of Physics, London

IOP Publishing, Temple Circus, Temple Way, Bristol, BS1 6HG, UK

US Office: IOP Publishing, Inc., 190 North Independence Mall West, Suite 601, Philadelphia, PA 19106, USA

Contents

Acknowledgement

This work was financially supported by the Hierarchical Green-Energy Materials (Hi-GEM) Research Center, from The Featured Areas Research Center Program within the framework of the Higher Education Sprout Project by the Ministry of Education (MOE) and the Ministry of Science and Technology (Grant number MOST 108-2112-M-006-016-MY3) in Taiwan.

Editor biographies

Thi Dieu Hien Nguyen

Thi Dieu Hien Nguyen achieved her PhD in Physics in July 2021 from National Cheng Kung University (NCKU), Taiwan. Currently, she works as a postdoctoral researcher at NCKU. Her academic research focuses on essential properties of 2D materials and anode, cathode, electrolyte battery materials using first-principles calculations.

Shih-Yang Lin

Shih-Yang Lin received his PhD in physics in 2015 from the National Cheng Kung University (NCKU), Taiwan. Since 2015, he has been a postdoctoral researcher at NCKU and NCU. Recently, he works as an exchange scholar in University of Houston. His scientific interests are focused on low-dimensional materials, battery materials, semiconducting material and their derivatives using first-principles calculations.

Hsien-Ching Chung

Hsien-Ching Chung received the PhD degree in physics from National Cheng Kung University, Tainan, Taiwan, in 2011. He is the RD Manager in Super Double Power Technology Co., Ltd, Taiwan. He has complete experience in the lithium-ion battery industry supply chain, from battery cell factory, pack factory to system integration factory. His recent interests focus on system integration and application of Li-ion battery energy storage systems. He also has rich experience in fundamental research. From 2011–2017, as a postdoctoral fellow, his main scientific interests in condensed matter physics include the electronic and optical properties of carbon-related materials, low-dimensional systems, and next-generation energy materials. He is the author of one review article, and more than 30 articles published in journals and conferences. Dr Chung was the Executive Secretary of Environment Safety and Policy Promotion Committee under the Taiwan Battery Association (TBA) (2018–2020), a member of the Institute of Electrical and Electronics Engineers (IEEE), a member of the IEEE Power and Energy Society (IEEE PES), a member of the American Physical Society (APS), and a member of the American Chemical Society (ACS).

Ngoc Thanh Thuy Tran

Ngoc Thanh Thuy Tran obtained her PhD in physics in 2017 from the National Cheng Kung University (NCKU), Taiwan. Afterward, she began to work as a postdoctoral researcher and then an assistant researcher at Hierarchical Green- Energy Materials (Hi-GEM) Research Center, NCKU. Her scientific interest is focused on the fundamental properties of 2D materials and rechargeable battery materials by means of the first-principles calculations.

Ming-Fa Lin

Ming-Fa Lin is a distinguished professor in the Department of Physics, National Cheng Kung University, Taiwan. He received his PhD in physics in 1993 from the National Tsing-Hua University, Taiwan. His main scientific interests focus on essential properties of carbon related materials and low-dimensional systems. He is a member of American Physical Society, American Chemical Society, and the Physical Society of Republic of China (Taiwan).

List of contributors

Hsien-Ching Chung
Super Double Power Technology Co., Ltd, Changhua City, Changhua County, Taiwan

Wen-Dung Hsu
Department of Materials Science and Engineering, National Cheng Kung University, Taiwan

Thi My Duyen Huynh
National Cheng Kung University, Tainan, Taiwan

Phung My Loan Le
Applied Physical Chemistry Laboratory, Ho Chi Minh University of Science (HCMUS), Ho Chi Minh city, Vietnam

Wei-Bang Li
National Cheng Kung University, Tainan, Taiwan

Ming-Fa Lin
National Cheng Kung University, Tainan, Taiwan

Shih-Yang Lin
National Cheng Kung University, Tainan, Taiwan

Hsin-Yi Liu
Center for Micro/Nano Science and Technology, National Cheng Kung University, Tainan, Taiwan

Thi Dieu Hien Nguyen
National Cheng Kung University, Tainan, Taiwan

Thi Han Nguyen
National Cheng Kung University, Tainan, Taiwan

Hai Duong Pham
National Kaohsiung University of Science and Technology, Tainan, Taiwan

Ngoc Thanh Thuy Tran
National Cheng Kung University, Tainan, Taiwan

Ming-Hsiu Tsai
National Cheng Kung University, Tainan, Taiwan

Khuong Dien Vo
National Cheng Kung University, Tainan, Taiwan

Chapter 1

Introduction

Thi Dieu Hien Nguyen, Shih-Yang Lin, Hsien-Ching Chung, Wei-Bang Li, Ngoc Thanh Thuy Tran, Nguyen Thi Han, Hsin-Yi Liu, Hai Duong Pham and Ming-Fa Lin

Energies, which come into existence in any environments through diverse configurations, accompany our everyday lives [1, 2]. They can be revealed in various forms: electromagnetic waves [3], solar light [4, 5], waterpower [3], black coal [3], mineral oils [6], natural gases [7], hydrogen atoms [3], nuclear reactions/fusion [8], biological materials [9, 10], and chemical batteries [11–14]. Furthermore, energies and matter could be efficiently transformed into one another [1]. For example, electric power is supplied for the charging processes in lithium-ion-based batteries, and then chemical energy is released to do work. Such equipment has become necessary products throughout modern society, e.g., electric vehicles/motorcycles/bicycles [15–17], cell phones [18, 19], spaceships [20], wireless studios [18], and power points [18]. High-performance batteries are achieved by using Li-ion and electron transport [18, 19], respectively, inside the cathode/electrolyte/anode materials and outside the copper/aluminum current collectors (figure 1.1). The maximum Li^+ currents, being displayed under wide voltage windows [11, 21], large temperature ranges [22] and optimal characteristic times [11, 12], are strongly required in commercial applications [11, 12, 18]. Apparently, the core materials will play critical roles in battery-created basic sciences [23], integral engineering [20], and mainstream products [15, 17]. Their main features have stirred giant responses in science [24], commercial [18] and industrial societies [18]. The current book is mainly focused on the development of the theoretical framework about diversified quasi-particle properties (charges and spin configurations; [25]), according to delicate Vienna *ab initio* simulation package (VASP) calculations and analyses. Many distinct cathode/electrolyte/anode systems are outstanding candidates in clearly illustrating the diverse phenomena of the geometric [25–27], electronic [25–27],

doi:10.1088/978-0-7503-4685-6ch1

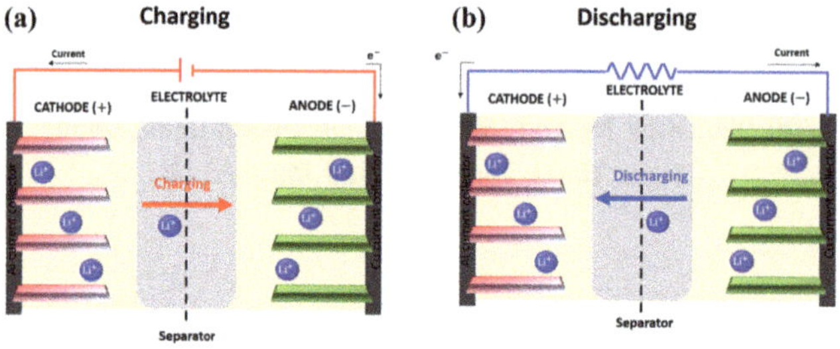

Figure 1.1. (a) Charging and (b) discharging processes of lithium-ion-related batteries.

magnetic [25], and optical properties [28]. Moreover, experimental syntheses [29–31] and examinations [32–34] are discussed in detail. In addition, there are some works on aluminum- [35–37] and iron-ion-based batteries ([38]; table 1.1), in which their diversified phenomena are under present investigations.

The basic science of theoretical research will be very useful in greatly enhancing the performance of various batteries [11–13, 18, 35, 36, 39] and their highly potential applications [11–15, 17–19, 35, 39]. The critical physical/chemical/material mechanisms, which are closely related to the efficiencies of cathodes, electrolytes and anodes, are absent according to the previous prediction works, such as, the significant roles of the highly non-uniform chemical environments under various moiré superlattices [26, 40, 41], multi-/single-orbital hybridizations in chemical bonds, diversified defects/a lot of intermediate states during charging and discharging processes, and steady currents due to ion transports. Theoretical frameworks, being suitable for systematic investigations, should be developed under current studies. From the physical viewpoints, the essential properties could be successfully explored by using first-principles simulations [27] and phenomenological models [42]. However, their strategies need to be strongly modified for most mainstream materials, e.g., a development of the theoretical framework suitable for core materials in ion-based batteries (details in this book; [26, 40, 41]). The former is available in studying the optimal geometries [26, 40, 41], electronic properties [26, 40, 41], magnetic configurations [25], mechanical strain [43], phonon spectra [44–46], optical absorptions [28], and Coulomb excitations [47]. There are some numerical calculations on the lattice symmetries and band structures of cathode/electrolyte/anode materials, especially for ternary lithium oxides [48], where their results might be short of accuracy, or the concise pictures are completely absent. Based on VASP calculations, the different physical and chemical properties, the geometric, electronic, magnetic, and optical ones, will be unified together to determine/identify/clarify the most important quasiparticle properties, the rich orbitals hybridizations and spin configurations. Very interestingly, the frequency-specific absorption peaks and orbital-dependent magnetic moments/spin density distribution, never appearing in the published works [41, 49–52], have been covered in the theoretical development. On the other side, the progress of phenomenological modes hardly grows in the presence of giant numerical barriers. For example, a Hermitian Hamiltonian matrix presents a very large dimension within a moiré superlattice, mainly owing to

Table 1.1. Various anodes, electrolytes, and cathodes in Li^+-based batteries.

	Materials	References
Anodes	TiO_2	[66–69]
	Graphite	[57, 369–373]
	Patterned Si	[70]
	Si film	[71]
	Carbon nanotubes	[61, 62]
	Carbon nanofibers	[374–376]
	Si nanowires	[72, 73]
	Si nanotubes	[20, 74]
	$Li_4Ti_5O_{12}$	[377–380]
	MoO_3	[75]
	SnO_2	[76]
	ZnO	[381]
	Fe_3O_4/carbon foam	[77]
	MnO	[78]
	Co_3O_4	[79]
	GaS_x	[80]
	MoS_2	[81]
Cathodes	V_2O_5	[100–102]
	$LiCoO_2$	[87, 88]
	Nano-$LiCoO_2$	[92]
	$LiMn_2O_4$	[89]
	$Li[Li_{0.20}Mn_{0.54}Ni_{0.13}Co_{0.13}]O_2$	[382]
	$LiNi_{1/3}Mn_{1/3}Co_{1/3}O_2$	[383, 384]
	$LiMn_{1.5}Ni_{0.5}O_4$	[385]
	$0.5Li_2MnO_3{\cdot}0.5LiNi_{0.375}Mn_{0.375}Co_{0.25}O_2$	[386]
	$Li_{1.2}Ni_{0.2}Mn_{0.6}O_2$	[387]
	$FePO_4$	[388]
	LiFe(Co/Ni)PO_4	[90, 91]
Solid-state electrolytes	Garnet ($Li_7La_3Zr_2O_{12}$)	[107]
	Perovskite ($Li_{3x}La_{2/3-x}TiO_3$)	[108]
	Na super-ionic conductor (NASICON)	[105]
	LIthium Super Ionic CONductor (LISICON)	[109]
	($LiMIV_2$)$(PO_4)_3$ (MIV = Ti, Zr, Ge, and Hf)	[389]
	$LiAlO_x$	[390]
	Li_3PO_4	[110]
	Lithium silicate	[391]
	Li (Ta/Nb)O_3	[392, 393]
	Li_3N	[106]
	$LiSiAlO_2$	[394]
	Sulfide (Li_4GeS_4, $Li_{10}GeP_2S_{12}$, Li_2S-P_2S_5)	[111]
	Argyrodite (Li_6PS_5X (X = Cl, Br, I))	[112]
	Anti-perovskite (Li_3OX (X = Cl, Br, I))	[113]
	LiSi/Ge/SnO	[40, 41, 230, 395]

(*Continued*)

Table 1.1. (*Continued*)

	Materials	References
Liquid electrolytes	l-Ethyl-3-methylimidazolium bis-(trifluoromethylsulfonyl)-imide (EMITFSI)	[117]
	1-allyl-3-vinyl imidazolium bis(trifluoromethanesulphonyl) imide ([AVIm][TFSI])	[118]
	N, *N*-diethyl-*N*-methyl-*N*-(2-methoxyethyl)ammonium TFSA (DEME–TFSA)	[119]
	N-Methyl-*N*-butylpyrrolidinium bis (trifluoromethylsufonyl)imide ($PYR_{14}TFSI$)	[115, 120, 121, 125, 129, 396, 397]
	N-methyl-*N*-propylpiperidinium bis (trifluoromethanesulfonyl) imide (PP_{13}–TFSI)	[116, 124, 130, 131]
	Pyrrolidinium nitrate ($PYRNO_3$)	[132]
	1-allyl-3-methylimidazolium bis(trifluoromethanesulfonyl)imide and 1-methyl-3-propylimidazolium bis (trifluoromethanesulfone) imide (AMImTFSI and $Im_{13}TFSI$)	[133]
	N, N-diethyl-N-methyl-N-(2-methoxyethyl)ammonium bis (trifluoromethylsulfonyl)imide (DMMATFSI)	[134]
	1-methyl-1-propylpiperidinium bis(trifluoro-methane sulfonyl)imide ($Pip_{13}TFSI$)	[123, 135, 136]
	methyl−methylcarboxymethyl pyrrolidinium bis-(trifluoromethane-sulfonyl)imide, N-butyl-N-methyl-pyrrolidinium bis(trifluoromethanesulfonyl) imide ($MMMPYRTFSIPYR_{14}TFSI$)	[137]
	1-methoxyethoxymethyl(tri-*n*-butyl)phosphonium bis (trifluoromethanesulfonyl)amide ($MEMBu_3PTFSI$)	[114]
	1-ethyl-3-methylimidazoliumbis(trifluoromethanesulfonyl) imide ($Im_{13}TFSI$)	[117, 127]
[Cation]$^+$ and the [anions]$^-$, $[BF_4]^-$ [122, 128], $[PF_6]^-$ [122, 398], $[TFO]^-$[122], $[TFSI]^-$[114–121, 123–125, 127, 129–137, 396, 397], $[FSI]^-$[117], $[HCOO]^-$ [122], $[MS]^-$ [122], and $[BNZ]^-$[122], $[FeCl_4]^-$ [399], $[NHSO_4]^-$ [400], $[MeSO_3]^-$ [401]		
Gel Polymer electrolytes	Poly (vinylidene fluoride) (PVdF)	[138]
	poly (ethylene oxide) (PEO)	[139]
	poly (vinylidene fluoride) hexafluoropropylene (PVdF-co-HFP)	[140, 147]
	poly(acrylonitrile) (PAN)	[141]
	poly (methyl methacrylate) (PMMA)	[142, 143]
	polystyrene-block-poly methyl methacrylate-block-polystyrene (PS-b-PMMA-b-PS)	[144]

the highly non-uniform chemical environment [25, 26, 40]. As a result, the tight-binding model [42], the static and dynamic Kubo formulas [53], the random-phase approximation [54], and the self-energy method [55], which are frequently utilized to investigate the diverse quasiparticle behaviors [25, 56], are almost absent in the above-mentioned research topics. How to combine two distinct frameworks becomes a new challenge. This might be useful in clarifying certain important issues.

Very interestingly, up-to-date experiments show that anode, electrolyte, and cathode materials of Li^+-based batteries have been successfully synthesized by the different methods, as clearly indicated in table 1.1 [details in 60–177]. To reach high-performance batteries [11, 12], the anode subsystems need to display a large capability of ion intercalation/adsorption [11, 12], high efficiency during charging and discharging processes [11, 12], outstanding cyclability [57, 58], low reactivity against electrolyte [11, 12], fast reaction rate [11, 12], low expense [11, 12], friendly environment [11, 12], and non-toxic operation [11, 12]. Bernal bulk graphite, a layered carbon-honeycomb lattice with well-behaved AB stacking [27], can serve as the high-efficiency anode of Li^+-based batteries and is predominantly utilized in commercial products [18, 23]. Lithium ions are electrochemically intercalated into the available spaces between graphitic sheets during the charging process and de-intercalated under the reverse one. A practical reversible capacity is greater than 360 mAh g^{-1} (about 372 mAh g^{-1} theoretically, [21]) through very high charging/discharging efficiencies [11, 21] . However, graphite intercalation compounds have huge backward factors due to volume expansion. There exist unusual carbon-related materials, such as, carbon nanofibers [59, 60] and coaxial carbon nanotubes [61, 62], where the single-walled graphene tubules are expected to exhibit reversible capacities about 300–600 mAh g^{-1} [63]. In addition to the mainstream condensed-matter systems, $Li_4Ti_5O_{12}$ is well known as a potential anode material for the next generation of Li^+-based batteries [64, 65]. In the current work, $Li_4Ti_5O_{12}$ is mainly focused on rich and unique essential properties with highly non-uniform environments, clearly revealing thermodynamic stability [64, 65], high cycle-life performance and safety [64, 65], compared to other anode candidates. The other materials are also available in anode electrodes, e.g., TiO_2 [66–69], patterned Si [70], Si film [71], Si nanowires [72, 73], Si nanotubes [74], MoO_3 [75], SnO_2 [76], ZnO [76], Fe_3O_4/carbon foam [77], MnO [78], Co_3O_4 [79], GaS_x [80], and MoS_2[81]. Obviously, according to the experimental [61, 82–84] and theoretical viewpoints [51, 85, 86], the significant differences between coaxial carbon and silicon nanotubes/nanoribbons are worthy of systematic examinations.

The most common cathode materials, which have been verified to be practicable in Li^+-driven batteries, cover $LiCoO_2$[87, 88], LiMnO [89], $LiFePO_4$ [90, 91] and lithium layered metal oxides [88], mainly owing to the very high charge capacity. Specifically, such complicated lithium oxide compounds contain the first-row transition metals (e.g., Mn, Fe, Co, and Ni with five kinds of 3d orbitals). The previous transport measurements clearly show that charge capacities of $LiCoO_2$/$LiMn_2O_4$ and variants/$LiFePO_4$ can reach 140/100–120/150–170 mAh g^{-1} [87–92]. The well-behaved cathode materials need to meet the following main requirements: high discharging voltages, large energy storage, long cycle lifetimes, prominent power densities [12, 93], light masses [12, 93], low self-discharging [12, 93] and absence of environmentally hazardous elements [12, 93]. Very interestingly, V_2O_5 is one of the potential materials for cathode electrode, which is successfully synthesized by certain different methods, such as, green route chemical synthesis [94], biological synthesis [95], sol–gel process/technology [96], chemical precipitation reactions [97], ultrasound-assisted technique [98], and thermal decompositions [99]. For the lithium-ion battery, V_2O_5 from the microwave-mediated green route clearly displays the specific capacity of 225 mAh g^{-1} under the specific 1.5–4.0 V voltage range [100–102].

Generally speaking, there exist three kinds of electrolyte materials between two electrodes, namely, the solid-state, liquid and gel structures (table 1.1). For the potential electrolyte candidates, their featured transport properties need to display large ionic conductivity of above 10^{-4} S cm^{-1} at room temperature [103], negligible electronic conductivity with a high ionic transference number [51, 103], and wide electrochemical stability windows [103, 104]. Many types of Li-ion solid electrolytes have been verified to possess the above-mentioned features through long-time experimental tests, such as, NaSiCoN [105], garnet [106, 107], perovskite [108], LiSiCoN [109], LiPON [110], Li_3N [106], sulfide [111], argyrodite [112], anti-perovskite [113] and so on. Among these systems, LISICON- and LiPON-types obviously show low ionic conductivities [~10^{-6}–10^{-5} S cm^{-1}], but are good at electrochemical stability. Li_3N-type exhibits reasonable ionic conductivities [~10^{-4} S cm^{-1}], but narrow electrochemical stability [106]. The sulfide- [111], argyrodite- [112] and anti-perovskite-type [113] solid electrolytes might be unstable under ambient atmosphere although they present high ionic conductivities, ~10^{-5}–10^{-2} S cm^{-1}, ~10^{-7}–10^{-2} S cm^{-1}, and ~10^{-4}–10^{-2} S cm^{-1}, respectively. In short, NASICON-, garnet- and perovskite-type solid electrolytes are promising for applications in solid-state batteries because of their performance in electrochemical stability and acceptable ionic conductivities (~10^{-4}–10^{-2} S cm^{-1}, ~10^{-4}–10^{-3} S cm^{-1}, ~10^{-5}–10^{-4} S cm^{-1}). Concerning battery safety, suitable selection of solid electrolytes must be careful, according to the great reduction of interface resistance. However, Li^+-based batteries are required to be encapsulated in avoiding contamination from the ambient environment, which could lead to a decrease in ionic conductivity and variation of the crystal structure. Solid electrolytes should greatly improve their properties and make compatible between solid electrolytes and electrodes.

Ionic liquid electrolytes (ILEs; details in table 1.1) have also stirred plenty of experimental [114, 115] and theoretical studies [116], mainly owing to safety advantages in rechargeable Li^+-based batteries. The most important features are that they are highly non-flammable, non-volatile, and conductive [117–121]. The characteristics of ILEs are electrochemically and thermally stable, being operated safely within wide-voltage cathodes at high temperatures [117, 122, 123]. The investigated ILEs are expected to present stable and large Li^+-ion current densities. On the other hand, ILEs need to overcome many drawbacks, such as, larger viscosity (e.g., a reduced total ionic conductivity at room temperature; [117, 124]), coexistent crystallization at low temperatures [117, 125], and temperature-dependent interfacial impedance at both the cathode and anode sides [93, 126]. In general, ILEs are constructed from the close combinations of cations and anions, for example, common cations might cover imidazolium, quaternary ammonium, pyrrolidinium and piperidinium, while popular anions are $[PF6]^-$ [127, 128], $[BF4]^-$ [122, 127], bis (trifluoromethanesulfonyl)imide ([TFSI]$^-$ [114, 116, 123, 124, 129–137]) and so on. Among the potential ILEs, 1-Ethyl-3-methylimidazolium bis(trifluoromethanesulfonyl)-imide (EMI-TFSI; [117]) has been frequently used with Li salt as an electrolyte for Li^+-based batteries due to its low viscosity property. In addition, there are many certain kinds of ILEs, including 1-allyl-3-vinyl imidazolium bis(trifluoromethanesulphonyl)imide [AVIm][TFSI]N [118], *N*, *N*-diethyl-*N*-methyl-*N*-(2-methoxyethyl)ammonium TFSA (DEME–TFSA;[119]), *N*-methyl-*N*-propylpiperidinium bis(trifluoromethanesulfonyl) imide (PP13–TFSI) [116, 124, 130], AMImTFSI and Im_{13}TFSI (1-allyl-3-methylimidazolium bis(trifluoromethanesulfonyl)imide and 1-methyl-3-propylimidazolium bis(trifluoromethanesulfone) imide) [133], DMMATFSI (N, N-diethyl-N-methyl-N-(2-methoxyethyl)ammonium is(trifluoromethylsulfonyl)imide) [134], etc. Besides, gel polymer electrolytes [138–144] could also be available in combination with ILEs for better efficiency. Generally speaking, gel polymer electrolytes (GPEs; [119–125]) possess several merits, e.g., no internal shorting [139, 145, 146], leakage of electrolytes [139, 146] and non-combustible reaction products at the electrode surface [139, 146]; the opposite of the case in ILEs. Most GPEs have been successfully synthesized through two steps, including the formation of a porous polymer membrane structure and its activation with a liquid organic electrolyte. Apparently, the outstanding GPEs are mainly determined by the intrinsic features of the polymer host. The specific polymers, polyethylene oxide (PEO; [139]), poly(vinylidene fluoride) (PVDF; [138]), poly(vinylidene fluoride-hexafluoropropylene) (PVDF-HFP; [140, 147]), poly(methyl methacrylate) (PMMA;[142, 143]), and polyacrylonitrile (PAN; [141]), have been frequently employed as unusual polymer hosts in gel polymer electrolytes.

Up-to-date experimental examinations show that there also exist aluminum- [148–150] and iron-ion-related batteries [39], especially for enough prominent research of the former. Aluminum-ion-based ones have been successfully applied in commercial products, such as, mobile devices [148, 149], hybrid electric vehicles (HEVs; [148, 149]), plug-in hybrid electric vehicles (PHEVs; [148, 149]), and electrical grids [148, 149]. Generally speaking, a typical aluminium-ion battery consists of a metallic anode of aluminum (table 1.2), an ionic liquid electrolyte, and a graphite-like cathode [37, 126, 151, 152]. Since aluminum-ion batteries are verified to be rechargeable, the Al-related molecular ions can generate energy due to transport from anode to cathode. In the recharging processes, the aluminum-related ions immigrate to the negative electrode (anode) and can exchange three electrons per ion. This clearly illustrates that the insertion of one Al^{3+} is equivalent to three Li^+ ions in conventional intercalation cathodes [148, 150]. The trivalent charge carrier in Al^{3+} is just merit and drawback, wherein the transfer of three-unit charges by one ion greatly enhances energy storage capacity [148, 153], but creates too strong electrostatic intercalations for the host materials under the well-defined electrochemical behavior [148, 153]. Very interestingly, an ionic liquid electrolyte (an ionically conductive medium) comes into existence by many compounds, for example, aluminum chloride and 1-ethyl-3-methylimidazolium chloride ($AlCl_3$/[EMIm]Cl; [148, 154]), aluminum chloride and 1-butyl-3-methyl imidazole salt chloride (AlCl3/[BMIm]Cl) [155], aluminum chloride and 1,3-di-n-butylimidazolium

Table 1.2. Various anodes, electrolytes, and cathodes in Al^{3+}-based batteries.

Anodes	Al	[152, 402]
Cathodes	graphite	[152]
	Mn_2O_4	[402]
	V_2O_5	[148, 155, 162]
	VO_2	[403]
	CFx	[152]
	Graphitic foam	[151]
	Pyrolytic graphite	[151]
	Graphic carbon paper	[36]
	$LiFePO_4$	[154]
	Mo_6S_8	[404]
	Ni_3S_2	[405]
	NiS	[406]
	S	[407]
	VCl_3	[408]
	CuHCF	[409]
	Polythiophene	[410]
	Polypyrrole	[410]
	FeS_2	[13, 411]
	FeS	[13, 159]
	Al	[163]
	Zn	[412]
	Cd	[412]
Electrolytes	Aluminum chloride and 1-ethyl-3-methylimidazolium chloride ($AlCl_3$/[EMIm]Cl)	[148, 154]
	Aluminum chloride and 1-butyl-3-methyl imidazole salt chloride and ($AlCl_3$/[BMIm]Cl)	[155]
	Aluminum chloride and 1,3-di-n-butylimidazolium Bromide ($AlCl_3$/[BIM]Br)	[152]
	$AlCl_3$–KCl–NaCl (acidic)	[156]
	$AlCl_3$–NaCl (acidic)	[157, 158]
	$AlCl_3$–NaCl–Al_2S_3 (basic)	[159]
	Alkali tetrahaloaluminate salts ($LiAlCl_4$–$NaAlCl_4$–$NaAlBr_4$–$KAlCl_4$)	[160]
	NaCl saturated $NaAlCl_4$ (acidic)	[161]
	Aluminum chloride and 1-methyl-3-ethylimidazolium chloride ($AlCl_3$–MEICl (acidic))	[162, 163]
	$AlCl_3$–dipropylsulfone–toluene	[164]
	$LiPF_6$–ethylene carbonate (EC)–ethyl methyl carbonate (EMC) ($LiPF_6$–EMC–VC)	[165]
	aluminum chloride-1-butyl-3-methylimidazolium chloride benzyl sulfoxide ($AlCl_3$–BMICl–benzyl sulfoxide)	[403]
	aluminum bromide and 1-ethyl-3-methylimidazolium chloride ($AlBr_3$–MEIBr (acidic))	[413]

bromide ($AlCl_3$/[BIM]Br; [152]), $AlCl_3$–KCl–NaCl (acidic; [156]), $AlCl_3$–NaCl (acidic; [157, 158]), $AlCl_3$–NaCl–Al_2S_3 (basic; [159]), alkali tetrahaloaluminate salts ($LiAlCl_4$/$NaAlCl_4$/$NaAlBr_4$/$KAlCl_4$;[160]), NaCl saturated $NaAlCl_4$ (acidic; [161]), aluminum chloride and 1-methyl-3-ethylimidazolium chloride ($AlCl_3$–MEICl (acidic); [162, 163]), $AlCl_3$–dipropylsulfone–toluene [164], $LiPF_6$–ethylene carbonate (EC)–ethyl methyl carbonate (EMC) ($LiPF_6$–EC–EMC; [165]), and so on. Among them, the mixing aluminium chloride [$AlCl_3$] and 1-ethyl-3-methylimidazolium chloride ((EMIm)Cl; [148]) or 1-butyl-3-methylimidazolium chloride ((BMIm) Cl; [155]) are predominant, being successfully synthesized at room temperature [166]. These ionic liquid electrolytes can provide the redox active chloroaluminate anions [$AlCl_4^-$ and $Al_2Cl_7^-$] in dominating the ion transport processes through the intercalation and deintercalation of electrode/anode/cathode. Currently, ionic liquid electrolytes can reach high ionic conductivities up to 10^{-3}–10^{-2} S cm^{-1} within the wide and stable electrochemical potential windows [150]. As for cathode materials of aluminum-ion batteries, they might come from many compounds, such as, graphite, Mn_2O_4, V_2O_5, VO_2, CF_x, graphitic foam pyrolytic graphite, graphic carbon paper, $LiFePO_4$, Mo_6S_8, Ni_3S_2, NiS, S, VCl_3, and so on (table 1.2). These systems should provide a high storage capacity to enhance the energy and power densities of aluminum-ion ones. However, most of the current cathode materials provide small discharge capacities (less than 100 mAh g^{-1}) and low electrode potentials (roughly lower than 1 eV, but only graphite with 2 eV [13]). In short, the development of high-performance aluminum-ion batteries might experience many high barriers, especially for pursuing efficient cathode materials. In addition, the iron-ion-based battery is also generated for the first time under a specific capacity of 207 mAh g^{-1} at 30 mA g^{-1}. Its huge advantage is that it is quite cheap, while the other characteristics require further high-resolution experimental examinations [39]. Very interestingly, this new type of battery mainly consists of mild steel, layered vanadium pentoxide (V_2O_5; [39]), and tetraethylene glycol dimethylether (TEGDME; [39]), respectively, corresponding to the anode, cathode, and electrolyte solvent. Obviously, the various materials will be combined together and tested to reach optimal battery performance (table 1.3). The important differences among lithium-, aluminium- and iron-ion-based batteries are worthy of systematic investigations in terms of experimental observations and theoretical simulations.

In addition to the above-mentioned experimental syntheses, high-resolution examination/identification/verification are required for the fundamental properties. The different spectroscopies are greatly developed to achieve diverse physical/chemical/material phenomena. First, the lattice symmetries can be confirmed from x-ray diffraction spectroscopy (details in chapter 2 section 2.1; [32, 167]), low-energy electron diffraction spectroscopy (LEED; [168]), scanning tunneling microscopy (STM, details in chapter 2

Table 1.3. Various anodes, electrolytes, and cathodes in Fe^{3+}-based batteries.

Anodes	Fe	[39]
Cathodes	V_2O_5	[39]
Electrolytes	tetraethylene glycol dimethylether (TEGDME)	[39]

section 2.2; [34, 169]), and tunneling electron microscopy (TEM; [170, 171]), being sensitive to the dimension-dependent crystal morphologies. The first equipment is very suitable for 3D core materials of ion-based batteries. The x-ray patterns have become the unique measurements in identifying the stable crystal structures before (after; [32, 167]) the charging/discharging processes. Most solid-state anode/cathode/electrolyte materials, as clearly shown in table 1.1, are observed to display the unusual crystal symmetries due to the various moiré superlattices (highly-nonuniform chemical environments; [26, 40]). However, this technique cannot resolve the intermediate crystal configurations during the chemical reactions. On the other hand, the other techniques are very useful in detecting nanoscaled morphologies, e.g., the STM measurements for the stackings and defects on a graphite surface [34, 172]. As for electronic properties, electronic energy spectra [173, 174] and van Hove singularities, are, respectively, observed by the angle-resolved photoemission spectroscopy (ARPES, details in chapter 3 section 3.3; [173, 175]) and scanning tunneling spectroscopy (details in chapter 2 section 2.2; [34]). Their examinations on graphite [34, 174] and graphite intercalation compounds [176, 177] provide reliable information about the low-lying occupied electronic states with the rich semi-metallic behaviors. The opposite is true for the detections of multi-component lithium oxides; that is, there are no experimental results to date. Good characterization of the very small first Brillouin zone is the main reason. Finally, the observations of optical excitations, which are performed by the reflection/absorption/transmission/photoluminescence spectroscopies are quite few, e.g., the rough measurements of excitonic photoluminescence spectra in Li_2GeO_3 [178, 179]. Systematic tests of optical properties are required under the current investigations [178, 179].

The quaternary-component lithium oxides, Li_xFePO_4 ($x = 1, 0.75, 0.5$ and 0.25), are outstanding samples for basic science research (rich and unique charge and spin phenomena; [180, 181]), as well as the frequently commercial products in cathodes of Li^+-based batteries [182, 183]. Such mainstream materials have been successfully synthesized by chemical methods in large-scale quantities [182, 184]. Furthermore, Li-concentration-diversified crystal structures are clearly verified from x-ray diffraction patterns [181, 184, 185]. Both experimental examinations [181, 182, 185, 186] and theoretical simulations [187–189] are consistent with each other in determining the orthorhombic lattice symmetries with distinct Li, Fe, P and O positions. There are many atoms in a unit cell, i.e., moiré superlattices [190] appear in any compounds. Very interestingly, this creates giant barriers in the further investigations of other essential properties, especially for high-precision measurements. For example, up-to-date experimental results show that their diverse band structures and van Hove singularities cannot be, respectively, detected by the angle-resolved photoemission spectroscopy (ARPES; details in chapter 3 section 3.3) and scanning tunneling spectroscopy (STS; details in chapter 3 section 3.2). On the other side, the first-principles method (VASP; details in chapter 2 section 2.1; [191, 192]), but not the phenomenological models (details in chapter 2 section 2.1), is able to fully explore optimal lattice symmetries [188], electronic properties [189], magnetic configurations [181], and even optical excitations (with excitonic effects). Most importantly, the intrinsic interactions, which mainly arise from quasiparticle charge distributions and spin arrangements, are the studying focus through the various physical and chemical quantities. The calculated results cover the real crystal structures, energy spectra and wave functions, spatial charge/spin density distributions, density of states, magnetic moments, with the atom-, orbital- and spin-projections. The thoroughly consistent analyses will clearly illustrate the active multi-orbital in Li–, Fe– and P–O bonds, the dominating ones in the specific valence and conduction energy range, the dominating ferromagnetic magnetism due to 3D orbital of Fe atoms, their spin-up dominance for the occupied states, and the opposite spin-down configuration responsible for the metallic behavior across the Fermi level. The theoretical framework of VASP-based simulations is never proposed in previous research. It is under the current systematic investigations.

It is well-known that graphite-related materials [193] could serve as anode and cathode in aluminum- and lithium-ion-based batteries, respectively. Their main merits cover the most stable configuration, an easy synthesis in chemical intercalation/de-intercalation [194], and a quite low cost. Very interestingly, pristine graphite might have AA (simple one), AB (Bernal), ABC (rhombohedral) and turbostratic stackings, mainly owing to the weak, but significant interlayer van der Waals interactions [195]. Such layered condensed-matter systems are rather suitable for exploring the 2D quasiparticle behaviors, e.g., the diverse magnetic quantization [196] and magneto-optical selection rules [197]. Among them, the Bernal system is frequently utilized in experimental measurements [6, 198], being extensively investigated for their essential properties both experimentally and theoretically. Each graphitic sheet is a honeycomb lattice with a rather strong carbon-sp^2 bonding [199], in which the π- and σ-electronic states are well characterized in the different energy ranges [200]. The spacing between two neighboring sheets becomes available in modulating the physical, chemical, and material phenomena. Up to now, many guest atoms/molecules/ions [201] have been successfully synthesized into graphite to form the various compounds. For example, alkali-atom graphite intercalation compounds are verified to reveal the stage-n structures, including LiC_{6n} and LiM_{8n} (M = K. Rb and Cs; n = 1, 2, 3; 4; [202]). On the other side, $AlCl_4^-$-ion/$AlCl_4$-molecule ones are examined to show stage-4 and stage-3 cases at room and lower temperatures, respectively. Stage-1 and stage-2 configurations, with higher intercalant concentrations, are unable to be synthesized in experimental laboratories. This might arise from the fact that it is quite difficult to build the periodical arrangements along the longitudinal z and transverse directions simultaneously for the large ions or molecules. A few theoretical studies of the first-principles method are mainly focused on stage-1 and stage-2 systems in terms of geometric and electronic properties. The critical features, being associated with the atom-dominated energy spectra and wave function within the specific energy ranges, the active multi-orbital hybridization in distinct chemical bonds, and atom- and orbital-decomposed van Hove singularities, will be thoroughly clarified by the delicate simulations and analyses. That the quasiparticle charges play the critical roles in the fundamental properties is clearly illustrated in chapter 5.

The well-known lithium-ion batteries (LIBs) are characterized by high and reversible capacity, excellent rate capacity and cycling stability [203–206]. Graphite intercalation compounds (GICs) are widely used as an electrode material for LIBs [207–212]. To improve the energy storage performance of LIBs, many approaches are developed to increase the interlayer distance between layered graphene and reduce their restacking as much as possible [208, 213]. The stage-1 $FeCl_3$-graphite intercalation compound possesses a rather large interlayer distance of 9.20 Å. The super big unit cell of $FeCl_3$, which consists of 50 C atoms, 8 Fe atoms and 24 Cl atoms, presents as a layered structure sandwiched between two graphitic layers. Up to now, a lot of theoretical and experimental studies have been applied to explore the essential and important properties of $FeCl_3$ graphite intercalation compounds. For instance, Raman spectroscopy and x-ray diffraction (XRD) are available for investigating the stage of configuration and the interlayer distance of lattice [214–216]. Based on the previous works, we use first-principles calculations on VASP to obtain more detailed and complete information, such as the band structure, the density of state and the spatial charge distribution. Furthermore, first-principles calculations reveal that the band structure of $FeCl_3$ graphite intercalation compound is similar to pure graphene but with a redshift in the low-lying energy range, i.e., the doping effect is carried out due to the charge transfer from graphitic layer to $FeCl_3$. The calculated results can display the dominating multi-orbital hybridizations in Fe, C, and Cl atoms. The better performance of $FeCl_3$ graphite intercalation compounds after intercalation, e.g., enhanced electrical conductivity, is attributed to their unique intercalation structures. The theoretical framework of $FeCl_3$ graphite intercalation compound is significant and helpful for in-depth investigation of the full features of graphite-related intercalation compound electrodes.

Among many potential materials for ion-based batteries, V_2O_5 possesses a specific geometric structure, in which one vanadium is surrounded by five oxygen atoms [102, 217]. The essential properties of this system will be examined from the hybridization point of view. The band structures, atom-dominated energy spectra and wave function, the specific multi-orbital hybridization in distinct chemical bonds, and atom- or orbital-decomposed van Hove singularities in density of states will be studied under first-principles calculations. Furthermore, the absorption spectra, energy loss spectrum and reflection coefficient can be utilized to supplement the essential properties of condensed matter systems. An optical plasmon mode is identified to be the coherent charge oscillations of V-[$4s$, $3d_{x^2-y^2}$, $3d_{xy}$, $3d_{yz}$, $3d_{xz}$, $3d_{z^2}$] and O-[$2s$, $2p_x$, $2p_y$, $2p_z$] orbitals. The 3D binary compound is decomposed by the experimental measurements [98, 101, 102]. The theoretical framework can combine with the phenological model [218, 219] and the experimental examination [100, 102, 220], is worthy of systematic investigation. Chapter 7 will provide the physical, chemical, and material pictures in understanding the fundamental properties of V_2O_5.

The Li_2SiO_3 compound has gained considerable attention because it satisfies many applicable standards of an electrolyte material in LIBs. For example, the Li_2SiO_3 compound with low-cost synthesis exhibits a reliable ionic conductivity (2.5×10^{-8} S.cm^{-1} at 25 °C) [221] and high mechanic strength, which is suitable for a fast and flexible thin-film ionic conductor [222], and a CO_2 fine absorption effect is evidence that this material can be utilized for gas capture [223]. Furthermore, many previous studies also indicated that the Li_2SiO_3 compound has also been actively explored as a luminescent material [224, 225] due to its exceptional optical properties. In addition, the polar orthorhombic symmetry of the orthorhombic structure suggests that the Li_2SiO_3 compound can be suitable for piezoelectric, pyroelectric, and electro-optic applications [226–229]. Up to now, Li_2SiO_3 compound crystals have been successfully synthesized by various methods [225, 228, 230–232]. Its essential physical/chemical properties, such as optimized geometry structures and surface morphologies, are usually examined by x-ray diffraction [224, 225, 233], scanning electronic microscopy (SEM) [234] and tunneling electronic microscope (TEM) [235, 236]. The energy spectrum depending on frequencies and wave vectors at the valence states and the van Hove singularities in the density of states are not achieved because of a lack of angle-resolved photoemission spectroscopy (ARPES) [175] and scanning tunneling spectroscopy (STS) measurements [237], respectively. Regarding the optical responses, photoluminescence, absorption, and transmission measurement have been done conducted [224, 225, 238]. However, they have only been investigated for doped Li_2SiO_3 or measured at high temperature, in which optical properties are strongly affected by surface defects. So, this material is based on the density functional theory calculation on the optimized geometric structure with position-dependent chemical bonding, the atom dominated energy spectrum at various energy ranges, the spatial charge distribution due to different orbitals, and the atom- and orbital-projected density of states associated with the overlap of orbitals. The specific orbital hybridizations will be used to interpret the onset of the optical frequency, stable excitonic states, a lot of prominent absorption structures, a very strong plasmon mode in terms of the dielectric functions, energy loss functions, absorption coefficients, and reflectance spectra under the distinct electric polarization. The predictions in chapter 8 require highly-resolved experimental measurements.

There are several different formulas for electrolytes, usually consisting of organic compounds, such as ethylene carbonate (EC), propylene carbonate (PC), vinylene carbonate (VC), diethyl carbonate (DEC), dimethyl carbonate (DMC), ethyl methyl carbonate (EMC), lithium salt and functional additives [239–241]. Apart from electrolyte materials, identifying suitable electrolyte additives also plays a significant role to enhance the performance of LIBs. After the first operating circle of LIB, the solid electrolyte interface (SEI) is formed. This SEI layer separates the cathode and anode but still allows ions to pass through. To prevent the films from becoming too restrictive, electrolyte additives are often added to enhance the performance of LIBs [242–246]. A good additive can help to enhance the cycle life of Li-ion and maintain a stable SEI film on the anode without side effects of the electrolyte reduction on the anode and oxidation on the cathode. Electron affinity (EA) and ionization energy (IE) are the two important parameters needed to be taken into account to verify the properties of electrolyte/electrolyte additives [247]. However, with a huge number of molecules waiting for testing, calculations or experimental measurements are not an effective way to deal with them. As an alternative approach, a search map for predicting the EA and IE values for LIB applications will be established based on artificial neural networks (ANN) [248].

Graphite is formed by periodical graphene layers arranged along the *z*-direction [249]. It has been widely used in our lives for a long time. There exist three kinds of typical stacking, namely AAA, ABA, and ABC stackings [250–256]. They are also known as the simple hexagonal [256], Bernal [252] and rhombohedral graphites [254]. The Bernal graphite comes from nature [252], while the simple hexagonal one is only fabricated in the laboratory [256]. The theoretical and experimental studies of these graphites reveal a lot of feature-rich essential properties, covering geometric configurations [249, 254], electronic properties [47, 257], optical spectra [249, 257–259], and magnetic quantization [260–262]. Pristine graphites are all semimetals due to the C-$2p_z$ orbitals interlayer atomic interactions. The weak but significant van der Waals interactions account for the interlayer attractiveness. On the other side, alkali atoms are suitable to intercalate into the graphite layers, resulting in a high electrical conductivity [263, 264]. Recently, stage-n alkali graphite intercalation compounds have been synthesized in many experiments [264–267]. Importantly, the stage-1 system of lithium intercalation graphite is LiC_6 [268–271], being different from that of Na, K, or Rb intercalation graphites [271, 272]. As for the Li^+ ions intercalated graphite, they might appear frequently in the charging and discharging processes of lithium-based batteries [272–274]. Both kinds of Li/Li^+ graphite intercalation compounds are predicted to present the distinct geometric/electronic properties. A systematic study on their essentials by using the first-principles method are needed.

A periodical table [274] is very helpful in fully understanding the various crystal symmetries, where they are greatly diversified by the concise mechanisms of multi-/single-element components [26, 27, 40], allotropes [275, 276], three-/two-/one-/zero-dimensions [26, 27, 40, 277–279], rich stacking configurations between two neighboring layers/nanotubes [27, 45], layer/nanotube numbers of layered/coaxial systems [27, 280], and planar/curved/scrolled/folded low-dimensional structures [281–283]. It is well known that the critical bondings [281], being closely associated with the atomic configurations, are responsible for the formations of symmetric crystals, e.g., a rather strong σ bonding of carbon-[2s, 2 p_x, 2 p_y] orbital hybridization in a honeycomb lattice [27, 281]. Very interestingly, many kinds of condensed-matter systems have been successfully generated in laboratories by physical [284] and chemical syntheses [284], such as, *n*-type and *p*-type graphite intercalation compounds [209, 285], silicone-related anodes in lithium-ion-based batteries [286, 287], lithium-, phosphorus- and sulfur-dependent compounds [112, 288], solar cells [289], hydrogen energy materials [290], as well as, II–VI/II–V/III–V/III–VI composite compounds [291–293]. A series of related materials are outstanding candidates in testing, identifying and verifying the theoretical developments of quasiparticle framework [25, 27, 281] and the experimental progress of high-precision measurements [175, 294, 295]. The up-to-date successful battery operations [16, 296], which are conducted under the sufficiently high efficiencies [296], could cover the common lithium-ion-based batteries [11, 12, 296], lithium sulfide batteries [297, 298], sodium-/potassium-/magnesium-associated ones [299–301], aluminum-ion-related batteries [13, 37], and iron-ion-dependent systems [39]. Generally speaking, there exist plenty of merits and certain serious drawbacks in each kind of battery. The systematic investigations on the basic sciences are almost absent, according to the previous works of theoretical [11, 13, 39, 297] and experimental searches [11, 13, 39, 297]. For example, the active orbital hybridizations of Na-/K-/Mg-metal graphite intercalation compounds are not presented in the published first-principles results [302–304]. The specific similarities and important differences among them will be thoroughly explored under the current chapter 11 in terms of geometric stabilities/symmetries [305, 306], band structures/wave functions [269, 306], spatial charge densities before and after intercalations [209, 269], and merged van Hove singularities under the various chemical environments [209, 305]. The similar quasiparticle viewpoints might be very suitable in studying all the green energy materials [307, 308]; that is, their rich and unique quasiparticle properties will be unified under such concise thoughts [209].

Resource depletion and environmental pollution have always been an important global issue. Developing green materials can solve both problems at the same time efficiently. Lithium-ion batteries (LIBs) are one of the optimal candidates due to their rechargeable features and large energy capacities, which can supply electric power for transportation [309]. Graphite materials have been applied as anode materials of LIBs for a long time due to their excellent cycling stability and flat discharge plateaus [310–312]. The theoretical specific capacity of graphite is 372 mAh g^{-1}. However, it has limits to provide energy for the constantly increasing demand. Therefore, investigating a new anode material with a large energy storage space has become a significant issue. Two-dimensional graphene [313–323] and one-dimensional carbon nanotubes [324–334] with their respective unique nanostructures exhibit a higher specific capacity than graphite. Especially for the graphene-based materials, the high reversible capacity can reach up to 1000 mAh g^{-1} [315, 316, 335–339]. Several studies have reported that graphene–carbon nanotube composites present a synergistic effect between two different graphitic components [321–323, 333]. Meanwhile, Si in batteries shows the extraordinarily high specific capacity of 4200 mAg^{-1}, far more than other various carbonaceous materials [340, 341]. However, the large volume expansion leads to a rapid loss of reversible capacity during cycling, which is a challenge posed for Si. Recent studies have overcome this problem with nanostructures, such as nanowires, nanoporous materials and nanotubes[286, 342–344]. Park *et al* have investigated the high reversible charge capacity of silicon nanotubes (3247 mAh g^{-1}) by increasing the surface area accessible to the electrolyte and deposited a carbon coating on the surface of the nanotubes [344].

Realizing the electronic properties of nanostructures is a significant step. In this chapter we will introduce single-wall armchair carbon and silicon nanotubes with different diameters and chiralities. Both kinds of nanotubes can have metallic and semiconducting features, dependent on their radius and chiralities. The geometric structure, energy bands, spatial charge distributions, and density of states are discussed in detail. Carbon and silicon nanotubes can be considered as rolled up graphene and silicone; however, the hybridization of (s, p_x, p_y, p_z) at low-lying energy for carbon and silicon nanotubes is totally different compared with graphene and silicene, respectively. The ground state energy E_0 obviously decreases with an increasing diameter, owing to the reduction of C–C or Si–Si bond length, buckling distance, and the curvature effect. The electronic properties are strongly dependent on the size and chiral angles. For example, armchair silicon nanotubes (m,m) (n = 2–11) present semiconducting behavior, while carbon nanotubes (m,m) (n = 3–12) exhibit a metallic one. The spatial charge distributions illustrate various chemical bonding for nanotubes that can be employed to investigate the multi-/single-orbital hybridizations. All the main characteristics are directly reflected in the projected density of states, which are used to fully understand the orbital-decomposition contributions.

As one of the most attractive anode materials, $Li_4Ti_5O_{12}$ possesses a lot of merits, e.g., the safety and cycling life for lithium-ion batteries. Currently, the ternary compound emerges into the commercial anode product for Li^+-ion based batteries [65, 345]. This core material shows rich and unique geometric, electronic, and optical properties [26, 50, 346]. The primitive cell has a huge number of atoms, which is known as moiré superlattices [26]. The geometric structure performs in a highly non-uniform environment, which is basically created from the Li–O and Ti–O bonds. Many significant electronic properties are presented such as band structures, atom-dominated energy spectrum, spatial charge density distributions, atom, the orbital-decomposed density of states [26]. The optical properties cover the dielectric functions, absorption spectrum, reflectance, and energy loss spectrum [347, 348]. The theoretical properties could be tested under the high-resolution experimental measurements. Many techniques can be used for investigating the whole structures, e.g., x-ray diffractions for measuring the lattice parameters, transmission electron microscopy (TEM) for morphology [171], angle-resolved photoemission spectroscopy (ARPES; details in chapter 3 section 3.3) and scanning tunneling spectroscopy (STS; details in chapter 3 section 3.2) for band structures examination along with van Hove singularities. Also, the tight connection between the electronic, optical properties in terms of multi-hybridizations is worth thorough investigation in chapter 13.

Transition metal dichalcogenides (TMDs) reveal the remarkable structural polymorphs which are indicated in the ranges of electronic properties ranging from semiconducting, semi-metallic to metallic [349]. 2D TMDs and monochalcogenides [350] are the most widely used anode material. They are expected to develop the capacity of modern LIBs, SIBs and KIBs by enhancing adsorption capacities and the diffusion rate of anode materials. In particular, MoS_2 [351] and WS_2 [352] have been studied considerably as anode in LIBs. In addition, these materials have also attracted significant interest as efficient electrode materials because of favorable electrochemical properties. [353] For example, WTe_2 [353] shows high first cycle specific charge capacity and low charge transfer resistance. Obviously, the efforts in researching are focused on improving cost feasible anode and cathode materials for batteries that correspond to the major portion of the final cost of a device. [354] Therefore, TMDs have become well-known and are being studied since some of them are naturally abundant. Among TMDs, group IV TMDs, particularly ZrX_2 and HfX_2, are candidates which are severely limited in previous reports for applications in the battery field. To clarify the characteristic of these materials, the detailed information about the geometry and electronic properties should be illustrated to find out the potential features for further research in the energy storage. These properties will be discussed in chapter 14.

The aforementioned chapters give simple fundamental knowledge and future research topics about materials in the secondary batteries based on the theoretical point of view. It is also important to realize the electric characteristic of the whole battery cell, made by the combination of anode/cathode materials, electrolytes, and separators. Chapter 15 provides a brief description of the standard charge/discharge process for understanding the electric characteristic of secondary batteries. Through the measurement based on the standard charge/discharge process, researchers/engineers can realize the performance and quality of the batteries, and further design high-performance battery systems according to the charge/discharge profiles. On the other hand, owing to the popularization of electric vehicles worldwide and the development of renewable energy supply, lithium-ion (Li-ion) batteries are widely used from small-scale personal mobile products to large-scale energy storage systems. Recently, the number of retired power batteries has largely increased, causing environmental protection threats and waste of resources. Since most of the retired power batteries still possess about 80% initial capacity, the second use of them becomes a possible route to solve the emergency problem. Safety and performance are important in using these second-use repurposed batteries. Underwriters Laboratories (UL), a global safety certification company, published the standard for evaluating the safety and performance of repurposed batteries, i.e., UL 1974. In chapter 15, the test procedures are designed according to UL 1974 for Li-ion batteries and the charge/discharge profile dataset of the $LiFePO_4$ repurposed batteries are provided.

In recent years, owing to the rapid development of renewable energy and the popularity of the electric vehicle, lithium-ion (Li-ion) batteries are intensively used, causing more fundamental research on high-power, high-density, long-life, and safe energy materials. In chapter 16, the engineering integrations on combining Li-ion battery cells to an energy storage system (ESS) are introduced. The battery management system (BMS) plays an important role in using the battery system efficiently and safely. The energy management system (EMS) can well-integrate the ESS and other systems, such as renewable power sources and loads. Current potential applications for power and stationary application on various size scales are demonstrated. Finally, the outlooks on the future of Li-ion batteries are given.

This book is organized as follows. Chapter 2 covers the theoretical frameworks of the first-principles method and the phenomenological models, especially for their separate developments and the direct combinations. The former is utilized to thoroughly clarify the diverse quasiparticle properties of cathode/electrolyte/anode materials in Li-, Al-, Fe-, K-, Na-, Mg-based batteries. By the delicate simulations and analyses, the significant multi-/single-orbital hybridizations and spin configurations can be obtained from the various physical quantities, the unusual/normal chemical bonds in a moiré superlattice/a small unit cell, the distinct atom dominances at the specific energy ranges of band structure, a whole charge/spin density distribution, the atom-, orbital- and spin-decomposed van Hove singularities, the atom- and orbital-generated magnetic moments, the featured absorption structures and the meta-stable excitons. Hope to establish the close relations between the specified orbital mixings and the frequency-dependent optical excitation channels. Theoretical predictions on the geometric, electronic, magnetic, and optical properties can be examined by the various spectroscopies, as clearly discussed in chapter 3. The high-resolution examinations on the theoretical predictions are achieved from the measurements of x-ray patterns [32, 167], occupied electronic energy bands [173, 174], the density of states across the Fermi level [34], optical reflection/absorption/transmission spectra [178, 179], photoluminescence [179] and net magnetic moments [355]. Only part of the experimental observations have been done up to now, and their full support for the predicted results is discussed in detail. More detections are required under the current investigations, e.g., the time-resolved x-ray diffraction peaks [32, 167] and the defect-enriched crystal structures [356].

The diverse physical/chemical/material phenomena are clearly indicated in the different materials through VASP simulations. Chapters 4–15 are, respectively, focused on the $LiFePO_4$ cathode material in Li^+-based batteries [182, 357], $AlCl_4^-/Al_2Cl_7^-$-graphite intercalation compounds of aluminum-ion-based battery cathodes [149, 150, 358], graphite-$FeCl_3$ intercalation compounds for iron-ion-based battery cathodes [359], vanadium pentoxide in iron-ion-based battery cathodes [39], ternary Li_2GeO_3/Li_2SiO_3 electrolytes in lithium-ion-based battery cathodes [40, 41], various liquid electrolytes in lithium-ion-based battery [360, 361], graphite alkali-atom-/ion-intercalated compounds of Li^+-related-based battery anodes [362], sodium-, potassium- and magnesium-related graphite intercalation compounds, silicon nanotubes: anodes of lithium-ion-based batteries [51], ternary $Li_4Ti_5O_{12}$ compound: Li^+-based battery anode [26, 347], and Bulk and monolayer transition metal dichalcogenides group IV MX_2 (M = Zr, Hf; X = S, Se, or Te) of Li^+-based battery as anode of Li^+-based battery [25, 363]. The above-mentioned condensed-matters can present rich and unique features, being clearly identified within the development of theoretical frameworks. As for engineering integrations [16, 19, 364] and potential applications [17, 365, 366] are investigated in chapter 16 [17, 365, 366]. Furthermore, chapter 16 includes detailed discussion on current and future battery products [11, 12, 39, 148, 149]. Full concluding remarks are made in chapter 17 for the whole content. A lot of open issues, which are strongly associated with the theoretical and experimental limits, are proposed in chapter 18, e.g., the establishments of the tight-binding model under moiré superlattices [25, 26, 40] and the generalized ion-transport theories [367, 368]. Also, some problems related to batteries, being suitable for research and teaching, are designed in chapter 19.

References

[1] Chu W, Calise F, Duić N, Østergaard P A, Vicidomini M and Wang Q 2020 Recent advances in technology, strategy and application of sustainable energy systems *Energies* **13** 5229

[2] Armaroli N and Balzani V 2007 The future of energy supply: challenges and opportunities *Angew. Chem. Int. Ed.* **46** 52–66

[3] Serway R and Jewett J 2004 *Physics for Scientists and Engineers* 6th edn (New York: Thomson)

[4] Şen Z 2004 Solar energy in progress and future research trends *Prog. Energy Combust. Sci.* **30** 367–416

[5] Cook T R, Dogutan D K, Reece S Y, Surendranath Y, Teets T S and Nocera D G 2010 Solar energy supply and storage for the legacy and nonlegacy worlds *Chem. Rev.* **110** 6474–502

[6] Aluyor E O and Ori-Jesu M 2009 Biodegradation of mineral oils–a review *Afr. J. Biotechnol.* **8** 915–20

[7] Kvenvolden K A and Lorenson T D 2001 The global occurrence of natural gas hydrate *GMS* **124** 3–18

[8] Bao J-D and Zhuo Y-Z 2003 Investigation on anomalous diffusion for nuclear fusion reactions *Phys. Rev.* C **67** 064606

[9] Meyers M A, Chen P-Y, Lin A Y-M and Seki Y 2008 Biological materials: structure and mechanical properties *Prog. Mater Sci.* **53** 1–206

[10] Bao G and Suresh S 2003 Cell and molecular mechanics of biological materials *Nat. Mater.* **2** 715–25

[11] Goodenough J B and Park K-S 2013 The Li-ion rechargeable battery: a perspective *J. Am. Chem. Soc.* **135** 1167–76

[12] Nitta N, Wu F, Lee J T and Yushin G 2015 Li-ion battery materials: present and future *Mater. Today* **18** 252–64

[13] Das S K, Mahapatra S and Lahan H 2017 Aluminium-ion batteries: developments and challenges *J. Mater. Chem.* A **5** 6347–67

[14] Ambroz F, Macdonald T J and Nann T 2017 Trends in aluminium-based intercalation batteries *Adv. Energy Mater.* **7** 1602093

[15] Higashimoto K, Homma H, Uemura Y, Kawai H, Saibara S and Hironaka K 2011 Automotive lithium-ion batteries *Hitachi Rev.* **60** 17–21 https://www.hitachi.com/rev/pdf/2011/r2011_01_103.pdf

[16] Lee S S, Kim T H, Hu S J, Cai W W and Abell J A 2010 Joining technologies for automotive lithium-ion battery manufacturing: a review *Int. Manufacturing Science and Engineering Conf.* **1** 541–9

[17] Barré A, Deguilhem B, Grolleau S, Gérard M, Suard F and Riu D 2008 A review on lithium-ion battery ageing mechanisms and estimations for automotive applications *J. Power Sources* **241** 680–9

[18] Pistoia G 2013 *Lithium-Ion Batteries: Advances and Applications* (Burlington, MA: Newnes)

[19] Deng D 2015 Li-ion batteries: basics, progress, and challenges *Energy Sci. Eng.* **3** 385–418

[20] Pearson C, Schrantz K and Neubauer J 2010 Lithium-ion space battery technology development and infusion *2010 IEEE Aerospace Conf.* (Piscataway, NJ: IEEE) pp 1–12

[21] Tarascon J-M and Armand M 2011 *Materials for Sustainable Energy: A Collection of Peer-Reviewed Research and Review Articles from Nature Publishing Group* (Singapore: World Scientific) pp 171–9

[22] Leng F, Tan C M and Pecht M 2015 Effect of temperature on the aging rate of Li ion battery operating above room temperature *Sci. Rep.* **5** 12967

[23] Scrosati B, Hassoun J and Sun Y-K 2011 Lithium-ion batteries. A look into the future *Energy Environ. Sci.* **4** 3287–95

[24] Vetter J, Novák P, Wagner M R, Veit C, Möller K-C, Besenhard J, Winter M, Wohlfahrt-Mehrens M, Vogler C and Hammouche A 2005 Ageing mechanisms in lithium-ion batteries *J. Power Sources* **147** 269–81

[25] Shih-Yang Lin H-Y L, Khanh Nguyen D, Thanh Thuy Tran N, Duong Pham H, Chang S-L, Lin C-Y and Lin M-F 2020 *Silicene-Based Layered Materials: Essential Properties* (Bristol: IOP Publishing)

[26] Nguyen T D H, Pham H D, Lin S Y and Lin M F 2020 Featured properties of Li^+-based battery anode: $Li_4Ti_5O_{12}$ *RSC Adv.* **10** 14071–9

[27] Tran N T T, Lin S-Y, Lin C-Y and Lin M-F 2017 *Geometric and Electronic Properties of Graphene-Related Systems: Chemical Bonding Schemes* (Boca Raton, FL: CRC Press)

[28] Dien V K, Pham H D, Tran N T T, Han N T, Huynh T M D, Nguyen T D H and Fa-Lin M 2020 Orbital-hybridization-created optical excitations in $Li_8Ge_4O_{12}$ arXiv:2009.02160

[29] Vogt P, Capiod P, Berthe M, Resta A, De Padova P, Bruhn T, Le Lay G and Grandidier B 2014 Synthesis and electrical conductivity of multilayer silicene *Appl. Phys. Lett.* **104** 021602

[30] Purwamargapratala Y, Sujatno A, Sabayu Y L and Kartini E 2019 Synthesis of $Li_4Ti_5O_{12}$ (LTO) by sol-gel method for lithium ion battery anode *IOP Conf. Series: Mater. Sci. Eng.* **553** 012062

[31] Geim A K and Novoselov K S 2010 *Nanoscience and Technology: A Collection of Reviews from Nature Journals* (Singapore: World Scientific) pp 11–19

[32] Li J and Sun J L 2017 Application of x-ray diffraction and electron crystallography for solving complex structure problems *Acc. Chem. Res.* **50** 2737–45

[33] Venkateswarlu M, Chen C, Do J, Lin C, Chou T-C and Hwang B 2005 Electrochemical properties of nano-sized $Li_4Ti_5O_{12}$ powders synthesized by a sol–gel process and characterized by X-ray absorption spectroscopy *J. Power Sources* **146** 204–8

[34] Ridene M, Girard J, Travers L, David C and Ouerghi A 2012 STM/STS investigation of edge structure in epitaxial graphene *Surf. Sci.* **606** 1289–92

[35] Lin M-C, Gong M, Lu B, Wu Y, Wang D-Y, Guan M, Angell M, Chen C, Yang J and Hwang B-J 2015 An ultrafast rechargeable aluminium-ion battery *Nature* **520** 324–8

[36] Sun H, Wang W, Yu Z, Yuan Y, Wang S and Jiao S 2015 A new aluminium-ion battery with high voltage, high safety and low cost *Chem. Commun.* **51** 11892–5

[37] Wang D-Y, Wei C-Y, Lin M-C, Pan C-J, Chou H-L, Chen H-A, Gong M, Wu Y, Yuan C and Angell M 2017 Advanced rechargeable aluminium ion battery with a high-quality natural graphite cathode *Nat. Commun.* **8** 1–7

[38] Saroja A P V K, Samantaray S S and Sundara R 2019 A room temperature multivalent rechargeable iron ion battery with an ether based electrolyte: a new type of post-lithium ion battery *Chem. Commun.* **55** 10416–9

[39] Saroja A, Samantaray S S and Sundara R 2019 A room temperature multivalent rechargeable iron ion battery with an ether based electrolyte: a new type of post-lithium ion battery *Chem. Commun.* **55** 10416–9

[40] Khuong Dien V, Thi Han N, Nguyen T D H, Huynh T M D, Pham H D and Lin M-F 2020 Geometric and electronic properties of Li_2GeO_3 *Front. Mater.* **7** 288

[41] Kuganathan N, Tsoukalas L and Chroneos A 2019 Defects, dopants and Li-ion diffusion in Li_2SiO_3 *Solid State Ionics* **335** 61–6

[42] Chen S C, Wu J Y, Lin C Y and Lin M F 2017 *Theory of Magnetoelectric Properties of 2D Systems* (Bristol: IOP Publishing)

[43] Liu K, Kong B, Liu W, Sun Y, Song M-S, Chen J, Liu Y, Lin D, Pei A and Cui Y 2018 Stretchable lithium metal anode with improved mechanical and electrochemical cycling stability *Joule* **2** 1857–65

[44] Gospodarev I, Kravchenko K, Syrkin E and Feodos'ev S 2009 Quasi-two-dimensional features in the phonon spectrum of graphite *Low Temp. Phys.* **35** 589–95
[45] Dubay O and Kresse G 2003 Accurate density functional calculations for the phonon dispersion relations of graphite layer and carbon nanotubes *Phys. Rev.* B **67** 035401
[46] Ramírez R and Herrero C 2019 Phonon dispersion in two-dimensional solids from atomic probability distributions *J. Chem. Phys.* **151** 224107
[47] Ho J-H, Lu C, Hwang C, Chang C and Lin M-F 2006 Coulomb excitations in AA-and AB-stacked bilayer graphites *Phys. Rev.* B **74** 085406
[48] Nguyen T D H, Pham H D, Lin S-Y and Lin M-F 2020 Featured properties of Li^+-based battery anode: $Li_4Ti_5O_{12}$ *RSC Adv.* **10** 14071–9
[49] Hafner J 2008 Ab-initio simulations of materials using VASP: density-functional theory and beyond *J. Comput. Chem.* **29** 2044–78
[50] Ouyang C, Zhong Z and Lei M 2007 *Ab initio* studies of structural and electronic properties of $Li_4Ti_5O_{12}$ spinel *Electrochem. Commun.* **9** 1107–12
[51] Durgun E, Tongay S and Ciraci S 2005 Silicon and III-V compound nanotubes: structural and electronic properties *Phys. Rev.* B **72** 075420
[52] Gao J F, Zhang J F, Liu H S, Zhang Q F and Zhao J J 2013 Structures, mobilities, electronic and magnetic properties of point defects in silicene *Nanoscale* **5** 9785–92
[53] Nakayama T and Shima H 1998 Computing the Kubo formula for large systems *Phys. Rev.* E **58** 3984
[54] Ren X, Rinke P, Joas C and Scheffler M 2012 Random-phase approximation and its applications in computational chemistry and materials science *J. Mater. Sci.* **47** 7447–71
[55] Potthoff M 2003 Self-energy-functional approach to systems of correlated electrons *Eur. Phys. J.* B **32** 429–36
[56] Tran N T T L, Lin S-Y, Lin C-Y and Lin M-F 2017 *Geometric and Electronic Properties of Graphene-Related Systems: Chemical Bonding Schemes* (Boca Raton, FL: CRC Press)
[57] Goriparti S, Miele E, De Angelis F, Di Fabrizio E, Zaccaria R P and Capiglia C 2014 Review on recent progress of nanostructured anode materials for Li-ion batteries *J. Power Sources* **257** 421–43
[58] Lu J, Chen Z, Pan F, Cui Y and Amine K 2018 High-performance anode materials for rechargeable lithium-ion batteries *Electrochem. Energy Rev.* **1** 35–53
[59] Al-Saleh M H and Sundararaj U 2009 A review of vapor grown carbon nanofiber/polymer conductive composites *Carbon* **47** 2–22
[60] Feng L, Xie N and Zhong J 2014 Carbon nanofibers and their composites: a review of synthesizing, properties and applications *Materials* **7** 3919–45
[61] Chen S, Chen P and Wang Y 2011 Carbon nanotubes grown *in situ* on graphene nanosheets as superior anodes for Li-ion batteries *Nanoscale* **3** 4323–9
[62] Liu C, Li F, Ma L P and Cheng H M 2010 Advanced materials for energy storage *Adv. Mater.* **22** E28–62
[63] Wang *et al* 2003 Preparation and characterization of carbon nanotubes for energy storage *J. Power Sources* **119** 16–23
[64] Zhao B, Ran R, Liu M and Shao Z 2015 A comprehensive review of $Li_4Ti_5O_{12}$-based electrodes for lithium-ion batteries: the latest advancements and future perspectives *Mater. Sci. Eng.: R Rep.* **98** 1–71
[65] Sandhya C P, John B and Gouri C 2014 Lithium titanate as anode material for lithium-ion cells: a review *Ionics* **20** 601–20

[66] Ban C, Xie M, Sun X, Travis J J, Wang G, Sun H, Dillon A C, Lian J and George S M 2013 Atomic layer deposition of amorphous TiO_2 on graphene as an anode for Li-ion batteries *Nanotechnology* **24** 424002

[67] Kim S-W, Han T H, Kim J, Gwon H, moon H-S, Kang S-W, Kim S O and Kang K 2009 Fabrication and electrochemical characterization of TiO_2 three-dimensional nanonetwork based on peptide assembly *ACS Nano* **3** 1085–90

[68] Cheah S K, Perre E, Rooth M, Fondell M, Hårsta A, Nyholm L, Boman M, Gustafsson T, Lu J and Simon P 2009 Self-supported three-dimensional nanoelectrodes for microbattery applications *Nano Lett.* **9** 3230–3

[69] Wang W, Tian M, Abdulagatov A, George S M, Lee Y-C and Yang R 2012 Three-dimensional Ni/TiO2 nanowire network for high areal capacity lithium ion microbattery applications *Nano Lett.* **12** 655–60

[70] He Y, Yu X, Wang Y, Li H and Huang X 2011 Alumina-coated patterned amorphous silicon as the anode for a lithium-ion battery with high Coulombic efficiency *Adv. Mater.* **23** 4938–41

[71] Xiao X, Lu P and Ahn D 2011 Ultrathin multifunctional oxide coatings for lithium ion batteries *Adv. Mater.* **23** 3911–5

[72] Kohandehghan A, Kalisvaart P, Cui K, Kupsta M, Memarzadeh E and Mitlin D 2013 Silicon nanowire lithium-ion battery anodes with ALD deposited TiN coatings demonstrate a major improvement in cycling performance *J. Mater. Chem.* A **1** 12850–61

[73] Lotfabad E M, Kalisvaart P, Cui K, Kohandehghan A, Kupsta M, Olsen B and Mitlin D 2013 ALD TiO_2 coated silicon nanowires for lithium ion battery anodes with enhanced cycling stability and coulombic efficiency *Phys. Chem. Chem. Phys.* **15** 13646–57

[74] Lotfabad E M, Kalisvaart P, Kohandehghan A, Cui K, Kupsta M, Farbod B and Mitlin D 2014 Si nanotubes ALD coated with TiO_2, TiN or Al_2O_3 as high performance lithium ion battery anodes *J. Mater. Chem.* A **2** 2504–16

[75] Riley L A, Cavanagh A S, George S M, Jung Y S, Yan Y, Lee S H and Dillon A C 2010 Conformal surface coatings to enable high volume expansion Li-ion anode materials *ChemPhysChem* **11** 2124–30

[76] Wang D, Yang J, Liu J, Li X, Li R, Cai M, Sham T-K and Sun X 2014 Atomic layer deposited coatings to significantly stabilize anodes for Li ion batteries: effects of coating thickness and the size of anode particles *J. Mater. Chem.* A **2** 2306–12

[77] Kang E, Jung Y S, Cavanagh A S, Kim G H, George S M, Dillon A C, Kim J K and Lee J 2011 Fe_3O_4 nanoparticles confined in mesocellular carbon foam for high performance anode materials for lithium-ion batteries *Adv. Funct. Mater.* **21** 2430–8

[78] Lipson A L, Puntambekar K, Comstock D J, Meng X, Geier M L, Elam J W and Hersam M C 2014 Nanoscale investigation of solid electrolyte interphase inhibition on Li-ion battery MnO electrodes via atomic layer deposition of Al_2O_3 *Chem. Mater.* **26** 935–40

[79] Liu W-W, Lau W-M and Zhang Y 2018 The electrochemical properties of Co_3O_4 as a lithium-ion battery electrode: a first-principles study *Phys. Chem. Chem. Phys.* **20** 25016–22

[80] Meng X, Libera J A, Fister T T, Zhou H, Hedlund J K, Fenter P and Elam J W 2014 Atomic layer deposition of gallium sulfide films using hexakis (dimethylamido) digallium and hydrogen sulfide *Chem. Mater.* **26** 1029–39

[81] Li S, Liu P, Huang X, Tang Y and Wang H 2019 Reviving bulky MoS_2 as an advanced anode for lithium-ion batteries *J. Mater. Chem.* A **7** 10988–97

[82] Dai H 2002 Carbon nanotubes: synthesis, integration, and properties *Acc. Chem. Res.* **35** 1035–44
[83] Castrucci P, Scarselli M, De Crescenzi M, Diociaiuti M, Chaudhari P S, Balasubramanian C, Bhave T M and Bhoraskar S 2006 Silicon nanotubes: synthesis and characterization *Thin Solid Films* **508** 226–30
[84] Iijima S, Ajayan P and Ichihashi T 1992 Growth model for carbon nanotubes *Phys. Rev. Lett.* **69** 3100
[85] Lin M-F and Shung K W-K 1994 Plasmons and optical properties of carbon nanotubes *Phys. Rev.* B **50** 17744
[86] Spataru C D, Ismail-Beigi S, Capaz R B and Louie S G 2007 *Carbon Nanotubes* (Berlin: Springer) pp 195–227
[87] Donders M, Arnoldbik W, Knoops H, Kessels W and Notten P 2013 Atomic layer deposition of $LiCoO_2$ thin-film electrodes for all-solid-state Li-ion micro-batteries *J. Electrochem. Soc.* **160** A3066
[88] He P, Yu H and Zhou H 2012 Layered lithium transition metal oxide cathodes towards high energy lithium-ion batteries *J. Mater. Chem.* **22** 3680–95
[89] Miikkulainen V, Ruud A, Østreng E, Nilsen O, Laitinen M, Sajavaara T and Fjellvåg H 2014 Atomic layer deposition of spinel lithium manganese oxide by film-body-controlled lithium incorporation for thin-film lithium-ion batteries *J. Phys. Chem.* C **118** 1258–68
[90] Liu J, Banis M N, Xiao B, Sun Q, Lushington A, Li R, Guo J, Sham T-K and Sun X 2015 Atomically precise growth of sodium titanates as anode materials for high-rate and ultralong cycle-life sodium-ion batteries *J. Mater. Chem.* A **3** 24281–8
[91] Fongy C, Gaillot A-C, Jouanneau S, Guyomard D and Lestriez B 2010 Ionic vs electronic power limitations and analysis of the fraction of wired grains in $LiFePO_4$ composite electrodes *J. Electrochem. Soc.* **157** A885
[92] Scott I D, Jung Y S, Cavanagh A S, Yan Y, Dillon A C, George S M and Lee S-H 2011 Ultrathin coatings on nano-$LiCoO_2$ for Li-ion vehicular applications *Nano Lett.* **11** 414–8
[93] Mekonnen Y, Sundararajan A and Sarwat A I 2016 A review of cathode and anode materials for lithium-ion batteries *SoutheastCon 2016* (Piscataway, NJ: IEEE) pp 1–6
[94] Talavera N, Navarro M, Sifontes A, Díaz Y, Villalobos H, Niño-Vega G, Boada-Sucre A and González I 2013 Green synthesis of nanosized vanadium pentoxide using *Saccharomyces cerevisiae* as biotemplate *Recent Res. Develop. Mater. Sci.* **10** 89
[95] Aliyu A, Garba S and Bognet O 2017 Green synthesis, characterization and antimicrobial activity of vanadium nanoparticles using leaf extract of *Moringa Oleifera Int. J. Chem. Sc.* **16** 231
[96] Chan Y-L, Pung S-Y and Sreekantan S 2014 Synthesis of V_2O_5 nanoflakes on PET fiber as visible-light-driven photocatalysts for degradation of RhB dye *J. Catal.* **2014** 370696
[97] Aslam M, Ismail I M, Salah N, Chandrasekaran S, Qamar M T and Hameed A 2015 Evaluation of sunlight induced structural changes and their effect on the photocatalytic activity of V_2O_5 for the degradation of phenols *J. Hazard. Mater.* **286** 127–35
[98] Karthik K, Nikolova M P, Phuruangrat A, Pushpa S, Revathi V and Subbulakshmi M 2020 Ultrasound-assisted synthesis of V_2O_5 nanoparticles for photocatalytic and antibacterial studies *Mater. Res. Innov.* **24** 229–34
[99] Saravanan R, Gupta V, Mosquera E and Gracia F 2014 Preparation and characterization of V_2O_5/ZnO nanocomposite system for photocatalytic application *J. Mol. Liq.* **198** 409–12

[100] Liu D, Liu Y, Pan A, Nagle K P, Seidler G T, Jeong Y-H and Cao G 2011 Enhanced lithium-ion intercalation properties of V_2O_5 xerogel electrodes with surface defects *J. Phys. Chem.* C **115** 4959–65

[101] Wang H g, Ma D l, Huang Y and Zhang X b 2012 Electrospun V_2O_5 nanostructures with controllable morphology as high-performance cathode materials for lithium-ion batteries *Chem. Eur. J.* **18** 8987–93

[102] Shevchuk V, Usatenko Y N, Demchenko P Y, Antonyak O and Serkiz R Y 2011 Nano-and micro-size V_2O_5 structures *Chem. Metals Alloys* **4** 67–71

[103] Zheng F, Kotobuki M, Song S, Lai M O and Lu L 2018 Review on solid electrolytes for all-solid-state lithium-ion batteries *J. Power Sources* **389** 198–213

[104] Ding Z, Li J, Li J and An C 2020 Interfaces: key issue to be solved for all solid-state lithium battery technologies *J. Electrochem. Soc.* **167** 070541

[105] Kimpa M, Mayzan M, Yabagi J, Nmaya M, Isah K and Agam M 2018 Review on material synthesis and characterization of Sodium (Na) super-ionic conductor (NASICON) *E&ES* **140** 012156

[106] Xu H, Li Y, Zhou A, Wu N, Xin S, Li Z and Goodenough J B 2018 Li_3N-modified garnet electrolyte for all-solid-state lithium metal batteries operated at 40 °C *Nano Lett.* **18** 7414–8

[107] Aaltonen T, Alnes M, Nilsen O, Costelle L and Fjellvåg H 2010 Lanthanum titanate and lithium lanthanum titanate thin films grown by atomic layer deposition *J. Mater. Chem.* **20** 2877–81

[108] Swamy D T, Babu K E and Veeraiah V 2013 Evidence for high ionic conductivity in lithium-lanthanum titanate $Li_{0.29}La_{0.57}TiO_3$ *Bull. Mater. Sci.* **19** 1115–9

[109] Deng Y, Eames C, Fleutot B, David R, Chotard J-N l, Suard E, Masquelier C and Islam M S 2017 Enhancing the lithium ion conductivity in lithium superionic conductor (LISICON) solid electrolytes through a mixed polyanion effect *ACS Appl. Mater. Interfaces* **9** 7050–8

[110] Hämäläinen J, Holopainen J, Munnik F, Hatanpää T, Heikkilä M, Ritala M and Leskelä M 2012 Lithium phosphate thin films grown by atomic layer deposition *J. Electrochem. Soc.* **159** A259

[111] Kamaya N, Homma K, Yamakawa Y, Hirayama M, Kanno R, Yonemura M, Kamiyama T, Kato Y, Hama S and Kawamoto K 2011 A lithium superionic conductor *Nat. Mater.* **10** 682–6

[112] Boulineau S, Courty M, Tarascon J-M and Viallet V 2012 Mechanochemical synthesis of Li-argyrodite Li_6PS_5X (X = Cl, Br, I) as sulfur-based solid electrolytes for all solid state batteries application *Solid State Ionics* **221** 1–5

[113] Zhao Y and Daemen L L 2012 Superionic conductivity in lithium-rich anti-perovskites *J. Am. Chem. Soc.* **134** 15042–7

[114] Usui H, Yamamoto Y, Yoshiyama K, Itoh T and Sakaguchi H 2011 Application of electrolyte using novel ionic liquid to Si thick film anode of Li-ion battery *J. Power Sources* **196** 3911–5

[115] Francis C F, Kyratzis I L and Best A S 2020 Lithium-ion battery separators for ionic-liquid electrolytes: a review *Adv. Mater.* **32** 1904205

[116] Plylahan N, Kerner M, Lim D-H, Matic A and Johansson P 2016 Ionic liquid and hybrid ionic liquid/organic electrolytes for high temperature lithium-ion battery application *Electrochim. Acta* **216** 24–34

[117] Kerner M, Plylahan N, Scheers J and Johansson P 2015 Ionic liquid based lithium battery electrolytes: fundamental benefits of utilising both TFSI and FSI anions? *Phys. Chem. Chem. Phys.* **17** 19569–81

[118] Wang Z, Cai Y, Wang Z, Chen S, Lu X and Zhang S 2013 Vinyl-functionalized imidazolium ionic liquids as new electrolyte additives for high-voltage Li-ion batteries *J. Solid State Electrochem.* **17** 2839–48
[119] Seki S, Ohno Y, Miyashiro H, Kobayashi Y, Usami A, Mita Y, Terada N, Hayamizu K, Tsuzuki S and Watanabe M 2008 Quaternary ammonium room-temperature ionic liquid/lithium salt binary electrolytes: electrochemical study *J. Electrochem. Soc.* **155** A421
[120] Appetecchi G B, Montanino M, Zane D, Carewska M, Alessandrini F and Passerini S 2009 Effect of the alkyl group on the synthesis and the electrochemical properties of N-alkyl-N-methyl-pyrrolidinium bis (trifluoromethanesulfonyl) imide ionic liquids *Electrochim. Acta* **54** 1325–32
[121] Yun Y S, Kim J H, Lee S-Y, Shim E-G and Kim D-W 2011 Cycling performance and thermal stability of lithium polymer cells assembled with ionic liquid-containing gel polymer electrolytes *J. Power Sources* **196** 6750–5
[122] Molinari N, Mailoa J P and Kozinsky B 2019 General trend of a negative Li effective charge in ionic liquid electrolytes *J. Phys. Chem. Lett.* **10** 2313–9
[123] Xiang H, Yin B, Wang H, Lin H, Ge X, Xie S and Chen C 2010 Improving electrochemical properties of room temperature ionic liquid (RTIL) based electrolyte for Li-ion batteries *Electrochim. Acta* **55** 5204–9
[124] Montanino M, Moreno M, Carewska M, Maresca G, Simonetti E, Presti R L, Alessandrini F and Appetecchi G 2014 Mixed organic compound-ionic liquid electrolytes for lithium battery electrolyte systems *J. Power Sources* **269** 608–15
[125] Kühnel R-S, Böckenfeld N, Passerini S, Winter M and Balducci A 2011 Mixtures of ionic liquid and organic carbonate as electrolyte with improved safety and performance for rechargeable lithium batteries *Electrochim. Acta* **56** 4092–9
[126] McKerracher R D, Holland A, Cruden A and Wills R G A 2019 Comparison of carbon materials as cathodes for the aluminium-ion battery *Carbon* **144** 333–41
[127] Hofmann A, Migeot M, Arens L and Hanemann T 2016 Investigation of ternary mixtures containing 1-ethyl-3-methylimidazolium bis (trifluoromethanesulfonyl) azanide, ethylene carbonate and lithium bis (trifluoromethanesulfonyl) azanide *Int. J. Mol. Sci.* **17** 670
[128] Denshchikov K, Izmaylova M, Zhuk A, Vygodskii Y, Novikov V and Gerasimov A 2010 1-Methyl-3-butylimidazolium tetraflouroborate with activated carbon for electrochemical double layer supercapacitors *Electrochim. Acta* **55** 7506–10
[129] Bi S, Banda H, Chen M, Niu L, Chen M, Wu T, Wang J, Wang R, Feng J and Chen T 2020 Molecular understanding of charge storage and charging dynamics in supercapacitors with MOF electrodes and ionic liquid electrolytes *Nat. Mater.* **19** 552–8
[130] Theivaprakasam S, MacFarlane D R and Mitra S 2015 Electrochemical studies of N-methyl N-propyl pyrrolidinium bis (trifluoromethanesulfonyl) imide ionic liquid mixtures with conventional electrolytes in $LiFePO_4$/Li cells *Electrochim. Acta* **180** 737–45
[131] Li H, Pang J, Yin Y, Zhuang W, Wang H, Zhai C and Lu S 2013 Application of a nonflammable electrolyte containing Pp13TFSI ionic liquid for lithium-ion batteries using the high capacity cathode material $Li[Li_{0.2}Mn_{0.54}Ni_{0.13}Co_{0.13}]O_2$ *RSC Adv.* **3** 13907–14
[132] Böckenfeld N, Willeke M, Pires J, Anouti M and Balducci A 2013 On the use of lithium iron phosphate in combination with protic ionic liquid-based electrolytes *J. Electrochem. Soc.* **160** A559
[133] Wang M, Shan Z, Tian J, Yang K, Liu X, Liu H and Zhu K 2013 Mixtures of unsaturated imidazolium based ionic liquid and organic carbonate as electrolyte for Li-ion batteries *Electrochim. Acta* **95** 301–7

[134] Hofmann A, Schulz M, Indris S, Heinzmann R and Hanemann T 2014 Mixtures of ionic liquid and sulfolane as electrolytes for Li-ion batteries *Electrochim. Acta* **147** 704–11
[135] Ababtain K, Babu G, Lin X, Rodrigues M-T F, Gullapalli H, Ajayan P M, Grinstaff M W and Arava L M R 2016 Ionic liquid–organic carbonate electrolyte blends to stabilize silicon electrodes for extending lithium ion battery operability to 100 °C *ACS Appl. Mater. Interfaces* **8** 15242–9
[136] Wang L and Byon H R 2013 N-Methyl-N-propylpiperidinium bis (trifluoromethanesulfonyl) imide-based organic electrolyte for high performance lithium–sulfur batteries *J. Power Sources* **236** 207–14
[137] Cao X, He X, Wang J, Liu H, Röser S, Rad B R, Evertz M, Streipert B, Li J and Wagner R 2016 High voltage $LiNi_{0.5}Mn_{1.5}O_4/Li_4Ti_5O_{12}$ lithium ion cells at elevated temperatures: carbonate-versus ionic liquid-based electrolytes *ACS Appl. Mater. Interfaces* **8** 25971–8
[138] Karuppasamy K, Rhee H W, Reddy P A, Gupta D, Mitu L, Polu A R and Shajan X S 2016 Ionic liquid incorporated nanocomposite polymer electrolytes for rechargeable lithium ion battery: a way to achieve improved electrochemical and interfacial properties *J. Ind. Eng. Chem.* **40** 168–76
[139] Malik R, Zhou F and Ceder G 2011 Kinetics of non-equilibrium lithium incorporation in $LiFePO_4$ *Nat. Mater.* **10** 587–90
[140] Xu K 2004 Nonaqueous liquid electrolytes for lithium-based rechargeable batteries *Chem. Rev.* **104** 4303–418
[141] Tamilarasan P and Ramaprabhu S 2013 Graphene based all-solid-state supercapacitors with ionic liquid incorporated polyacrylonitrile electrolyte *Energy* **51** 374–81
[142] Li Y, Wong K W, Dou Q and Ng K M 2016 A single-ion conducting and shear-thinning polymer electrolyte based on ionic liquid-decorated PMMA nanoparticles for lithium-metal batteries *J. Mater. Chem.* A **4** 18543–50
[143] Rao M, Geng X, Liao Y, Hu S and Li W 2012 Preparation and performance of gel polymer electrolyte based on electrospun polymer membrane and ionic liquid for lithium ion battery *J. Membr. Sci.* **399** 37–42
[144] Kitazawa Y, Iwata K, Imaizumi S, Ahn H, Kim S Y, Ueno K, Park M J and Watanabe M 2014 Gelation of solvate ionic liquid by self-assembly of block copolymer and characterization as polymer electrolyte *Macromolecules* **47** 6009–16
[145] Long L, Wang S, Xiao M and Meng Y 2016 Polymer electrolytes for lithium polymer batteries *J. Mater. Chem.* A **4** 10038–69
[146] Li X, Qian K, He Y-B, Liu C, An D, Li Y, Zhou D, Lin Z, Li B and Yang Q-H 2017 A dual-functional gel-polymer electrolyte for lithium ion batteries with superior rate and safety performances *J. Mater. Chem.* A **5** 18888–95
[147] Singh V K and Singh R K 2015 Development of ion conducting polymer gel electrolyte membranes based on polymer PVdF-HFP, BMIMTFSI ionic liquid and the Li-salt with improved electrical, thermal and structural properties *J. Mater. Chem.* C **3** 7305–18
[148] Jayaprakash N, Das S and Archer L 2011 The rechargeable aluminum-ion battery *Chem. Commun.* **47** 12610–2
[149] Zhang Y, Liu S Q, Ji Y J, Ma J M and Yu H J 2018 Emerging nonaqueous aluminum-ion batteries: challenges, status, and perspectives *Adv. Mater.* **30** 170630
[150] Zhu G, Angell M, Pan C-J, Lin M-C, Chen H, Huang C-J, Lin J, Achazi A J, Kaghazchi P and Hwang B-J 2019 Rechargeable aluminum batteries: effects of cations in ionic liquid electrolytes *RSC Adv.* **9** 11322–30

[151] Levitin G, Yarnitzky C and Licht S 2002 Fluorinated graphites as energetic cathodes for nonaqueous Al batteries *Electrochem. Solid-State Lett.* **5** A160

[152] Rani J V, Kanakaiah V, Dadmal T, Rao M S and Bhavanarushi S 2013 Fluorinated natural graphite cathode for rechargeable ionic liquid based aluminum–ion battery *J. Electrochem. Soc.* **160** A1781

[153] Zhang K Q, Kirlikovali K O, Suh J M, Choi J W, Jang H W, Varma R S, Farha O K and Shokouhimehr M 2020 Recent advances in rechargeable aluminum-ion batteries and considerations for their future progress *ACS Appl. Energy Mater.* **3** 6019–35

[154] Sun X-G, Fang Y, Jiang X, Yoshii K, Tsuda T and Dai S 2016 Polymer gel electrolytes for application in aluminum deposition and rechargeable aluminum ion batteries *Chem. Commun.* **52** 292–5

[155] Wang H, Bai Y, Chen S, Luo X, Wu C, Wu F, Lu J and Amine K 2015 Binder-free V_2O_5 cathode for greener rechargeable aluminum battery *ACS Appl. Mater. Interfaces* **7** 80–4

[156] Holleck G L 1972 The reduction of chlorine on carbon in $AlCl_3$-KCl-NaCl melts *J. Electrochem. Soc.* **119** 1158

[157] Li Q and Bjerrum N J 2002 Aluminum as anode for energy storage and conversion: a review *J. Power Sources* **110** 1–10

[158] Jiang T, Brym M C, Dubé G, Lasia A and Brisard G 2006 Electrodeposition of aluminium from ionic liquids: part I—electrodeposition and surface morphology of aluminium from aluminium chloride ($AlCl_3$)–1-ethyl-3-methylimidazolium chloride ([EMIm] Cl) ionic liquids *Surf. Coat. Technol.* **201** 1–9

[159] Takami N and Koura N 1988 Anodic sulfidation of FeS electrode in a NaCl saturated $AlCl_3$-NaCl melt *Electrochim. Acta* **33** 1137–42

[160] Hjuler H, Von Winbush S, Berg R W and Bjerrum N 1989 A novel inorganic low melting electrolyte for secondary aluminum-nickel sulfide batteries *J. Electrochem. Soc.* **136** 901

[161] Berrettoni M, Tossici R, Zamponi S, Marassi R and Mamantov G 1993 A cyclic voltammetric study of the electrochemical behavior of NiS_2 in molten NaCl saturated $NaAlCl_4$ melts *J. Electrochem. Soc.* **140** 969

[162] Reed L D and Menke E 2013 The roles of V_2O_5 and stainless steel in rechargeable Al–ion batteries *J. Electrochem. Soc.* **160** A915

[163] Dymek C Jr, Williams J, Groeger D and Auborn J 1984 An aluminum acid-base concentration cell using room temperature chloroaluminate ionic liquids *J. Electrochem. Soc.* **131** 2887

[164] Chiku M, Takeda H, Matsumura S, Higuchi E and Inoue H 2015 Amorphous vanadium oxide/carbon composite positive electrode for rechargeable aluminum battery *ACS Appl. Mater. Interfaces* **7** 24385–9

[165] Zhang X, Tang Y, Zhang F and Lee C S 2016 A novel aluminum–graphite dual-ion battery *Adv. Energy Mater.* **6** 1502588

[166] Fredlake C P, Crosthwaite J M, Hert D G, Aki S N and Brennecke J F 2004 Thermophysical properties of imidazolium-based ionic liquids *J. Chem. Eng. Data* **49** 954–64

[167] Speakman S A 2013 Introduction to x-ray powder diffraction data analysis *Center for Mater. Sci. Eng. at MIT*

[168] VanHove M A, Weinberg W H and Chan C-M 2012 *Low-Energy Electron Diffraction: Experiment, Theory and Surface Structure Determination* vol 6 (Cham: Springer)

[169] Chen C J 1993 *Introduction to Scanning Tunneling Microscopy* vol 4 (Oxford: Oxford University Press)

[170] Reimer L 2013 *Transmission Electron Microscopy: Physics of Image Formation and Microanalysis* vol 36 (Berlin: Springer)
[171] Hayat M E 2012 *Basic Techniques for Transmission Electron Microscopy* (Amsterdam: Elsevier)
[172] Wang Z, Xiang R, Shi Q, Yang J, Wang X, Hou J and Chen J 2006 Modeling STM images in graphene using the effective-mass approximation *Phys. Rev.* B **74** 125417
[173] Lv B, Qian T and Ding H 2019 Angle-resolved photoemission spectroscopy and its application to topological materials *Nat. Rev. Phys.* **1** 609–26
[174] Zhang C, Li Y, Pei D, Liu Z and Chen Y 2020 Angle-resolved photoemission spectroscopy study of topological quantum materials *Annu. Rev. Mater. Res.* **50** 131–53
[175] Damascelli A 2004 Probing the electronic structure of complex systems by ARPES *Phys. Scr.* **2004** 61
[176] Jishi R and Dresselhaus M 1992 Superconductivity in graphite intercalation compounds *Phys. Rev.* B **45** 12465
[177] Belash I, Bronnikov A, Zharikov O and Pal'nichenko A 1989 Superconductivity of graphite intercalation compound with lithium C_2Li *Solid State Commun.* **69** 921–3
[178] Trukhin A 1997 Self-trapped excitons in SiO_2, GeO_2, Li_2GeO_3, $AlPO_4$ and $GaPO_4$ *Mater. Sci. Forum* (Freienbach: Trans Tech Publ.) pp 531–6
[179] Trukhin A N, Rogulis U and Spingis M 1997 Self-trapped exciton in Li_2GeO_3 *J. Lumin.* **72** 890–2
[180] Castro L, Dedryvere R, El Khalifi M, Lippens P-E, Bréger J, Tessier C and Gonbeau D 2010 The spin-polarized electronic structure of $LiFePO_4$ and $FePO_4$ evidenced by in-lab XPS *J. Phys. Chem.* C **114** 17995–8000
[181] Kang B and Ceder G 2009 Battery materials for ultrafast charging and discharging *Nature* **458** 190–3
[182] Yuan L-X, Wang Z-H, Zhang W-X, Hu X-L, Chen J-T, Huang Y-H and Goodenough J B 2011 Development and challenges of $LiFePO_4$ cathode material for lithium-ion batteries *Energy Environ. Sci.* **4** 269–84
[183] Yamada A, Chung S and Hinokuma K Optimized $LiFePO_4$ for lithium battery cathodes *J. Electrochem. Soc.* **148** A224
[184] Andersson A S, Kalska B, Häggström L and Thomas J O 2000 Lithium extraction/insertion in $LiFePO_4$: an x-ray diffraction and Mössbauer spectroscopy study *Solid State Ionics* **130** 41–52
[185] Wang L, He X, Sun W, Wang J, Li Y and Fan S 2012 Crystal orientation tuning of $LiFePO_4$ nanoplates for high rate lithium battery cathode materials *Nano Lett.* **12** 5632–6
[186] Ong S P, Wang L, Kang B and Ceder G 2008 Li–Fe–P–O_2 phase diagram from first principles calculations *Chem. Mater.* **20** 1798–807
[187] Nakayama M, Yamada S, Jalem R and Kasuga T 2016 Density functional studies of olivine-type $LiFePO_4$ and $NaFePO_4$ as positive electrode materials for rechargeable lithium and sodium ion batteries *Solid State Ionics* **286** 40–4
[188] Nagpure S C, Babu S, Bhushan B, Kumar A, Mishra R, Windl W, Kovarik L and Mills M 2011 Local electronic structure of $LiFePO_4$ nanoparticles in aged Li-ion batteries *Acta Mater.* **59** 6917–26
[189] Xia Y, Zhang W, Huang H, Gan Y, Tian J and Tao X 2011 Self-assembled mesoporous $LiFePO_4$ with hierarchical spindle-like architectures for high-performance lithium-ion batteries *J. Power Sources* **196** 5651–8

[190] Kresse G and Furthmüller J 1996 Efficient iterative schemes for *ab initio* total-energy calculations using a plane-wave basis set *Phys. Rev.* B **54** 11169
[191] Kresse G and Joubert D 1999 From ultrasoft pseudopotentials to the projector augmented-wave method *Phys. Rev.* B **59** 1758
[192] Ohzuku T, Takehara Z and Yoshizawa S 1978 A graphite compound as cathode for rechargeable nonaqueous lithium battery *Denki Kagaku oyobi Kogyo Butsuri Kagaku* **46** 438–41
[193] Sutter P, Sadowski J T and Sutter E A 2010 Chemistry under cover: tuning metal–graphene interaction by reactive intercalation *J. Am. Chem. Soc.* **132** 8175–9
[194] Bucko T, Hafner J, Lebegue S and Angyán J G 2010 Improved description of the structure of molecular and layered crystals: *ab initio* DFT calculations with van der Waals corrections *J. Phys. Chem.* A **114** 11814–24
[195] Goerbig M 2011 Electronic properties of graphene in a strong magnetic field *Rev. Mod. Phys.* **83** 1193
[196] Huang Y, Lin M-F and Chang C 2008 Landau levels and magneto-optical properties of graphene ribbons *J. Appl. Phys.* **103** 073709
[197] Rong Z Y and Kuiper P 1993 Electronic effects in scanning tunneling microscopy: Moiré pattern on a graphite surface *Phys. Rev.* B **48** 17427
[198] Allen M J, Tung V C and Kaner R B 2010 Honeycomb carbon: a review of graphene *Chem. Rev.* **110** 132–45
[199] Johnston D 1955 The structure of the π-band of graphite *Proc. R. Soc. Lond.* A **227** 349–58
[200] Watanabe K, Kondow T, Soma M, Onishi T and Tamaru K 1973 Molecular-sieve type sorption on alkali graphite intercalation compounds *Proc. R. Soc. Lond.* A **333** 51–67
[201] Hartwigsen C, Witschel W and Spohr E 1997 Ab initio study of structural properties of stage-1 alkali graphite intercalation compounds *Ber. Bunsenges. Phys. Chem.* **101** 859–62
[202] Sathiya M, Rousse G, Ramesha K, Laisa C, Vezin H, Sougrati M T, Doublet M-L, Foix D, Gonbeau D and Walker W 2013 Reversible anionic redox chemistry in high-capacity layered-oxide electrodes *Nat. Mater.* **12** 827–35
[203] Song J, Shin D W, Lu Y, Amos C D, Manthiram A and Goodenough J B 2012 Role of oxygen vacancies on the performance of Li $[Ni_{0.5-x}Mn_{1.5+x}]O_4$ ($x = 0$, 0.05, and 0.08) spinel cathodes for lithium-ion batteries *Chem. Mater.* **24** 3101–9
[204] Ventosa E, Mei B, Xia W, Muhler M and Schuhmann W 2013 TiO_2 (B)/anatase composites synthesized by spray drying as high performance negative electrode material in Li-Ion batteries *ChemSusChem* **6** 1312–5
[205] Singh D P, Mulder F M, Abdelkader A M and Wagemaker M 2013 Facile micro templating $LiFePO_4$ electrodes for high performance Li-ion batteries *Adv. Energy Mater.* **3** 572–8
[206] Su D S and Schlögl R 2010 Nanostructured carbon and carbon nanocomposites for electrochemical energy storage applications *ChemSusChem* **3** 136–68
[207] Wu Z-S, Zhou G, Yin L-C, Ren W, Li F and Cheng H-M 2012 Graphene/metal oxide composite electrode materials for energy storage *Nano Energy* **1** 107–31
[208] Natori A, Ohno T and Oshiyama A 1985 Work function of alkali-atom adsorbed graphite *J. Phys. Soc. Jpn.* **54** 3042–50
[209] Li W-B, Lin S-Y, Tran N T T, Lin M-F and Lin K-I 2020 Essential geometric and electronic properties in stage-n graphite alkali-metal-intercalation compounds *RSC Adv.* **10** 23573–81

[210] Bhauriyal P, Mahata A and Pathak B 2017 The staging mechanism of $AlCl_4$ intercalation in a graphite electrode for an aluminium-ion battery *Phys. Chem. Chem. Phys.* **19** 7980–9

[211] Wu M, Xu B, Chen L and Ouyang C 2016 Geometry and fast diffusion of $AlCl_4$ cluster intercalated in graphite *Electrochim. Acta* **195** 158–65

[212] Wang C, Li D, Too C O and Wallace G G 2009 Electrochemical properties of graphene paper electrodes used in lithium batteries *Chem. Mater.* **21** 2604–6

[213] Ni Z, Wang H, Kasim J, Fan H, Yu T, Wu Y H, Feng Y and Shen Z 2007 Graphene thickness determination using reflection and contrast spectroscopy *Nano Lett.* **7** 2758–63

[214] Wang Y, Ni Z, Shen Z, Wang H and Wu Y 2008 Interference enhancement of Raman signal of graphene *Appl. Phys. Lett.* **92** 043121

[215] Kerber S, Barr T, Mann G, Brantley W, Papazoglou E and Mitchell J 1998 The complementary nature of x-ray photoelectron spectroscopy and angle-resolved x-ray diffraction part II: analysis of oxides on dental alloys *J. Mater. Eng. Perform.* **7** 334–42

[216] Shklover V, Haibach T, Ried F, Nesper R and Novak P 1996 Crystal structure of the product of Mg^{2+} insertion into V_2O_5 single crystals *J. Solid State Chem.* **123** 317–23

[217] Foulkes W M C and Haydock R 1989 Tight-binding models and density-functional theory *Phys. Rev.* B **39** 12520

[218] Kittel C 1996 *Introduction to Solid State Physics* vol 402 (New York: Wiley)

[219] Sakai Y and Ehara S 1999 Surface structure and electronic states of V_2O_5 surface studied by scanning tunneling microscopy/spectroscopy *Jpn. J. Appl. Phys.* **38** 2944

[220] Nakagawa A, Kuwata N, Matsuda Y and Kawamura J 2010 Characterization of stable solid electrolyte lithium silicate for thin film lithium battery *J. Phys. Soc. Jpn.* **79** 98–101

[221] Shi L, Zhu T, Gao G, Zhang X, Wei W, Liu W and Ding S 2018 Highly stretchable and transparent ionic conducting elastomers *Nat. Commun.* **9** 1–7

[222] Watanabe M, Inoi J, Kim S W, Kaneko T, Toda A, Sato M, Uematsu K, Toda K, Koide J and Toda M 2017 Synthesis of Li_2SiO_3 using novel water-assisted solid state reaction method *J. Ceram. Soc. Jpn.* **125** 472–5

[223] Alemi A, Khademinia S and Sertkol M 2015 Part III: lithium metasilicate (Li_2SiO_3)—mild condition hydrothermal synthesis, characterization and optical properties *Int. Nano Lett.* **5** 77–83

[224] Alemi A and Khademinia S 2015 Part I: lithium metasilicate (Li_2SiO_3)—mild condition hydrothermal synthesis, characterization, and optical properties *Int. Nano Lett.* **5** 15–20

[225] Sakiyama H, Yamasaki M, Suzuki T, Watanabe Y, Ito R, Ohnishi A and Nishida Y 2006 Formation of polar crystals of a chiral dinuclear nickel (II) complex and their pyroelectric effect and electro-optic effect *Inorg. Chem. Commun.* **9** 18–20

[226] Ballato J and Haertling G H 2004 Piezoelectric and electro-optic *Ceramic Mater. Electron.* **68** 207

[227] Recker K and Wallrafen F 1984 Crystal growth and dielectric, piezoelectric, electrooptic, elastic and thermoelastic properties of orthorhombic lithium hydrogen selenite, $LiHSeO_3$ *Z. Kristallogr. Cryst. Mater.* **169** 249–56

[228] Lewis M H 2013 *Glasses and Glass-Ceramics* (Berlin: Springer)

[229] Habasaki J and Okada I 1992 Molecular dynamics study of Li_2SiO_3 in the liquid and glassy states *Mol. Simul.* **8** 179–95

[230] Tang T, Chen P, Luo W, Luo D and Wang Y 2012 Crystalline and electronic structures of lithium silicates: a density functional theory study *J. Nucl. Mater.* **420** 31–8

[231] Zulueta Y A, Froeyen M and Nguyen M T 2017 Structural properties and mechanical stability of lithium-ion based materials. A theoretical study *Comput. Mater. Sci.* **136** 271–9

[232] Richet P, Mysen B O and Andrault D 1996 Melting and premelting of silicates: Raman spectroscopy and x-ray diffraction of Li_2SiO_3 and Na_2SiO_3 *Phys. Chem. Miner.* **23** 157–72

[233] Ortiz-Landeros J, Contreras-García M, Gomez-Yañez C and Pfeiffer H 2011 Surfactant-assisted hydrothermal crystallization of nanostructured lithium metasilicate (Li_2SiO_3) hollow spheres:(I) synthesis, structural and microstructural characterization *J. Solid State Chem.* **184** 1304–11

[234] Li X and Yang H 2014 Morphology-controllable Li_2SiO_3 nanostructures *CrystEngComm* **16** 4501–7

[235] Furusawa S, Tsurui T and Shimizu K 2014 TEM observation and ionic conductivity study of Li_2SiO_3 thin-film on sapphire substrate *Key Eng. Mater.* **596** 15–20

[236] Feenstra R M 1994 Scanning tunneling spectroscopy *Surf. Sci.* **299** 965–79

[237] Haussühl S, Liebertz J and Stähr S 1982 Single crystal growth and pyroelectric, dielectric, piezoelectric, elastic, and thermoelastic properties of orthorhombic Li_2SiO_3, Li_2GeO_3, and Na_2GeO_3 *Cryst. Res. Technol.* **17** 521–6

[238] Foss C E L, Svensson A M, Gullbrekken Ø, Sunde S and Vullum-Bruer F 2018 Temperature effects on performance of graphite anodes in carbonate based electrolytes for lithium ion batteries *J. Energy Storage* **17** 395–402

[239] Gireaud L, Grugeon S, Laruelle S, Pilard S and Tarascon J-M 2005 Identification of Li battery electrolyte degradation products through direct synthesis and characterization of alkyl carbonate salts *J. Electrochem. Soc.* **152** A850

[240] Leggesse E G and Jiang J-C 2012 Theoretical study of the reductive decomposition of 1, 3-propane sultone: SEI forming additive in lithium-ion batteries *RSC Adv.* **2** 5439–46

[241] Liu X, Zhou J, Xu Z and Wang Y 2020 Atomic thermodynamics and microkinetics of the reduction mechanism of electrolyte additives to facilitate the formation of solid electrolyte interphases in lithium-ion batteries *RSC Adv.* **10** 16302–12

[242] Zhang S S 2006 A review on electrolyte additives for lithium-ion batteries *J. Power Sources* **162** 1379–94

[243] Li S, Luo Z, Li L, Hu J, Zou G, Hou H and Ji X 2020 Recent progress on electrolyte additives for stable lithium metal anode *Energy Storage Mater.* **32** 306–19

[244] Eshetu G G and Figgemeier E 2019 Confronting the challenges of next-generation silicon anode-based lithium-ion batteries: role of designer electrolyte additives and polymeric binders *ChemSusChem* **12** 2515–39

[245] Gao H, Maglia F, Lamp P, Amine K and Chen Z 2017 Mechanistic study of electrolyte additives to stabilize high-voltage cathode–electrolyte interface in lithium-ion batteries *ACS Appl. Mater. Interfaces* **9** 44542–9

[246] Halls M D and Tasaki K 2010 High-throughput quantum chemistry and virtual screening for lithium ion battery electrolyte additives *J. Power Sources* **195** 1472–8

[247] Wu B, Han S, Shin K G and Lu W 2018 Application of artificial neural networks in design of lithium-ion batteries *J. Power Sources* **395** 128–36

[248] Bernal J D 1924 The structure of graphite *Proc. R. Soc. Lond.* A **106** 749–73

[249] Warner J H, Rümmeli M H, Gemming T, Büchner B and Briggs G A D 2009 Direct imaging of rotational stacking faults in few layer graphene *Nano Lett.* **9** 102–6

[250] Coletti C, Forti S, Principi A, Emtsev K, Zakharov A, Daniels K, Daas B, Chandrashekhar M, MacDonald A and Polini M 2014 Revealing the electronic band structure of quasi-free trilayer graphene on SiC(0001) *MRS Online Proceedings Library Archive* **1693** 159–67

[251] Beckett R and Croft R 1952 The structure of graphite oxide *J. Phys. Chem.* **56** 929–35

[252] Kopnin N, Ijäs M, Harju A and Heikkilä T 2013 High-temperature surface superconductivity in rhombohedral graphite *Phys. Rev.* B **87** 140503

[253] Yang Y, Zou Y-C, Woods C R, Shi Y, Yin J, Xu S, Ozdemir S, Taniguchi T, Watanabe K and Geim A K 2019 Stacking order in graphite films controlled by van der Waals technology *Nano Lett.* **19** 8526–32

[254] Liu Z, Suenaga K, Harris P J and Iijima S 2009 Open and closed edges of graphene layers *Phys. Rev. Lett.* **102** 015501

[255] Lee J-K, Lee S-C, Ahn J-P, Kim S-C, Wilson J I and John P 2008 The growth of AA graphite on (111) diamond *J. Chem. Phys.* **129** 234709

[256] Wallace P R 1947 The band theory of graphite *Phys. Rev.* **71** 622

[257] Amelinckx S and Delavignette P 1960 Electron optical study of basal dislocations in graphite *J. Appl. Phys.* **31** 2126–35

[258] Lin M, Huang C and Chuu D 1997 Plasmons in graphite and stage-1 graphite intercalation compounds *Phys. Rev.* B **55** 13961

[259] Matsui T, Kambara H, Niimi Y, Tagami K, Tsukada M and Fukuyama H 2005 STS observations of Landau levels at graphite surfaces *Phys. Rev. Lett.* **94** 226403

[260] Fujita M, Wakabayashi K, Nakada K and Kusakabe K 1996 Peculiar localized state at zigzag graphite edge *J. Phys. Soc. Jpn.* **65** 1920–3

[261] Červenka J, Katsnelson M and Flipse C 2009 Room-temperature ferromagnetism in graphite driven by two-dimensional networks of point defects *Nat. Phys.* **5** 840–4

[262] De S, King P J, Lotya M, O'Neill A, Doherty E M, Hernandez Y, Duesberg G S and Coleman J N 2010 Flexible, transparent, conducting films of randomly stacked graphene from surfactant-stabilized, oxide-free graphene dispersions *Small* **6** 458–64

[263] Inagaki M 1989 Applications of graphite intercalation compounds *J. Mater. Res.* **4** 1560–8

[264] Caragiu M and Finberg S 2005 Alkali metal adsorption on graphite: a review *J. Phys. Condens. Matter* **17** R995

[265] Watanabe K, Kondow T, Onishi T and Tamaru K 1972 Molecular-sieve type sorption on alkali graphites *Chem. Lett.* **1** 477–8

[266] Bower C, Suzuki S, Tanigaki K and Zhou O 1998 Synthesis and structure of pristine and alkali-metal-intercalated single-walled carbon nanotubes *Appl. Phys.* A **67** 47–52

[267] Hightower A, Ahn C, Fultz B and Rez P 2000 Electron energy-loss spectrometry on lithiated graphite *Appl. Phys. Lett.* **77** 238–40

[268] Selloni A, Chen C and Tosatti E 1988 Scanning tunneling spectroscopy of graphite and intercalates *Phys. Scr.* **38** 297

[269] Nobuhara K, Nakayama H, Nose M, Nakanishi S and Iba H 2013 First-principles study of alkali metal-graphite intercalation compounds *J. Power Sources* **243** 585–7

[270] Wang Z, Ratvik A P, Grande T and Selbach S M 2015 Diffusion of alkali metals in the first stage graphite intercalation compounds by vdW-DFT calculations *RSC Adv.* **5** 15985–92

[271] Rangappa D, Sone K, Kudo T and Honma I 2010 Directed growth of nanoarchitectured $LiFePO_4$ electrode by solvothermal synthesis and their cathode properties *J. Power Sources* **195** 6167–71

[272] Kanno R, Shirane T, Kawamoto Y, Takeda Y, Takano M, Ohashi M and Yamaguchi Y 1996 Synthesis, structure, and electrochemical properties of a new lithium iron oxide, $LiFeO_2$, with a corrugated layer structure *J. Electrochem. Soc.* **143** 2435

[273] Takahashi M, Tobishima S, Takei K and Sakurai Y 2001 Characterization of $LiFePO_4$ as the cathode material for rechargeable lithium batteries *J. Power Sources* **97** 508–11

[274] Anastas P T and Zimmerman J B 2019 The periodic table of the elements of green and sustainable chemistry *Green Chem.* **21** 6545–66

[275] Hirsch A 2010 The era of carbon allotropes *Nat. Mater.* **9** 868–71

[276] Enyashin A N and Ivanovskii A L 2011 Graphene allotropes *Phys. Status Solidi* **248** 1879–83

[277] Mittal J and Lin K 2017 Carbon nanotube-based interconnections *J. Mater. Sci.* **52** 643–62

[278] Shyu L F and Lin M F 2002 Electronic and optical properties of narrow-gap carbon nanotubes *J. Phys. Soc. Jpn.* **71** 1820–3

[279] Omacrsawa E 2012 *Perspectives of Fullerene Nanotechnology* (Cham: Springer)

[280] Siddiqi M and Kishek R 2019 A predictive model for multipactor discharge in coaxial systems based on chaos theory *IEEE Trans. Electron Devices* **66** 4403–7

[281] Lin S Y T, Chang N T T, Su S L and Lin, M. F. W P 2018 *Structure- and Adatom-Enriched Essential Properties of Graphene Nanoribbons* (Boca Raton, FL: CRC Press)

[282] Yin W-J, Xie Y-E, Liu L-M, Chen Y-P, Wang R-Z, Wei X-L, Zhong J-X and Lau L 2013 Atomic structure and electronic properties of folded graphene nanoribbons: a first-principles study *J. Appl. Phys.* **113** 173506

[283] Li H, Li M and Kang Z 2018 Mechanics of the scrolling and folding of graphene *Nanotechnology* **29** 245604

[284] Notarianni M, Liu J, Vernon K and Motta N 2016 Synthesis and applications of carbon nanomaterials for energy generation and storage *Beilstein J. Nanotechnol.* **7** 149–96

[285] Meng X, Tongay S, Kang J, Chen Z, Wu F, Li S-S, Xia J-B, Li J and Wu J 2013 Stable p- and n-type doping of few-layer graphene/graphite *Carbon* **57** 507–14

[286] Chan C K, Peng H, Liu G, McIlwrath K, Zhang X F, Huggins R A and Cui Y 2008 High-performance lithium battery anodes using silicon nanowires *Nat. Nanotechnol.* **3** 31–5

[287] Zuo X, Zhu J, Müller-Buschbaum P and Cheng Y-J 2017 Silicon based lithium-ion battery anodes: a chronicle perspective review *Nano Energy* **31** 113–43

[288] Umeshbabu E, Zheng B and Yang Y 2019 Recent progress in all-solid-state Lithium –Sulfur batteries using high Li-ion conductive solid electrolytes *Electrochem. Energy Rev.* **2** 199–230

[289] Xue R, Zhang J, Li Y and Li Y 2018 Organic solar cell materials toward commercialization *Small* **14** 1801793

[290] Sherif S A, Goswami D Y, Stefanakos E L and Steinfeld A 2014 *Handbook of Hydrogen Energy* (Boca Raton, FL: CRC Press)

[291] Adachi S 2009 *Properties of Semiconductor Alloys: Group-IV, III-V and II-VI Semiconductors* vol 28 (New York: Wiley)

[292] Madelung O 1982 Semiconductors-physics of II-VI and I-VII compounds, semi-magnetic semiconductors *Landolt-Bornstein-Numerical Data and Fundamental Relationships in Science and Technology* vol 17 (Cham: Springer) p 255

[293] Ohno H, Munekata H, Penney T, Von Molnar S and Chang L 1993 Partial ferromagnetic order in p-Type (In, Mn) as diluted magnetic III-V semiconductors *Mater. Sci. Forum* **117–118** 297–302

[294] Kordyuk A 2014 ARPES experiment in fermiology of quasi-2D metals *Low Temp. Phys.* **40** 286–96

[295] Bussolotti F, Chi D, Goh K J, Huang Y L and Wee A T 2020 *2D Semiconductor Materials and Devices* (Amsterdam: Elsevier) pp 199–220

[296] Scrosati W A v S B 2002 *Advances in Lithium-Ion Batteries* (Cham: Springer)

[297] Li G, Chen Z and Lu J 2018 Lithium-sulfur batteries for commercial applications *Chem* **4** 3–7

[298] Yang L, Li Q, Wang Y, Chen Y, Guo X, Wu Z, Chen G, Zhong B, Xiang W and Zhong Y 2020 A review of cathode materials in lithium-sulfur batteries *Ionics* **26** 5299–318

[299] Kubota K, Dahbi M, Hosaka T, Kumakura S and Komaba S 2018 Towards K-ion and Na-ion batteries as 'Beyond Li-Ion' *Chem. Rec.* **18** 459–79

[300] Yabuuchi N, Kubota K, Dahbi M and Komaba S 2014 Research development on sodium-ion batteries *Chem. Rev.* **114** 11636–82

[301] Yoo H D, Shterenberg I, Gofer Y, Gershinsky G, Pour N and Aurbach D 2013 Mg rechargeable batteries: an on-going challenge *Energy Environ. Sci.* **6** 2265–79

[302] Li Y, Lu Y, Adelhelm P, Titirici M-M and Hu Y-S 2019 Intercalation chemistry of graphite: alkali metal ions and beyond *Chem. Soc. Rev.* **48** 4655–87

[303] Zhou W and Sit P H-L 2020 First-principles understanding of the staging properties of the graphite intercalation compounds towards dual-ion battery applications *ACS Omega* **5** 18289–300

[304] Calandra M and Mauri F 2006 Possibility of superconductivity in graphite intercalated with alkaline earths investigated with density functional theory *Phys. Rev.* B **74** 094507

[305] Hembram K P, Jung H, Yeo B C, Pai S J, Lee H J, Lee K-R and Han S S 2016 A comparative first-principles study of the lithiation, sodiation, and magnesiation of black phosphorus for Li-, Na-, and Mg-ion batteries *Phys. Chem. Chem. Phys.* **18** 21391–7

[306] Okamoto Y 2014 Density functional theory calculations of alkali metal (Li, Na, and K) graphite intercalation compounds *J. Phys. Chem.* C **118** 16–9

[307] Oh J-M, Biswick T T and Choy J-H 2009 Layered nanomaterials for green materials *J. Mater. Chem.* **19** 2553–63

[308] Thakur V K 2013 *Green Composites from Natural Resources* (Boca Raton, FL: CRC Press)

[309] Bhardwaj T, Antic A, Pavan B, Barone V and Fahlman B D 2010 Enhanced electrochemical lithium storage by graphene nanoribbons *J. Am. Chem. Soc.* **132** 12556–8

[310] Chen J and Cheng F 2009 Combination of lightweight elements and nanostructured materials for batteries *Acc. Chem. Res.* **42** 713–23

[311] Bruce P G, Scrosati B and Tarascon J M 2008 Nanomaterials for rechargeable lithium batteries *Angew. Chem. Int. Ed.* **47** 2930–46

[312] Kaskhedikar N A and Maier J 2009 Lithium storage in carbon nanostructures *Adv. Mater.* **21** 2664–80

[313] Novoselov K S, Geim A K, Morozov S V, Jiang D, Zhang Y, Dubonos S V, Grigorieva I V and Firsov A A 2004 Electric field effect in atomically thin carbon films *Science* **306** 666–9

[314] He L, Lu J Q and Jiang H 2009 Controlled carbon-nanotube junctions self-assembled from graphene nanoribbons *Small* **5** 2802–6

[315] Wang G, Shen X, Yao J and Park J 2009 Graphene nanosheets for enhanced lithium storage in lithium ion batteries *Carbon* **47** 2049–53
[316] Lian P, Zhu X, Liang S, Li Z, Yang W and Wang H 2010 Large reversible capacity of high quality graphene sheets as an anode material for lithium-ion batteries *Electrochim. Acta* **55** 3909–14
[317] Chen S, Chen P, Wu M, Pan D and Wang Y 2010 Graphene supported Sn–Sb@ carbon core–shell particles as a superior anode for lithium ion batteries *Electrochem. Commun.* **12** 1302–6
[318] Chen S Q and Wang Y 2010 Microwave-assisted synthesis of a Co_3O_4–graphene sheet-on-sheet nanocomposite as a superior anode material for Li-ion batteries *J. Mater. Chem.* **20** 9735–9
[319] Murugan A V, Muraliganth T and Manthiram A 2009 Rapid, facile microwave-solvothermal synthesis of graphene nanosheets and their polyaniline nanocomposites for energy strorage *Chem. Mater.* **21** 5004–6
[320] Gwon H, Kim H-S, Lee K U, Seo D-H, Park Y C, Lee Y-S, Ahn B T and Kang K 2011 Flexible energy storage devices based on graphene paper *Energy Environ. Sci.* **4** 1277–83
[321] Tung V C, Chen L-M, Allen M J, Wassei J K, Nelson K, Kaner R B and Yang Y 2009 Low-temperature solution processing of graphene–carbon nanotube hybrid materials for high-performance transparent conductors *Nano Lett.* **9** 1949–55
[322] Yu D and Dai L 2010 Self-assembled graphene/carbon nanotube hybrid films for supercapacitors *J. Phys. Chem. Lett.* **1** 467–70
[323] Fan Z, Yan J, Zhi L, Zhang Q, Wei T, Feng J, Zhang M, Qian W and Wei F 2010 A three-dimensional carbon nanotube/graphene sandwich and its application as electrode in supercapacitors *Adv. Mater.* **22** 3723–8
[324] Schnorr J M and Swager T M 2011 Emerging applications of carbon nanotubes *Chem. Mater.* **23** 646–57
[325] Wang G, Ling Y, Qian F, Yang X, Liu X-X and Li Y 2011 Enhanced capacitance in partially exfoliated multi-walled carbon nanotubes *J. Power Sources* **196** 5209–14
[326] Zhang D, Zhao Y, Goodenough J B, Lu Y, Chen C, Wang L and Liu J 2011 Exfoliation from carbon nanotubes versus tube size on lithium insertion *Electrochem. Commun.* **13** 125–8
[327] Zhang H, Cao G and Yang Y 2009 Carbon nanotube arrays and their composites for electrochemical capacitors and lithium-ion batteries *Energy Environ. Sci.* **2** 932–43
[328] Wang Y, Wu M, Jiao Z and Lee J Y 2009 Sn@ CNT and Sn@ C@ CNT nanostructures for superior reversible lithium ion storage *Chem. Mater.* **21** 3210–5
[329] Han Y, Wang Y, Wang Y, Jiao L, Yuan H and Liu S 2011 Hydrothermal synthesis and electrochemical properties of cobalt–carbon nanotubes nanocomposite *Electrochim. Acta* **56** 3258–63
[330] Welna D T, Qu L, Taylor B E, Dai L and Durstock M F 2011 Vertically aligned carbon nanotube electrodes for lithium-ion batteries *J. Power Sources* **196** 1455–60
[331] Wang Y and Lee J Y 2006 One-step, confined growth of bimetallic tin–antimony nanorods in carbon nanotubes grown *In Situ* for reversible Li^+ ion storage *Angew. Chem. Int. Ed.* **45** 7039–42
[332] Benvidi A, Kakoolaki P, Zare H R and Vafazadeh R 2011 Electrocatalytic oxidation of hydrazine at a Co(II) complex multi-wall carbon nanotube modified carbon paste electrode *Electrochim. Acta* **56** 2045–50

[333] Su Q, Liang Y, Feng X and Müllen K 2010 Towards free-standing graphene/carbon nanotube composite films via acetylene-assisted thermolysis of organocobalt functionalized graphene sheets *Chem. Commun.* **46** 8279–81

[334] Liu S, Zhu J, Mastai Y, Felner I and Gedanken A 2000 Preparation and characteristics of carbon nanotubes filled with cobalt *Chem. Mater.* **12** 2205–11

[335] Yoo E, Kim J, Hosono E, Zhou H-s, Kudo T and Honma I 2008 Large reversible Li storage of graphene nanosheet families for use in rechargeable lithium ion batteries *Nano Lett.* **8** 2277–82

[336] Pan D, Wang S, Zhao B, Wu M, Zhang H, Wang Y and Jiao Z 2009 Li storage properties of disordered graphene nanosheets *Chem. Mater.* **21** 3136–42

[337] Wu Z-S, Ren W, Xu L, Li F and Cheng H-M 2011 Doped graphene sheets as anode materials with superhigh rate and large capacity for lithium ion batteries *ACS Nano* **5** 5463–71

[338] Chen S, Bao P, Xiao L and Wang G 2013 Large-scale and low cost synthesis of graphene as high capacity anode materials for lithium-ion batteries *Carbon* **64** 158–69

[339] Wang X, Weng Q, Liu X, Wang X, Tang D-M, Tian W, Zhang C, Yi W, Liu D and Bando Y 2014 Atomistic origins of high rate capability and capacity of N-doped graphene for lithium storage *Nano Lett.* **14** 1164–71

[340] Pollak E, Geng B, Jeon K-J, Lucas I T, Richardson T J, Wang F and Kostecki R 2010 The interaction of Li^+ with single-layer and few-layer graphene *Nano Lett.* **10** 3386–8

[341] Lee E and Persson K A 2012 Li absorption and intercalation in single layer graphene and few layer graphene by first principles *Nano Lett.* **12** 4624–8

[342] Cui L-F, Yang Y, Hsu C-M and Cui Y 2009 Carbon–silicon core–shell nanowires as high capacity electrode for lithium ion batteries *Nano Lett.* **9** 3370–4

[343] Kim H and Cho J 2008 Superior lithium electroactive mesoporous Si@ Carbon core–shell nanowires for lithium battery anode material *Nano Lett.* **8** 3688–91

[344] Park M-H, Kim M G, Joo J, Kim K, Kim J, Ahn S, Cui Y and Cho J 2009 Silicon nanotube battery anodes *Nano Lett.* **9** 3844–7

[345] Yi T-F, Yang S-Y and Xie Y 2015 Recent advances of $Li_4Ti_5O_{12}$ as a promising next generation anode material for high power lithium-ion batteries *J. Mater. Chem.* A **3** 5750–77

[346] Liu Y, Lian J, Sun Z, Zhao M, Shi Y and Song H 2017 The first-principles study for the novel optical properties of $LiTi_2O_4$, $Li_4Ti_5O_{12}$, $Li_2Ti_2O_4$ and $Li_7Ti_5O_{12}$ *Chem. Phys. Lett.* **677** 114–9

[347] Liu H, Zhu Z, Huang J, He X, Chen Y, Zhang R, Lin R, Li Y, Yu S and Xing X 2019 Elucidating the limit of Li insertion into the spinel $Li_4Ti_5O_{12}$ *ACS Mater. Lett.* **1** 96–102

[348] Ozen S, Senay V, Pat S and Korkmaz S 2016 Optical, morphological properties and surface energy of the transparent $Li_4Ti_5O_{12}$ (LTO) thin film as anode material for secondary type batteries *J. Phys. D-Appl. Phys.* **49** 105303

[349] Huang H H, Fan X F, Singh D J and Zheng W T 2020 Recent progress of TMD nanomaterials: phase transitions and applications *Nanoscale* **12** 1247–68

[350] Karmakar S, Chowdhury C and Datta A 2016 Two-dimensional group IV monochalcogenides: anode materials for Li-ion batteries *J. Phys. Chem.* C **120** 14522–30

[351] Liu H, Su D W, Zhou R F, Sun B, Wang G X and Qiao S Z 2012 Highly ordered mesoporous MoS_2 with expanded spacing of the (002) crystal plane for ultrafast lithium ion storage *Adv. Energy Mater.* **2** 970–5

[352] Liu H, Su D W, Wang G X and Qiao S Z 2012 An ordered mesoporous WS_2 anode material with superior electrochemical performance for lithium ion batteries *J. Mater. Chem.* **22** 17437–40
[353] Soares D M and Singh G 2020 Superior electrochemical performance of layered WTe_2 as potassium-ion battery electrode *Nanotechnology* **31** 455406
[354] Soares D M, Mukherjee S and Singh G 2020 TMDs beyond MoS_2 for electrochemical energy storage *Chem.-Eur. J.* **26** 6320–41
[355] Ziarkowska W, Nowicki M and Charubin T 2019 Magnetic moment measurement stand *Int. Conf. Mechatronics* pp 178–85
[356] Raebiger H 2010 Theory of defect complexes in insulators *Phys. Rev.* B **82** 073104
[357] Guzmán G, Vazquez-Arenas J, Ramos-Sánchez G, Bautista-Ramírez M and González I 2017 Improved performance of $LiFePO_4$ cathode for Li-ion batteries through percolation studies *Electrochim. Acta* **247** 451–9
[358] Wang H, Gu S, Bai Y, Chen S, Zhu N, Wu C and Wu F 2015 Anion-effects on electrochemical properties of ionic liquid electrolytes for rechargeable aluminum batteries *J. Mater. Chem.* A **3** 22677–86
[359] Zhao W, Tan P H, Liu J and Ferrari A C 2011 Intercalation of few-layer graphite flakes with $FeCl_3$: Raman determination of Fermi level, layer by layer decoupling, and stability *J. Am. Chem. Soc.* **133** 5941–6
[360] Zhang S S 2013 Liquid electrolyte lithium/sulfur battery: fundamental chemistry, problems, and solutions *J. Power Sources* **231** 153–62
[361] Li Q, Chen J, Fan L, Kong X and Lu Y 2016 Progress in electrolytes for rechargeable Li-based batteries and beyond *Green Energy Environ.* **1** 18–42
[362] Lin S-Y, Li W-B, Tran N T T, Hsu W-D, Liu H-Y and Fa-Lin M 2019 Essential properties of Li/Li^+ graphite intercalation compounds *Green Energy Materials Handbook* (Boca Raton, FL: CRC Press)
[363] Matthes L, Pulci O and Bechstedt F 2014 Optical properties of two-dimensional honeycomb crystals graphene, silicene, germanene, and tinene from first principles *New J. Phys.* **16** 105007
[364] Saw L H, Ye Y and Tay A A 2016 Integration issues of lithium-ion battery into electric vehicles battery pack *J. Clean. Prod.* **113** 1032–45
[365] Debe M K 2012 Electrocatalyst approaches and challenges for automotive fuel cells *Nature* **486** 43–51
[366] Lin M, Cheng Y and Yen S 2005 Evaluation of PEMFC power systems for UPS base station applications *J. Power Sources* **140** 346–9
[367] Zhu X, Ong C S, Xu X, Hu B, Shang J, Yang H, Katlakunta S, Liu Y, Chen X and Pan L 2013 Direct observation of lithium-ion transport under an electrical field in Li_xCoO_2 nanograins *Sci. Rep.* **3** 1–8
[368] Fong K D, Self J, Diederichsen K M, Wood B M, McCloskey B D and Persson K A 2019 Ion transport and the true transference number in nonaqueous polyelectrolyte solutions for lithium ion batteries *ACS Cent. Sci.* **5** 1250–60
[369] Jung Y S, Cavanagh A S, Riley L A, Kang S H, Dillon A C, Groner M D, George S M and Lee S H 2010 Ultrathin direct atomic layer deposition on composite electrodes for highly durable and safe Li-ion batteries *Adv. Mater.* **22** 2172–6
[370] Wang X and Yushin G 2015 Chemical vapor deposition and atomic layer deposition for advanced lithium ion batteries and supercapacitors *Energy Environ. Sci.* **8** 1889–904

[371] Wang H-Y and Wang F-M 2013 Electrochemical investigation of an artificial solid electrolyte interface for improving the cycle-ability of lithium ion batteries using an atomic layer deposition on a graphite electrode *J. Power Sources* **233** 1–5

[372] Liu Y, Wang Y M, Yakobson B I and Wood B C 2014 Assessing carbon-based anodes for lithium-ion batteries: a universal description of charge-transfer binding *Phys. Rev. Lett.* **113** 028304

[373] Raccichini R, Varzi A, Passerini S and Scrosati B 2015 The role of graphene for electrochemical energy storage *Nat. Mater.* **14** 271–9

[374] Yoon S-H, Park C-W, Yang H, Korai Y, Mochida I, Baker R and Rodriguez N M 2004 Novel carbon nanofibers of high graphitization as anodic materials for lithium ion secondary batteries *Carbon* **42** 21–32

[375] Yan X, Teng D, Jia X, Yu Y and Yang X 2013 Improving the cyclability and rate capability of carbon nanofiber anodes through in-site generation of SiO_x-rich overlayers *Electrochim. Acta* **108** 196–202

[376] Yue H, Li F, Yang Z, Tang J, Li X and He D 2014 Nitrogen-doped carbon nanofibers as anode material for high-capacity and binder-free lithium ion battery *Mater. Lett.* **120** 39–42

[377] Snyder M Q, Trebukhova S A, Ravdel B, Wheeler M C, DiCarlo J, Tripp C P and DeSisto W J 2007 Synthesis and characterization of atomic layer deposited titanium nitride thin films on lithium titanate spinel powder as a lithium-ion battery anode *J. Power Sources* **165** 379–85

[378] Leonidov I, Leonidova O, Perelyaeva L, Samigullina R, Kovyazina S and Patrakeev M 2003 Structure, ionic conduction, and phase transformations in lithium titanate $Li_4Ti_5O_{12}$ *Phys. Solid State* **45** 2183–8

[379] Liu J, Li X, Cai M, Li R and Sun X 2013 Ultrathin atomic layer deposited ZrO_2 coating to enhance the electrochemical performance of $Li_4Ti_5O_{12}$ as an anode material *Electrochim. Acta* **93** 195–201

[380] Ahn D and Xiao X 2011 Extended lithium titanate cycling potential window with near zero capacity loss *Electrochem. Commun.* **13** 796–9

[381] Lee J-H, Hon M-H, Chung Y-W and Leu C 2011 The effect of TiO_2 coating on the electrochemical performance of ZnO nanorod as the anode material for lithium-ion battery *Appl. Phys.* A **102** 545–50

[382] Jung Y S, Cavanagh A S, Yan Y, George S M and Manthiram A 2011 Effects of atomic layer deposition of Al_2O_3 on the Li $[Li_{0.20}Mn_{0.54}Ni_{0.13}Co_{0.13}]O_2$ cathode for lithium-ion batteries *J. Electrochem. Soc.* **158** A1298

[383] Riley L A, Van Atta S, Cavanagh A S, Yan Y, George S M, Liu P, Dillon A C and Lee S-H 2011 Electrochemical effects of ALD surface modification on combustion synthesized $LiNi1/3Mn_{1/3}Co_{1/3}O_2$ as a layered-cathode material *J. Power Sources* **196** 3317–24

[384] Li X, Liu J, Banis M N, Lushington A, Li R, Cai M and Sun X 2014 Atomic layer deposition of solid-state electrolyte coated cathode materials with superior high-voltage cycling behavior for lithium ion battery application *Energy Environ. Sci.* **7** 768–78

[385] Deng H, Nie P, Luo H, Zhang Y, Wang J and Zhang X 2014 Highly enhanced lithium storage capability of $LiNi_{0.5}Mn_{1.5}O_4$ by coating with Li_2TiO_3 for Li-ion batteries *J. Mater. Chem.* A **2** 18256–62

[386] Bettge M, Li Y, Sankaran B, Rago N D, Spila T, Haasch R T, Petrov I and Abraham D P 2013 Improving high-capacity $Li_{1.2}Ni_{0.15}Mn_{0.55}Co_{0.1}O_2$-based lithium-ion cells by modifiying the positive electrode with alumina *J. Power Sources* **233** 346–57

[387] Zhang X, Belharouak I, Li L, Lei Y, Elam J W, Nie A, Chen X, Yassar R S and Axelbaum R L 2013 Structural and electrochemical study of Al_2O_3 and TiO_2 coated $Li_{1.2}Ni_{0.13}Mn_{0.54}Co_{0.13}O_2$ cathode material using ALD *Adv. Energy Mater.* **3** 1299–307

[388] Gandrud K B, Pettersen A, Nilsen O and Fjellvåg H 2013 High-performing iron phosphate for enhanced lithium ion solid state batteries as grown by atomic layer deposition *J. Mater. Chem.* A **1** 9054–9

[389] Thangadurai V and Weppner W 2006 Recent progress in solid oxide and lithium ion conducting electrolytes research *Ionics* **12** 81–92

[390] Comstock D J and Elam J W 2013 Mechanistic study of lithium aluminum oxide atomic layer deposition *J. Phys. Chem.* C **117** 1677–83

[391] Tomczak Y, Knapas K, Sundberg M, Leskelä M and Ritala M 2013 *In situ* reaction mechanism studies on lithium hexadimethyldisilazide and ozone atomic layer deposition process for lithium silicate *J. Phys. Chem.* C **117** 14241–6

[392] Liu J, Banis M N, Li X, Lushington A, Cai M, Li R, Sham T-K and Sun X 2013 Atomic layer deposition of lithium tantalate solid-state electrolytes *J. Phys. Chem.* C **117** 20260–7

[393] Østreng E, Sønsteby H H, Sajavaara T, Nilsen O and Fjellvåg H 2013 Atomic layer deposition of ferroelectric $LiNbO_3$ *J. Mater. Chem.* C **1** 4283–90

[394] Perng Y-C, Cho J, Sun S Y, Membreno D, Cirigliano N, Dunn B and Chang J P 2014 Synthesis of ion conducting $Li_xAl_ySi_zO$ thin films by atomic layer deposition *J. Mater. Chem.* A **2** 9566–73

[395] Musa N, Woo H, Teo L and Arof A 2017 Optimization of Li_2SnO_3 synthesis for anode material application in Li-ion batteries *Mater. Today Proc.* **4** 5169–77

[396] Ivanov S, Cheng L, Wulfmeier H, Albrecht D, Fritze H and Bund A 2013 Electrochemical behavior of anodically obtained titania nanotubes in organic carbonate and ionic liquid based Li ion containing electrolytes *Electrochim. Acta* **104** 228–35

[397] Kim G-T, Appetecchi G B, Carewska M, Joost M, Balducci A, Winter M and Passerini S 2010 UV cross-linked, lithium-conducting ternary polymer electrolytes containing ionic liquids *J. Power Sources* **195** 6130–7

[398] Chang H-H, Chang C-K, Tsai Y-C and Liao C-S 2012 Electrochemically synthesized graphene/polypyrrole composites and their use in supercapacitor *Carbon* **50** 2331–6

[399] Xu L, Xia J, Xu H, Yin S, Wang K, Huang L, Wang L and Li H 2014 Reactable ionic liquid assisted solvothermal synthesis of graphite-like C_3N_4 hybridized α-Fe_2O_3 hollow microspheres with enhanced supercapacitive performance *J. Power Sources* **245** 866–74

[400] Feng L, Wang K, Zhang X, Sun X, Li C, Ge X and Ma Y 2018 Flexible solid-state supercapacitors with enhanced performance from hierarchically graphene nanocomposite electrodes and ionic liquid incorporated gel polymer electrolyte *Adv. Funct. Mater.* **28** 1704463

[401] Anouti M, Couadou E, Timperman L and Galiano H 2012 Protic ionic liquid as electrolyte for high-densities electrochemical double layer capacitors with activated carbon electrode material *Electrochim. Acta* **64** 110–7

[402] Menictas C, Skyllas-Kazacos M and Lim T M 2014 *Advances in Batteries for Medium and Large-Scale Energy Storage: Types and Applications* (Amsterdam: Elsevier)

[403] Wang W, Jiang B, Xiong W, Sun H, Lin Z, Hu L, Tu J, Hou J, Zhu H and Jiao S 2013 A new cathode material for super-valent battery based on aluminium ion intercalation and deintercalation *Sci. Rep.* **3** 1–6

[404] Geng L, Lv G, Xing X and Guo J 2015 Reversible electrochemical intercalation of aluminum in Mo_6S_8 *Chem. Mater.* **27** 4926–9

[405] Wang S, Yu Z, Tu J, Wang J, Tian D, Liu Y and Jiao S 2016 A novel aluminum-ion battery: Al/$AlCl_3$-[EMIm] Cl/Ni_3S_2@ graphene *Adv. Energy Mater.* **6** 1600137

[406] Yu Z, Kang Z, Hu Z, Lu J, Zhou Z and Jiao S 2016 Hexagonal NiS nanobelts as advanced cathode materials for rechargeable Al-ion batteries *Chem. Commun.* **52** 10427–30

[407] Xia S, Zhang X-M, Huang K, Chen Y-L and Wu Y-T 2015 Ionic liquid electrolytes for aluminium secondary battery: influence of organic solvents *J. Electroanal. Chem.* **757** 167–75

[408] Suto K, Nakata A, Murayama H, Hirai T, Yamaki J-i and Ogumi Z 2016 Electrochemical properties of Al/vanadium chloride batteries with $AlCl_3$-1-ethyl-3-methylimidazolium chloride electrolyte *J. Electrochem. Soc.* **163** A742

[409] Reed L, Ortiz S, Xiong M and Menke E 2015 A rechargeable aluminum-ion battery utilizing a copper hexacyanoferrate cathode in an organic electrolyte *Chem. Commun.* **51** 14397–400

[410] Hudak N S 2014 Chloroaluminate-doped conducting polymers as positive electrodes in rechargeable aluminum batteries *J. Phys. Chem.* C **118** 5203–15

[411] Koura N 1980 A preliminary investigation for an Al/$AlCl_3$–NaCl/FeS_2 secondary cell *JElS* **127** 1529–31

[412] Reynolds G and Dymek C Jr 1985 Primary and secondary room temperature molten salt electrochemical cells *J. Power Sources* **15** 109–18

[413] Tsuda T, Kokubo I, Kawabata M, Yamagata M, Ishikawa M, Kusumoto S, Imanishi A and Kuwabata S 2014 Electrochemical energy storage device with a Lewis Acidic $AlBr_3$−1-ethyl-3-methylimidazolioum bromide room-temperature ionic liquid *J. Electrochem. Soc.* **161** A908

Chapter 2

Theoretical frameworks on geometric, electronic, magnetic, and optical properties

Thi Dieu Hien Nguyen, Vo Khuong Dien, Hai Duong Pham, Shih-Yang Lin, Hsien-Ching Chung, Wei-Bang Li, Ngoc Thanh Thuy Tran, Nguyen Thi Han, Hsin-Yi Liu and Ming-Fa Lin

Theoretical developments are very important in green energy materials, since they are capable of providing scientific viewpoints, especially for the critical mechanisms in solving the mainstream issues of ion-based batteries. A whole framework covers the numerical simulations and/or the phenomenological models, being introduced in sections 2.1 and 2.2, respectively. Both of them are expected to propose concise physical/chemical/material pictures in fully understanding the diversified essential properties. The current book is focused on the former, while the latter can be found in the other books [10–12]. The relations with the various experimental spectroscopies are also discussed in chapter 3.

2.1 Numerical simulations: first-principles methods and others, including physical and chemical viewpoints

Theoretical strategy, which is built from first-principles simulations [1, 2], is focused on examining the critical orbital hybridizations and spin arrangements (the basic quasiparticle properties; [3, 4]) in chemical bonds from geometric symmetries, electronic properties, magnetic configurations, and optical absorption spectra without/with many-body effects. This theoretical framework, being tested on a lot of core materials of ion-based batteries [5–8], is able to clearly identify the significant orbital hybridizations associated with the frequency-dependent electronic excitations. It covers the detailed examinations of position-dependent chemical bonds atom dominance from band structure, the overall orbital hybridizations through charge

doi:10.1088/978-0-7503-4685-6ch2

density distributions, the energy-range-related ones under the atom- and orbital-projected van Hove singularities, and the specific orbital hybridization for each prominent absorption structure. Most importantly, the close combinations of the different results are required for fully consistent identifications. These delicate calculations and analyses are very complex, being never achieved in the previous studies. In addition to charge densities, quasiparticle properties, atom-dependent spin arrangements, have been considered by the spin-orbital couplings and the on-site-spin bi-particle interactions. The atom- and orbital-generated magnetic configurations are thoroughly examined from the net magnetic moments, the spin densities, and the spin-split electronic properties [9, 10]. The strong competition/cooperation between chemical bondings and magnetic configurations will dominate whether the ferromagnetic/anti-ferromagnetic/non-magnetic configurations [10–12] would come to exist in condensed-matter systems.

The diversified phenomena, being created by the various quasiparticles [3, 4, 13], are the mainstream topics of basic science research. The essential properties [10, 12] and functionality-based potential applications [5–8] in any condensed-matter systems are dominated by them. For example, electrons, phonons, and photons are pure quasiparticles with the dual features of particle and wave packet [14]. Furthermore, their composites under the same/different subsystems become new ones, e.g., plasmons [14], polarons [14], and polaritons [14]. In general, quantum and classical pictures are available in characterizing the fundamental quasiparticle behaviors. There exist plenty of elastic and inelastic scattering events due to the quasiparticle propagation. The intrinsic and significant interactions mainly arise from the charges and spins, belonging to the single- and bi-particle forms (details in sections 2.1 and 2.2). It is well known that the latter have no exact energy spectra and wave functions for the distinct numerical simulations [15–17] and phenomenological models [18, 19]. Apparently, the approximate techniques are frequently utilized in the theoretical predictions and even the experimental analyses, in which the merits and drawbacks are discussed in the following paragraphs. Very interestingly, the quasiparticle behaviors are expected to be strongly and even/dramatically changed by the external gate voltages, static and uniform magnetic fields, position- and time-dependent Coulomb fields and electromagnetic waves. How to directly include the important couplings between quasiparticles and applied modulations in the first-principles calculations are worthy of a thorough investigation.

2.1.1 Optimal lattice symmetries and electronic properties

The single- and bi-particle interactions between quasi-particles (both chemical bonds and spin configurations) play critical roles in determining the geometric and electronic properties of condensed-matter systems. The Coulomb potentials are experienced by the electron quasiparticles, mainly from the carrier distribution and the periodical ions. It is well known that the classical electron–electron Coulomb interactions are inversely proportional to the distance of two separate charges [20]. Very interestingly, the static and dynamic charge screenings are able to generate many-body interactions beyond the classical Coulomb-law form [21, 22]. Any materials possess the exchange and correlation energies, in which the former and the latter, respectively, represent the first- and other-order Coulomb corrections through the Feynman diagrams of quasiparticle self-energy [23]. Generally speaking, the former has an analytic formula [24, 25], but the latter cannot be solved exactly [20, 24]. A lot of approximate models are developed to get well-behaved single-particle/quasiparticle properties, such as, study focuses on the pair distribution function [20, 26], the normalization constant [14], the induced charge density due to impurity [14], and the quasiparticle energy spectra/decay rates [27]. Apparently, the exact correlation energy is thoroughly absent even under a specific carrier density [20, 24]. This will create a very high barrier in the first-principles simulations. Such an issue could be solved when the many-particle interaction energies are well characterized by the charge-density-dependent forms [21, 22]. The up-to-date theoretical studies show that it is successfully overcome by the generalized gradient approximation for uniform electron gases (GGA, or Perdew–Burke–Ernzerhof form; [28]). It should be noticed that the highly non-uniform systems might present serious drawbacks in the theoretical predictions. For example, the important modifications on the interlayer van der Waals interactions need to be covered in layered graphene systems/graphene nanoribbons/carbon nanotubes [12]/[11]/[29]. Moreover, it is very difficult to delicately simulate the periodical crystal potential with the frequently electronic scatterings at any positions and times. In general, this ionic Coulomb potential is generated by the core valence electrons and protons of the same/different atoms in the mono-/multi-element systems. Furthermore, its exact form is very sensitive to the atomic/molecular components, so that the direct experimental examinations on the theoretical predictions are absent or impossible [30–32]. The optimal pseudopotential could be obtained from a lot of numerical tests [10, 33] and thus dominates the calculated band structures and wave functions.

All the theoretical developments, being focused on the ground state energies/the stable lattices of any condensed-matter systems, clearly indicate the suitable simulation methods [34], but not the phenomenological ones [34]. For example, the current book fully explores the essential properties of cathode/electrolyte/anode materials using the Vienna *ab initio* simulation package (VASP) calculations [1, 2]. It is well known that numerical calculations have become one of the very powerful tools in investigating the diverse physical/chemical/materials phenomena. Very interestingly, as to the electron-related quasiparticles, how to solve the reliable Schrödinger equation, with the single- and many-particle interactions (discussed in the previous paragraph), is the first-step investigation. Different approximations/viewpoints [35, 36] have been developed to achieve stable geometric structures, and reasonable electronic properties. Very obviously, the first-principles simulations are a dominating method for the fundamentals properties of periodic systems, i.e., they are frequently utilized to study the geometric, electronic, magnetic, and optical properties (details in sections 2.1.1 and 2.1.2). They are rather efficient in solving the unusual electron–electron and electron–ion interactions through a series of approximations and simplifications to get the quasiparticle energy spectra and wave packets. For recently decades, the VASP [1, 2] can evaluate an approximate solution to the many-body Schrödinger equation, either within density functional theory (DFT), solving the Kohn–Sham equations (the flow chart in figure 2.1), or within the Hartree–Fock (HF) approximation, solving the Roothaan equations [37]. In this work, the rich and unique essential properties are fully investigated by first-principles DFT using VASP. The exchange-correlation energy is calculated from the Perdew–Burke–Ernzerhof (PBE) functional under GGA [38]. Furthermore, the projector-augmented wave pseudopotentials are available in dealing with the electron–crystal scatterings [39]. The wave functions are the linear superposition of a plane–wave complete set within a maximum energy cutoff of 500 eV, where they would be tested to present a very high accuracy. All the atoms during the relaxation processes are subjected to their position adjustments in order to have an optimized structure with the lowest total ground state energy. In general, the convergence condition for one full relaxation is to set the Hellmann–Feynman force on each atom lower than 0.01 eV/Å and the total energy difference of $\Delta E_t < 0.01$ eV. Consequently, the most stable configuration in a unit cell will be used as an initial input for the further evaluations of other fundamental properties (figure 2.2).

In order to understand the noninteracting kinetic energy functional, Kohn–Sham theory provides the energy expression which collects for a single-particle solution. This energy can be expressed under the form

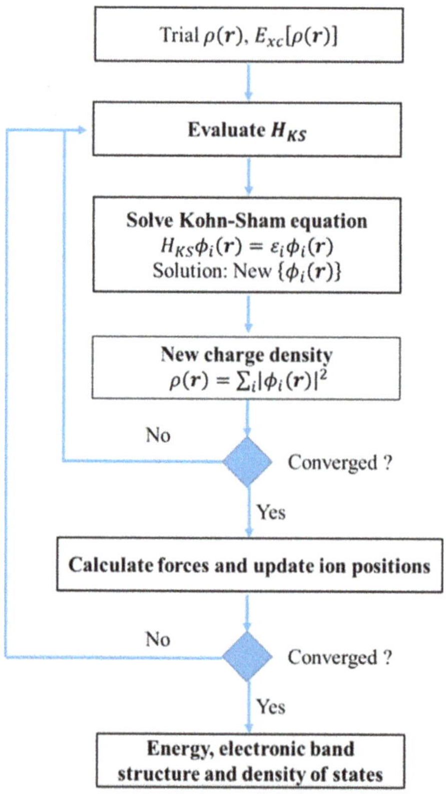

Figure 2.1. Flow chart for the VASP simulations of the optimal lattice symmetries, electronic energy spectra, wave functions, charge densities, and atom- and orbital-projected van Hove singularities.

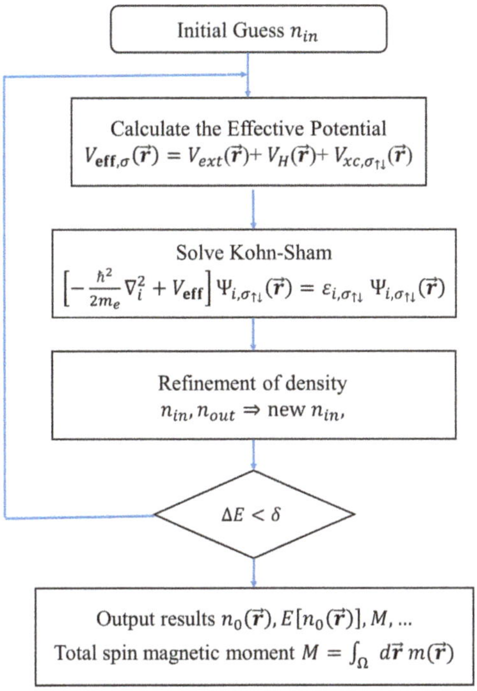

Figure 2.2. Similar plot to figure 2.1, but shown for the calculations of spin-dependent magnetic properties (atom- and orbital-projected magnetic moments, spin-split band structures, spin configurations and spin-decomposed density of states).

$$E_T[n] = \int V_{\text{ext}}(\vec{r})n(\vec{r})d^3r + \frac{1}{2}\iint \frac{n(\vec{r})n(\vec{r}')}{|\vec{r} - \vec{r}'|}d^3rd^3r' + T[n] + E_{xc}[n]$$

where noninteracting electrons ($T[n]$), external potential $V_{\text{ext}}(r)$, and charge density $n(r)$ are given in the formula. It is noted that the second term presents Coulomb energy and the last term shows the exchange-correlation energy. In practical use, the Kohn–Sham equations require the exact form of the exchange-correlation energy functional. However, it might face difficulty with an unknown mathematical form of E_{xc}. Therefore, DFT and other approximations of exchange- correlation are used to overcome this problem. Some exchange-correlation functionals are proposed by Perdew, known as 'Jacob's ladder' [40], providing varying accuracy with the development of the order. The first approach is the local-density approximation (LDA), in which a real inhomogeneous system is divided into infinitesimal volumes, and the electron density in each of the volumes is taken to be constant. The exchange-correlation energy for the density within each volume is then assumed to be the exchange-correlation energy $E_{xc}^{LDA}[n] = \int n(\vec{r})\epsilon_{xc}^{unif}(n(\vec{r}))dr$ obtained from the uniform electron gas for that density. Thus, the total exchange-correlation energy of the system can be written as where ϵ_{xc}^{unif} is the exchange-correlation energy per particle of the interacting uniform electron gas of density $n(r)$. In practice, exchange and correlation are calculated $E_{xc}^{LDA}[n] = E_x^{LDA}[n] + E_c^{LDA}[n]$ separately. Apparently, DFT can express the analytical expression for the exchange energy in the exact form of $E_x^{LDA}[n] = -\frac{3}{4}\left(\frac{3}{\pi}\right)^{1/3} \int n^{\frac{4}{3}}(r)dr$. There are several expressions for the correlation energy which have been obtained by fitting to the results of accurate Monte Carlo calculations (QMC) calculations of the Uniform Electron Gas (UEG). The next approach is GGA, which improves in the accuracy of LDA. This calculation depends not just on the value of the density at a point (as in the LDA case) but also on its gradient $E_{xc}^{GGA}[n] = \int n(\vec{r})\epsilon_{xc}(n(\vec{r})\ |\nabla n(\vec{r})\ |)dr$. In general, most GGA functionals are constructed in the form of a correction term which is added to the LDA functional. According to higher accuracy, the second derivative (the Laplacian) will be added in Meta-GGA (MGGA) functionals. These fourth generation functionals add 'exact exchange' calculated from the hybrid functionals functional to some conventional treatment of DFT exchange and correlation. The philosophy behind the hybrid functional is simple and rooted in the 'adiabatic connection' formula, which is a rigorous *ab initio* formula for the exchange-correlation energy of DFT [125]. One conventional expression of this formula is $E_{xc} = \int_0^1 U_{xc}^{\lambda} d\lambda$ where λ is an interelectronic coupling strength parameter that switches on the 1/ |ri−rj | Coulomb repulsion between electrons and U_{xc}^{λ} is the potential energy of exchange and correlation at λ. This formula connects the non-interacting reference system with the fully interacting one at density $n(r)$. Recognizing that the non-interacting $\lambda = 0$ limit is nothing more than HF exchange, it is expected that exact exchange must play a role in 'better' exchange-correlation functionals and thus LDA and GGA exchange and correlation functionals are mixed with a fraction of HF exchange. Hybrids use HSE. To date, different approaches to the exchange-correlation potential have been implemented: LDA, GGA hybrids such as HSE, Meta-GGAs, VdW-functionals: Grimme D2 and D3, Tkatchenko–Scheffler, vdW-DF more advanced methods GW, ACFDT in RPA (adiabatic connection fluctuation dissipation theorem in the random phase approximation, and MP2 (second order Møller–Plesset). The above methods will play a critical role in comprehending the noninteracting kinetic energy functional of condensed matter systems. Also, the close relations between the many-body Coulomb interactions and the carrier density distribution deserve a detailed examination.

Up to now, the standard VASP calculations have been successful in exploring the diversified phenomena [1, 2] and providing the full information about the potential applications [5, 6]. Specifically, the theoretical framework, which is thoroughly supported by the fundamental physical/chemical/material properties, is developed to achieve their concise mechanisms. The systematic investigations, which are closely related to the quasiparticle charges (the single-/multi-orbital hybridizations in chemical bonds), are conducted on graphene-related emergent materials, such as layered graphene systems [12], graphene nanoribbons [11], and silicene-based condensed matters [10]. Three books clearly show that the geometric symmetries, band structures/wave packets, spatial charge densities and atom- and orbital-decomposed van Hove singularities are greatly diversified by the various dimensionalities [10–12], planar/curved structures [10–12], quantum-confinement boundaries [10, 12], stacking configurations [10, 12], layer/nanotube numbers [11, 12], adatom absorptions [10, 12], host-atom substitutions [10, 12], heterojunction effects between 2D material and substrate [10], and composite compounds of distinct 2D subsystems [10, 12]. In addition, layered germanene/tinene/plumbene systems [41]/[42]/[43] are under current study. According to the theoretical predictions [10, 12], both carbon and silicon atoms, which belong to the group-IV elements, can create very active chemical environments. A lot of the mainstream materials are chosen for the delicate evaluations and analyses, in which they clearly illustrate the concise mechanisms in determining the essential properties. That is to say, the π and σ bondings, the full sp^3 bonding, or their perturbation mixings with the dominating π one, are responsible for the featured properties. For example, the orthogonal π and σ bondings only lead to the planar honeycomb lattices in all the layered graphene systems/graphene nanoribbons, such as, the AA- [12, 44, 45], AB- [12, 44], ABC- [12, 46] and AAB-stacked graphene-sp^2 systems [12, 47]. The significant van der Waals interactions, the interlayer carbon-$2p_z$-orbital hybridizations [12, 48], can strongly modify the low-lying valence and conduction energy subbands [12, 48], e.g., the linear [12, 49], parabolic [12] or oscillatory dispersion relations [12, 49] due to the intralayer and interlayer π bondings. Furthermore, the π-electronic states are well distinguished from the σ-electronic ones, mainly covering their initiated states at the different symmetry points and their whole band widths within the distinct energy ranges [33, 49]. On the other hand, the sp^2-dominant carbon honeycomb lattices can present the curved/folded/scrolled surface structures [11, 50, 51], as successfully synthesized under the various experimental methods. Such nonplanar carbon lattices are identified to possess the misorientations of $2p_z$ orbitals and the weak, but significant sp^3-orbital hybridizations, according to the accurate calculations of charge density distributions. For example, very narrow [4, 0]/[5, 0] zigzag carbon nanotubes are predicted to be metals from the orbital-projected density of states [52–54], in which the van Hove singularities will show the drastic variations on the π-bonding strengths and four-orbital emergences. In most cases, the π and σ bands are roughly separated from each other. Very interestingly, the strong competition between sp^2 and sp^3 chemical bondings in honeycomb lattices even creates the significant bucklings of A and B sublattice. The non-orthogonal π and σ bondings, which survive in silicene/germanene/tinene/plumbene systems [10]/[41]/[13, 42]/[43], are responsible for the anti-crossing energy spectra. More importantly, the 2D energy bands are effectively changed into the 1D ones through the totally different van Hove singularities. When the interlayer multi-orbital interactions are further taken into consideration, both π and σ -electronic features almost disappear at the whole energy ranges. Such phenomena are obviously revealed in bilayer silicene/germanene systems with various stacking configurations [10, 55–57], but not graphene ones [12, 45, 58–60]. The main reason lies in the largely enhanced bucklings and chemical bondings. It becomes meaningless for the identification of π and σ -electronic states.

The chemical modifications, chemisorptions [61, 62] and substitutions [63], are very efficient in diversifying the fundamental properties [61–63] and enhancing the potential applications [64]. There exists a lot of theoretical and experimental research [65–67]. The up-to-date results clearly show that the first-principles simulations are the only and unique manner in unifying the various quasiparticle phenomena, since they are capable of evaluating the full physical/chemical/material quantities to examine and identify the concise orbital hybridizations in various chemical bonds. In general, layered graphene systems [12], cylindrical carbon nanotubes [29, 68], graphene nanoribbons [11] and silicene/germanene/tinene/plumbene [10]/[41]/[13, 42]/[43] have the active surface environment with the rich dangling bonds, at least, due to the outmost p_z orbitals. Through detailed investigations in three books [10–12], the slightly modified/strongly distorted/totally destroyed π bondings and the creations of single-/multi-orbital hybridizations (related to the four orbitals of group-IV elements) frequently come to exist in the distinct chemical bonds. Very interestingly, the alkali-adatoms on carbon-honeycomb lattices, with the hollow-site positions, indicate the weak, but significant chemical bondings [33, 69]. This leads to the roughly rigid Dirac-cone structure with an obvious blue-shift Fermi level, i.e., the π bondings behave well on a planar lattice under the almost full charge transfer. The bridge- and top-site chemisorptions can greatly change the orbital hybridizations, e.g., the multi-orbital hybridizations in C–O bonds of graphene oxides [33, 70] and the sp^3-bonding hydrogenated graphene systems (the five-orbital bonding, accompanied with a buckled structure; [71, 72], respectively). The complex and unique orbital hybridizations, which survive in silicene-based emergent materials are clearly predicted from the accurate VASP calculations (details in Shih Yang's book [10]). For example, boron-, carbon- and nitrogen-substituted silicene systems, are predicted to exhibit hybridized π and σ bondings under the orthogonal feature [63], similar behaviors through strong distortion of the former [63], and fully mixed sp^3 bondings [63]. Three types of four-orbital hybridizations are thoroughly examined from optimal lattice symmetries, energy spectra, charge density distributions, and van Hove singularities.

Compared with group-IV layered materials [73, 74], the ternary/multi-component lithium oxides [75, 76] present totally different physical/chemical/material phenomena. Generally speaking, the latter have moiré superlattices in the unit cells, where there are many atoms or chemical bonds. How to unify the complicated energy spectra and wave packets might induce high barriers for the current numerical calculations. The position-dependent chemical bonds, which generate the highly non-uniform environment, are responsible for the critical orbital hybridizations and thus determine the various fundamental properties. Specifically, the observable fluctuations of chemical bond lengths are discussed thoroughly for their roles. By analyzing a lot of valence and conduction subbands, the distinct atom dominances could be further classified into specific energy ranges. The spatial charge density distributions are able to provide all the orbital hybridizations. Moreover, the atom- and orbital-decomposed densities of states are utilized to examine the specific orbital mixings through the merged van Hove singularities. Such special structures, as frequently observed within the valence- and conduction-state ranges, will be directly reflected in other essential properties, e.g., vertical optical excitations (details in section 2.1.3). However, their accurate examinations by scanning tunneling microscopy (details in chapter 3 section 3.3) cannot be done up to now [32], since the bulk properties are totally different from the surface ones. In addition, the spin configurations are also included in certain cases (details in section 2.1.2).

2.1.2 Stable magnetic configurations

In addition to charge, another quasiparticle property, spin configuration, will play a critical role in the intrinsic magnetic properties. It is well known that fermions and bosons are well distinguished from their half-integer and integer spins, respectively, e.g., $s = 1/2$ and $s = 0$ for electrons and phonons. It is well known that the spin-created magnetic properties are totally different between isolated atoms/molecules and periodical materials since the latter has more complicated single- and many-particle spin interactions.

Electronic energy spectra and wave functions of condensed-matter systems might depend on the spin-up and spin-down states, being revealed as the spin-dependent energy gap [14], band splitting, net magnetic moment, spatial spin-density distribution, and spin-decomposed van Hove singularities. Their experimental examinations could be finished by the spin-polarized spectroscopies, e.g., the SP-STM, SP-STS (details in chapter 3 section 3.3) and SP-ARPES (section 3.4), respectively, for the nanoscaled morphology [77, 78], the spin-projected density of states [79], and spin-split or spin-degenerate occupied energy bands [80]. According to the previous theoretical predictions, two kinds of spin-induced interactions, spin-orbital couplings, and spin-on-site Coulomb scatterings, frequently come to exist in determining the featured magnetic properties. The former, which is mainly due to the crystal Coulomb potential experienced, belongs to the relativistic interaction under the single-particle coupling form. Very interestingly, for many mono-element layered materials, this significant magnetic coupling has been identified to greatly diversify the low-energy essential properties, such as, the diverse band structures in silicene [10, 41]/germanene [41]/tinene [13, 42]/plumbene [43]/antimonene [81, 82] /bismuthene systems [82].

However, it cannot lead to the ferromagnetic [41, 42] and anti-ferromagnetic configurations. On the other hand, the latter presents a very layer on-site bi-particle Coulomb interaction. This has been thoroughly discussed through the tight-binding Hubbard model [18, 19, 34], being very useful in qualitatively understanding the strong competition/cooperation of charge- and spin-dependent interactions. The spin-generated Coulomb interactions are fully covered in the current VASP simulations. In short, whether the various magnetic phenomena appear in mainstream materials is determined by the orbital-hybridization- and spin-related ground state energy.

In many-body physics, the magnetic ground-state properties are defined by the wave function or determined by the density alternatively. DFT calculations are able to evaluate the magnetic moment or magnetic states of the system such as ferromagnetic, antimagnetic, or nonmagnetic [10–12]. The flow-chart of figure 2.1 demonstrates the process of calculating the charge and spin-density distribution. In this work, the total magnetization is determined by $M = \int_{\Omega} d\vec{r} m(\vec{r})$ with respect to the stable geometric structure (after optimization process). It is clear that the results are highly dependent on the exchange-correlation potential such that it is really difficult to find the exact solution. In the local spin density approximation (LSDA), the exchange correlation energy can be assumed in the form $E_{xc} = \int n(\mathbf{r})\varepsilon_{xc}[n_{\uparrow}(\vec{r}) - n_{\downarrow}(\vec{r})]$. This approximation has good performance when calculating the potential of the spin-polarized electron gas. However, in many cases, the GGA provides better prediction and approaches experimental results.

The systematic VASP simulations on graphene- [12], graphene-nanoribbon-[11] and silicene-related systems [10] clearly show the diverse magnetic phenomena. Very interestingly, the open zigzag edges in carbon honeycomb lattices (zigzag graphene nanoribbons; [11]) are able to present anti-ferromagnetic configuration across the ribbon center and ferromagnetic along on the same side. Furthermore, the different spin arrangements could be modulated by the variation of adatom concentration and configuration. For example, the alkali adatoms above graphene nanoribbon lead to the full charge transfers (the 1D metals; [11, 50]), as well as the drastic changes of spin density distributions (reduction, enhancement and absence of spin-up and spin-down states). More rich and unique spin density distributions are predicted to be revealed in the curved/scrolled/folded graphene nanoribbons [11, 50, 51] and AA- and AB-stacked bilayer ones [12, 44, 45], with the significant interlayer atomic interactions, e.g., the dominating ferromagnetism and/or anti-ferromagnetism along the specific direction. The zigzag boundaries, non-planar structure and stacking configuration can drive the magnetic properties in carbon-atom lattices, while such atoms are unable to create them. The combined ground state energy, which is due to the intralayer and interlayer orbital hybridizations and the spin-dependent on-site Coulomb interactions, will determine the charge and spin density distributions. While the buckling of honeycomb lattice appears in silicene-related systems [10, 41, 42], the significant sp^3 is stabilized by the coupling of multi-orbital hybridizations and spin–orbital interactions. This will lead to unusual behaviors, where the distortion and separation of Dirac-cone structure and the rise of σ-electronic valence bands grow in the increment of atomic number. For example, tinene/plumbene/antimonene/bismuthene possess low-lying π and σ bands [13, 42, 43, 82]. The magnetic properties could also be introduced by transition and rare-earth metal atoms through chemical modifications, e.g., Fe/Co/Ni on monolayer silicene [10] and silicene/Ag(111) [83–85]. Most importantly, the atom- and orbital-dependent hybridizations and spin states are simultaneously examined from all the VASP calculation results. The similar analyses could be generalized to the magnetic materials, as done for ternary LiFe/Co/NiO compounds [86–88]/[89–91]/[92, 93].

2.1.3 Vertical optical excitations

The special optical properties, which appear in various condensed-matter materials, are frequently utilized in highly potential applications, e.g., optical sensors [94], modulators [94] and energy storages [94]. The main characteristics are clearly revealed in the optical gap (threshold excitation frequency; [94, 95]), prominent absorption structures, reflectance, absorption coefficient, refractive index, and energy loss function with charge screenings. The position-dependent chemical environment can create the strongly anisotropic excitation spectra, being sensitive to the directions of electric polarizations. Moreover, the many-body effects, the excitonic ones [23, 96], might play an important role in determining the featured optical properties, e.g., the stable exciton bound states in greatly reducing the initial absorption frequencies. How to use the first-principles results for examining the critical orbital hybridizations is the main study focus. The cathode, electrolyte and anode materials of ion-based batteries are chosen as the typical systems to test the developed theoretical framework.

A condensed-matter system is easily modulated by a time- and position-dependent electromagnetic field, in which there exist significant couplings between charges and propagating waves (electrons and photons). The carrier dynamics and wave dispersion relations, which are, respectively, solved by quantum scattering theories and Maxwell equations, determine all the features of optical properties. Electrons are vertically excited from the occupied states to the unoccupied ones with the same wave vector since photons can only provide sufficient energies under the normal cases (the long wavelength limit). Such carriers effectively screen the perturbed electric field, create the induced current density, and thus build the electric polarization. The charge screening ability is fully characterized by the transverse dielectric function (a dimensionless ϵ ($q \to 0,\omega$)), as done for the macroscopic dielectric constant due to the static Coulomb-field perturbation [18, 97]. That is, its definition is the specific ratio of the electric fields before and after the charge screenings under the (momentum, frequency) spaces.

As a result, its real and imaginary parts, $Re(\epsilon(\omega))$ and $Im(\epsilon(\omega))$, can define all the optical physical quantities through the long wavelength limit (further discussions; [14, 18]). Very obviously, how to calculate and get the suitable and reliable dielectric function in the absence/presence of many-body effects should be the most important focus. To evaluate the complex dielectric function, experimental measurements and theoretical calculations present the different manners. The former is able to examine the frequency-dependent reflectance/absorption/transmission spectra, being insufficient in determining the real- and imaginary-part response functions simultaneously $Re(\epsilon(\omega))$-1 and $Re(\epsilon(\omega))$. However, the measured results could be supported by the following Kramers–Kronig relations

Equation (2.1); relations using $Re(\epsilon$-1) and $Im(\epsilon$-1)

$$
\begin{aligned}
Re\frac{\epsilon(\omega)}{\epsilon_0} &= 1 + \frac{2}{\pi}P\int_0^\infty \frac{Im\frac{\epsilon(\omega')}{\epsilon_0}}{\omega' - \omega}d\omega' \\
Im\frac{\epsilon(\omega)}{\epsilon_0} &= -\frac{1}{\pi}P\int_0^\infty \frac{\left[Re\frac{\epsilon(\omega')}{\epsilon_0} - 1\right]}{\omega' - \omega}d\omega'
\end{aligned}
\tag{2.1}
$$

Any physical responses come to exist after the external perturbations. therefore, they are well-behaved functions of excitation frequencies in the presence of de-excitation mechanisms [14, 23]. While the absorption frequency is further regarded as a complex variable, the real- and imaginary-part response functions are examined to be associated with each other. That is, doing the principle-value integration of $\frac{Re[\epsilon - 1]}{Im[\epsilon - 1]}$ can achieve $\frac{Im[\epsilon - 1]}{Re[\epsilon - 1]}$. Furthermore, the $\int_0^\infty$ integration is enough because of the even/odd property in $\frac{Re[\epsilon - 1]}{Im[\epsilon - 1]}$. It should be noticed that very wide frequency ranges are required in the various optical spectroscopies. This might induce the main drawbacks for estimating the frequency-dependent dielectric function. How to overcome such frequent issues could utilize the full assistance of the empirical models, e.g., the high-frequency plasmon-mode approximation [14, 98].

The dynamic Kubo formula [99], without/with the many-body effects, is frequently utilized to explore optical excitations. To satisfy the carrier dynamics under an electromagnetic wave, the canonical momentum of electron consists of the kinetic and magnetic-field ones, in which the latter is proportional to field strength and charge quantity [100]. The dominating Hamiltonian perturbation belongs to the electric-dipole moment. Its square of the inner product between the initial and final states will dominate the scattering probability of each available transition channel and thus even determine the well-behaved optical selection rules [14, 101]. How to evaluate it exactly through the suitable energy spectra and wave functions might cover the significant difficulties of the theoretical predictions. That is, the close combinations of the tight-binding model and the Kubo formula would be very important in efficiently exploring the diversified excitation phenomena of emergent materials (the detailed development in our books [18, 19, 97, 102]). In addition to absorption intensity, the joint density of states, being generated by the occupied and unoccupied electronic states, represents the available number of excitation channels. These two critical factors are responsible for the dielectric function under the single-particle pictures [14, 103], as revealed in much previous research. But for semi-conducting or insulation materials, the excitonic effects come to exist very often under the suitable environment, e.g., the great reduction of threshold excitation frequency in a wide-gap system [23, 104], and certain prominent absorption peaks at specific frequencies lower than the single-particle ones [23, 104]. Excitons are stable/meta-stable bound states of the excited electrons and holes since there exists a significant attractive Coulomb potential between them. Their binding strength is rather sensitive to the charge screenings of the other carriers. The creation of two-particle states and the dielectric screenings need to be included in all the scattering Feynman diagrams [23]. This corresponds to the standard Bethe–Salpeter equation [105, 106], as done in the current VASP calculations (the flow chart in figure 2.3).

The various optical properties are clearly characterized under the four Maxwell equations of electric and magnetic fields and the related boundary conditions [100]. Under the Coulomb gauge of $\nabla . A = 0$, the electromagnetic and Coulomb fields, respectively, possess the transverse and longitudinal electric components. As for condensed-matter systems without free current densities and magnetic configurations [14, 100], the significant couplings between an electromagnetic wave and charge carriers are well described through the Helmholtz equation

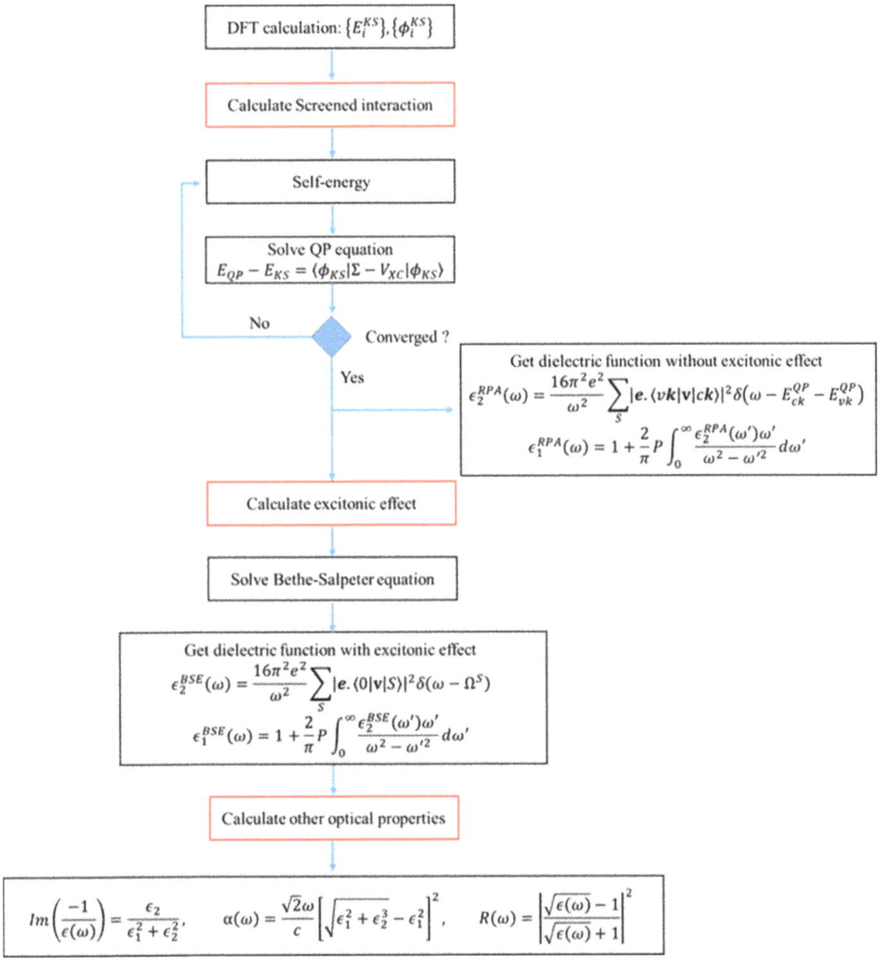

Figure 2.3. First-principles calculations on the optical properties in the presence/absence of excitonic effects, covering the real- and imaginary-part dielectric functions, absorption coefficients, reflectance spectra, and energy loss functions.

$$c^2\nabla^2 E = \frac{\partial^2 D}{\partial t^2} \tag{2.2}$$

This equation corresponds to the strong dependences of position vectors and time. By the further Fourier transforms into the (momentum, frequency)-space, one can achieve the dispersion relation

$$c^2k^2 = \epsilon(\omega)\omega^2 \tag{2.3}$$

The wave vector of an electromagnetic wave agrees with the close relation of $k = \frac{\omega\sqrt{\epsilon(\omega)}}{c}$. Its value might be real or complex, being mainly determined by the imaginary-part dielectric function (being sensitive to the available excitation channels). In general, $\sqrt{\epsilon(\omega)} = n + i\frac{c}{\omega}\alpha$; therefore, the complex k is utilized to define the refractive index and the absorption coefficient simultaneously. The inverse of α is the characteristic propagation length in which an electromagnetic wave can penetrate a condensed-matter system. Very interestingly, the various kinds of valence/conduction electrons are able to efficiently screen the electromagnetic wave at the different frequency ranges, or such carriers can generate the diversified optical excitations. The significant couplings, with an electromagnetic wave absorption, directly determine the reflectance, absorption, and transmission spectra [94, 95].

When an electromagnetic wave is incident on a sample along the normal direction, both electric and magnetic fields are situated at the transverse plane [100]. The electric/magnetic field is classified into three components: incident, reflected and transmitted ones. There exist two continuous boundary conditions for the tangential electric and magnetic fields. By using Faraday's and modified Ampère's laws, the reflectance, being characterized by the square ratio of the reflected versus incident electric fields, is expressed as

$$R(\omega) = \left| \frac{\sqrt{\epsilon(\omega)} - 1}{\sqrt{\epsilon(\omega)} + 1} \right|^2 \tag{2.4}$$

The nearly total reflectance comes to exist, while the dielectric function presents a purely negative value within the specific frequency range [100]. The very large R is very useful in fully understanding the forbidden frequency range for which an electromagnetic wave only declines quickly. Similar behavior is revealed in the low-frequency reflectance due to the conduction electrons, being frequently described by the Drude model [14]. The rapid decrease of reflectance, the plasmon edge, appears before and after the frequency of coherent charge oscillations [14]. On the other hand, the optical excitations thoroughly disappear for any semiconductor within an optical gap. The imaginary-part dielectric function is vanishing, and its real-part one can provide a low reflectance, mainly owing to free propagation in the absence of energy loss.

It is well known that the imaginary-part dielectric function and the energy loss function, $Im(\epsilon)$ and $Im\left[\frac{-1}{\epsilon}\right]$, respectively, are the available optical/Coulomb excitations under the single-particle and collective pictures. The former and the latter are responsible for the electron–hole pair excitations and the plasmon modes at the various frequency ranges. That is, all screening behaviors due to all the kinds of valence/conductions are obviously covered in the electron energy loss spectra corresponding to the directly measured results (EELS in [107, 108]). Such important differences have been very successfully clarified from a lot of experimental [108–110] and theoretical research [14]. For example, Taft and Philipp can measure the high-resolution reflectance spectrum of the semimetallic Bernal graphite within a broad frequency range of $\omega<26$ eV [111]. Such a wide absorption range is very sufficient and reliable in fully understanding the optical excitations associated with the free conduction electrons, the valence π electrons (the carbon-$2p_z$-orbitals), and the deeper-energy σ ones (the C-(2s, $2p_x$, $2p_y$)-orbitals). These quasiparticle behaviors directly reflect the energy-dependent band structure. That is, they are respectively, dominated by the weak, but significant inter-layer van der Waals interactions (the interlayer $2p_z$-orbital hybridizations), the important intralayer nearest-neighbor hopping integrals (the intralayer π bondings), and the intralayer σ-electron multi-orbital hybridizations. Using the delicate analyses on $Im(\epsilon)$ and $Im\left[\frac{-1}{\epsilon}\right]$, the first, second and third kinds of quantum quasiparticles are clearly identified to, respectively, appear in (I) $\omega<0.8$ eV, (II) 0.8 eV $<\omega<9$ eV, and (III) $\omega>9$ eV Furthermore, the π and σ plasmon modes under long wavelengths come to exist at 7.0 and 25.0 eVs, respectively.

In this book, one of the main focuses is to fully explore the optical properties in cathode/electrolyte/anode materials of ion-based batteries through first-principles calculations. By using the non-local Kohn–Sham equation with the bi-particle Coulomb interactions (details in section 2.1; [112]], the wave-vector-dependent energy spectra and wave functions are available in further calculating the dielectric function in the absence of excitonic effects. But when the significant Coulomb attractions between the photon-induced electron and hole pairs take part in all the intermediate excitation processes, the standard Bethe and Salpeter equation, as clearly indicated in figure 2.3, needs to be solved for the excitonic effects. How to analyze their strong modifications on the various optical properties might be relatively easy for condensed-matter systems with few valence and conduction energy subbands, e.g., the excitonic effects in stage-n graphite intercalation compounds [113].

The theoretical and experimental analyses, which are mainly conducted on the important absorption structures of ternary lithium oxides (core materials of batteries in chapter 1), would present serious challenges. Such compounds have a lot of atoms in a unit cell and thus many energy subbands within the greatly reduced first Brillouin zone [75]. Furthermore, both electric-polarization directions and excitonic effects are able to affect the intensities of absorption structures. In general, all the prominent absorption structures might be due to the stable bound excitons and the band-edge states with the joint van Hove singularities. The former frequently appear before the single-particle threshold absorption frequency, e.g., the initial strong peaks of the 1s, 2s and 3s excitons [23]. Very interestingly, the complicated optical excitation processes, which directly determine two kinds of pronounced absorption structures, might become open issues in identifying the specific transition channels. How to reach the excellent predictions on the unusual optical properties will be solved in the different chapters under the suitable physical pictures, e.g., the delicate analyses on unique absorption spectra of Li_2SiO_3 compounds in the absence and presence of excitonic effects (details in chapter 8). Most importantly, the critical orbital hybridizations in chemical bonds, being associated with the band-edge states, are found to be meaningful in dominating the various absorption peaks through the specific absorption frequencies. A broader theoretical framework, which consists of the geometric, electronic, magnetic, and optical properties, will show a successful development under the unified viewpoints. Furthermore, the diversified chemical bondings are clearly revealed in the cathode, electrolyte, and anode materials of ion-based batteries.

2.2 Phenomenological models: generalized tight-binding model/ effective-mass approximation, Kubo formula, random-phase approximation and others

Generally speaking, the phenomenological models are able to provide the qualitative and quantitative viewpoints for the various physical/chemical/material properties, especially the concise pictures [18, 19] and the critical mechanisms [18, 19]. They become more powerful in understanding the high-symmetry emergent materials, such as, the random-phase approximation and self-energy method about 3D/2D/1D electron gases [114, 115]/[116]/[117], layered graphene systems [118], and carbon nanotubes [119]. A lot of mainstream condensed-matter systems are successfully synthesized under distinct physical and chemical methods; therefore, the greatly diversified phenomena need to be fully explored through the delicate development of the theoretical framework. Up to now, this is clearly revealed in the systematic investigations, as done for 3D graphites [120], 2D graphene-related systems [18, 19], and 1D carbon nanotubes [52] under the single-particle and many-body theories. Detailed discussions are illustrated in the following sections.

2.2.1 The generalized tight-binding model for electronic properties under any external fields

Both effective-mass approximation and tight-binding model are available in the investigation of the low-lying essential properties of emergent materials. In general, the former is the reduced form of the latter [18, 97], while the low-energy quasiparticle properties are dominated by the electronic states near the specific valleys. For example, up-to-date studies show that this approximate model might be successful in few-layer graphene systems under well-behaved interlayer $2p_z$-orbital interactions [18, 19, 97]. The electronic energy spectra and wave functions, which are initiated from the K/K* valleys, are greatly diversified by the layer number and stacking configuration. There exist a Dirac-cone structure, three pairs of vertical Dirac cones, one pair of slightly distorted Dirac cone and two pairs of parabolic bands, and one pair of partially flat/sombrero-shape/linear energy bands, respectively, in monolayer [121], AAA- [121], ABA- [121, 122] and ABC-trilayer graphenes [121–123]. Similar studies are suitable and reliable for monolayer silicene and germanene with weak, but significant spin–orbital interaction. On the other side, this method would become too cumbersome to deal with the heavy atomic systems, mainly owing to the significant couplings of the multi-orbital hybridizations [41, 74] and spin–orbital interactions [41, 74]. The previous calculations, as clearly done by the tight-binding model [19] and the first-principles method [10] have shown the simultaneous emergences of the π and σ-electronic energy bands associated with the K and Γ valleys, e.g., those of monolayer tinene/plumbene/antimonene/bismuthene [13, 42, 43, 81, 82]. The state degeneracy and the irregular energy dispersions can completely terminate the further magnetic quantization from the effective-mass model. As a result, it is impossible to thoroughly comprehend the featured Landau levels in ABC- and AAB-stacked layered graphenes, bilayer silicene/germanene systems [47, 121, 123]/[97, 124], and single-layer tinene/plumbene/antimonene/bismuthene materials [13, 42, 43, 81, 82].

Whether the tight-binding model is reliable within the whole energy spectra is mainly determined by the critical factors, the various hopping integrals of the neighboring orbitals [18, 19] and the spin-dependent interactions [18, 19]. In general, such parameters are obtained from a good fitting with the low-lying band structures of the first-principles calculations [10, 12]. For example, there exist different interlayer $2p_z$-orbital hybridizations in AAA- [97], ABA- [122], ABC- [97, 122] and AAB-stacked graphenes [97]. Only the last one is a narrow-gap semiconductor, while the others are semimetals in the presence of important band overlaps. Specifically, to achieve three pairs of valence and conduction bands with the oscillatory/sombrero-shape/parabolic dispersions in trilayer AAB stacking, one intralayer and nine interlayer $2p_z$-orbital hopping integrals are utilized in the tight-binding model. This method is also successful in three kinds of well-stacked bulk graphites, simple [102, 125], Bernal [126, 127] and rhombohedral ones [128, 129] through the diversified band structures and free carrier densities. The similar $2p_z$-orbital model, which is modified by their misorientation along the azimuthal direction (on the cylindrical surface) is suitable for sufficient large carbon nanotubes [11]. More importantly, the open boundary condition, the non-uniform hopping integrals, and the Hubbard-like on-site spin interactions need to be included in the 1D graphene nanoribbons [130, 131]. The up-to-date predictions, both $3p_z$/$4p_z$-orbital hybridizations (π bondings and spin-orbital interactions) are enough for understanding the essential properties of monolayer silicene/germanene systems, such as, their electronic properties [41, 74], optical absorption spectra [13, 42], and Coulomb excitations and deexcitations [18, 42]. However, for heavy group-IV and group-V systems with the seriously buckled honeycomb lattices, the sp^3 bondings and the strong spin–orbital couplings are capable of creating the featured energy spectra, e.g., the low-lying π and σ bands at the distinct valleys and their anti-crossings in tinene/plumbene/antimonene/bismuthene [13, 42, 43, 81, 82]. On the other hand, the large barriers of numerical calculations might be introduced in the tight-binding model through the highly non-uniform chemical bonds in a moiré superlattice. A lot of hopping integrals, which are associated with the position-, atom- and orbital-dependent chemical bonds, must be determined simultaneously. It should be noticed that no suitable parameters are revealed in the previous theoretical studies for the ternary lithium oxide compounds (cathode/electrolyte/anode materials of ion-based batteries; [5, 6]). This will become the open issue in near-future investigations.

In general, the electronic properties are easily modulated by the external electric and magnetic fields, especially for the latter [19]. The up-to-date theoretical predictions on the magnetic quantization phenomena show that the generalized tight-binding model is able to fully explore the various condensed-matter systems, with the different dimensionalities [18, 19, 97], stacking configurations [18, 19, 97], layer numbers [18, 19, 97], multi-orbital hybridizations [18, 19, 97], and spin–orbital couplings [18, 19, 97], such as, the diverse Landau levels in trilayer AAA- [97], ABA- [122]. ABC- [122, 132] and AAB-stacked graphene materials [47, 97]. This developed framework is also reliable for the non-uniform and/or composite fields, covering the spatially modulated magnetic or electric fields [47, 122, 133], and a uniform magnetic field combined with the former [47, 122, 123]. However, the effective-mass approximation is only suitable for high-symmetry systems in the presence of simple energy dispersions (e.g., trilayer AAA [97] and ABA stackings [97, 122, 133]), but not reliable through the degenerate constant-energy loops (e.g., ABC and AAB ones). Furthermore, first-principles calculations cannot deal with the super-large moiré superlattice due to the vector potential (**A**(**r**)). A monolayer graphene is chosen to fully illustrate the magnetic-field effects on the pristine Hamilton. The generalized tight-binding model is based on two equivalent A and B sublattices of a honeycomb lattice. In addition to the original period of crystal potential, **A**(**r**), being due to a uniform perpendicular magnetic field ($B_{z\ z}$ in figure 2.4), presents another **B**-dependent one. The greatly enlarged unit cell, which is taken to be a large rectangle along the armchair direction, just covers one flux quantum ($\phi_0 = \frac{hc}{e}$; [100]) through the $2R_B$-hexagons. That is, $2R_B\phi = \phi_0$, where a uniform magnetic field penetrates a carbon hexagon using the flux of ϕ. As a result, the magnetic Hamiltonian is a $4R_B \times 4R_B$ Hermitian matrix (details in [134]). Its diagonalization can achieve the unusual Landau-level spectra and wave functions. The valence and conduction bands of Dirac-cone band structures are magnetically quantized into $E^{c,\,v}(n^{c,\,v}) = v_F\sqrt{n^{c,\,v}B_z}$. V_F is the Fermi velocity, the slope of the linear Dirac cone [134]. Very interestingly, the quantum number, $n^{c,\,v} = 0, 1, 2$, is determined by the zero-point number of the spatial probability distribution under the dominating sublattices. This viewpoint could be generalized to any layered graphene-related systems (details in several books [18, 97, 102]), in which the various interlayer orbital hybridizations and spin–orbital couplings are taken into account simultaneously. Bulk graphite [102, 135] and graphite alkali-intercalation compounds [102], very popular cathode materials of Li^+-based batteries, are the outstanding layered samples in clearly illustrating the dimension- and doping-diversified magneto-electronic properties [102, 122, 132]. However, the magnetic quantization in 3D ternary lithium oxides might become an open issue since it is almost impossible to

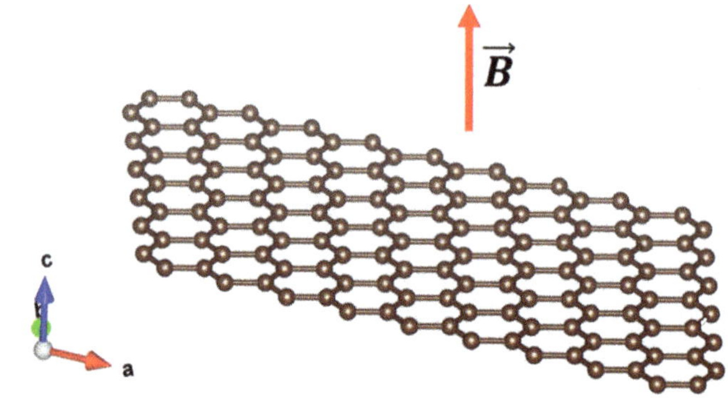

Figure 2.4. A monolayer honeycomb lattice under a uniform perpendicular magnetic field with a moiré superlattice.

solve under too complicated hopping integrals (rather complex orbital hybridizations in position-dependent chemical bonds; [75]). Most importantly, the generalized tight-binding model is successful in directly combining with the other methods for the other essential properties (to be discussed later).

2.2.2 The static and dynamic Kubo formulas

It is well known that the static and dynamic Kubo formulas are, respectively, available in exploring the transport and optical excitations for electrons of condensed-matter systems. The latter will be recovered to the former under the limit of $\omega \rightarrow 0$, and then it is very useful for studying quantum Hall effects of layered materials after the direct combination with the generalized tight-binding model, e.g., the unusual Hall conductivities in sliding bilayer/trilayer graphene systems [136]/[137] and bilayer silicene [138]. Very interestingly, the selection rules of quantum transport properties, which might survive between the initial and final Landau-level scatterings, could be thoroughly examined under the numerical analyses. The above-mentioned physical models are mainly focused on quasiparticle behaviors. How to develop a theoretical framework suitable for the ion transports of batteries will become an urgent issue, since concise pictures and critical mechanisms are absent up to now.

The optical properties of the coupled electromagnetic fields and electrons are discussed in section 2.1.3 thoroughly. Compared with the first-principles calculations, the Kubo formula and the generalized tight-binding model are able to deal with absorption spectra very efficiently in the absence of excitonic effects. For example, the theoretical predictions can explain the experimental measurements [131, 139] on absorption peaks of trilayer ABA- [122, 133] and ABC-stacked [122, 123, 132] graphene systems, in which the joint van Hove singularities are well identified. In addition, the electric dipole moment between the initial and final states is evaluated within the gradient approximation [140], being consistent with the sublattice bases of the generalized tight-binding model. Moreover, up-to-date studies clearly show that their direct combination is the only unique framework reliable for the magneto-optical properties. Furthermore, the numerical technique needs to be greatly enhanced in overcoming the huge magnetic Hamiltonian matrix (details in [18, 97]). The systematic investigations have been finished for various graphene systems [10] and AA-, AB- and ABC-stacked graphites [102]. Very interestingly, there exist diverse magneto-absorption structures and magneto-optical selection rules in emergent layered materials, e.g., the rich and unique phenomena in sliding bilayer graphenes [19, 141], tinene [19], phosphorene [19] and bismuthene [142]. Whether the diversified magneto-optical properties could survive in ternary lithium oxides [75] is worthy of closer examination.

The direct combination of the static Kubo formula and the generalized tight-binding model is very useful in comprehending the magnetic transport properties, especially for the quantum Hall effects. in graphene [136, 143, 144] and silicene systems [145, 146]. This unified framework has been utilized to explore the diverse quantum conductivities of 2D layered structures, such as, monolayer graphene/

silicene, sliding bilayer graphene systems with various stacking configurations, AAA-, ABA-, ABC- and AAB-stacked graphenes [122, 123, 132, 133], and AB-bt bilayer silicene [147]. The theoretical predictions are consistent with the up-to-date Hall measurements [138]. Very interestingly, the magnetically quantized Landau levels in few-layer graphenes, as discussed earlier in section 2.2.1, can generate rich and unique transport properties [137]. Within the linear response, the transverse Hall conductivities are evaluated from the static Kubo formula [99], in which the velocity matrix elements are evaluated from the gradient approximation (the above-mentioned discussions in the dynamic Kubo formula; [99]). The distinct stacking configurations and layer numbers are shown to greatly diversify the Hall effects. The quantum conductivities strongly depend on the Fermi energy (free carrier density) and magnetic-field strength. The diverse features include the specific non-integer conductivities, the integer conductivities with the different steps, the splitting-created reduction and complexity of quantum conductivity, a vanishing or non-zero conductivity at the neutral point, and the well-like, staircase, composite, and abnormal plateau structures in the field-dependences. Such stacking-dependent characteristics principally arise from the crossing, anti-crossing and splitting magneto-electronic energy spectra and three kinds of quantized modes (the well-behaved, perturbed and undefined Landau levels; [56]). Whether the Hall quantum transports would be revealed in cathode/electrolyte/anode materials requires closer examination. As for the ion transports within batteries, how to develop a fully theoretical framework through the quantum models of many-particle scatterings is the mainstream focus [148, 149].

2.2.3 The modified RPA/self-energy method for layered systems

In addition to electromagnetic waves, Coulomb fields are very efficient in perturbing condensed-matter systems (examining the screening abilities of free and valence charges through the various electron-mediated apparatuses [150], such as, the available sources in transmission (details in our books) and reflection [18]) electron energy loss spectroscopies. Under the Coulomb gauge (details in section 2.1.3), Coulomb and electromagnetic fields, respectively, belong to the longitudinal and transverse ones [18]. Both of them are capable of generating significant similarities and differences in the single-particle and collective excitations, being characterized by the longitudinal and transverse dielectric functions [18]. The former induces the inelastic electron–electron Coulomb interactions under the conservation of the transferred momentum and frequency $[q, \omega]$, in which q is vanishing during the vertically optical excitations. That is, the Coulomb fields carry sufficient momentum and energy at any times and positions, so that they could be transformed into intermediated excited quasiparticles and plasma waves [18]. The diversified electronic excitations/de-excitations, which are defined in the $[q, \omega]$-excitation phase diagrams [23], are easily observed in a lot of related materials, such as the various electron–hole excitations and plasmon modes in 3D graphites/graphite intercalation compounds [120, 125, 151], 2D graphene systems with distinct staking configurations [18, 151], 1D coaxial carbon nanotubes [68, 152], graphene nanoribbons [131, 153], and 0D carbon fullerenes [154]. Similar phenomena could be observed in silicon- [155] and germanium-atom materials [19]. How to fully explore and understand the rich and unique Coulomb excitations in the well-known and emergent materials requires systematic investigations both theoretically and experimentally. As to theoretical models, all the intrinsic interactions and external fields need to be taken into account together within a unified theoretical framework [19]. This viewpoint, as successfully done for graphene-related systems (details in books [11, 12]), is finished by the direct combination of the tight-binding model and the modified random-phase approximation/self-energy method (RPA discussed later; [116–118]).

RPA, which has been thoroughly examined under low dimensionalities [118], is very efficient and suitable in determining the featured Coulomb-field excitations. For example, RPA is identified to be reliable for 2D/1D [115, 116, 156] electron gases, and quantum wells/wires [157, 158]. Also, it is available in investigating the diverse electronic excitations in layered graphene systems with distinct layer numbers and stacking configurations [12]. Most importantly, the complex interlayer $2p_z$-orbital hybridizations, the interlayer Coulomb interactions, and the external electric and magnetic fields are required to be covered within the calculation models. The modified RPA, which is based on the layer-dependent A and B sublattices [97], is proposed to thoroughly comprehend the intralayer and interlayer electric polarizations/charge oscillations/single-particle and collective excitations [19, 97]. It is fully consistent with the generalized tight-binding model even in the presence of steady external fields. According to the theoretical predictions [19, 97] and part of experimental measurements [143, 159, 160], the diversified excitation phenomena are clearly revealed in three types of graphites [102] (simple, Bernal and rhombohedral ones; [127, 129]), layered graphene systems, and multi-wall carbon nanotubes [161]. For example, AAA-, ABA-, ABC- and AAB-stacked graphenes show the stacking- and doping-enriched electron–hole pair regions/boundaries, and acoustic and optical plasmon modes without/with the significant Landau dampings. Furthermore, bilayer AA and AB stackings possess the featured magneto-plasmon modes. But when the single-orbital hybridizations are replaced by the multi-orbital ones, the Coulomb scattering matrix elements become very complicated, e.g., those in 2D tinene/plumbene/antimony/bismuthene [13, 42, 43, 82]. In addition, they are expected to play a critical role in the elementary excitations of ternary lithium oxides. How to get reliable electron–electron interactions due to the band-structure effects is an open issue. This is under current investigation.

The Coulomb-field excitations would become the effective channels of quasiparticle decay rates [18, 101]; that is, the inelastic electron–electron scatterings play critical roles in the lifetimes of the excited electrons and holes [97]. In general, the self-energy method, which is done under the screened exchange energy [114, 158], is frequently utilized in exploring quasi-particle properties. The up-to-date calculations on quasiparticle lifetimes clearly show great progress for comprehending diverse phenomena, such as the dimensionality-diversified temperature and energy dependences in 3D [14], 2D [162] and 1D [163] electron gas. As for emergent layered materials [44, 47, 97], how to unify the intralayer and interlayer interactions together (e.g., hopping integrals, Coulomb interactions and spin-orbital couplings) in a theoretical framework is a very interesting study topic. This is finished in the latest book, as examined for 1D coaxial carbon nanotubes. Furthermore, few-layer graphene systems are under current investigation. The modified self-energy method, corresponding to layered structures, should be fully tested using many mainstream 2D systems, for example, AA and AB bilayer graphenes, with totally different energy dispersions [44, 132], are expected to exhibit the unusual wave-vector dependences about their exciting quasiparticle states. In short, the developed framework about the quasiparticle properties is under test for more popular materials [97]. The high-resolution experimental measurements are urgently required to examine the carrier dynamics. The screening abilities of various carriers might dominate the transport performance of Li^+-based batteries [89, 148, 149]. For example, there exist close relations between an electron cloud in the cathode and Li^+ ions within graphite intercalation compound [164, 165] through the significant charge screenings [164, 165].

References

[1] Hafner J 2008 Ab-initio simulations of materials using VASP: density-functional theory and beyond *J. Comput. Chem.* **29** 2044–78

[2] Lozhnikov V E, Mamonov A V, Borzilov V O, Mamonova M V, Prudnikov P V, Sorokin A A and Baksheev G G 2019 Estimating the performance of *Ab Initio* calculation by VASP on openpower high performance system CEUR Workshop Proc. pp 24–9

[3] Van Neck D, Verdonck S, Bonny G, Ayers P and Waroquier M 2006 Quasiparticle properties in a density-functional framework *Phys. Rev.* A **74** 042501

[4] Christensen R S, Levinsen J and Bruun G M 2015 Quasiparticle properties of a mobile impurity in a Bose–Einstein condensate *Phys. Rev. Lett.* **115** 160401

[5] Nitta N, Wu F, Lee J T and Yushin G 2015 Li-ion battery materials: present and future *Mater. Today* **18** 252–64

[6] Goodenough J B and Park K-S 2013 The Li-ion rechargeable battery: a perspective *J. Am. Chem. Soc.* **135** 1167–76

[7] Das S K, Mahapatra S and Lahan H 2017 Aluminium-ion batteries: developments and challenges *J. Mater. Chem.* A **5** 6347–67

[8] Lin M-C, Gong M, Lu B, Wu Y, Wang D-Y, Guan M, Angell M, Chen C, Yang J and Hwang B-J 2015 An ultrafast rechargeable aluminium-ion battery *Nature* **520** 324–8

[9] Sarmah A and Hobza P 2017 Understanding the spin-dependent electronic properties of symmetrically far-edge doped zigzag graphene nanoribbon from a first principles study *RSC Adv.* **7** 46604–14

[10] Shih-Yang Lin H-Y L, Nguyen D K, Thanh Thuy Tran N, Duong Pham H, Chang S-L, Lin C-Y and Lin M-F 2020 *Silicene-Based Layered Materials: Essential Properties* (Bristol: IOP Publishing)

[11] Lin S Y T, T N T, Chang S L, Su W P and Lin M F 2018 *Structure- and Adatom-Enriched Essential Properties of Graphene Nanoribbons* (Boca Raton, FL: CRC Press)

[12] Tran N T T, Lin S-Y, Lin C-Y and Lin M-F 2017 *Geometric and Electronic Properties of Graphene-Related Systems: Chemical Bonding Schemes* (Boca Raton, FL: CRC Press)

[13] Matthes L, Pulci O and Bechstedt F 2013 Massive Dirac quasiparticles in the optical absorbance of graphene, silicene, germanene, and tinene *J. Phys.-Condens. Matter* **25** 395305

[14] Kittel C 1996 *Introduction to Solid State Physics* (New York: Wiley) p 402
[15] Meyer J, Tombers M, van Wüllen C, Niedner-Schatteburg G, Peredkov S, Eberhardt W, Neeb M, Palutke S, Martins M and Wurth W 2015 The spin and orbital contributions to the total magnetic moments of free Fe, Co, and Ni clusters *J. Chem. Phys.* **143** 104302
[16] Jiles D 2015 *Introduction to Magnetism and Magnetic Materials* (Boca Raton, FL: CRC Press)
[17] Mattis D C 2012 *The Theory of Magnetism I: Statics and Dynamics* vol 17 (Berlin: Springer)
[18] Chiun-Yan Lin J-Y W, Chiu C-W and Lin M-F 2019 *Coulomb Excitations and Decays in Graphene-Related Systems* (Boca Raton, FL: CRC Press)
[19] Chen S, Wu J, Lin C and Lin M 2016 Theory of magnetoelectric properties of 2d systems *New J. Phys.* **18** 103024
[20] Tsuda N, Nasu K, Fujimori A and Siratori K 2000 *Electronic Conduction in Oxides* (Berlin: Springer) pp 119–55
[21] Albareda G, Suñé J and Oriols X 2009 Many-particle Hamiltonian for open systems with full coulomb interaction: application to classical and quantum time-dependent simulations of nanoscale electron devices *Phys. Rev.* B **79** 075315
[22] Kira M and Koch S W 2006 Many-body correlations and excitonic effects in semiconductor spectroscopy *Prog. Quantum Electron.* **30** 155–296
[23] Mahan G D 2013 *Many-Particle Physics* (Berlin: Springer)
[24] Kohn W and Sham L J 1965 Self-consistent equations including exchange and correlation effects *Phys. Rev.* **140** A1133
[25] Chachiyo T and Chachiyo H 2020 Simple and accurate exchange energy for density functional theory *Molecules* **25** 3485
[26] Hedin L, Lundqvist B and Lundqvist S 1970 Beyond the one-electron approximation: density of states for interacting electrons *J. Res. Natl. Bur. Stand. Sec. A, Phys. Chem.* **74** 417
[27] Nechaev I, Echenique P M and Chulkov E V 2010 Inelastic decay rate of quasiparticles in a two-dimensional spin–orbit coupled electron system *Phys. Rev.* B **81** 195112
[28] Ernzerhof M and Scuseria G E 1999 Assessment of the Perdew–Burke–Ernzerhof exchange-correlation functional *J. Chem. Phys.* **110** 5029–36
[29] Orellana W, Miwa R and Fazzio A 2003 First-principles calculations of carbon nanotubes adsorbed on Si (001) *Phys. Rev. Lett.* **91** 166802
[30] Lv B, Qian T and Ding H 2019 Angle-resolved photoemission spectroscopy and its application to topological materials *Nat. Rev. Phys.* **1** 609–26
[31] Baltes H, Yacoby Y, Pindak R, Clarke R, Pfeiffer L and Berman L 1997 Measurement of the x-ray diffraction phase in a 2D crystal *Phys. Rev. Lett.* **79** 1285–8
[32] Bussolotti F, Chi D, Goh K J, Huang Y L and Wee A T 2020 *2D Semiconductor Materials and Devices* (Amsterdam: Elsevier) pp 199–220
[33] Tran N T T L, Lin S Y and Lin, M. F. C Y 2017 *Geometric and Electronic Properties of Graphene-Related Systems: Chemical Bonding Schemes* (Boca Raton, FL: CRC Press)
[34] Foulkes W M C and Haydock R 1989 Tight-binding models and density-functional theory *Phys. Rev.* B **39** 12520
[35] Grimme S, Antony J, Ehrlich S and Krieg H 2010 A consistent and accurate *ab initio* parametrization of density functional dispersion correction (DFT-D) for the 94 elements H-Pu *J. Chem. Phys.* **132** 154104

[36] Grimme S, Ehrlich S and Goerigk L 2011 Effect of the damping function in dispersion corrected density functional theory *J. Comput. Chem.* **32** 1456–65
[37] Echenique P and Alonso J L 2007 A mathematical and computational review of Hartree–Fock SCF methods in quantum chemistry *Mol. Phys.* **105** 3057–98
[38] Perdew J P, Burke K and Ernzerhof M 1996 Generalized gradient approximation made simple *Phys. Rev. Lett.* **77** 3865
[39] Rangel T, Caliste D, Genovese L and Torrent M 2016 A wavelet-based projector augmented-wave (PAW) method: reaching frozen-core all-electron precision with a systematic, adaptive and localized wavelet basis set *Comput. Phys. Commun.* **208** 1–8
[40] Perdew J P and Schmidt K 2001 Jacob's ladder of density functional approximations for the exchange-correlation energy *AIP Conf. Proc.* (College Park, MD: American Institute of Physics) pp 1–20
[41] Roome N J and Carey J D 2014 Beyond graphene: stable elemental monolayers of silicene and germanene *ACS Appl. Mater. Interfaces* **6** 7743–50
[42] Matthes L, Pulci O and Bechstedt F 2014 Optical properties of two-dimensional honeycomb crystals graphene, silicene, germanene, and tinene from first principles *New J. Phys.* **16** 105007
[43] Das D K, Sarkar J and Singh S K 2018 Effect of sample size, temperature and strain velocity on mechanical properties of plumbene by tensile loading along longitudinal direction: a molecular dynamics study *Comput. Mater. Sci.* **151** 196–203
[44] Ho J-H, Lu C, Hwang C, Chang C and Lin M-F 2006 Coulomb excitations in AA-and AB-stacked bilayer graphites *Phys. Rev.* B **74** 085406
[45] Laref A, Alsagri M, Alay-e-Abbas S M, Laref S, Huang H, Xiong Y, Yang J, Khandy S A, Rai D P and Varshney D 2020 Electronic structure and optical characteristics of AA stacked bilayer graphene: a first principles calculations *Optik* **206** 163755
[46] Yelgel C 2016 Electronic structure of abc-stacked multilayer graphene and trigonal warping: a first principles calculation *J. Phys. Conf. Ser.* **707** 012022
[47] Do T-N, Lin C-Y, Lin Y-P, Shih P-H and Lin M-F 2015 Configuration-enriched magneto-electronic spectra of AAB-stacked trilayer graphene *Carbon* **94** 619–32
[48] Berland K, Cooper V R, Lee K, Schröder E, Thonhauser T, Hyldgaard P and Lundqvist B I 2015 van der Waals forces in density functional theory: a review of the vdW-DF method *Rep. Prog. Phys.* **78** 066501
[49] Birowska M, Milowska K and Majewski J 2011 Van der Waals density functionals for graphene layers and graphite *Acta Phys. Pol.* A **120** 845–8
[50] Yin W-J, Xie Y-E, Liu L-M, Chen Y-P, Wang R-Z, Wei X-L, Zhong J-X and Lau L 2013 Atomic structure and electronic properties of folded graphene nanoribbons: a first-principles study *J. Appl. Phys.* **113** 173506
[51] Chang S, Chiang C, Lin S and Lin M 2016 Feature-rich Geometric and Electronic Properties of Carbon Nanoscrolls arXiv: 1607.00452
[52] Dresselhaus M, Dresselhaus G and Saito R 1995 Physics of carbon nanotubes *Carbon* **33** 883–91
[53] Bulusheva L, Okotrub A, Romanov D and Tomanek D 1998 Electronic structure of (n, 0) zigzag carbon nanotubes: cluster and crystal approach *J. Phys. Chem.* A **102** 975–81
[54] Benguediab S, Tounsi A, Zidour M and Semmah A 2014 Chirality and scale effects on mechanical buckling properties of zigzag double-walled carbon nanotubes *Compos. Part B: Eng.* **57** 21–4

[55] Fu H, Zhang J, Ding Z, Li H and Meng S 2014 Stacking-dependent electronic structure of bilayer silicene *Appl. Phys. Lett.* **104** 131904

[56] Liu H, Lin S-Y and Wu J-y 2020 Stacking-configuration-enriched essential properties of bilayer graphenes and silicenes *J. Chem. Phys.* **153** 154707

[57] Hastuti D P, Amalia W, Priska Z and Nurwantoro P 2020 First-principles density-functional-theory calculations of formation and dissociation energies in germanene multi-vacancies *Mater. Today Commun.* **22** 100754

[58] Castro Neto A H, Guinea F, Peres N M R, Novoselov K S and Geim A K 2009 The electronic properties of graphene *RvMP* **81** 109–62

[59] Wong J-H, Wu B-R and Lin M-F 2012 Strain effect on the electronic properties of single layer and bilayer graphene *J. Phys. Chem.* C **116** 8271–7

[60] Lin C-Y, Do T-N, Huang Y-K and Lin M-F 2017 *Optical Properties of Graphene in Magnetic and Electric Felds* (Bristol: IOP Publishing)

[61] Li Y, Ren J-C, Zhang R-Q, Lin Z and Van Hove M A 2012 Atomic nitrogen chemisorption on graphene with extended line defects *J. Mater. Chem.* **22** 21167–72

[62] Margulis V and Muryumin E 2008 Theory of chemisorption of a single hydrogen atom on graphene *Carbon Nanomaterials in Clean Energy Hydrogen Systems* (Cham: Springer) p 183

[63] Pham H D, Nguyen T D H, Vo K D, Huynh T M D and Lin M-F 2020 Rich essential properties of boron, carbon, and nitrogen substituted germanenes *Appl. Phys. Express* **13** 085502

[64] Nalwa H S 2001 *Handbook of Surfaces and Interfaces of Materials, Five-Volume Set* (Amsterdam: Elsevier)

[65] Karlický F, Lepetit B and Lemoine D 2014 Quantum modelling of hydrogen chemisorption on graphene and graphite *J. Chem. Phys.* **140** 124702

[66] Yuan S, Bao J L, Wang L, Xia Y, Truhlar D G and Wang Y 2016 Graphene-supported nitrogen and boron rich carbon layer for improved performance of lithium–sulfur batteries due to enhanced chemisorption of lithium polysulfides *Adv. Energy Mater.* **6** 1501733

[67] Lv R, Li Q, Botello-Méndez A R, Hayashi T, Wang B, Berkdemir A, Hao Q, Elías A L, Cruz-Silva R and Gutiérrez H R 2012 Nitrogen-doped graphene: beyond single substitution and enhanced molecular sensing *Sci. Rep.* **2** 586

[68] Lin M-F and Shung K W-K 1994 Plasmons and optical properties of carbon nanotubes *Phys. Rev.* B **50** 17744

[69] Praveen C, Piccinin S and Fabris S 2015 Adsorption of alkali adatoms on graphene supported by the Au/Ni (111) surface *Phys. Rev.* B **92** 075403

[70] Thanh Thuy Tran N, Lin S-Y, Glukhova O E and Lin M-F 2015 π-bonding-dominated energy gaps in graphene oxides arXiv:1512.08693

[71] Boukhvalov D, Katsnelson M and Lichtenstein A 2008 Hydrogen on graphene: electronic structure, total energy, structural distortions and magnetism from first-principles calculations *Phys. Rev.* B **77** 035427

[72] Huang H-C, Lin S-Y, Wu C-L and Lin M-F 2016 Configuration-and concentration-dependent electronic properties of hydrogenated graphene *Carbon* **103** 84–93

[73] Galbiati M, Motta N, De Crescenzi M and Camilli L 2019 Group-IV 2D materials beyond graphene on nonmetal substrates: challenges, recent progress, and future perspectives *Appl. Phys. Rev.* **6** 041310

[74] Vogt P 2018 *Silicene, Germanene and Other Group IV 2D Materials* (Frankfurt: Beilstein-Institute)

[75] Nguyen T D H, Pham H D, Lin S Y and Lin M F 2020 Featured properties of Li^+-based battery anode: $Li_4Ti_5O_{12}$ *RSC Adv.* **10** 14071–9

[76] Khuong Dien V, Thi Han N, Nguyen T D H, Huynh T M D, Pham H D and Lin M-F 2020 Geometric and electronic properties of Li_2GeO_3 *Front. Mater.* **7** 288

[77] Schwarz A, Bode M and Wiesendanger R 2007 Scanning probe techniques: MFM and SP-STM *Handbook of Magnetism and Advanced Magnetic Materials* (Cham: Springer)

[78] Kubetzka A, Bode M, Pietzsch O and Wiesendanger R 2002 Spin-polarized scanning tunneling microscopy with antiferromagnetic probe tips *Phys. Rev. Lett.* **88** 057201

[79] Wiesendanger R 2007 Mapping spin structures on the atomic scale *Europhys. News* **38** 16–21

[80] Kuroda K, Yaji K, Harasawa A, Noguchi R, Kondo T, Komori F and Shin S 2018 Experimental methods for spin-and angle-resolved photoemission spectroscopy combined with polarization-variable laser *J. Vis. Exp.* **136** e57090

[81] Singh D, Gupta S K, Sonvane Y and Lukačević I 2016 Antimonene: a monolayer material for ultraviolet optical nanodevices *J. Mater. Chem.* C **4** 6386–90

[82] Kecik D, Özçelik V O, Durgun E and Ciraci S 2019 Structure dependent optoelectronic properties of monolayer antimonene, bismuthene and their binary compound *Phys. Chem. Chem. Phys.* **21** 7907–17

[83] Lin C-L, Arafune R, Kawahara K, Tsukahara N, Minamitani E, Kim Y, Takagi N and Kawai M 2012 Structure of silicene grown on Ag (111) *Appl. Phys. Express* **5** 045802

[84] Feng L, Yabuoshi K, Sugimoto Y, Onoda J, Fukuda M and Ozaki T 2018 Structural identification of silicene on the Ag (111) surface by atomic force microscopy *Phys. Rev.* B **98** 195311

[85] Takagi N, Lin C L and Arafune R 2016 *Silicene* (Berlin: Springer) pp 143–65

[86] Buyukyazi M and Mathur S 2015 3D nanoarchitectures of a-$LiFeO_2$ and alpha-$LiFeO_2$/C nanofibers for high power lithium-ion batteries *Nano Energy* **13** 28–35

[87] Li J, Li J, Luo J, Wang L and He X 2011 Recent advances in the $LiFeO_2$-based materials for Li-ion batteries *Int. J. Electrochem. Sci.* **6** 1550–61

[88] Boufelfel A 2013 Electronic structure and magnetism in the layered $LiFeO_2$: DFT+U calculations *J. Magn. Magn. Mater.* **343** 92–8

[89] Andriyevsky B, Doll K and Jacob T 2014 Electronic and transport properties of $LiCoO_2$ *Phys. Chem. Chem. Phys.* **16** 23412–20

[90] Khatun F, Gafur M, Ali M, Islam M and Sarker M 2014 Impact of lithium composition on structural, electronic and optical properties of lithium cobaltite prepared by solid-state reaction *J. Sci. Res.* **6** 217–31

[91] Donders M, Arnoldbik W, Knoops H, Kessels W and Notten P 2013 Atomic layer deposition of $LiCoO_2$ thin-film electrodes for all-solid-state Li-ion micro-batteries *J. Electrochem. Soc.* **160** A3066

[92] Molenda J, Wilk P and Marzec J 2002 Structural, electrical and electrochemical properties of $LiNiO_2$ *Solid State Ionics* **146** 73–9

[93] Daré A-M, Hayn R and Richard J-L 2003 Orbital and spin exchange in $LiNiO_2$ *EPL (Europhys. Lettl)* **61** 803

[94] Singh J 2006 *Optical Properties of Condensed Matter and Applications* vol 6 (New York: Wiley)

[95] Fox M 2001 *Optical Properties of Solids* (Oxford: Oxford University Press)

[96] Mak K F, Shan J and Heinz T F 2011 Seeing many-body effects in single-and few-layer graphene: observation of two-dimensional saddle-point excitons *Phys. Rev. Lett.* **106** 046401

[97] Lin C-Y, Ho C-H, Wu J-Y, Do T-N, Shih P-H, Lin S-Y and Lin M-F 2019 *Diverse Quantization Phenomena in Layered Materials* (Boca Raton, FL: CRC Press)
[98] Zhou F, Liu F, Xiao L, Cui K, Feng X, Zhang W and Huang Y 2017 Extending the frequency range of surface plasmon polariton mode with meta-material *Nano-Micro Lett.* **9** 9
[99] Nakayama T and Shima H 1998 Computing the Kubo formula for large systems *Phys. Rev.* E **58** 3984
[100] Jackson J D 2007 *Classical Electrodynamics* (New York: Wiley)
[101] Fromer N, Schüller C, Lai C, Chemla D, Perakis I, Driscoll D and Gossard A 2002 Coulomb correlations in a two-dimensional electron gas in large magnetic fields *Phys. Rev.* B **66** 205314
[102] Lin C-Y, Chen R-B, Ho Y-H and Lin M-F 2017 *Electronic and Optical Properties of Graphite-Related Systems* (Boca Raton, FL: CRC Press)
[103] Fox M 2002 *Optical Properties of Solids* (College Park, MD: American Association of Physics Teachers)
[104] Wu S, Cheng L and Wang Q 2017 Excitonic effects and related properties in semiconductor nanostructures: roles of size and dimensionality *Mater. Res. Express* **4** 085017
[105] Blase X, Duchemin I, Jacquemin D and Loos P-F 2020 The Bethe–Salpeter equation formalism: from physics to chemistry *J. Phys. Chem. Lett.* **11** 7371–82
[106] Leng X, Jin F, Wei M and Ma Y 2016 GW method and Bethe–Salpeter equation for calculating electronic excitations *WIREs* **6** 532–50
[107] Nguyen-Truong H T 2014 Energy-loss function including damping and prediction of plasmon lifetime *J. Electron. Spectrosc. Relat. Phenom.* **193** 79–85
[108] Egerton R F 2008 Electron energy-loss spectroscopy in the TEM *Rep. Prog. Phys.* **72** 016502
[109] Xin H L, Dwyer C, Muller D A, Zheng H and Ercius P 2013 Scanning confocal electron energy-loss microscopy using valence-loss signals *Microsc. Microanal.* **19** 1036–49
[110] Iakoubovskii K, Mitsuishi K, Nakayama Y and Furuya K 2008 Thickness measurements with electron energy loss spectroscopy *Microsc. Res. Tech.* **71** 626–31
[111] Taft E and Philipp H 1965 Optical properties of graphite *Phys. Rev.* **138** A197
[112] Yu H S, Li S L and Truhlar D G 2016 Perspective: Kohn-Sham density functional theory descending a staircase *J. Chem. Phys.* **145** 130901
[113] Lin W-B, Tran N T T and Lin S-Y 2019 Diverse fundamental properties in stage-n graphite alkali-intercalation compounds: anode materials of Li^+-based batteries arXiv:2001.02042
[114] Gould T, Ruzsinszky A and Perdew J P 2019 Simple self-interaction correction to random-phase-approximation-like correlation energies *Phys. Rev.* A **100** 022515
[115] Ren X, Rinke P, Joas C and Scheffler M 2012 Random-phase approximation and its applications in computational chemistry and materials science *J. Mater. Sci.* **47** 7447–71
[116] Schulze H-J, Schuck P and Van Giai N 2000 Two-dimensional electron gas in the random-phase approximation with exchange and self-energy corrections *Phys. Rev.* B **61** 8026
[117] Grüneis A, Marsman M, Harl J, Schimka L and Kresse G 2009 Making the random phase approximation to electronic correlation accurate *J. Chem. Phys.* **131** 154115
[118] Olsen T and Thygesen K S 2013 Random phase approximation applied to solids, molecules, and graphene-metal interfaces: from van der Waals to covalent bonding *Phys. Rev.* B **87** 075111
[119] Sakai H, Suzuura H and Ando T 2003 Effective-mass theory of electron correlations in band structure of semiconducting carbon nanotubes *J. Phys. Soc. Jpn.* **72** 1698–705
[120] Belenkov E and Ali-Pasha V 2011 3D-graphite structure *Crystallogr. Rep.* **56** 101–6

[121] Lin C-Y, Do T-N, Huang Y-K and Lin M-F 2017 *Optical Properties of Graphene in Magnetic and Electric Fields* (Bristol: IOP Publishing)
[122] Yuan S, Roldán R and Katsnelson M I 2011 Landau level spectrum of ABA-and ABC-stacked trilayer graphene *Phys. Rev.* B **84** 125455
[123] Lin Y-P, Lin C-Y, Ho Y-H, Do T-N and Lin M-F 2015 Magneto-optical properties of ABC-stacked trilayer graphene *Phys. Chem. Chem. Phys.* **17** 15921–7
[124] Liu H-Y, Lin S-Y and Wu J-Y 2020 Stacking-configuration-enriched essential properties of bilayer graphenes and silicenes *J. Chem. Phys.* **153** 154707
[125] Shyu F-L and Lin M-F 2000 Plasmons and optical properties of semimetal graphite *J. Phys. Soc. Jpn.* **69** 3781–4
[126] Lai Y, Ho J, Chang C and Lin M-F 2008 Magnetoelectronic properties of bilayer Bernal graphene *Phys. Rev.* B **77** 085426
[127] Li T and Lin M 2013 Electronic properties of bilayer Bernal graphene in a modulated magnetic field *J. Magn. Magn. Mater.* **348** 61–7
[128] Chang C, Wang J, Lu C, Huang Y, Lin M-F and Chen R 2008 Optical properties of simple hexagonal and rhombohedral few-layer graphenes in an electric field *J. Appl. Phys.* **103** 103109
[129] Chiu C-W, Huang Y-C, Chen S-C, Lin M-F and Shyu F-L 2011 Low-frequency electronic and optical properties of rhombohedral graphite *Phys. Chem. Chem. Phys.* **13** 6036–42
[130] Chung H-C, Chang C-P, Lin C-Y and Lin M-F 2016 Electronic and optical properties of graphene nanoribbons in external fields *Phys. Chem. Chem. Phys.* **18** 7573–616
[131] Wu J, Chiu Y-H, Lien J and Lin M-F 2009 The effects of the modulated magnetic fields on electronic structures of graphene nanoribbons *J. Nanosci. Nanotechnol.* **9** 3193–200
[132] Lu L C, Chang C P, Huang Y C, Ho J H, Hwang C C and Lin M F 2007 Electronic properties of AA-and ABC-stacked few-layer graphites *J. Phys. Soc. Jpn.* **76** 024701
[133] Chiun-Yan L, Ching-Hong H, Jhao-Ying W and Ming-Fa L 2020 Unusual electronic excitations in ABA trilayer graphene *Sci. Rep.* **10** 11106
[134] Konschuh S, Gmitra M and Fabian J 2010 Tight-binding theory of the spin–orbit coupling in graphene *Phys. Rev.* B **82** 245412
[135] Kumar A, Poumirol J, Escoffier W, Goiran M, Raquet B and Broto J 2011 Electronic properties of graphene, few-layer graphene, and bulk graphite under very high magnetic field *Int. J. Nanosci.* **10** 43–7
[136] Barlas Y, Yang K and MacDonald A 2012 Quantum Hall effects in graphene-based two-dimensional electron systems *Nanotechnology* **23** 052001
[137] Do T-N, Chang C-P, Shih P-H, Wu J-Y and Lin M-F 2017 Stacking-enriched magneto-transport properties of few-layer graphenes *Phys. Chem. Chem. Phys.* **19** 29525–33
[138] Do T-N, Gumbs G, Shih P-H, Huang D and Lin M-F 2019 Valley-and spin-dependent quantum Hall states in bilayer silicene *Phys. Rev.* B **100** 155403
[139] Yin L-J, Shi L-J, Li S-Y, Zhang Y, Guo Z-H and He L 2019 High-magnetic-field tunneling spectra of ABC-stacked trilayer graphene on graphite *Phys. Rev. Lett.* **122** 146802
[140] List N H, Kauczor J, Saue T, Jensen H J A and Norman P 2015 Beyond the electric-dipole approximation: a formulation and implementation of molecular response theory for the description of absorption of electromagnetic field radiation *J. Chem. Phys.* **142** 244111
[141] Yang L, Deslippe J, Park C-H, Cohen M L and Louie S G 2009 Excitonic effects on the optical response of graphene and bilayer graphene *Phys. Rev. Lett.* **103** 186802

[142] Chen S-C, Wu J-Y and Lin M-F 2018 Feature-rich magneto-electronic properties of bismuthene *New J. Phys.* **20** 062001
[143] Zhang Y, Tan Y-W, Stormer H L and Kim P 2005 Experimental observation of the quantum Hall effect and Berry's phase in graphene *Nature* **438** 201–4
[144] Jiang Z, Zhang Y, Tan Y-W, Stormer H and Kim P 2007 Quantum Hall effect in graphene *Solid State Commun.* **143** 14–9
[145] Liu C-C, Feng W and Yao Y 2011 Quantum spin Hall effect in silicene and two-dimensional germanium *Phys. Rev. Lett.* **107** 076802
[146] Ezawa M 2012 Quantum Hall effects in silicene *J. Phys. Soc. Jpn.* **81** 064705
[147] Do T-N, Gumbs G, Shih P-H, Huang D, Chiu C-W, Chen C-Y and Lin M-F 2019 Peculiar optical properties of bilayer silicene under the influence of external electric and magnetic fields *Sci. Rep.* **9** 1–15
[148] Georén P and Lindbergh G 2001 Characterisation and modelling of the transport properties in lithium battery polymer electrolytes *Electrochim. Acta* **47** 577–87
[149] Krachkovskiy S A, Bazak J D, Werhun P, Balcom B J, Halalay I C and Goward G R 2016 Visualization of steady-state ionic concentration profiles formed in electrolytes during Li-ion battery operation and determination of mass-transport properties by *in situ* magnetic resonance imaging *J. Am. Chem. Soc.* **138** 7992–9
[150] Vilian A E, Kim W, Park B, Oh S Y, Kim T, Huh Y S, Hwangbo C K and Han Y-K 2019 Efficient electron-mediated electrochemical biosensor of gold wire for the rapid detection of C-reactive protein: a predictive strategy for heart failure *Biosens. Bioelectron.* **142** 111549
[151] Trevisanutto P E, Holzmann M, Côté M and Olevano V 2009 High-energy excitonic effects in graphite and graphene arXiv:0909.1682
[152] Shyu F-L, Chang C-P, Chen R-B, Chiu C-W and Lin M-F 2003 Magnetoelectronic and optical properties of carbon nanotubes *Phys. Rev.* B **67** 045405
[153] Chen S, Wang T, Lee C and Lin M-F 2008 Magneto-electronic properties of graphene nanoribbons in the spatially modulated electric field *Phys. Lett.* A **372** 5999–6002
[154] Notarianni M, Liu J, Vernon K and Motta N 2016 Synthesis and applications of carbon nanomaterials for energy generation and storage *Beilstein J. Nanotechnol.* **7** 149–96
[155] Thorgrimsson B, McJunkin T, MacQuarrie E, Coppersmith S and Eriksson M 2020 The effect of external electric fields on silicon with superconducting gallium nano-precipitates *J. Appl. Phys.* **127** 215102
[156] Ruzsinszky A, Constantin L A and Pitarke J 2016 Kernel-corrected random-phase approximation for the uniform electron gas and jellium surface energy *Phys. Rev.* B **94** 165155
[157] Kushwaha M S 2008 Analyzing the existence of magnetoroton excitations in magnetized quantum wires *Phys. Rev.* B **78** 153306
[158] Faleev S V and Stockman M I 2002 Self-consistent random-phase approximation for interacting electrons in quantum wells and intersubband absorption *Phys. Rev.* B **66** 085318
[159] Zhang L, Zhang Y, Camacho J, Khodas M and Zaliznyak I 2011 The experimental observation of quantum Hall effect of l = 3 chiral quasiparticles in trilayer graphene *Nat. Phys.* **7** 953–7
[160] Giesbers A, Ponomarenko L, Novoselov K, Geim A, Katsnelson M, Maan J and Zeitler U 2009 Gap opening in the zeroth Landau level of graphene *Phys. Rev.* B **80** 201403
[161] Bao H, Ruan X and Fisher T S 2010 Optical properties of ordered vertical arrays of multi-walled carbon nanotubes from FDTD simulations *Opt. Express* **18** 6347–59

[162] Korotyeyev V, Kochelap V and Varani L 2012 Wave excitations of drifting two-dimensional electron gas under strong inelastic scattering *J. Appl. Phys.* **112** 083721
[163] Sharma A, Kaur K, Garg V and Moudgil R K 2018 Exchange-correlation effects in a finite-temperature quasi-one-dimensional electron gas *Phys. Status Solidi (B)* **255** 1800174
[164] Li Y, Lu Y, Adelhelm P, Titirici M-M and Hu Y-S 2019 Intercalation chemistry of graphite: alkali metal ions and beyond *Chem. Soc. Rev.* **48** 4655–87
[165] Wang L, Zhu Y, Guo C, Zhu X, Liang J and Qian Y 2014 Ferric chloride-graphite intercalation compounds as anode materials for li-ion batteries *ChemSusChem* **7** 87–91

IOP Publishing

Chapter 3

Experimental measurements

Shih-Yang Lin, Vo Khuong Dien, Wei-Bang Li, Thi Dieu Hien Nguyen, Hsien-Ching Chung, Ngoc Thanh Thuy Tran, Nguyen Thi Han, Hsin-Yi Liu, Hai Duong Pham and Ming-Fa Lin

Simultaneous developments of theoretical predictions and experimental measurements can lead to full progress of basic science research [1–10]. The current Vienna *ab initio* simulation package (VASP) simulations are mainly focused on optimal lattice symmetries [11, 12], spatial charge distributions [13, 14], van Hove singularities in density of states [15, 16], valence and conduction energy subbbands [11, 17, 18], optical excitation structures [19, 20], and magnetic moments. The calculated results are, respectively, examined by high-resolution observations through x-ray diffractions/low-energy electron diffractions (LEED; [21–27]), scanning tunneling microscopy/x-ray scatterings (STM; [28–35]), scanning tunneling spectroscopy (STS; [1, 36–42]), angle-resolved photoemission spectroscopy (ARPES; [43–46]), optical reflection/absorption/transmission/photoluminescence spectroscopies [47–49]/[50–52] /[49]/[52], and SQUID [53, 54]. Up-to-date distinct techniques and their significant contributions to diverse quasiparticle properties are thoroughly discussed in the following paragraphs.

3.1 X-ray diffractions

It is well known that x-ray diffraction is very powerful in detecting crystal structures of condensed-matter systems since the first discovery of x-rays by Roentgen in 1895 [55]. Later, Bragg established the x-ray elastic scattering theory [56], confirmed by the experimental observation by Laue in 1912 [57, 58]. According to the quantum theory of electron–photon elastic scatterings [59, 60] (figure 3.1), the incident photon beam, with the initial wave vector of **k**, experiences the significant interactions from the spatial charge density distribution and thus scatters along the different direction

doi:10.1088/978-0-7503-4685-6ch3

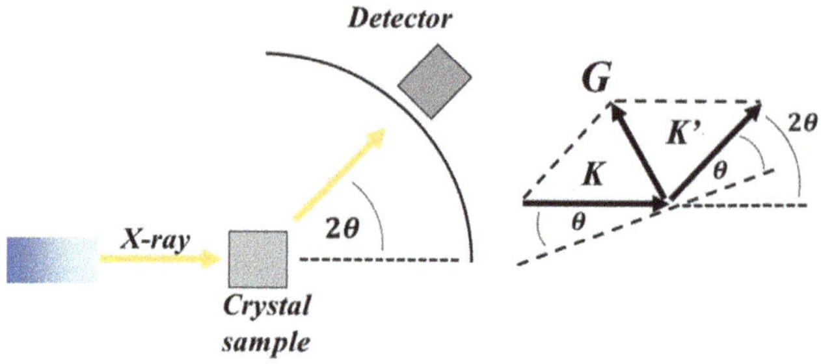

Figure 3.1. The elastic scatterings of charge density due to a photon beam.

of **k**′ through the specific energy conservation. As a result of the periodical property, the diffraction peaks are characterized by **k**′- **k** = **G**, where **G** is the reciprocal lattice vector. That is to say, the diffracted peak structures, coming into existence at the different θs (the incident angle), are mainly dominated by the suitable **G**s. The scattering form factor, being due to atoms/molecules in a primitive unit cell, is determined by their spatial charge density distributions. How to utilize the x-ray measurements in evaluating the real carrier densities becomes an important issue. For example, a 3D sodium chloride, which consists of a double fcc structure [61], has been successfully identified to present the almost fully ionic configurations [Na^+ and Cl^-] with the spherical charge distributions. To date, the VASP simulations, which are conducted on the multi-/single-orbital hybridizations in chemical bonds (details in this book), are very accurate for any crystal structures. The calculated results, as well as the optimal lattice symmetry, are available in realizing the x-ray peaks in terms of diffraction conditions and spectral intensities [59, 60]. Most importantly, the close cooperation of theoretical and experimental studies is able to thoroughly identify the significant orbital hybridizations in establishing moiré superlattices/high-symmetry lattices, e.g., those in ternary lithium oxide compounds [62–64].

There are many well-established synchrotron radiation centers around the world, with the high-resolution and continuous light resources, e.g., Taiwan National Synchrotron Radiation Research Center (NSRRC) at Hsinchu [65–67]. Such standard laboratories are capable of providing stable x-ray diffraction measurements. The incident photons present the highly modulated wavelengths comparable to or smaller than the lattice constants, generate the elastic scatterings with a periodic charge density, and then lead to the constructive or destructive diffraction patterns. During its propagation, an electromagnetic wave experiences elastically scattering events from the whole carrier density due to the orbital hybridizations in chemical bonds, being the so-called Thomson scatterings [68–70]. All the valence orbitals, which behave like a Hertz dipole through the frequency of propagating photons, serve as a source of dipole radiation. The wavelength λ of x-rays remains unchanged for the Thomson scatterings. According to the theoretical predictions, the spectral intensity at distance R is estimated to be $I(R) = I_0(r_e/R)(1 + \cos^2 2\theta/2)$ in the absence of atomic nuclei, in which $r_e = \left(\frac{\mu_0}{4\pi}\right)\left(\frac{e^2}{m}\right)$ is the classical radius of an electron (2.818×10^{-15} m) [71]. In addition, the nuclei components, with heavy masses, will only yield a less than 10^{-6} contribution, compared with that of electrons. Figure 3.1 clearly illustrates an observation for a simple cubic crystal through sufficient irradiation of monochromatic x-rays with a wavelength of λ. Each atom has six neighboring ones at a lattice constant of a and the angle between two close chemical bonds is always 90°. One can characterize the critical scattering vector $\mathbf{G} = \mathbf{k}\text{-}\mathbf{k}'$, the wave vector difference between the incident and reflected electromagnetic waves. By using the geometry of scattering vector construction, the diffraction condition is $G = 2k\sin\theta$ (θ the incident angle related to $\mathbf{k}$). $\mathbf{G}$ at $I(R)\rightarrow m$ ax is further achieved from three Laue conditions [72]. Obviously, the available **G**s are mainly determined by the crystal symmetries and the various atoms/molecules in a unit cell, since the total scattering intensities strongly depend on the whole charge distributions related to the latter. Both specific scattering events and spectral intensities could be identified by high-resolution x-ray diffractions. The detailed comparisons of VASP calculations and experimental measurements are thus expected to thoroughly clarify the multi-/single-orbital orbital hybridizations in chemical bonds.

A lot of high-resolution x-ray diffractions clearly illustrate that high-symmetry lattices or moiré superlattices are revealed in various anode/electrolyte/cathode materials of ion-based batteries. For example, the stage-1 lithium graphite intercalation compound, which frequently serves as a commercial anode, is verified to present a layered LiC_6 structure with lithium- and carbon-atom honeycomb lattices [73–76]. Furthermore, the interlayer distance between two AA-stacked graphitic sheets is 3.706 Å [74], being consistent with the VASP result of 3.815 Å [77] (details in chapter 10). Similar phenomena can be found in alkali-doped graphite systems, e.g., stage-1 KC_8, NaC_8, RbC_8, and CsC_8 under a specific AB stacking. As for lithium oxide compounds, their unusual x-ray diffraction patterns obviously show certain **G**-dependent peak structures, with the various intensities and shapes. It should be noticed that Rietveld refinement [78] has been successfully utilized to characterize the rich and unique crystal structures of anode/electrolyte/cathode materials using the 2 θ-related diffraction peaks. The core components of Li^+-based batteries, $Li_4Ti_5O_{12}$, Li_2GeO_3/Li_2SiO_3 and $LiFe/Co/NiO_2$ (details in chapters 12, 8 and 4), respectively, exhibit the diversified crystal phases: the cubic spinel structure, the orthorhombic Cmc21, and the orthorhombic Pmnm [79]/ hexagonal R 3 m [80]/ hexagonal R 3 m [81]. Furthermore, their lattice constants of [a, b, c] are examined to be [9.24(8), 5.5016(4), 4.848 57(24)] [in the unit of Å], [9.6024(8), 5.5016(4), 4.848 57(24)], and [9.6024(8), 5.5016(4), 4.848 57(24)], being very close to those of VASP simulations [9.6024(8), 5.5016(4), = 4.848 57(24)], [9.612, 5.462, 4.874], and [9.6024 (8), 5.5016(4), c = 4.848 57(24)]. On the other hand, the experimental estimations cannot provide very useful information about the position-dependent bond lengths and the spatial charge density distributions. By the delicate evaluations and analyses, the theoretical predictions on the optimal geometries obviously display the observable fluctuations of the same chemical bonds in moiré superlattices (more than 5%). This indicates the highly non-uniform chemical environment which might be responsible for driving the transport of Li^+-ions inside the specific electrolytes. The modifications of Rietveld refinement [78] could be required in further experimental examinations. Moreover, the calculated results present the multi-/single-orbital mixing behaviors between two neighboring atoms. Such carrier distributions are enough in identifying the significant hybridizations. Very interestingly, how to link the main features of the measured x-ray diffraction peaks (the θ dependence, intensity, and form) and the theoretical calculations of the atom-created form factor will become an emergent issue in thoroughly understanding the critical chemical bondings.

3.2 STM and STS: nano-morphology and van Hove singularities

The core materials of ion-based batteries can present unusual nano-crystal structures and/or surface morphologies [82], and lattice-symmetry-induced van Hove singularities [83, 84], e.g., layered honeycomb lattices in graphite/graphite alkali-atom intercalation compounds with semi-metallic/metallic behaviors [85, 86]/[87, 88]. The close relations between the geometric and electronic properties are very interesting in basic science research, as frequently observed in other physical/chemical/material phenomena. They can be directly clarified by the simultaneous scanning-tunneling observations. For example, it is famous that the hexagonal arrangement along the single-walled carbon nanotube axis and its density of states across the Fermi level was clearly detected in the measured STM/STS spectra of the various gate-voltage-dependent scanning tunneling currents in 1998 [89]. Similar high-resolution examinations have been successfully applied to 1D graphene nanoribbons [90, 91] and silicon nanotubes [92]. The designs and operation principles for STM/STS are introduced in the following paragraphs.

A lot of experimental examinations clearly indicate that high-resolution scanning tunneling microscopy (STM; [93, 94]) is the most important technique in identifying the nanoscale surface topographies since it was first discovered by Binnig and Rohrer in 1982 [33, 95, 96]. Its highly accurate measurements have been verified to be very powerful in accounting for the crystal structures of emergent 2D, 1D and 0D lattices [91, 97, 98]. This unusual apparatus can accurately characterize the surface structures along both lateral and vertical atomic resolutions, in which the successful identifications cover the lattice constants/bond lengths [28, 29], arrangement chiralities [30, 31], planar/curved forms [32, 99], step edges [100], dangling bonds/local vacancies [101], dislocations/impurities [102, 103], chemi-adsorbed/substituted adatoms, molecules and functional groups [104, 105], and nanoclusters/nanoislands [106]. The STM instrument, as clearly indicated figure 3.2(a), consists of a conducting/semiconducting solid surface and a rather sharp metal tip [33]. Their several-angstroms spacing can be manipulated by the specific piezoelectric feedback devices [107]. By using a bias voltage between surface and tip, an extremely weak, but observable quantum current [34] arises from the well-test tunneling effect [108], e.g., the precise detections on tunneling currents below several nAs. Its observation strongly depends on distance, and the exponential decay is approximately utilized to analyze the height-dependent quantum current [109]. Very interestingly, this current transports from the occupied electronic states of tip through the unoccupied ones of sample by using a positive bias voltage ($V > 0$; [110, 111]), or vice versa. Most

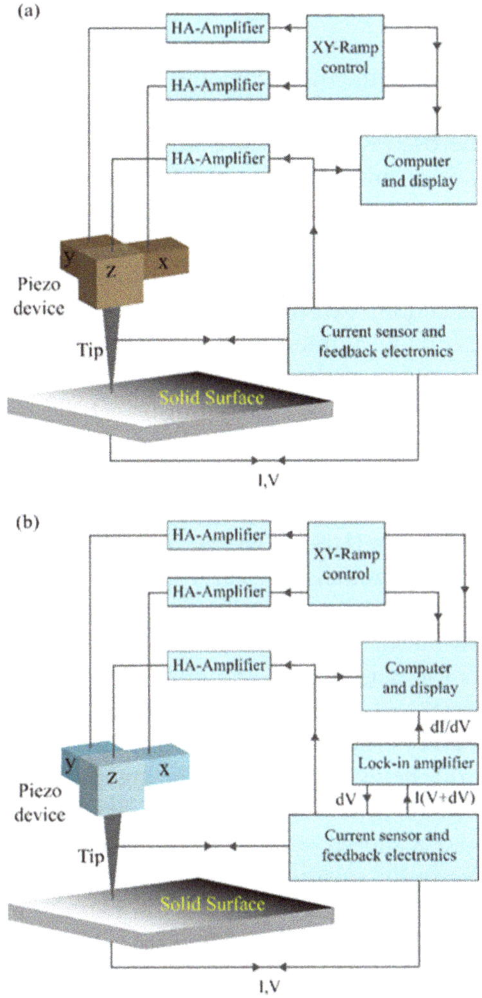

Figure 3.2. The optical excitations in the simultaneous creations of the excited valence holes and conduction electrons.

importantly, the tunneling quantum current is capable of serving as a characteristic response. The frequently utilized mode, a constant tunneling quantum current versus a specific voltage [112, 113], is operated to detect a local nanostructure very accurately under a piezoelectric device. Simultaneously, a direct combination with a precise scanning device, the total information of surface morphology can be presented at the preselected positions through the well-defined cases [114]. In general, the response of surface charges hardly relies on the background effects, mainly owing to an ultra-high vacuum environment. The best spatial resolution of STM measurements can reach ~ 0.1 Å. It should be noticed that STM and x-ray diffraction measurements, respectively, provide atom images and charge density distributions (details in section 3.1). Specifically, the latter includes the effective orbital hybridizations of chemical bonds within a crystal. Whether the strongly modified STM/STS could estimate the carrier densities becomes an open issue. In addition to STM, other experimental techniques are available in fully characterizing the unique geometries of the low-dimensional emergent materials, such as, TEM (tunneling electron microscopy; [115–124]), atomic force microscopy (AFM; [93, 115–129]), low-energy electron diffraction (LEED, [130–137]), and reflection high-energy electron diffraction (RHEED; [138–143]). For example, their examinations are responsible for the various stacking configurations [144], buckled/folded/scrolled lattices [145]/[146]/[147], and the complicated heterojunctions [148].

Up-to-date STM detections can directly identify the optimal surface structures under nanoscale precision. Very interestingly, a lot of experimental measurements [114, 149–151] and theoretical analyses [152–154], which are conducted on carbon-sp^2 condensed-matter systems, have clearly illustrated the rich and unique crystal structures. The mainstream organic materials include 3D bulk graphites/graphite intercalation compounds [155, 156]/[157, 158], 2D layered graphenes [159–161], 1D carbon nanotubes/graphene nanoribbons [38, 162–164]/[150, 165–167], and 0D carbon fullerenes/onions/disks [168–170]/[171, 172]/[173, 174]. The direct verifications of the diverse crystal symmetries clearly indicate the strong and complicated relations among the planar honeycomb lattice, the finite-width quantum confinement, the flexible/non-buckled geometry, and the significant modification of chemical bondings. As for Bernal bulk graphite [175], the STM observations have confirmed the 2D networks of local defects on the surface [176], and the pyridinic-nitrogen and graphitic-N structures [177]. Very interestingly, the electrostatic-manipulation STM [178], being performed on a highly oriented pyrolytic graphite surface to overcome the weak, but significant van der Waals interactions, dramatically leads to a continuous and large-range shift of the top layer. That is to say, this method is very effective in the creation of various stacking configurations, as done for an AAB-stacked trilayer graphene [179–181]. The atomic-scale examinations have successfully shown ABA- [182, 183], ABC- [182–184], AAB-stacked [152–154], turbostratic few-layer graphene systems [185], the corrugated substrate and buffer graphene layer [186, 187], the rippled graphene nanoislands [106, 188], and the adatom distribution configurations on the planar/curved graphene surfaces [104]. In addition, the AA-stacked graphenes [144, 189] are almost absent in the experimental synthesis and analysis. It is a scientific breakthrough in 1998 [190, 191]: the simultaneous identifications for the geometric and electronic properties of the single-walled carbon nanotubes. That both radius and chirality can determine the metallic and semiconducting behaviors (the theoretical predictions; [192–194]) is verified by the high-resolution STM/STS measurements (discussed later again). Similar investigations are generalized to graphene nanoribbons [90, 166] and silicon nanotubes [92, 195] with the chiral open edges and buckled A/B sublattices, respectively. In short, carbon atoms, with four active orbitals, have generated the rich allotropes with the unusual lattice symmetries.

Scanning tunneling spectroscopy (STS; [1, 36–42]), being designed in figure 3.2(b), could be regarded as extended STM under more delicate operations and analyses in the absence of feedback devices. This apparatus can characterize the energy-dependent van Hove singularities centered about the Fermi level [196] and thus specify the orbital hybridizations of chemical bonds within the different energy ranges, based on the close relationship between tunneling quantum current and bias voltage [197]. Under the same tip-height [198], the various bias voltages [*V*'s] are available in driving the sensitive responses of quantum currents [*I*'s]; furthermore, the position-dependent *I*–*V* curves are taken from the accurate operations on a whole surface. The local surface information has been covered in the modified electronic conductance spectrum. In general, the measured data strongly depend to surface electronic states, when the tunneling quantum current is qualitatively approximated to be proportional to the constant density of states of a sharp tip [100, 199]. The normalized differential conductance of the tip-surface tunneling junction, being defined as $[dI/dV]/[I/V]$ in the more precise form, can account for the significant density of states in the surface nanostructure. That is, its functional relation with bias voltage is characterized as the energy-dependent density of states. Most importantly, the coexistent noise within the measured differential conductance could be greatly reduced by using a lock-in amplifier through a small AC dV modulation [200]. The up-to-now STS measurements clearly show the highest resolution about several pAs [201]. In addition, they could be further utilized to identify the spin-split density of states after the successful observations of magnetic tips [202].

High-resolution STS detections are capable of identifying the main features of van Hove singularities (their energies, numbers, intensities, and structures; [203–210]) and the semiconducting/semi-metallic/metallic behaviors [203, 204, 211]. Such measurements are quite useful in distinguishing the diversified phenomena of the monoelement-related emergent materials [212]. From theoretical viewpoints, the delicate VASP calculations and analyses (details in chapters 4–13) can reach an important conclusion: the concise relations between the specific van Hove singularities and orbital mixings within certain energy ranges. Specifically, the systematic STS investigations are very successful for verifying the geometry-, dimension- and modification-enriched electronic properties in carbon-sp^2-related systems, such as, 1D carbon nanotubes [1, 213], graphene nanoribbons [203, 204], few-layer graphenes [205, 206], adatom-adsorbed graphenes [207, 208] and graphite [209, 210]. As for the single-walled seamless carbon nanotubes [214], a breakthrough of STS measurement [214] displays a lot of asymmetric peaks that clearly indicate the chirality- and radius-created band gaps and energy spacings between two neighboring subbands. Even under the obvious curvature effects on narrow cylindrical surfaces (the $2p_z$-orbital misorientation and the significant sp^3 hybridization; [215]), the single-/multi-wall armchair nanotubes [33, 216] are the 1D metals with finite densities of states at the Fermi level (enough high free carrier density). According to the VASP calculations and analyses [13, 217], whether the π and σ bondings are perpendicular to each other could be distinguished from the emergence of the $2p_z$- and [2s, $2p_x$, $2p_y$]-decomposed van Hove singularities. The composite examinations of STS and STM show other outstanding quantum-confinement effects [218, 219]. The main features of planar graphene nanoribbons [109, 220], the width- and edge-structure-dominated energy gaps and the square-root divergences of asymmetric peaks/

plateau structures in 1D parabolic/linear energy bands, are clearly observed through the precisely defined boundary forms. The diverse phenomena are expected to appear in silicon nanotube/nanoribbons [92, 195, 221]/[222]. Very interestingly, more features of the low-energy van Hove singularities are revealed in layered and adatom-doped graphene systems, such as, a linear energy dependence vanishing at the Fermi level for monolayer honeycomb lattice (a zero-gap semiconductor; [223, 224]), the logarithmic peak structures due to the saddle points in twisted bilayer graphenes [225, 226], a gate-voltage-created band gap for bilayer AB and tri-layer ABC stackings [227, 228], a significant peak just at the Fermi level associated with the localized surface states in tri-layer and penta-layer ABC stackings [229], a sharp dip structure at E_F surrounded by a pair of asymmetric peaks for tri-layer AAB stacking (a very narrow gap in the presence of low-lying constant-energy loops; [230]), and a red-shift Dirac point related to the *n*-type electron doping of Bi adatoms [231]. Finally, the measured density of states for bulk Bernal graphite shows a finite value near E_F characteristic of a 3D semimetal [232] and the broadening of prominent π- and π^*-peak structures at the deeper/higher state energies [233]. Such results directly reflect the maximum interlayer atomic interactions. Also noticed is that the well-defined π and σ band widths in graphene-related systems, only with the orthogonal characteristics, can be thoroughly detected by further STS examinations. Similar STS measurements on graphite alkali-atom/alkali-ion intercalation compounds are available in verifying the n-type metals/semimetals [234]/[235]. On the other side, there are no STS measurements for the multi-component lithium oxides, e.g., the absence of experimental van Hove singularities in $Li_4Ti_5O_{12}$ [236], $LiGe_2O_3$ [237], and $LiFePO_4$ [7]. The main difficulties might lie in wide band gaps and too many valence and conduction subbands.

3.3 ARPES: occupied energy spectra and widths

Very interestingly, the high-resolution measurements of the angle-resolved photoemission spectroscopy (ARPES; [238]) can directly examine the wave-vector-dependent electronic energy spectra and widths for the occupied states, but not the unoccupied ones. That is, they are the only way to thoroughly identify the isotropic or anisotropic behaviors of the occupied/excited quasiparticle states. It should be noticed that the energy dependence of the unoccupied states could be verified from the full examination of scanning tunneling spectroscopy (STS; [239]). The up-to-date ARPES measurements are very successful in comprehending the low-lying occupied energy bands for most mainstream materials, especially for those in 2D few-layer condensed-matter systems without/with substrate, such as the unusual energy dispersions of layered graphenes/silicenes/germanenes [240]/[241]/[242]. In addition, the ARPES tests on 1D carbon nanotubes are thoroughly absent because of the identification of the well-defined wave vector along the tubular axis (discussed later). Generally speaking, ARPES measurements and VASP simulations are compared with each other, leading to the identification of a low-energy band structure [243]. Furthermore, the theoretical predictions of the tight-binding model [244] only play minor roles in the detailed comparisons.

Certain significant issues come to exist during the delicate analyses, when the ARPES measurements have been conducted on the available materials. Both intermediate excitation processes (all the elastic and inelastic scatterings; [245, 246]) and sample surface (the destruction of the conservation for the perpendicular momentum; [247, 248]) are able to generate very large challenges in verifying the rich and unique quasiparticle properties. The incident photons, which carry the enough high momenta and energies, directly excite the occupied electronic states within a condensed-matter system into the unoccupied ones using them. In general, such quantized quasiparticles have created the excited valence holes and conduction electrons simultaneously (figure 3.3). The excited states will experience an infinite intermediate configuration, since the scattering resources must survive at any positions and times, e.g., the electron–electron Coulomb interactions [249], the electron–photon couplings under a finite temperature [250], and the carrier-defect scatterings [251], being associated with the intrinsic environments. If they overcome the quantum confinement of an open boundary, the momenta and energies of the scattered electron beam could be measured through the angle-resolved counter [252]. Only the carriers near a sample surface take part in the photoemission processes, in which the characteristic depths are roughly estimated to ~10–20 Å [252]. This fact is very useful in analyzing the measured ARPES. Very interestingly, the rich and unique low-lying band structures, which correspond to 2D few-layer emergent materials without/with substrates, could be identified through the delicate analyses. For example, monolayer and bilayer silicene on Ag[111] are fully explored for their band structures using about the four-layer substrate.

It is very popular that the ARPES chamber is directly combined with the setup of sample growth for fully testing the *in situ* quasi-particle properties [253]. After enough efficient illumination through the soft x-ray beam on the specific materials

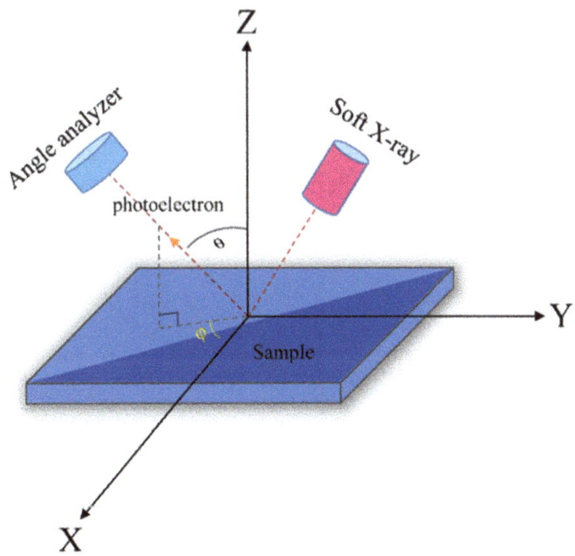

Figure 3.3. The setup plots of (a) STM and (b) STS.

(figure 3.3), the electric-dipole perturbations induce the dramatic transitions between the occupied and unoccupied electronic states (figure 3.4), the intermediate configurations of the excited quasiparticles come to exist under the frequently scattering events, and then they escape an open boundary by overcoming the surface work function [252]. Within an angle-resolved (momentum, energy) analyzer, the total momenta of photoelectrons are estimated using an electron-gas model, in which they are decomposed into the parallel and perpendicular components associated with the polar and azimuthal angles (θ and φ in figure 3.4). The former is conserved during the photoemission/escaping process, but the latter cannot satisfy the conservation law because of the breaking of translation symmetry. In principle, the ARPES examinations are more reliable for layered structures with weak energy bands along the perpendicular direction [k_z]. Maybe, the non-conservation barrier could be overcome under the featured energy dispersions of the significant k_z-dependence. For example, successful analyses have been conducted on the 3D band structure in the AB-stacked Bernal graphite by the specific energy spectra at $k_z = 0$ and zone boundary [254]. It should be noticed that more careful examinations are required for 3D complex band structures in ternary lithium oxides (predictions in this book). Very interestingly, the high-resolution ARPES measurements also provide the distribution widths of the occupied quasiparticle energies. Their inverses are responsible for the wave-vector-dependent quasiparticle lifetimes, being dominated by both Coulomb decays [255] and electron–phonon interactions [256]. How to greatly improve energy and momentum resolutions will play critical roles in fully investigating the many-body mechanisms, e.g., the best precisions of ~1 meV and 0.1° in the UV range [257].

The recent ARPES measurements, which are mainly focused on layered emergent materials, are very successful in examining/identifying/verifying the diverse quasiparticle energy spectra and lifetimes, such as, the rich and unique quasiparticle properties in group-IV [44, 258] and group-V 2D systems [259, 260]. After a lot of delicate growth [261], measurements [44] and analyses [14, 262], the graphene-related systems, with the dominating sp^2 bondings, have been shown to exhibit dimension-, layer-number-, stacking-configuration-, adatom-/molecule-chemisorption-, guest-atom-substitution-, and substrate-diversified band structures. The main features cover the [k_x, k_y, k_z]-dependent 3D energy dispersions in a semi-metallic Bernal graphite [263], the blue-shift Fermi levels of alkali-doped graphite

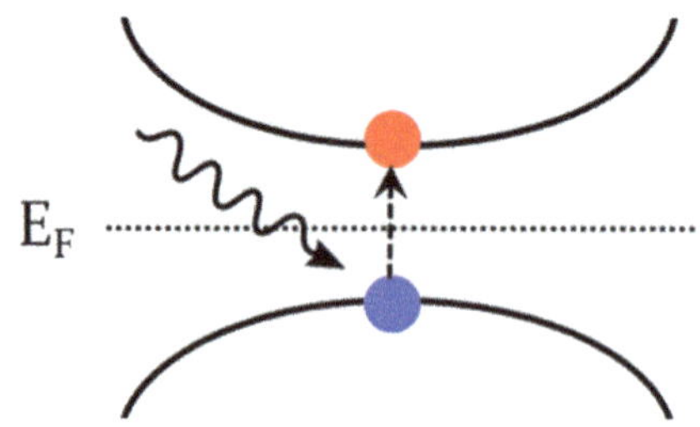

Figure 3.4. The apparatus design for angle-resolved photoemission spectroscopy with an incident photon beam and a scattered electron beam.

intercalation compounds [264], 1D parabolic subbands and their gaps of semiconducting graphene nanoribbons [265], an isotropic and zero-gap Dirac cone for monolayer graphene [266], two pairs of parabolic energy dispersions under bilayer AB stacking [267], the modified dispersions of linear and parabolic bands in twisted bilayer graphenes [268], the linear and parabolic bands in trilayer graphene with ABA stacking [254], the partially flat, sombrero-shaped and linear energy bands/the linear, parabolic and parabolic ones in trilayer ABC/ABA stacking [269, 270]/[246], the semiconductor-metal transitions/the enlarged band gaps through chemisorption and substitution [14, 271, 272]; the heterojunction-bonding-created dramatic changes of band structures for layered graphene structures on different substrates [43, 273]. In addition, the direct ARPES tests, which are required on the angular-momentum-decoupled valence subbands of coaxial carbon nanotubes (e.g., the theoretical predictions in [274]), have not been finished to date since their discovery in 1991 by Iijima [275]. How to characterize the 1D wave vector along the nanotube axis very accurately might be the most important barrier [276, 277].

As for non-carbon-sp^2 bonding materials, similar ARPES experiments are available in fully exploring the diverse energy subbands below the chemical potential, covering the other group-IV- and group-V-related ones. The unusual occupied band structures are clearly observed in layered silicenes [278, 279], germanenes [242, 280], tinenes, plumbenes [281], phosphorenes [282–284], antimonenes [285], and bismuthenes [286, 287]. Very interestingly, the significant chemical bondings with the various substrates are generated through the epitaxial growths, being consistent with the predicted VASP simulations [287, 288]. For example, silicene sheets on Ag(111) could display a 0.6-eV band gap under the separated valence and conduction Dirac points near the K/K' valleys. However, a very narrow gap (< 10 meV) is expected to appear under the free-standing case by the spin–orbital coupling (the first-principles calculations and the tight-binding model; [289, 290]). The significant band-gap difference has indicated the unique Si–Ag multi-orbital hybridizations at heterojunction boundary [291, 292]. The modified multi-Dirac-cone energy spectra are shown in few-layer germanene on Au(111) [280, 293]. On the other hand, the quasi-graphene band structures could be thoroughly broken by the prominent multi-orbital hybridizations at the mixed boundary, e.g., the full modification of band structure in silicene/ZrB_2 (0001) [294, 295]. A bilayer tinene on Bi_2Te_3 (111) is examined to exhibit hole-like metallic behavior at the Γ point, being in sharp contrast with the first-principles results in the absence of a substrate [296, 297]. Moreover, the gapless property is detected to survive in Pb monolayer/Au(111) [298], in which the critical mechanism is deduced to be very strong adlayer-substrate chemical bondings. In short, the significant multi-orbital hybridizations could survive in a pristine system, as well as the heterojunction with substrate. Their critical roles, which dominate the main features of the low-lying quasiparticle energy spectra [14], could be verified by the high-resolution ARPES examinations.

The ARPES measurements seem to be useless for most cathode/electrolyte/anode materials in ion-based batteries, while the multi-component compounds have 3D moiré superlattices as primitive unit cells. Specifically, the ternary lithium oxides, being frequently utilized in high-performance batteries [299–303] possess a lot of atom- and position-dependent chemical bonds, such as, the well-known LiFe/Co/Ni/Si/Ge/Sn/TiO compounds (details in table 1.1 of chapter 1; [236, 237, 304, 305]). For example, a $Li_8Ge_4O_{12}$ compound presents 32 Li–O and 16 Ge–O bonds within a unit cell, with strong fluctuations about their lengths. Li, Ge and O atoms, respectively, 1, 4 and 4 active orbitals, leading to very pronounced multi-orbital mixings. Obviously, this system has shown a complex and unusual band structure, in which the principal features cover a large indirect band gap, the high asymmetry of valence hole and conduction electron energy spectra, the strong anisotropy, and the various dispersion relations. The featured quasiparticle spectrum clearly illustrates the mission impossible in ARPES examinations (no experimental reports up to now). How to overcome the non-conservation of the perpendicular momenta and the characterization of energy subbands at the high-symmetry point will be open issues in near-future experimental investigations.

The decay rates of the excited valence/conduction holes are directly detected from the wave-vector-dependent distribution widths of ARPES measurements. The examined results are very reliable for few-layer emergent materials, as successfully conducted on potassium-adsorbed graphenes [306]. The critical factors, which dominate the observed resolutions, cover the good conservation of the perpendicular momentum, the easy characterization of the $[x,y]$-plane wave vectors (the high-symmetry points), and the concise band structure. For example, according to the first-principle calculations [307, 308], the alkali-doped graphenes are predicted to the n-type metals, in which the blue shift of the Fermi level and the slightly separated Dirac-cone structure appear within the approximately rigid-band approximation. The spectral intensities, being measured along the high-symmetry directions through the distinct alkali-adatom concentrations absorbed on monolayer graphene, clearly reveal the (k_x, k_y)-related linewidths. They are able to provide the doping-induced momentum distribution curves (MDCs; [309, 310]). Generally speaking, there exist the Lorentzian peaks centered about the quasi-particle energies; furthermore, their full widths related to the half-maximum intensity are characterized as the inelastic scattering rate. By using the self-energy method within the screened exchange form [311], the single-particle electron–hole excitations and coherent plasma oscillations of free carriers, as well as the electron–phonon interactions at finite temperatures, could account for the rich energy/wave vector dependences of the MDC linewidths. Obviously, the theoretical and experimental studies on the excited quasiparticles are absent for the ternary lithium oxides because of too complicated cases. However, there are certain research works on graphite intercalation compounds [312–314]. In addition, five kinds of femtosecond optical spectroscopies are designed for the specific quasiparticle lifetimes by examining photoelectron distribution [315, 316], optical absorption [317, 318]/transmission [319, 320]/reflectivity [321, 322], respectively, corresponding to metallic and semiconducting materials. In short, the accurate ARPES measurements at very low temperatures are able to provide the full information for the Coulomb-dominated quasiparticle energies and MDCs and thus test the theoretical calculations.

3.4 Optical spectroscopies: reflectance, absorption, transmission, and photoluminescence

The various high-resolution optical spectroscopies on the distinct excitation phenomena [323–325] have been successfully developed for examining/identifying/verifying the diverse excitation phenomena in the mainstream materials. For example, the theoretical predictions of VASP calculations are mainly focused on anode/electrolyte/cathode/ materials of ion-based batteries (details in this book; [236, 237, 304]). The frequency-dependent dielectric functions, reflectance spectra, absorption coefficients and energy loss functions, with/without the excitonic effects [326, 327], are thoroughly done using delicate evaluations and analyses. Most importantly, the close relations between the orbital hybridizations and specific sorption structures are clearly identified from the unified physical and chemical properties. The concise and critical mechanisms, which dominate all the excitation behaviors under an electromagnetic wave perturbation, require the systematic detections from the optical reflection/absorption/transmission/ photoluminescence spectroscopies [328]/[329]/[330]/[331]. The up-to-date apparatuses, being operated under the different cases, are briefly introduced in the following paragraphs. The progress of experimental examinations is also discussed in detail.

A broad-band light source, which covers the frequency ranges of NIR, VIS and UV (roughly for $0\ \text{eV} \leqslant \omega \leqslant 20\ \text{eV}$; [332]), is available in measuring the ω-dependent reflectance spectrum. Furthermore, a PC-driven miniaturized spectrometer is operated within the corresponding spectral range, as clearly indicated in figure 3.5(a). The incident electromagnetic wave initiated from the light source is guided to a reflection probe through the specific fibers of A1–A6, being vertically transmitted into a current sample. In general, this condensed-matter system could be regarded as the stacking of transparent and semi-transparent films above an appropriate reflective or transmitting substrate. The latter might be a silicon wafer and glass slide. Under simultaneous observation, the reflection probe collects all the reflected light using fiber B in figure 3.5(a). The continuous reflectance spectrum is achieved from the high-precision ratio between the intensities of reflected and incident electromagnetic fields (details in equation (2.4). That is, a normally incident beam from the light source interacts with valence/conduction carriers within a sample and creates a reflectance signal that is continued by the spectrometer. The reflectance measurements have been successfully done on a lot of well-known materials, e.g., those for bulk graphite [333, 334] and carbon nanotube bundles [335, 336]. However, how to analyze the complicated absorption structures and critical mechanisms associated with moiré superlattices becomes a new challenge (discussed with core materials of Li^+-based batteries later). Such an open issue is never solved in the previous examinations [236, 237, 304, 326, 327] since it mainly arises from the unique electronic energy spectra with plenty of valence and conduction subbands (too rich joint van Hove singularities). The wide-frequency measurements of the reflectance spectrum, which is conducted on the AB-stacked graphite by a normal incident electromagnetic wave, present an excellent example in clearly illustrating the strong relations between the experimental and theoretical

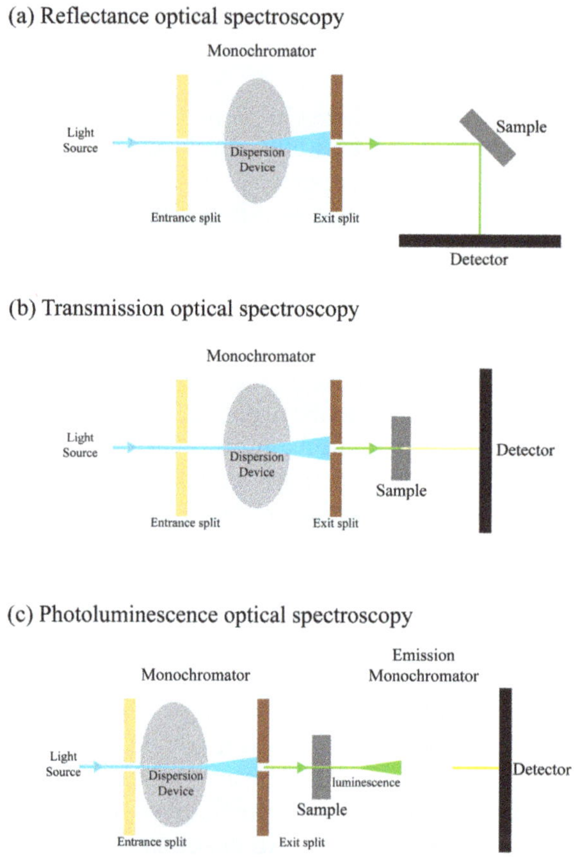

Figure 3.5. Optical spectroscopies in measuring the frequency-dependent (a) reflectance, (b) transmission and (c) photoluminescence spectra.

viewpoints, as well as the optical excitations due to the well-characterized π and σ chemical bondings. Many years ago, Taft and Philipp [334] successfully detected its reflectance within the frequency range of 0 eV < ω < 26 eV, being enough for electronic states structure associated with carbon-[2s, $2p_x$, $2p_y$, $2p_z$] orbitals (independent of the 1s-core levels). To determine the real- and imaginary-part dielectric function at the long wave-length limit, the Kramers–Kronig principle-value integrations under the perturbation-response physical events are utilized in the estimations of Re[ε(ω)]. Furthermore, the energy loss function of Im[−1/ε (ω)] can comprehend the whole charge screening ability. Very interestingly, this work presents a detailed examination of the sum rules arising from the bare and screened response functions and Im[ε(ω)]. In addition, some errors are introduced in the bare/screened response functions, since it is mission impossible to measure an optical spectrum within an infinite frequency range. Apparently, the featured optical properties cover the unusual reflectance curve at about ω→0, 5, 15 eVs, the maximum π-electronic excitations/the initial σ-electronic transition at 4.5 eV/15 eV indicated from Im[ε(ω)] (the single-particle excitation spectrum), and the 7-eV π plasmon and 25-eV π+σ plasmons in energy loss spectrum. Good separation of π and σ bondings shows their orthogonal behaviors and thus carbon-2p_z-dominated interlayer atomic interactions/low-lying band structure. Similar orbital-hybridization phenomena could be found in graphene-related sp^2-bonding systems [14]. On the other hand, the low-frequency reflectance spectra in n-/p-type graphite intercalation compounds clearly show the dramatic changes because of the blue/redshift of the Fermi level (the greatly enhanced free conduction-electron/valence-hole density; [337–339]/[340, 341]). Their interband threshold absorption frequency is about $2E_F$. Optical examinations are one of the effective methods in estimating the doping effects.

When a sample is sufficiently thin, its frequency-dependent transmission spectrum could be experimentally examined within a wide ω-range, as shown in figure 3.5(b). Under the simultaneous observations of reflectance and transmittance, the absorption spectrum will be determined through a simple conservation law of the total incident intensity. Very interestingly, the transmission/absorption spectroscopy is very suitable for fully exploring the diverse low-frequency optical properties in emergent layered materials, such as those in group-IV [48] and group-V [342] 2D systems. Their high-resolution measurements can directly clarify the diversified electronic energy spectra due to the unusual interactions in terms of layer numbers [343, 344], stacking configurations [345–347], and external magnetic/electric fields [348]. As for few-layer graphene systems, the theoretical predictions (the phenomenological models in [255, 348]) agree with the experimental results, in which the main features of optical excitations, the form, intensity, number and frequency of special absorption structures, are very sensitive to the crystal symmetries of the composite honeycomb lattices. It is well known that the measured absorbance/transmittance arises from a finite-thickness sample with quantum-confinement effects. Extra boundary conditions are expected to induce the dimensionality-crossover phenomena. This should be covered in further theoretical and experimental developments.

In addition to vertically optical excitations, the further decays of excited electrons and holes are very interesting research topics [255, 349] and play critical roles in application devices [350–352]. The many-particle quasiparticle dynamics can be directly examined by the high-resolution photoluminescence spectroscopy under a low enough temperature. Apparently, this belongs to a contactless and non-destructive method, being very suitable in detecting the main features of stable and metastable excitons. Figure 3.5(c) clearly shows the initial optical excitations by an incident electromagnetic wave with a specific frequency above the threshold one. The weakly or strongly coupled excitons experience a lot of inelastic scatterings, e.g., the electron–electron Coulomb [230, 248, 249] and electron–phonon interactions [353, 354]. And then, the excited quasiparticles could further decay into the single-particle states by light emission or another non-radiative de-excitation. The former is collected by an optical detector. Both absorbance and photoluminescence, which come to exist within a narrow-frequency range, are simultaneously examined for evaluating the bound-state energies and their lifetimes of excitons. Another direct combination of photoluminescence and reflectance spectra is very useful in determining the many-body and single-particle threshold absorption frequencies. Such examinations can distinguish the important differences among the excitonic bound-state energy [355–357], the optical gap [358–360], and the band gap [361, 362]. When the time-resolved pump and probe are applied to a condensed-matter system, the rich optical excitation channels could be thoroughly explored for their characteristics. This manner is helpful in examining the theoretical viewpoints in chapter 2, e.g., the various absorption peaks due to the different multi-/single-orbital hybridization in chemical bonds (details in chapter 8).

It should be noticed that the simultaneous measurements of the frequency-dependent reflectance and transmission spectra on finite-width samples [332, 363, 364] will open an interesting study topic, the width-diversified optical transitions with/without excitonic effects [326, 327]. The vertical optical excitations, which appear under the single-particle or many-body processes [348, 365], are well characterized by the dynamic charge screening abilities of all the carriers in a sample. That is, any optical properties are closely related to the real- and imaginary-part dielectric function, such as reflectance, absorption length, transmission coefficient and energy loss function (details in section 2.1]. Most importantly, $\varepsilon(\omega)$ is very sensitive to the sample width and thus the quantum-confinement effect, since it is mainly determined by the energy subbands and wave functions [366–368]. This critical factor is expected to display the dramatic transformation during the variation of a sample thickness. For example, the transverse dielectric functions of layered graphene systems are predicted to strongly depend on the layer numbers and stacking configurations [369, 370]. Their exact calculations require the band-structures effects [371, 372]. From theoretical viewpoints, the continuous optical processes of reflection/absorption and absorption/transmission appear at the reflected and transmitted surfaces, respectively [332, 373]. The phenomenological models, with infinite scattering processes of finite-width films, are absent in the previous theoretical predictions. They could be explored for a full understanding about the unique dimensionality-crossover phenomena [374, 375]. The theoretical framework of optical properties, being assisted by quantum confinements, are worthy of systematic investigation in near-future studies.

References

[1] Wilder J W, Venema L C, Rinzler A G, Smalley R E and Dekker C 1998 Electronic structure of atomically resolved carbon nanotubes *Nature* **391** 59–62

[2] Clauss W, Bergeron D and Johnson A 1998 Atomic resolution STM imaging of a twisted single-wall carbon nanotube *Phys. Rev.* B **58** R4266

[3] Venema L, Meunier V, Lambin P and Dekker C 2000 Atomic structure of carbon nanotubes from scanning tunneling microscopy *Phys. Rev.* B **61** 2991

[4] Osváth Z, Koós A, Horváth Z, Gyulai J, Benito A and Martınez M *et al* 2002 Arc-grown Y-branched carbon nanotubes observed by scanning tunneling microscopy (STM) *Chem. Phys. Lett.* **365** 338–42

[5] Osváth Z, Vértesy G, Tapasztó L, Wéber F, Horváth Z and Gyulai J *et al* 2005 Atomically resolved STM images of carbon nanotube defects produced by Ar^+ irradiation *Phys. Rev.* B **72** 045429

[6] Wang Z, Poncharal P and De Heer W 2000 Measuring physical and mechanical properties of individual carbon nanotubes by *in situ* TEM *J. Phys. Chem. Solids* **61** 1025–30
[7] Yamada A, Chung S-C and Hinokuma K 2001 Optimized $LiFePO_4$ for lithium battery cathodes *J. Electrochem. Soc.* **148** A224
[8] Kang B and Ceder G 2009 Battery materials for ultrafast charging and discharging *Nature* **458** 190–93
[9] Harinipriya S, Diwakar V D and Subramanian V R 2008 Performance characteristics of cathode materials for lithium-ion batteries: a Monte Carlo strategy *J. Electrochem. Soc.* **155** A875
[10] Malik R, Zhou F and Ceder G 2011 Kinetics of non-equilibrium lithium incorporation in $LiFePO_4$ *Nat. Mater.* **10** 587–90
[11] Nakayama M, Yamada S, Jalem R and Kasuga T 2016 Density functional studies of olivine-type $LiFePO_4$ and $NaFePO_4$ as positive electrode materials for rechargeable lithium and sodium ion batteries *Solid State Ionics* **286** 40–4
[12] Ong S P, Wang L, Kang B and Ceder G 2008 Li–Fe–P–O_2 phase diagram from first principles calculations *Chem. Mater.* **20** 1798–807
[13] Lin S-Y, Tran N T T, Chang S-L, Su W-P and Lin M-F 2018 *Structure-and Adatom-Enriched Essential Properties of Graphene Nanoribbons* (Boca Raton, FL: CRC Press)
[14] Tran N T T, Lin S-Y, Lin C-Y and Lin M-F 2017 *Geometric and Electronic Properties of Graphene-Related Systems: Chemical Bonding Schemes* (Boca Raton, FL: CRC Press)
[15] Chowdhury S, Majumdar A and Jana D 2017 Search for magnetism in transition metal atoms doped tetragonal graphene: a DFT approach *J. Magn. Magn. Mater.* **441** 523–30
[16] Nagpure S C, Babu S, Bhushan B, Kumar A, Mishra R and Windl W *et al* 2011 Local electronic structure of $LiFePO_4$ nanoparticles in aged Li-ion batteries *Acta Mater.* **59** 6917–26
[17] Castro L, Dedryvere R, El Khalifi M, Lippens P-E, Bréger J and Tessier C *et al* 2010 The spin-polarized electronic structure of $LiFePO_4$ and $FePO_4$ evidenced by in-lab XPS *J. Phys. Chem.* C **114** 17995–8000
[18] Zhou Z, Steigerwald M, Hybertsen M, Brus L and Friesner R A 2004 Electronic structure of tubular aromatic molecules derived from the metallic (5, 5) armchair single wall carbon nanotube *J. Am. Chem. Soc.* **126** 3597–607
[19] Benrekia A, Benkhettou N, Nassour A, Driz M, Sahnoun M and Lebegue S 2012 Structural, electronic and optical properties of cubic $SrTiO_3$ and $KTaO_3$: *Ab initio* and GW calculations *Physica* B **407** 2632–36
[20] Rani P, Dubey G S and Jindal V 2014 DFT study of optical properties of pure and doped graphene *Physica* E **62** 28–35
[21] Germer L and Hartman C 1960 Improved low energy electron diffraction apparatus *Rev. Sci. Instrum.* **31** 784
[22] Prutton M and Tokutaka H 1969 A UHV evaporator for leed and Auger emission studies *Thin Solid Films* **3** 411–16
[23] Fujinaga Y 1979 LEED and AES analyses of Cu-Pd alloy surface prepared by evaporation and annealing *Surf. Sci.* **84** 1–16
[24] Onchi M and Farnsworth H 1968 Adsorption and oxidation of carbon monoxide on {100} nickel: II. Combined LEED and mass spectrometer measurements *Surf. Sci.* **11** 203–15
[25] Chang C C 1971 Auger electron spectroscopy *Surf. Sci.* **25** 53–79

[26] Sojka F, Meissner M, Zwick C, Forker R and Fritz T 2013 Determination and correction of distortions and systematic errors in low-energy electron diffraction *Rev. Sci. Instrum.* **84** 015111

[27] Sojka F, Meissner M, Zwick C, Forker R, Vyshnepolsky M and Klein C *et al* 2013 To tilt or not to tilt: correction of the distortion caused by inclined sample surfaces in low-energy electron diffraction *Ultramicroscopy* **133** 35–40

[28] Hordvik A 1966 Bond-length/dihedral-angle and bond-length/bondorder relationships for sulphur (II)-sulphur (II) bonds *Acta Chem. Scand.* **20** 1885–91

[29] Clair S, Pons S, Seitsonen A P, Brune H, Kern K and Barth J V 2004 STM study of terephthalic acid self-assembly on Au (111): hydrogen-bonded sheets on an inhomogeneous substrate *J. Phys. Chem.* B **108** 14585–90

[30] De Feyter S, Grim P C, Rücker M, Vanoppen P, Meiners C and Sieffert M *et al* 1998 Expression of chirality by achiral coadsorbed molecules in chiral monolayers observed by STM *Angew. Chem. Int. Ed.* **37** 1223–26

[31] Yablon D G, Giancarlo L C and Flynn G W 2000 Manipulating self-assembly with achiral molecules: an STM study of chiral segregation by achiral adsorbates *J. Phys. Chem.* B **104** 7627–35

[32] Kong X-H, Yang Y-L, Lei S-B and Wang C 2008 On the topography multiplicity of non-planar titanyl (IV) phthalocyanine molecules and the STM imaging mechanism *Surf. Sci.* **602** 684–92

[33] Binnig G, Rohrer H, Gerber C and Weibel E 1982 Surface studies by scanning tunneling microscopy *Phys. Rev. Lett.* **49** 57

[34] Pascual J, Méndez J, Gómez-Herrero J, Baró A, Garcia N and Binh V T 1993 Quantum contact in gold nanostructures by scanning tunneling microscopy *Phys. Rev. Lett.* **71** 1852

[35] Tománek D, Louie S G, Mamin H J, Abraham D W, Thomson R E and Ganz E *et al* 1987 Theory and observation of highly asymmetric atomic structure in scanning-tunneling-microscopy images of graphite *Phys. Rev.* B **35** 7790

[36] Olk C H and Heremans J P 1994 Scanning tunneling spectroscopy of carbon nanotubes *J. Mater. Res.* **9** 259–62

[37] Carroll D, Redlich P, Ajayan P, Charlier J-C, Blase X and De Vita A *et al* 1997 Electronic structure and localized states at carbon nanotube tips *Phys. Rev. Lett.* **78** 2811

[38] Venema L, Janssen J, Buitelaar M, Wildöer J, Lemay S and Kouwenhoven L *et al* 2000 Spatially resolved scanning tunneling spectroscopy on single-walled carbon nanotubes *Phys. Rev.* B **62** 5238

[39] Ouyang M, Huang J-L, Cheung C L and Lieber C M 2001 Atomically resolved single-walled carbon nanotube intramolecular junctions *Science* **291** 97–100

[40] Janssen J, Lemay S, Kouwenhoven L and Dekker C 2002 Scanning tunneling spectroscopy on crossed carbon nanotubes *Phys. Rev.* B **65** 115423

[41] LeRoy B J, Lemay S, Kong J and Dekker C 2004 Scanning tunneling spectroscopy of suspended single-wall carbon nanotubes *Appl. Phys. Lett.* **84** 4280–82

[42] LeRoy B J, Kong J, Pahilwani V, Dekker C and Lemay S 2005 Three-terminal scanning tunneling spectroscopy of suspended carbon nanotubes *Phys. Rev.* B **72** 075413

[43] Zhou S Y, Gweon G-H, Fedorov A, First, de P, De Heer W and Lee D-H *et al* 2007 Substrate-induced bandgap opening in epitaxial graphene *Nat. Mater.* **6** 770–75

[44] Ohta T, Bostwick A, McChesney J L, Seyller T, Horn K and Rotenberg E 2007 Interlayer interaction and electronic screening in multilayer graphene investigated with angle-resolved photoemission spectroscopy *Phys. Rev. Lett.* **98** 206802
[45] Mucha-Kruczyński M, Tsyplyatyev O, Grishin A, McCann E, Fal'ko V I and Bostwick A *et al* 2008 Characterization of graphene through anisotropy of constant-energy maps in angle-resolved photoemission *Phys. Rev.* B **77** 195403
[46] Sprinkle M, Siegel D, Hu Y, Hicks J, Tejeda A and Taleb-Ibrahimi A *et al* 2009 First direct observation of a nearly ideal graphene band structure *Phys. Rev. Lett.* **103** 226803
[47] Weightman P, Martin D, Cole R and Farrell T 2005 Reflection anisotropy spectroscopy *Rep. Prog. Phys.* **68** 1251
[48] John R and Merlin B 2017 Optical properties of graphene, silicene, germanene, and stanene from IR to far UV–A first principles study *J. Phys. Chem. Solids* **110** 307–15
[49] Karrai K and Warburton R J 2003 Optical transmission and reflection spectroscopy of single quantum dots *Superlattices Microstruct.* **33** 311–37
[50] Kazaoui S, Minami N, Jacquemin R, Kataura H and Achiba Y 1999 Amphoteric doping of single-wall carbon-nanotube thin films as probed by optical absorption spectroscopy *Phys. Rev.* B **60** 13339
[51] Kazaoui S, Minami N, Yamawaki H, Aoki K, Kataura H and Achiba Y 2000 Pressure dependence of the optical absorption spectra of single-walled carbon nanotube films *Phys. Rev.* B **62** 1643
[52] Gangopadhyay P, Kesavamoorthy R, Bera S, Magudapathy P, Nair K and Panigrahi B *et al* 2005 Optical absorption and photoluminescence spectroscopy of the growth of silver nanoparticles *Phys. Rev. Lett.* **94** 047403
[53] Finkler A, Segev Y, Myasoedov Y, Rappaport M L, Ne'eman L and Vasyukov D *et al* 2010 Self-aligned nanoscale SQUID on a tip *Nano Lett.* **10** 1046–49
[54] Allman M S, Altomare F, Whittaker J D, Cicak K, Li D and Sirois A *et al* 2010 rf-SQUID-mediated coherent tunable coupling between a superconducting phase qubit and a lumped-element resonator *Phys. Rev. Lett.* **104** 177004
[55] Mould R F 1995 The early history of x-ray diagnosis with emphasis on the contributions of physics 1895-1915 *Phys. Med. Biol.* **40** 1741
[56] Bragg W L 1913 The structure of some crystals as indicated by their diffraction of X-rays *Proc. R. Soc. Lond.* A **89** 248–77
[57] Bragg W L 1912 The specular reflection of x-rays *Nature* **90** 410
[58] Eckert M 2012 Disputed discovery: the beginnings of x-ray diffraction in crystals in 1912 and its repercussions *Acta Crystallogr., Sect. A: Found. Crystallogr.* **68** 30–9
[59] Florescu V and Gavrila M 1976 Elastic scattering of photons by K-shell electrons at high energies *Phys. Rev.* A **14** 211
[60] Birkholz M 2006 *Thin Film Analysis by X-Ray Scattering* (New York: Wiley)
[61] Vermolen E, Kuijk A, Filion L, Hermes M, Thijssen J and Dijkstra M *et al* 2009 Fabrication of large binary colloidal crystals with a NaCl structure *Proc. Natl Acad. Sci.* **106** 16063–7
[62] Whittingham M S 1976 The role of ternary phases in cathode reactions *J. Electrochem. Soc.* **123** 315
[63] Tarascon J, Wang E, Shokoohi F, McKinnon W and Colson S 1991 The spinel phase of $LiMn_2O_4$ as a cathode in secondary lithium cells *J. Electrochem. Soc.* **138** 2859

[64] Godshall N 1986 Lithium transport in ternary lithium-copper-oxygen cathode materials *Solid State Ionics* **18** 788–93

[65] Song Y-F, Chang C-H, Liu C-Y, Chang S-H, Jeng U-S and Lai Y-H *et al* 2007 X-ray beamlines for structural studies at the NSRRC superconducting wavelength shifter *J. Synchrotron Radiat.* **14** 320–25

[66] Liu D-G, Chang C-H, Liu C-Y, Chang S-H, Juang J-M and Song Y-F *et al* 2009 A dedicated small-angle x-ray scattering beamline with a superconducting wiggler source at the NSRRC *J. Synchrotron Radiat.* **16** 97–104

[67] Kung P-J, Lau W, Lee A-p and Leou K-C 2016 Development of a high resolution beam position monitor for NSRRC VUV/THz FEL *7th Int. Particle Accelerator Conf. (IPAC'16), Busan, Korea, May 8–13* 2016 pp 2680–2

[68] Stehle P 1963 High-intensity limit of Thomson scattering *JOSA* **53** 1003

[69] Fiocco G and Thompson E 1963 Thomson scattering of optical radiation from an electron beam *Phys. Rev. Lett.* **10** 89

[70] Cormack A 1954 Interference of Rayleigh and nuclear Thomson scattering *Phys. Rev.* **94** 1397

[71] Prince E and Wilson A J C 2004 *International Tables for Crystallography* (Dordrecht: Kluwer)

[72] Amorós J L 2012 *The Laue Method* (Amsterdam: Elsevier)

[73] Rydberg H, Dion M, Jacobson N, Schröder E, Hyldgaard P and Simak S *et al* 2003 Van der Waals density functional for layered structures *Phys. Rev. Lett.* **91** 126402

[74] Persson K, Hinuma Y, Meng Y S, Van der Ven A and Ceder G 2010 Thermodynamic and kinetic properties of the Li-graphite system from first-principles calculations *Phys. Rev.* B **82** 125416

[75] Magerl A, Zabel H and Anderson I 1985 In-plane jump diffusion of Li in LiC_6 *Phys. Rev. Lett.* **55** 222

[76] Chou M, Cohen M L and Louie S G 1986 Theoretical Compton profiles of graphite and LiC_6 *Phys. Rev.* B **33** 6619

[77] Li W-B, Lin S-Y, Tran N T T, Lin M-F and Lin K-I 2020 Essential geometric and electronic properties in stage-n graphite alkali-metal-intercalation compounds *RSC Adv.* **10** 23573–81

[78] Rietveld H M 1969 A profile refinement method for nuclear and magnetic structures *J. Appl. Crystallogr.* **2** 65–71

[79] Li J, Li J, Luo J, Wang L and He X 2011 Recent advances in the $LiFeO_2$-based materials for Li-ion batteries *Int. J. Electrochem. Sci.* **6** 1550–61

[80] Inaba M, Iriyama Y, Ogumi Z, Todzuka Y and Tasaka A 1997 Raman study of layered rock-salt $LiCoO_2$ and its electrochemical lithium deintercalation *J. Raman Spectrosc.* **28** 613–17

[81] Ohzuku T, Ueda A and Nagayama M 1993 Electrochemistry and structural chemistry of $LiNiO_2$ (R3m) for 4 volt secondary lithium cells *J. Electrochem. Soc.* **140** 1862

[82] Aurbach D, Gofer Y and Langzam J 1989 The correlation between surface chemistry, surface morphology, and cycling efficiency of lithium electrodes in a few polar aprotic systems *J. Electrochem. Soc.* **136** 3198

[83] Brihuega I, Mallet P, González-Herrero H, De Laissardière G T, Ugeda M and Magaud L *et al* 2012 Unraveling the intrinsic and robust nature of van Hove singularities in twisted

bilayer graphene by scanning tunneling microscopy and theoretical analysis *Phys. Rev. Lett.* **109** 196802
[84] Li G, Luican A, Dos Santos J L, Neto A C, Reina A and Kong J *et al* 2010 Observation of Van Hove singularities in twisted graphene layers *Nat. Phys.* **6** 109–13
[85] Land T, Michely T, Behm R, Hemminger J and Comsa G 1992 STM investigation of single layer graphite structures produced on Pt (111) by hydrocarbon decomposition *Surf. Sci.* **264** 261–70
[86] Batra I P, Garcia N, Rohrer H, Salemink H, Stoll E and Ciraci S 1987 A study of graphite surface with STM and electronic structure calculations *Surf. Sci.* **181** 126–38
[87] Kelty S P and Lieber C M 1989 Atomic-resolution scanning-tunneling-microscopy investigations of alkali-metal–graphite intercalation compounds *Phys. Rev.* B **40** 5856
[88] Kelty S P, Lu Z and Lieber C M 1991 Scanning-tunneling-microscopy investigations of ternary graphite intercalation compounds *Phys. Rev.* B **44** 4064
[89] Alvarado S, Seidler P, Lidzey D and Bradley D 1998 Direct determination of the exciton binding energy of conjugated polymers using a scanning tunneling microscope *Phys. Rev. Lett.* **81** 1082
[90] Tapasztó L, Dobrik G, Lambin P and Biro L P 2008 Tailoring the atomic structure of graphene nanoribbons by scanning tunnelling microscope lithography *Nat. Nanotechnol.* **3** 397–401
[91] Cai J, Ruffieux P, Jaafar R, Bieri M, Braun T and Blankenburg S *et al* 2010 Atomically precise bottom-up fabrication of graphene nanoribbons *Nature* **466** 470–73
[92] De Crescenzi M, Castrucci P, Scarselli M, Diociaiuti M, Chaudhari P S and Balasubramanian C *et al* 2005 Experimental imaging of silicon nanotubes *Appl. Phys. Lett.* **86** 231901
[93] Hapala P, Kichin G, Wagner C, Tautz F S, Temirov R and Jelínek P 2014 Mechanism of high-resolution STM/AFM imaging with functionalized tips *Phys. Rev.* B **90** 085421
[94] Wendt S, Schaub R, Matthiesen J, Vestergaard E K, Wahlström E and Rasmussen M D *et al* 2005 Oxygen vacancies on TiO_2 (1 1 0) and their interaction with H_2O and O_2: a combined high-resolution STM and DFT study *Surf. Sci.* **598** 226–45
[95] Binnig G and Rohrer H 2000 Scanning tunneling microscopy *IBM J. Res. Dev.* **44** 279
[96] Binnig G, Rohrer H, Gerber C and Weibel E 1982 Vacuum tunneling *Phys. B*+ C **109** 2075–77
[97] Howells S, Chen T, Gallagher M, Sarid D, Lichtenberger D L and Wright L L *et al* 1992 High resolution images of single C60 molecules on gold (111) using scanning tunneling microscopy *Surf. Sci.* **274** 141–46
[98] Stolyarova E, Rim K T, Ryu S, Maultzsch J, Kim P and Brus L E *et al* 2007 High-resolution scanning tunneling microscopy imaging of mesoscopic graphene sheets on an insulating surface *Proc. Natl Acad. Sci.* **104** 9209–12
[99] Zhang Y, Gao T, Gao Y, Xie S, Ji Q and Yan K *et al* 2011 Defect-like structures of graphene on copper foils for strain relief investigated by high-resolution scanning tunneling microscopy *ACS Nano* **5** 4014–22
[100] Niimi Y, Matsui T, Kambara H, Tagami K, Tsukada M and Fukuyama H 2006 Scanning tunneling microscopy and spectroscopy of the electronic local density of states of graphite surfaces near monoatomic step edges *Phys. Rev.* B **73** 085421
[101] Yamauchi T, Takahara Y and Narita N 2004 STM/STS measurements of adatom vacancies on Si (111)-7x7 reconstruction surface *J. Jpn I. Met. Mater.* **68** 904–7

[102] Merte L R, Knudsen J, Grabow L C, Vang R T, Lægsgaard E and Mavrikakis M *et al* 2009 Correlating STM contrast and atomic-scale structure by chemical modification: vacancy dislocation loops on FeO/Pt (1 1 1) *Surf. Sci.* **603** L15–18

[103] De la Figuera J, González M, García-Martínez R, Rojo J, Hernán O and de Parga A V *et al* 1998 STM characterization of extended dislocation configurations in Au (001) *Phys. Rev.* B **58** 1169

[104] Tilinin I, Rose M, Dunphy J, Salmeron M and Van Hove M 1998 Identification of adatoms on metal surfaces by STM: experiment and theory *Surf. Sci.* **418** 511–20

[105] Björk J, Matena M, Dyer M S, Enache M, Lobo-Checa J and Gade L H *et al* 2010 STM fingerprint of molecule–adatom interactions in a self-assembled metal–organic surface coordination network on Cu (111) *Phys. Chem. Chem. Phys.* **12** 8815–21

[106] Cai T, Song Z, Rodriguez J A and Hrbek J 2004 Preparation and structural characterization of RuS_2 nanoislands on Au (111) *J. Am. Chem. Soc.* **126** 8886–87

[107] Bonnail N, Tonneau D, Jandard F, Capolino G-A and Dallaporta H 2004 Variable structure control of a piezoelectric actuator for a scanning tunneling microscope *IEEE Trans. Ind. Electron.* **51** 354–63

[108] Grifoni M and Hänggi P 1998 Driven quantum tunneling *Phys. Rep.* **304** 229–354

[109] Söde H, Talirz L, Gröning O, Pignedoli C A, Berger R and Feng X *et al* 2015 Electronic band dispersion of graphene nanoribbons via Fourier-transformed scanning tunneling spectroscopy *Phys. Rev.* B **91** 045429

[110] Heinze S, Blügel S, Pascal R, Bode M and Wiesendanger R 1998 Prediction of bias-voltage-dependent corrugation reversal for STM images of bcc (110) surfaces: W (110), Ta (110), and Fe (110) *Phys. Rev.* B **58** 16432

[111] Cometto F, Frank K, Stel B, Arisnabarreta N, Kern K and Lingenfelder M 2017 The STM bias voltage-dependent polymorphism of a binary supramolecular network *Chem. Commun.* **53** 11430–32

[112] Reiss G, Schneider F, Vancea J and Hoffmann H 1990 Scanning tunneling microscopy on rough surfaces: deconvolution of constant current images *Appl. Phys. Lett.* **57** 867–69

[113] Hellenthal C, Heimbuch R, Sotthewes K, Kooij E S and Zandvliet H J 2013 Determining the local density of states in the constant current STM mode *Phys. Rev.* B **88** 035425

[114] Klusek Z, Kozlowski W, Waqar Z, Datta S, Burnell-Gray J and Makarenko I *et al* 2005 Local electronic edge states of graphene layer deposited on Ir (1 1 1) surface studied by STM/CITS *Appl. Surf. Sci.* **252** 1221–27

[115] Monthioux M, Smith B, Burteaux B, Claye A, Fischer J and Luzzi D 2001 Sensitivity of single-wall carbon nanotubes to chemical processing: an electron microscopy investigation *Carbon* **39** 1251–72

[116] Lourie O and Wagner H 1998 Transmission electron microscopy observations of fracture of single-wall carbon nanotubes under axial tension *Appl. Phys. Lett.* **73** 3527–29

[117] Sharma R and Iqbal Z 2004 *In situ* observations of carbon nanotube formation using environmental transmission electron microscopy *Appl. Phys. Lett.* **84** 990–92

[118] Park C, Crooks R E, Siochi E J, Harrison J S, Evans N and Kenik E 2003 Adhesion study of polyimide to single-wall carbon nanotube bundles by energy-filtered transmission electron microscopy *Nanotechnology* **14** L11

[119] Lv R, Cui T, Jun M S, Zhang Q, Cao A and Su D S *et al* 2011 Open-ended, N-doped carbon nanotube–graphene hybrid nanostructures as high-performance catalyst support *Adv. Funct. Mater.* **21** 999–1006

[120] Colomer J-F, Henrard L, Van Tendeloo G, Lucas A and Lambin P 2004 Study of the packing of double-walled carbon nanotubes into bundles by transmission electron microscopy and electron diffraction *J. Mater. Chem.* **14** 603–6

[121] Zhu Y, Li L, Zhang C, Casillas G, Sun Z and Yan Z *et al* 2012 A seamless three-dimensional carbon nanotube graphene hybrid material *Nat. Commun.* **3** 1–7

[122] Nguyen D D, Tai N-H, Chen S-Y and Chueh Y-L 2012 Controlled growth of carbon nanotube–graphene hybrid materials for flexible and transparent conductors and electron field emitters *Nanoscale* **4** 632–38

[123] Suzuki S, Bower C and Zhou O 1998 In-situ TEM and EELS studies of alkali–metal intercalation with single-walled carbon nanotubes *Chem. Phys. Lett.* **285** 230–34

[124] Nethravathi C, Viswanath B, Michael J and Rajamath M 2012 Hydrothermal synthesis of a monoclinic VO_2 nanotube–graphene hybrid for use as cathode material in lithium ion batteries *Carbon* **50** 4839–46

[125] Garcıa R and Perez R 2002 Dynamic atomic force microscopy methods *Surf. Sci. Rep.* **47** 197–301

[126] Giessibl F J 2003 Advances in atomic force microscopy *Rev. Mod. Phys.* **75** 949

[127] Hiesgen R and Friedrich K A 2011 *Atomic Force Microscopy* (Boca Raton, FL: CRC Press)

[128] Meyer E 1992 Atomic force microscopy *Prog. Surf. Sci.* **41** 3–49

[129] Rugar D and Hansma P 1990 Atomic force microscopy *Phys. Today* **43** 23–30

[130] Hanbücken M, Neddermeyer H and Rupieper P 1982 Low energy electron diffraction, Auger electron spectroscopy and angle-resolved photoemission from silver films on Si (100) and Si (111) *Thin Solid Films* **90** 37–42

[131] Hagen D, Nieuwenhuys B, Rovida G and Somorjai G 1976 Low-energy electron diffraction, Auger electron spectroscopy, and thermal desorption studies of chemisorbed CO and O_2 on the (111) and stepped [6 (111)×(100)] iridium surfaces *Surf. Sci.* **57** 632–50

[132] Lagally M G and Martin J A 1983 Instrumentation for low-energy electron diffraction *Rev. Sci. Instrum.* **54** 1273–88

[133] Lander J and Morrison J 1963 Scattering factors and other properties of low-energy electron diffraction *J. Appl. Phys.* **34** 3517–35

[134] Held G 2011 Low-energy electron diffraction (LEED) *Surface and Thin Film Analysis* (Weinheim: Wiley) pp 93–109

[135] Lander J, Morrison J and Unterwald F 1962 Improved design and method of operation of low energy electron diffraction equipment *Rev. Sci. Instrum.* **33** 782–83

[136] Scheibner E, Germer L and Hartman C 1960 Apparatus for direct observation of low-energy electron diffraction patterns *Rev. Sci. Instrum.* **31** 112–14

[137] Mitchell K 1973 Low-energy electron diffraction *Contemp. Phys.* **14** 251–71

[138] Wang G-C and Lu T-M 2013 *RHEED Transmission Mode and Pole Figures: Thin Film and Nanostructure Texture Analysis* (Berlin: Springer)

[139] Wang G-C and Lu T-M 2014 RHEED transmission mode and RHEED pole figure *RHEED Transmission Mode and Pole Figures* (Berlin: Springer) pp 73–106

[140] Braun W 1999 *Applied RHEED: Reflection High-Energy Electron Diffraction during Crystal Growth* (Berlin: Springer)

[141] Dabrowska-Szata M 2003 Analysis of RHEED pattern from semiconductor surfaces *Mater. Chem. Phys.* **81** 257–59

[142] Ichimiya A, Cohen P I and Cohen P I 2004 *Reflection High-Energy Electron Diffraction* (Cambridge: Cambridge University Press)

[143] Haeni J, Theis C D and Schlom D G 2000 RHEED intensity oscillations for the stoichiometric growth of $SrTiO_3$ thin films by reactive molecular beam epitaxy *J. Electroceram.* **4** 385–91

[144] Borysiuk J, Sołtys J and Piechota J 2011 Stacking sequence dependence of graphene layers on SiC (000 1−)—experimental and theoretical investigation *J. Appl. Phys.* **109** 093523

[145] Ruiz-García M, Bonilla L L and Prados A 2016 STM-driven transition from rippled to buckled graphene in a spin-membrane model *Phys. Rev.* B **94** 205404

[146] Zhang J, Xiao J, Meng X, Monroe C, Huang Y and Zuo J-M 2010 Free folding of suspended graphene sheets by random mechanical stimulation *Phys. Rev. Lett.* **104** 166805

[147] Schaper A K, Hou H, Wang M, Bando Y and Golberg D 2011 Observations of the electrical behaviour of catalytically grown scrolled graphene *Carbon* **49** 1821–28

[148] Nguyen G D, Tsai H-Z, Omrani A A, Marangoni T, Wu M and Rizzo D J *et al* 2017 Atomically precise graphene nanoribbon heterojunctions from a single molecular precursor *Nat. Nanotechnol.* **12** 1077

[149] Biedermann L B, Bolen M L, Capano M A, Zemlyanov D and Reifenberger R G 2009 Insights into few-layer epitaxial graphene growth on 4*H*-SiC (0001) substrates from STM studies *Phys. Rev.* B **79** 125411

[150] Marchini S, Günther S and Wintterlin J 2007 Scanning tunneling microscopy of graphene on Ru (0001) *Phys. Rev.* B **76** 075429

[151] Wang Z, Xiang R, Shi Q, Yang J, Wang X and Hou J *et al* 2006 Modeling STM images in graphene using the effective-mass approximation *Phys. Rev.* B **74** 125417

[152] Bocquet M-L and Wang B 2010 Metal–organic interaction probed by first principles STM simulations *Prog. Surf. Sci.* **85** 435–59

[153] Martínez-Galera A J, Nicoara N, Martínez J I, Dappe Y J, Ortega J and Gómez-Rodríguez J M 2014 Imaging molecular orbitals of PTCDA on graphene on Pt (111): electronic structure by STM and first-principles calculations *J. Phys. Chem.* C **118** 12782–88

[154] Blanco J M, González C, Jelínek P, Ortega J, Flores F and Pérez R 2004 First-principles simulations of STM images: from tunneling to the contact regime *Phys. Rev.* B **70** 085405

[155] Duo I, Fujishima A and Comninellis C 2003 Electron transfer kinetics on composite diamond (sp^3)–graphite (sp^2) electrodes *Electrochem. Commun.* **5** 695–700

[156] Unwin P R, Güell A G and Zhang G 2016 Nanoscale electrochemistry of sp^2 carbon materials: from graphite and graphene to carbon nanotubes *Acc. Chem. Res.* **49** 2041–48

[157] Inoshita T, Nakao K and Kamimura H 1977 Electronic structure of potassium-graphite intercalation compound: C8K *J. Phys. Soc. Jpn.* **43** 1237–43

[158] Siebentritt S, Pues R, Rieder K-H and Shikin A M 1997 Surface phonon dispersion in graphite and in a lanthanum graphite intercalation compound *Phys. Rev.* B **55** 7927

[159] Ferrari A C, Meyer J, Scardaci V, Casiraghi C, Lazzeri M and Mauri F *et al* 2006 Raman spectrum of graphene and graphene layers *Phys. Rev. Lett.* **97** 187401

[160] Hashimoto A, Suenaga K, Gloter A, Urita K and Iijima S 2004 Direct evidence for atomic defects in graphene layers *Nature* **430** 870–73

[161] Lauffer P, Emtsev K, Graupner R, Seyller T, Ley L and Reshanov S *et al* 2008 Atomic and electronic structure of few-layer graphene on SiC (0001) studied with scanning tunneling microscopy and spectroscopy *Phys. Rev.* B **77** 155426

[162] Andreoni W 2000 *The Physics of Fullerene-Based and Fullerene-Related Materials* (Boston, MA: Springer)

[163] Dresselhaus M S, Dresselhaus G, Saito R and Jorio A 2005 Raman spectroscopy of carbon nanotubes *Phys. Rep.* **409** 47–99
[164] Yarotski D A, Kilina S V, Talin A A, Tretiak S, Prezhdo O V and Balatsky A V *et al* 2009 Scanning tunneling microscopy of DNA-wrapped carbon nanotubes *Nano Lett.* **9** 12–7
[165] Han M Y, Özyilmaz B, Zhang Y and Kim P 2007 Energy band-gap engineering of graphene nanoribbons *Phys. Rev. Lett.* **98** 206805
[166] Talirz L, Söde H, Dumslaff T, Wang S, Sanchez-Valencia J R and Liu J *et al* 2017 On-surface synthesis and characterization of 9-atom wide armchair graphene nanoribbons *ACS Nano* **11** 1380–88
[167] Jiao L, Zhang L, Wang X, Diankov G and Dai H 2009 Narrow graphene nanoribbons from carbon nanotubes *Nature* **458** 877–80
[168] Balch A L and Olmstead M M 1998 Reactions of transition metal complexes with fullerenes (C60, C70, etc.) and related materials *Chem. Rev.* **98** 2123–66
[169] Sakurai T, Wang X-D, Xue Q, Hasegawa Y, Hashizume T and Shinohara H 1996 Scanning tunneling microscopy study of fullerenes *Prog. Surf. Sci.* **51** 263–408
[170] Oberdörster E 2004 Manufactured nanomaterials (fullerenes, C60) induce oxidative stress in the brain of juvenile largemouth bass *Environ. Health Perspect.* **112** 1058–62
[171] Banhart F and Ajayan P M 1996 Carbon onions as nanoscopic pressure cells for diamond formation *Nature* **382** 433–35
[172] Sano N, Wang H, Chhowalla M, Alexandrou I and Amaratunga G 2001 Synthesis of carbon 'onions' in water *Nature* **414** 506–7
[173] Wang J, Martinez T, Yaniv D R and McCormick L D 1991 Scanning tunneling microscopic investigation of surface fouling of glassy carbon surfaces due to phenol oxidation *J. Electroanal. Chem. Interfacial Electrochem.* **313** 129–40
[174] Yaniv D R and McCormick L D 1991 Novel STM-EC center for *in situ* imaging under potentiostatic/galvanostatic control: electrodeposition of copper and cadmium on carbon substrates *Electroanalysis* **3** 103–10
[175] Bernal J D 1924 The structure of graphite *Proc. R. Soc. Lond.* A **106** 749–73
[176] Rong Z Y 1994 Extended modifications of electronic structures caused by defects: scanning tunneling microscopy of graphite *Phys. Rev.* B **50** 1839
[177] Lin Y-C, Teng P-Y, Yeh C-H, Koshino M, Chiu P-W and Suenaga K 2015 Structural and chemical dynamics of pyridinic-nitrogen defects in graphene *Nano Lett.* **15** 7408–13
[178] Xu P, Yang Y, Qi D, Barber S, Schoelz J and Ackerman M *et al* 2012 Electronic transition from graphite to graphene via controlled movement of the top layer with scanning tunneling microscopy *Phys. Rev.* B **86** 085428
[179] Nelson F, Diebold A C and Hull R 2010 Simulation study of aberration-corrected high-resolution transmission electron microscopy imaging of few-layer-graphene stacking *Microsc. Microanal.* **16** 194
[180] Rong Z Y and Kuiper P 1993 Electronic effects in scanning tunneling microscopy: moiré pattern on a graphite surface *Phys. Rev.* B **48** 17427
[181] Campanera J, Savini G, Suarez-Martinez I and Heggie M 2007 Density functional calculations on the intricacies of moiré patterns on graphite *Phys. Rev.* B **75** 235449
[182] Hattendorf S, Georgi A, Liebmann M and Morgenstern M 2013 Networks of ABA and ABC stacked graphene on mica observed by scanning tunneling microscopy *Surf. Sci.* **610** 53–8

[183] Yan K, Peng H, Zhou Y, Li H and Liu Z 2011 Formation of bilayer bernal graphene: layer-by-layer epitaxy via chemical vapor deposition *Nano Lett.* **11** 1106–10

[184] Pierucci D, Sediri H, Hajlaoui M, Girard J-C, Brumme T and Calandra M *et al* 2015 Evidence for flat bands near the Fermi level in epitaxial rhombohedral multilayer graphene *ACS Nano* **9** 5432–39

[185] Garlow J A, Barrett L K, Wu L, Kisslinger K, Zhu Y and Pulecio J F 2016 Large-area growth of turbostratic graphene on Ni (111) via physical vapor deposition *Sci. Rep.* **6** 19804

[186] Fiori S, Murata Y, Veronesi S, Rossi A, Coletti C and Heun S 2017 Li-intercalated graphene on SiC (0001): an STM study *Phys. Rev.* B **96** 125429

[187] Goler S, Coletti C, Piazza V, Pingue P, Colangelo F and Pellegrini V *et al* 2013 Revealing the atomic structure of the buffer layer between SiC (0001) and epitaxial graphene *Carbon* **51** 249–54

[188] De Parga A V, Calleja F, Borca B, Passeggi M Jr, Hinarejos J and Guinea F *et al* 2008 Periodically rippled graphene: growth and spatially resolved electronic structure *Phys. Rev. Lett.* **100** 056807

[189] Liu Z, Suenaga K, Harris P J and Iijima S 2009 Open and closed edges of graphene layers *Phys. Rev. Lett.* **102** 015501

[190] Delaney P, Choi H J, Ihm J, Louie S G and Cohen M L 1998 Broken symmetry and pseudogaps in ropes of carbon nanotubes *Nature* **391** 466–68

[191] Hone J, Ellwood I, Muno M, Mizel A, Cohen M L and Zettl A *et al* 1998 Thermoelectric power of single-walled carbon nanotubes *Phys. Rev. Lett.* **80** 1042

[192] Krupke R, Hennrich F, Löhneysen H. v. and Kappes M M 2003 Separation of metallic from semiconducting single-walled carbon nanotubes *Science* **301** 344–47

[193] Umeno Y, Kitamura T and Kushima A 2004 Theoretical analysis on electronic properties of zigzag-type single-walled carbon nanotubes under radial deformation *Comput. Mater. Sci.* **30** 283–87

[194] Barone V and Scuseria G E 2004 Theoretical study of the electronic properties of narrow single-walled carbon nanotubes: beyond the local density approximation *J. Chem. Phys.* **121** 10376–79

[195] Lin X, Lu J and Zhu H 2017 The stability and electronic properties of a new allotrope of silicene and silicon nanotube *Superlattices Microstruct.* **101** 480–87

[196] Mintmire J and White C 1998 Universal density of states for carbon nanotubes *Phys. Rev. Lett.* **81** 2506

[197] Kubby J A and Boland J J 1996 Scanning tunneling microscopy of semiconductor surfaces *Surf. Sci. Rep.* **26** 61–204

[198] Moresco F, Meyer G, Rieder K-H, Tang H, Gourdon A and Joachim C 2001 Low temperature manipulation of big molecules in constant height mode *Appl. Phys. Lett.* **78** 306–8

[199] Passoni M, Donati F, Bassi A L, Casari C S and Bottani C E 2009 Recovery of local density of states using scanning tunneling spectroscopy *Phys. Rev.* B **79** 045404

[200] Yin Y, Zech M, Williams T L, Wang X, Wu G and Chen X *et al* 2009 Scanning tunneling spectroscopy and vortex imaging in the iron pnictide superconductor $BaFe_{1.8}Co_{0.2}As_2$ *Phys. Rev. Lett.* **102** 097002

[201] Ji S-H, Zhang T, Fu Y-S, Chen X, Ma X-C and Li J *et al* 2008 High-resolution scanning tunneling spectroscopy of magnetic impurity induced bound states in the superconducting gap of Pb thin films *Phys. Rev. Lett.* **100** 226801

[202] Kleiber M, Bode M, Ravlić R, Tezuka N and Wiesendanger R 2002 Magnetic properties of the Cr (001) surface studied by spin-polarized scanning tunneling spectroscopy *J. Magn. Magn. Mater.* **240** 64–9
[203] Son Y-W, Cohen M L and Louie S G 2006 Half-metallic graphene nanoribbons *Nature* **444** 347–49
[204] Son Y-W, Cohen M L and Louie S G 2006 Energy gaps in graphene nanoribbons *Phys. Rev. Lett.* **97** 216803
[205] Koh Y K, Bae M-H, Cahill D G and Pop E 2010 Heat conduction across monolayer and few-layer graphenes *Nano Lett.* **10** 4363–68
[206] Leon V, Quintana M, Herrero M A, Fierro J L, de la Hoz A and Prato M *et al* 2011 Few-layer graphenes from ball-milling of graphite with melamine *Chem. Commun.* **47** 10936–38
[207] Lv C, Xue Q, Xia D, Ma M, Xie J and Chen H 2010 Effect of chemisorption on the interfacial bonding characteristics of graphene– polymer composites *J. Phys. Chem.* C **114** 6588–94
[208] Sluiter M H and Kawazoe Y 2003 Cluster expansion method for adsorption: application to hydrogen chemisorption on graphene *Phys. Rev.* B **68** 085410
[209] Bennett A J, McCarroll B and Messmer R P 1971 Molecular orbital approach to chemisorption. II. Atomic H, C, N, O, and F on graphite *Phys. Rev.* B **3** 1397
[210] Volpe M and Cleri F 2003 Chemisorption of atomic hydrogen in graphite and carbon nanotubes *Surf. Sci.* **544** 24–34
[211] Chen Z, Lin Y-M, Rooks M J and Avouris P 2007 Graphene nano-ribbon electronics *Physica* E **40** 228–32
[212] Bhimanapati G R, Lin Z, Meunier V, Jung Y, Cha J and Das S *et al* 2015 Recent advances in two-dimensional materials beyond graphene *ACS Nano* **9** 11509–39
[213] Odom T W, Huang J-L, Kim P and Lieber C M 1998 Atomic structure and electronic properties of single-walled carbon nanotubes *Nature* **391** 62–4
[214] Xue Y and Datta S 1999 Fermi-level alignment at metal-carbon nanotube interfaces: application to scanning tunneling spectroscopy *Phys. Rev. Lett.* **83** 4844
[215] Yakobson B, Campbell M, Brabec C and Bernholc J 1997 Tensile strength, atomistics of fracture, and C-chain unraveling in carbon nanotubes *J. Comput.-Aided Mater. Design* **3** 173
[216] Yoshioka H and Odintsov A A 1999 Electronic properties of armchair carbon nanotubes: bosonization approach *Phys. Rev. Lett.* **82** 374
[217] Lin T, Zhang W-D, Huang J and He C 2005 A DFT study of the amination of fullerenes and carbon nanotubes: reactivity and curvature *J. Phys. Chem.* B **109** 13755–60
[218] Hou J, Wang B, Yang J, Wang K, Lu W and Li Z *et al* 2003 Disorder and suppression of quantum confinement effects in Pd nanoparticles *Phys. Rev. Lett.* **90** 246803
[219] Wolf O, Dasog M, Yang Z, Balberg I, Veinot J and Millo O 2013 Doping and quantum confinement effects in single Si nanocrystals observed by scanning tunneling spectroscopy *Nano Lett.* **13** 2516–21
[220] Chang S-L, Lin S-Y, Lin S-K, Lee C-H and Lin M-F 2014 Geometric and electronic properties of edge-decorated graphene nanoribbons *Sci. Rep.* **4** 6038
[221] Park M-H, Kim M G, Joo J, Kim K, Kim J and Ahn S *et al* 2009 Silicon nanotube battery anodes *Nano Lett.* **9** 3844–47

[222] De Padova P, Quaresima C, Ottaviani C, Sheverdyaeva P M, Moras P and Carbone C *et al* 2010 Evidence of graphene-like electronic signature in silicene nanoribbons *Appl. Phys. Lett.* **96** 261905

[223] Cocco G, Cadelano E and Colombo L 2010 Gap opening in graphene by shear strain *Phys. Rev.* B **81** 241412

[224] Lv R and Terrones M 2012 Towards new graphene materials: doped graphene sheets and nanoribbons *Mater. Lett.* **78** 209–18

[225] Moon P and Koshino M 2013 Optical absorption in twisted bilayer graphene *Phys. Rev.* B **87** 205404

[226] Uchida K, Furuya S, Iwata J-I and Oshiyama A 2014 Atomic corrugation and electron localization due to Moiré patterns in twisted bilayer graphenes *Phys. Rev.* **B 90** 155451

[227] Lui C H, Li Z, Mak K F, Cappelluti E and Heinz T F 2011 Observation of an electrically tunable band gap in trilayer graphene *Nat. Phys.* **7** 944–47

[228] Chuang Y-C, Wu J-Y and Lin M-F 2013 Electric field dependence of excitation spectra in AB-stacked bilayer graphene *Sci. Rep.* **3** 1368

[229] Wang J, Wang Z, Zhang R, Zheng Y, Chen L and Wang S *et al* 2018 A first-principles study of the electrically tunable band gap in few-layer penta-graphene *Phys. Chem. Chem. Phys.* **20** 18110–16

[230] Lin C-Y, Huang B-L, Ho C-H, Gumbs G and Lin M-F 2018 Geometry-diversified Coulomb excitations in trilayer AAB stacking graphene *Phys. Rev.* **B 98** 195442

[231] Aktürk O Ü and Tomak M 2010 Bismuth doping of graphene *Appl. Phys. Lett.* **96** 081914

[232] García N, Esquinazi P, Barzola-Quiquia J and Dusari S 2012 Evidence for semiconducting behavior with a narrow band gap of Bernal graphite *New J. Phys.* **14** 053015

[233] FL S and Lin M F 2001 Low-frequency π-electronic excitations of simple hexagonal graphite *J. Phys. Soc. Jpn.* **70** 897–901

[234] Goncher S J, Zhao L, Pasupathy A N and Flynn G W 2013 Substrate level control of the local doping in graphene *Nano Lett.* **13** 1386–92

[235] Li W-B, Lin S-Y, Tran N T T, Lin K-I and Lin M-F 2020 Essential electronic properties on stage-1 Li/Li^+-graphite-intercalation compounds for different concentrations arXiv:2006.12055

[236] Nguyen T D H, Pham H D, Lin S-Y and Lin M-F 2020 Featured properties of Li^+-based battery anode: $Li_4Ti_5O_{12}$ *RSC Adv.* **10** 14071–79

[237] Dien V K, Han N T, Nguyen T D H, Huynh T M D, Pham H D and Lin M-F 2020 Geometric and electronic properties of Li_2GeO_3 arXiv:2009.02154

[238] Kordyuk A 2015 Pseudogap from ARPES experiment: three gaps in cuprates and topological superconductivity *Low Temp. Phys.* **41** 319–41

[239] Matt C E, Pirie H, Soumyanarayanan A, He Y, Yee M M and Chen P *et al* 2020 Consistency between ARPES and STM measurements on SmB 6 *Phys. Rev.* **B 101** 085142

[240] Hwang E and Sarma S D 2008 Quasiparticle spectral function in doped graphene: electron-electron interaction effects in ARPES *Phys. Rev.* **B 77** 081412

[241] Avila J, De Padova P, Cho S, Colambo I, Lorcy S and Quaresima C *et al* 2013 Presence of gapped silicene-derived band in the prototypical (3×3) silicene phase on silver (111) surfaces *J. Phys. Condens. Matter* **25** 262001

[242] Lin C-H, Huang A, Pai W W, Chen W-C, Chen T-Y and Chang T-R *et al* 2018 Single-layer dual germanene phases on Ag (111) *Phys. Rev. Mater.* **2** 024003

[243] Cui J, Chen C, He W, Avila J, Zhao L-D and Asensio M C *et al* 2018 Large enhancement of electrical transport properties of SnS in the out-of-plane direction by n-type doping: a combined ARPES and DFT study *J. Mater. Chem.* A **6** 24588–94
[244] Zhong Y, Chen Y, Guan J, Zhao J, Rao Z and Tang C *et al* 2018 Extraction of tight binding parameters from *in situ* ARPES on the continuously doped surface of cuprates *Sci. China Phys. Mech.* **61** 127403
[245] Novikov D 2007 Elastic scattering theory and transport in graphene *Phys. Rev.* B **76** 245435
[246] Lin C-Y, Ho C-H, Wu J-Y and Lin M-F 2020 Unusual electronic excitations in ABA trilayer graphene *Sci. Rep.* **10** 11106
[247] Shih P-H, Chiu Y-H, Wu J-Y, Shyu F-L and Lin M-F 2017 Coulomb excitations of monolayer germanene *Sci. Rep.* **7** 40600
[248] Lin C-Y, Lee M-H and Lin M-F 2018 Coulomb excitations in ABC-stacked trilayer graphene *Phys. Rev.* B **98** 041408
[249] Shih P-H, Chiu C-W, Wu J-Y, Do T-N and Lin M-F 2018 Coulomb scattering rates of excited states in monolayer electron-doped germanene *Phys. Rev.* B **97** 195302
[250] Van D, Tuan N Q and Khanh 2013 Plasmon modes of double-layer graphene at finite temperature *Physica* E **54** 267–72
[251] Anno Y, Takeuchi M, Matsuoka M, Takei K, Akita S and Arie T 2017 Effect of defect-induced carrier scattering on the thermoelectric power of graphene *Appl. Phys. Lett.* **110** 263501
[252] Dudin P, Lacovig P, Fava C, Nicolini E, Bianco A and Cautero G *et al* 2010 Angle-resolved photoemission spectroscopy and imaging with a submicrometre probe at the SPECTROMICROSCOPY-3.2 L beamline of Elettra *J. Synchrotron Radiat.* **17** 445–50
[253] Ren Y, Chen S, Cai W, Zhu Y, Zhu C and Ruoff R S 2010 Controlling the electrical transport properties of graphene by *in situ* metal deposition *Appl. Phys. Lett.* **97** 053107
[254] Sugawara K, Yamamura N, Matsuda K, Norimatsu W, Kusunoki M and Sato T *et al* 2018 Selective fabrication of free-standing ABA and ABC trilayer graphene with/without Dirac-cone energy bands *NPG Asia Mater.* **10** e466
[255] Lin C-Y, Wu J-Y, Chiu C-W and Lin M-F 2019 *Coulomb Excitations and Decays in Graphene-Related Systems* (Boca Raton, FL: CRC Press)
[256] Park C-H, Giustino F, Cohen M L and Louie S G 2008 Electron–phonon interactions in graphene, bilayer graphene, and graphite *Nano Lett.* **8** 4229–33
[257] Ishida Y, Jung J, Kim M, Kwon J, Kim Y and Chung D *et al* 2020 Work function seen with sub-meV precision through laser photoemission *Commun. Phys.* **3** 1–8
[258] Sadeddine S, Enriquez H, Bendounan A, Das P K, Vobornik I and Kara A *et al* 2017 Compelling experimental evidence of a Dirac cone in the electronic structure of a 2D Silicon layer *Sci. Rep.* **7** 44400
[259] Ling X, Wang H, Huang S, Xia F and Dresselhaus M S 2015 The renaissance of black phosphorus *Proc. Natl Acad. Sci.* 112 4523–30
[260] Zheng X, Chen R, Shi G, Zhang J, Xu Z and Jiang T 2015 Characterization of nonlinear properties of black phosphorus nanoplatelets with femtosecond pulsed Z-scan measurements *Opt. Lett.* **40** 3480–83
[261] Losurdo M, Giangregorio M M, Capezzuto P and Bruno G 2011 Graphene CVD growth on copper and nickel: role of hydrogen in kinetics and structure *Phys. Chem. Chem. Phys.* **13** 20836–43

[262] Boukhvalov D, Katsnelson M and Lichtenstein A 2008 Hydrogen on graphene: electronic structure, total energy, structural distortions and magnetism from first-principles calculations *Phys. Rev.* B **77** 035427

[263] Partoens B and Peeters F 2006 From graphene to graphite: electronic structure around the K point *Phys. Rev.* B **74** 075404

[264] Sugawara K, Kanetani K, Sato T and Takahashi T 2011 Fabrication of Li-intercalated bilayer graphene *AIP Adv.* **1** 022103

[265] Senkovskiy B V, Usachov D Y, Fedorov A V, Haberer D, Ehlen N and Fischer F R *et al* 2018 Finding the hidden valence band of $N = 7$ armchair graphene nanoribbons with angle-resolved photoemission spectroscopy *2D Mater.* **5** 035007

[266] Yakovkin I N 2016 Dirac cones in graphene, interlayer interaction in layered materials, and the band gap in MoS_2 *Crystals* **6** 143

[267] Junaid M and Witjaksono G 2019 Analysis of band gap in AA and Ab stacked bilayer graphene by Hamiltonian tight binding method Proc. of the 2019 IEEE Int. Conf. on Sensors and Nanotechnology, Penang Island, Malaysia pp 24–5

[268] Zou L, Po H C, Vishwanath A and Senthil T 2018 Band structure of twisted bilayer graphene: emergent symmetries, commensurate approximants, and Wannier obstructions *Phys. Rev.* B **98** 085435

[269] Lin Y-P, Wang J, Lu J-M, Lin C-Y and Lin M-F 2014 Energy spectra of ABC-stacked trilayer graphene in magnetic and electric fields *RSC Adv.* **4** 56552–60

[270] Ho C-H, Chang C-P and Lin M-F 2016 Evolution and dimensional crossover from the bulk subbands in ABC-stacked graphene to a three-dimensional Dirac cone structure in rhombohedral graphite *Phys. Rev.* B **93** 075437

[271] Shih P-H, Do T-N, Gumbs G, Huang D, Pham T P and Lin M-F 2020 Magneto-transport properties of B-, Si-and N-doped graphene *Carbon* **160** 211–18

[272] Shih P-H, Do T-N, Gumbs G and Lin M-F 2020 Electronic and optical properties of doped graphene *Phys.* E **118** 113894

[273] Kim S, Ihm J, Choi H J and Son Y-W 2008 Origin of anomalous electronic structures of epitaxial graphene on silicon carbide *Phys. Rev. Lett.* **100** 176802

[274] Mishra B K and Ashok B 2018 Coaxial carbon nanotubes: from springs to ratchet wheels and nanobearings *Mater. Res. Express* **5** 075023

[275] Iijima S 1991 Helical microtubules of graphitic carbon *Nature* **354** 56–8

[276] Li T-S and Lin M-F 2006 Electronic properties of carbon nanotubes under external fields *Phys. Rev.* B **73** 075432

[277] Lin M F and Chuu D S 1998 Electronic states of toroidal carbon nanotubes *J. Phys. Soc. Jpn.* **67** 259–63

[278] De Padova P, Avila J, Resta A, Razado-Colambo I, Quaresima C and Ottaviani C *et al* 2013 The quasiparticle band dispersion in epitaxial multilayer silicene *J. Phys. Condens. Matter* **25** 382202

[279] Noguchi E, Sugawara K, Yaokawa R, Hitosugi T, Nakano H and Takahashi T 2015 Direct observation of Dirac cone in multilayer silicene intercalation compound $CaSi_2$ *Adv. Mater.* **27** 856–60

[280] Dávila M E and Le Lay G 2016 Few layer epitaxial germanene: a novel two-dimensional Dirac material *Sci. Rep.* **6** 20714

[281] Klimovskikh I I, Otrokov M M, Voroshnin V Y, Sostina D, Petaccia L and Di Santo G *et al* 2017 Spin–orbit coupling induced gap in graphene on Pt (111) with intercalated Pb monolayer *ACS Nano* **11** 368–74

[282] Kou L, Chen C and Smith S C 2015 Phosphorene: fabrication, properties, and applications *J. Phys. Chem. Lett.* **6** 2794–805

[283] Zhang W, Enriquez H, Tong Y, Bendounan A, Kara A and Seitsonen A P *et al* 2018 Epitaxial synthesis of blue phosphorene *Small* **14** 1804066

[284] Zhang W, Enriquez H, Tong Y, Bendounan A, Kara A and Seitsonen A P *et al* 2018 Blue phosphorene: epitaxial synthesis of blue phosphorene *Small* **14** 1870249

[285] Lei T, Li J-M, Li F-S, Wang J-O, Ibrahim K and Zhang K 2019 Anisotropic electronic structure of antimonene *Appl. Phys. Lett.* **115** 221602

[286] Claessen R, Stuehler R, Reis F and Schaefer J 2018 One-dimensional physics in the edge states of the high-temperature quantum spin Hall system bismuthene on SiC (0001) *APS* **2018** E08. 008

[287] Reis F, Li G, Dudy L, Bauernfeind M, Glass S and Hanke W *et al* 2017 Bismuthene on a SiC substrate: a candidate for a high-temperature quantum spin Hall material *Science* **357** 287–90

[288] Bhuvaneswari R, Maria J P, Nagarajan V and Chandiramouli R 2019 DFT study on the interaction properties of V-series nerve agent molecules on novel bismuthene nanotubes *J. Inorg. Organomet. Polym. Mater.* **29** 2226–36

[289] Liu C-C, Jiang H and Yao Y 2011 Low-energy effective Hamiltonian involving spin–orbit coupling in silicene and two-dimensional germanium and tin *Phys. Rev.* B **84** 195430

[290] Wu C-H 2018 Tight-binding model and *ab initio* calculation of silicene with strong spin–orbit coupling in low-energy limit arXiv:1804.01695

[291] Kara A, Enriquez H, Seitsonen A P, Voon L L Y, Vizzini S and Aufray B *et al* 2012 A review on silicene—new candidate for electronics *Surf. Sci. Rep.* **67** 1–18

[292] Sone J, Yamagami T, Aoki Y, Nakatsuji K and Hirayama H 2014 Epitaxial growth of silicene on ultra-thin Ag (111) films *New J. Phys.* **16** 095004

[293] Endo S, Kubo O, Nakashima N, Iwaguma S, Yamamoto R and Kamakura Y *et al* 2017 Germanene on Al (111) grown at nearly room temperature *Appl. Phys. Express* **11** 015502

[294] Fleurence A, Gill T, Friedlein R, Sadowski J, Aoyagi K and Copel M *et al* 2016 Single-domain epitaxial silicene on diboride thin films *Appl. Phys. Lett.* **108** 151902

[295] Lee C-C, Fleurence A, Yamada-Takamura Y, Ozaki T and Friedlein R 2014 Band structure of silicene on zirconium diboride (0001) thin-film surface: convergence of experiment and calculations in the one-Si-atom Brillouin zone *Phys. Rev.* B **90** 075422

[296] Chen M, Peng J-P, Zhang H-M, Wang L-L, He K and Ma X-C *et al* 2012 Molecular beam epitaxy of bilayer Bi (111) films on topological insulator Bi_2Te_3: a scanning tunneling microscopy study *Appl. Phys. Lett.* **101** 081603

[297] Lei T, Jin K-H, Zhang N, Zhao J-L, Liu C and Li W-J *et al* 2016 Electronic structure evolution of single bilayer Bi (1 1 1) film on 3D topological insulator $Bi_2Se_xTe_{3-x}$ surfaces *J. Phys. Condens. Matter* **28** 255501

[298] Crepaldi A, Pons S, Frantzeskakis E, Calleja F, Etzkorn M and Seitsonen A *et al* 2013 Combined ARPES and STM study of Pb/Au (111) Moiré structure: one overlayer, two symmetries *Phys. Rev.* B **87** 115138

[299] Guo M, Li K, Liu L, Zhang H, Guo W and Hu X *et al* 2019 Manganese-based multi-oxide derived from spent ternary lithium-ions batteries as high-efficient catalyst for VOCs oxidation *J. Hazard. Mater.* **380** 120905

[300] Hou X, Zhu G, Niu X, Dai Z, Yin Z and Dong Q *et al* 2017 Ternary transition metal oxide derived from Prussian blue analogue for high-performance lithium ion battery *J. Alloys Compd.* **729** 518–25

[301] Ren J, Li R, Liu Y, Cheng Y, Mu D and Zheng R *et al* 2017 The impact of aluminum impurity on the regenerated lithium nickel cobalt manganese oxide cathode materials from spent LIBs *New J. Chem.* **41** 10959–65

[302] Ding C, Wang L, Zhou W, Wang D, Du Y and Wen G 2018 New design on Li-ion battery anode of ternary complex metal/metal oxide@ CNT: a case study of hierarchical NiCo-$NiCo_2O_4$@ CNTs *Chem. Eng. J.* **353** 340–49

[303] Huang Y, Li Y, Huang R and Yao J 2019 Ternary $Fe_2O_3/Fe_3O_4/FeCO_3$ composite as a high-performance anode material for lithium-ion batteries *J. Phys. Chem.* C **123** 12614–22

[304] Han N T, Dien V K, Tran N T T, Nguyen D K, Su W-P and Lin M-F 2020 First-principles studies of electronic properties in lithium metasilicate (Li_2SiO_3) arXiv:2001.07128

[305] Dai Y-M, Wang Y-F and Chen C-C 2018 Synthesis and characterization of magnetic $LiFe_5O_8$-$LiFeO_2$ as a solid basic catalyst for biodiesel production *Catal. Commun.* **106** 20–4

[306] Pervan P and Lazić P 2017 Adsorbed or intercalated: Na on graphene/Ir (111) *Phys. Rev. Mater.* **1** 044202

[307] Khademi A, Sajadi E, Dosanjh P, Bonn D, Folk J A and Stöhr A *et al* 2016 Alkali doping of graphene: the crucial role of high-temperature annealing *Phys. Rev.* B **94** 201405

[308] Lee C K, Seo J G, Kim H J, Hong S J, Song G and Ahn C *et al* 2019 Versatile and tunable electrical properties of doped nonoxidized graphene using alkali metal chlorides *ACS Appl. Mater. Interfaces* **11** 42520–27

[309] Arakane T, Sato T, Takahashi T, Fujii T and Asamitsu A 2011 Angle-resolved photoemission study of the doping evolution of a three-dimensional Fermi surface in Na_xCoO_2 *New J. Phys.* **13** 043021

[310] Neupane M, Xu S-Y, Alidoust N, Sankar R, Belopolski I and Sanchez D S *et al* 2015 Surface versus bulk Dirac state tuning in a three-dimensional topological Dirac semimetal *Phys. Rev.* B **91** 241114

[311] Tan T S, Kas J and Rehr J 2018 Coulomb-hole and screened exchange in the electron self-energy at finite temperature *Phys. Rev.* B **98** 115125

[312] Lenchuk O, Adelhelm P and Mollenhauer D 2019 New insights into the origin of unstable sodium graphite intercalation compounds *Phys. Chem. Chem. Phys.* **21** 19378–90

[313] Li Y, Lu Y, Adelhelm P, Titirici M-M and Hu Y-S 2019 Intercalation chemistry of graphite: alkali metal ions and beyond *Chem. Soc. Rev.* **48** 4655–87

[314] Xu J, Dou Y, Wei Z, Ma J, Deng Y and Li Y *et al* 2017 Recent progress in graphite intercalation compounds for rechargeable metal (Li, Na, K, Al)-ion batteries *Adv. Sci.* **4** 1700146

[315] Tong X-M 2019 Photoelectron angular distribution of atoms in pulsed XUV and IR fields *Phys. Rev.* A **99** 043422

[316] Zhang C, Andersson T, Förstel M, Mucke M, Arion T and Tchaplyguine M *et al* 2013 The photoelectron angular distribution of water clusters *J. Chem. Phys.* **138** 234306

[317] Bialkowski S E 1996 *Photothermal Spectroscopy Methods for Chemical Analysis* (New York: Wiley)

[318] Dawlaty J M, Shivaraman S, Strait J, George P, Chandrashekhar M and Rana F *et al* 2008 Measurement of the optical absorption spectra of epitaxial graphene from terahertz to visible *Appl. Phys. Lett.* **93** 131905

[319] Soukoulis C M 2012 *Photonic Band Gap Materials* (Berlin: Springer)

[320] Wooten F 2013 *Optical Properties of Solids* (New York: Academic)

[321] Beal A, Liang W and Hughes H 1976 Kramers-Kronig analysis of the reflectivity spectra of 3R-WS_2 and 2H-WSe_2 *J. Phys.* C **9** 2449

[322] Prasankumar R P and Taylor A J 2016 *Optical Techniques for Solid-State Materials Characterization* (Boca Raton, FL: CRC Press)

[323] Hollas J M 2013 *High Resolution Spectroscopy* (Oxford: Butterworth-Heinemann)

[324] Popova M, Chukalina E, Erofeev D, Jablunovskis A, Gudim I and Malkin B 2020 High-resolution optical spectroscopy and modeling of spectral and magnetic properties of multiferroic $ErFe_3(BO_3)_4$ *Phys. Rev.* B **101** 205108

[325] Tkachenko N V 2006 *Optical Spectroscopy: Methods and Instrumentations* (Amsterdam: Elsevier)

[326] Dien V K, Pham H D, Tran N T T, Han N T, Huynh T M D and Nguyen T D H *et al* 2020 Orbital-hybridization-created optical excitations in $Li_8Ge_4O_{12}$ arXiv:2009.02160

[327] Han N T, Dien V K and Lin M-F 2020 Excitonic effects in the optical spectra of lithium metasilicate (Li_2SiO_3) arXiv:2010.11621

[328] Kortüm G 2012 *Reflectance Spectroscopy: Principles, Methods, Applications* (Berlin: Springer)

[329] Yano J and Yachandra V K 2009 X-ray absorption spectroscopy *Photosynth. Res.* **102** 241–54

[330] Ahn C C 2006 *Transmission Electron Energy Loss Spectrometry in Materials Science and the EELS Atlas* (New York: Wiley)

[331] Gilliland G 1997 Photoluminescence spectroscopy of crystalline semiconductors *Mater. Sci. Eng.: R: Rep.* **18** 99–399

[332] Fox M 2002 *Optical Properties of Solids* (College Park, MD: American Association of Physics Teachers)

[333] Philipp H R 1977 Infrared optical properties of graphite *Phys. Rev.* B **16** 2896

[334] Taft E and Philipp H 1965 Optical properties of graphite *Phys. Rev.* **138** A197

[335] Kuhlmann U, Jantoljak H, Pfänder N, Journet C, Bernier P and Thomsen C 1999 Infrared reflectance of single-walled carbon nanotubes *Synth. Met.* **103** 2506–7

[336] Zeng H, Jiao L, Xian X, Qin X, Liu Z and Cui X 2008 Reflectance spectra of individual single-walled carbon nanotubes *Nanotechnology* **19** 045708

[337] Giovannetti G, Khomyakov P A, Brocks G, Karpan V V, van den Brink J and Kelly P J 2008 Doping graphene with metal contacts *Phys. Rev. Lett.* **101** 026803

[338] Park H Y, Jung W S, Kang D H, Jeon J, Yoo G and Park Y *et al* 2016 Extremely low contact resistance on graphene through n-Type doping and edge contact design *Adv. Mater.* **28** 864–70

[339] Wei P, Liu N, Lee H R, Adijanto E, Ci L and Naab B D *et al* 2013 Tuning the Dirac point in CVD-grown graphene through solution processed n-type doping with 2-(2-methoxy-phenyl)-1, 3-dimethyl-2, 3-dihydro-1 H-benzoimidazole *Nano Lett.* **13** 1890–97

[340] Kepaptsoglou D, Hardcastle T P, Seabourne C R, Bangert U, Zan R and Amani J A *et al* 2015 Electronic structure modification of ion implanted graphene: the spectroscopic signatures of p-and n-type doping *ACS Nano* **9** 11398–407

[341] Wu M, Cao C and Jiang J 2010 Light non-metallic atom (B, N, O and F)-doped graphene: a first-principles study *Nanotechnology* **21** 505202

[342] Allaoui I, Benyoussef A and Kenz A E 2020 Optoelectronic properties of pristine antimonene, phosphorene and their binary compound *Comput. Condens. Matter* **24** e00488

[343] Lin C-Y and Lin M-F 2019 Magneto-electronic properties of twisted bilayer graphene system arXiv:1907.08858

[344] Lin C-Y and Lin M-F 2019 Electronic and optical properties of stacking-configuration-modulated bilayer graphene in electric and magnetic fields arXiv:1907.08785

[345] Ho Y-H, Wu J-Y, Chen R-B, Chiu Y-H and Lin M-F 2010 Optical transitions between Landau levels: AA-stacked bilayer graphene *Appl. Phys. Lett.* **97** 101905

[346] Huang B-L, Chuu C-P and Lin M-F 2019 Asymmetry-enriched electronic and optical properties of bilayer graphene *Sci. Rep.* **9** 1–12

[347] Lin Y-P, Lin C-Y, Ho Y-H, Do T-N and Lin M-F 2015 Magneto-optical properties of ABC-stacked trilayer graphene *Phys. Chem. Chem. Phys.* **17** 15921–27

[348] Lin C-Y, Do T-N, Huang Y-K and Lin M-F 2017 *Optical Properties of Graphene in Magnetic and Electric Fields* (Bristol: IOP Publishing)

[349] Shiha P-H, Chiub C-W, Wuc J-Y, Dob T-N and Lina M-F 2017 Coulomb scattering rates of excited states in germanene arXiv:1710.05281

[350] Otsuji T, Popov V and Ryzhii V 2014 Active graphene plasmonics for terahertz device applications *J. Phys.* D **47** 094006

[351] Low T and Avouris P 2014 Graphene plasmonics for terahertz to mid-infrared applications *ACS Nano* **8** 1086–101

[352] Winnerl S, Mittendorff M, König-Otto J C, Schneider H, Helm M and Winzer T *et al* 2017 Ultrafast processes in graphene: from fundamental manybody interactions to device applications *Ann. Phys.* **529** 1700022

[353] Giustino F, Cohen M L and Louie S G 2007 Electron-phonon interaction using Wannier functions *Phys. Rev.* B **76** 165108

[354] Giustino F, Yates J R, Souza I, Cohen M L and Louie S G 2007 Electron-phonon interaction via electronic and lattice Wannier functions: superconductivity in boron-doped diamond reexamined *Phys. Rev. Lett.* **98** 047005

[355] Rohlfing M and Louie S G 1998 Excitonic effects and the optical absorption spectrum of hydrogenated Si clusters *Phys. Rev. Lett.* **80** 3320

[356] Yang L, Cohen M L and Louie S G 2007 Excitonic effects in the optical spectra of graphene nanoribbons *Nano Lett.* **7** 3112–15

[357] Yang L, Deslippe J, Park C-H, Cohen M L and Louie S G 2009 Excitonic effects on the optical response of graphene and bilayer graphene *Phys. Rev. Lett.* **103** 186802

[358] Arima T, Tokura Y and Torrance J 1993 Variation of optical gaps in perovskite-type 3*d* transition-metal oxides *Phys. Rev.* B **48** 17006

[359] Vasiliev I, Öğüt S and Chelikowsky J R 2001 *Ab initio* absorption spectra and optical gaps in nanocrystalline silicon *Phys. Rev. Lett.* **86** 1813

[360] Wang Y, Qian D, Cui Y, Zhang H, Hou J and Vandewal K *et al* 2018 Optical gaps of organic solar cells as a reference for comparing voltage losses *Adv. Energy Mater.* **8** 1801352

[361] Muscat J, Wander A and Harrison N 2001 On the prediction of band gaps from hybrid functional theory *Chem. Phys. Lett.* **342** 397–401

[362] Zaanen J, Sawatzky G and Allen J 1985 Band gaps and electronic structure of transition-metal compounds *Phys. Rev. Lett.* **55** 418

[363] Kumar V, Sharma S K, Sharma T and Singh V 1999 Band gap determination in thick films from reflectance measurements *Opt. Mater.* **12** 115–19

[364] McCarthy D E 1965 The reflection and transmission of infrared materials: III, spectra from 2 μ to 50 μ *Appl. Opt.* **4** 317–20

[365] Lin C-Y, Chen R-B, Ho Y-H and Lin M-F 2017 *Electronic and Optical Properties of Graphite-Related Systems* (Boca Raton, FL: CRC Press)

[366] Hybertsen M S and Louie S G 1987 *Ab initio* static dielectric matrices from the density-functional approach. I. Formulation and application to semiconductors and insulators *Phys. Rev.* B **35** 5585

[367] Levine Z H and Louie S G 1982 New model dielectric function and exchange-correlation potential for semiconductors and insulators *Phys. Rev.* B **25** 6310

[368] Zhao X, Wei C, Yang L and Chou M 2004 Quantum confinement and electronic properties of silicon nanowires *Phys. Rev. Lett.* **92** 236805

[369] Li X L, Han W P, Wu J B, Qiao X F, Zhang J and Tan P H 2017 Layer-number dependent optical properties of 2D materials and their application for thickness determination *Adv. Funct. Mater.* **27** 1604468

[370] Wang Y, Ni Z, Liu L, Liu Y, Cong C and Yu T *et al* 2010 Stacking-dependent optical conductivity of bilayer graphene *ACS Nano* **4** 4074–80

[371] Blase X, Rubio A, Louie S G and Cohen M L 1995 Quasiparticle band structure of bulk hexagonal boron nitride and related systems *Phys. Rev.* B **51** 6868

[372] Qiu D Y, Coh S, Cohen M L and Louie S G 2020 Comparison of G W band structure to semiempirical approach for an FeSe monolayer *Phys. Rev.* B **101** 235154

[373] Kittel C and McEuen P 1976 *Introduction to Solid State Physics* (New York: Wiley)

[374] Pärnaste M, Marcellini M, Holmström E, Bock N, Fransson J and Eriksson O *et al* 2007 Dimensionality crossover in the induced magnetization of Pd layers *J. Phys. Condens. Matter* **19** 246213

[375] Pinto N, Rezvani S J, Perali A, Flammia L, Milošević M V and Fretto M *et al* 2018 Dimensional crossover and incipient quantum size effects in superconducting niobium nanofilms *Sci. Rep.* **8** 1–12

Chapter 4

Diversified quasiparticle properties of four-component Li_xFePO_4 cathode materials in Li^+-based batteries: charges and spins

Shih-Yang Lin, Thi Dieu Hien Nguyen, Hsien-Ching Chung, Wei-Bang Li, Ngoc Thanh Thuy Tran, Nguyen Thi Han, Hsin-Yi Liu, Hai Duong Pham and Ming-Fa Lin

In general, the essential physical properties of quasiparticles in any condensed-matter systems are dominated by the interacting charge distributions and spin configurations, according to both numerical simulations [1, 2] and phenomenological models [3, 4]. Certain core materials in ion-based batteries possess these two critical factors, such as, four-component Li_xFePO_4 compounds with variable crystal structures [5–7], and three-component ones LiFeO/LiCoO/LiNiO [8, 9]/[10, 11]/[12, 13]. Such materials, which have displayed commercialized applications [14], are expected to be outstanding samples in basic science research [15]. The main reason lies in the rich and unique crystal structures (the various meta-stable moiré superlattices; [6, 16]). The Vienna *ab initio* simulation package (VASP)-based theoretical frameworks are developed in the current book [1, 2], in which they are able to provide the full information about the diverse physical/chemical/material phenomena. Most important, the critical mechanisms, the multi-/single-orbital hybridizations in distinct chemical bonds and the spin configurations due to different atoms/orbitals [1], will be thoroughly examined from all the available quantities through delicate calculations and analyses. It should be noticed that compared to the well-stacked graphene-related systems [17], the current ones present highly non-uniform chemical environments and thus very difficult simulations. How to overcome the various theoretical barriers is one of the significant works. On the other hand, those associated with the phenomenological models only become open issues because of the absence of reliable Hamiltonians (details in chapter 17).

doi:10.1088/978-0-7503-4685-6ch4

In addition to the commercial applications/the higher-performance cathode materials of Li^+-based batteries [8, 10, 12], the multi-component transition-metal lithium oxide compounds are very suitable for basic science research. The numerical simulation methods are only efficient for identifying the diverse quasiparticle particle properties, according to the previous theoretical studies [18, 19]. The phenomenological models are almost absent, since all the stable/meta-stable crystal structures belong to moiré superlattices with many atoms/chemical bonds in primitive unit cells [20, 21]. Each highly non-uniform chemical environment can be successfully obtained from first-principles calculations, as done for $LiFePO_4$, as well as, $LiFeO_2$/$LiCoO_2$/$LiNiO_2$ [22]/[23, 24]/[24], The theoretical predictions are focused on the semiconducting or metallic behaviors, and the ferromagnetic, anti-ferromagnetic or non-configurations by examining the band structures, net magnetic moments and spin-split densities of states. Whether such results are reliable requires the careful checks. Apparently, the delicate calculations and analyses, being almost absent to date, should be very useful in fully understanding the featured essential properties, where they include the atom-dominated energy spectrum within the specific energy ranges, the spatial charge distributions, the atom- and orbital-created magnetic moments, the spin-up and spin-down arrangements, and the energy-dependent emergences of the atom-, orbital- and spin-created van Hove singularities. As for the intermediate crystal structures during the charging and discharging processes [25, 26], the dramatic transformation of geometric symmetries remains the mainstream issue. Dynamic Monte Carlo simulations have been utilized to comprehend the Li^+-ion release/intercalation of Li_xFePO_4 compounds [27–32]. The most important conclusion is the strong correlation between the optimal chemical environment and the Li^+-ion diffusion. How to characterize and examine the dynamic phenomena due to the large variation of Li^+ ions deserves systematic investigations, e.g., the x-diversified fundamental properties [33] and the significant roles of magnetic configurations on ion transports [34].

The Li_xFePO_4 compounds have become the commercial cathode products of Li^+-based batteries since the outstanding transport measurements by Goodenough and co-workers in 1997 [35]. Such four-components are thoroughly explored through the initial delicate growths for the different lithium concentrations [36], the x-ray diffractions due to the various crystal structures [7], and the stable current densities in the charging and discharging processes [5, 25]. Lithium iron phosphate compounds are successfully synthesized by the direct solid-state reaction among the stoichiometric amounts of Fe-acetate, ammonium phosphate, and lithium carbonate, covering Li_xFePO_4 of $x = 0.25$, 0.50, and 0.75 under an inert atmosphere. The accurate x-ray powder diffraction technique is available in identifying the distinct crystal structures [37, 38]. Very interestingly, the reversible release of Li^+-ions from $LiFePO_4$ and intercalation into $FePO_4$ is efficiently operated under 3.5 V by 0.05 mA cm^{-2}. This clearly illustrates an excellent cathode candidate for the cathode of a low-power and rechargeable lithium battery, being inexpensive, nontoxic, and environmentally benign. Furthermore, the specific capacity is 100 to 110 mAh g^{-1} even with the incomplete extraction and insertion. It is comparable to those of the theoretical predictions (~ 170 mAh g^{-1}; [39]). To greatly improve transport performance, tremendous efforts are focused on various experimental techniques in getting the most efficient Li_xFePO_4 compounds, such as direct chemical reactions in the presence of high-surface area carbon-black [40], mechanical activation for carbon-coated lithium iron phosphate [41, 42], one-pot solvothermal synthesis [43, 44], *in situ* combinatorial method [45], and *in situ* pyrolysis and catalytic graphitization [46]. Its essential properties can be measured by x-ray diffractometer [47], field emission scanning electron microscopy (FE-SEM; [48]), high-resolution transmission electron microscopy (HRTEM; [49]), selected area electron diffraction (SAED; [50]), and energy dispersive x-ray spectroscopy (EDX; [51]). Since the spin-created magnetic configurations are very active in transition-metal-relate compounds [52], the experimental examinations are required to clarify the diverse magneto-electronic properties. For example, the spin-polarized band structures of $LiFePO_4$ are detected by x-ray photoelectron spectroscopy (XPS; [53]).

According to first-principles simulations [18, 19], certain meta-stable crystal structures of four-component Li_xFePO_4 compounds (x = 1.0, 0.75, 0.5; 0.25) are chosen as model systems to clearly illustrate the diversified quasiparticle phenomena (the various charge and spin behaviors;[1]). High-resolution numerical evaluations are required for important steps (details in chapter 2), such as the sufficient conditions for the maximum plane-wave energies, the wave-vector points in the greatly reduced first Brillouin zone, and the almost vanishing net forces on each atom. All the useful physical quantities, which are utilized to achieve the concise mechanisms and pictures, cover the optimal geometries (numbers of distinct atoms, atomic positions and bond lengths within moiré superlattices), single-/multi-atom-dominated energy subbands with/without spin splittings at specific energy ranges, whole orbital hybridizations related to the spatial charge distributions, atom-, orbital- and spin-projected van Hove singularities due to the specific energy spectra/ranges, atom- and orbital-induced magnetic moments, and spin-up and spin-down arrangements [1, 2]. The diverse quasiparticle behaviors, which are driven by the variation of lithium concentration, are fully explored by delicate analyses, such as sensitive x-dependences of the fundamental properties. This will be very useful in further studies of intermediate crystal structures [16, 50]. Detailed comparisons with the ternary partners [10, 13, 22] and other cathodes [54–56] are thoroughly made in terms of the geometric, electronic and magnetic properties.

4.1 Stable/meta-stable crystal structures of Li_xFePO_4 compounds

Li_xFePO_4 (x = 1.0, 0.75, 0.5 and 025) can present rich and unique crystal structures, being achieved from the delicate VASP simulations and analyses. Such 3D materials possess moiré superlattice with many atoms in a unit cell so that the numerical calculations need to have highly precise evaluation during the optimal processes, especially for the requirement of an almost vanishing net force for each atom. To fully explore the spin-dependent magnetic configurations, GGA+U is utilized to deal with the single- and many-particle intrinsic interactions. This simulation method principally covers the frequent electron–ion scatterings due to the atom-dependent crystal potentials [57], the complicated exchange and correlation energies

Table 4.1. The details of VASP simulations.

VASP detail	Run Type: GGA+U # of K-points: 20×16×8 Pseudopotentials: Li_sv/Fe_pv/P/O Energy Cutoff: 520 eV U Values of Fe: 5.3 eV

arising from the spatial charge distribution, and the Hubbard-like on-site bi-particle ones associated with spin-up and spin-down states. The first and second terms are evaluated with well-characterized generalized gradient approximation (GGA) (details in chapter 2; [57]), and the third one is closely related to the maximum one-site Coulomb interaction. Pseudopotentials have been built for the periodical Li/Fe/P/O atoms. A significant value of U~5.4 eV is only created by the specific iron atoms (table 4.1) and vanishing in other non-transition-metal ones. A complete set of plane waves, with the cutoff kinetic energy of 520 eV, is reliable in fully exploring the featured wave packets of quasiparticle estates. Moreover, the wave-vector points, 20×16×8, are reliable to reach the relaxation status the non-zero force below 0.01 eV $Å^{-1}$.

The quaternary Li_xFePO_4 compounds present highly non-uniform physical and chemical environments, according to the delicate VASP results in tables 4.2(a) and (b) and figure 4.1. Such complicated systems belong to the orthorhombic lattices with the different lattice constants along three perpendicular axes and the point group of mmm (table 4.2(a)). Furthermore, they possess position-dependent and anisotropic chemical bonds, in which there are only stable oxidization bonds, namely, the Li–, Fe– and P–O bonds (table 4.2(b)). There are six, six and four neighboring oxygen atoms, respectively, in Li–, Fe– and P–O bonds. This is independent of x. However, the fluctuations of their bond lengths are significant and observable. In general, the featured crystal structures are very sensitive to the lithium concentration, as clearly illustrated in figures 4.1(a)–(d). For example, concerning $x = 1$, such material exhibits the lattice constant of 4.74/6.06/10.41 Å along $x/y/z$-axis, as well as the Li–/Fe–/P–O-range 2.098–2.203/2.087–2.219/1.536–1.572 Å. Each bond length directly reflects the binding strength of distinct orbitals (discussed again in charge density of figure 4.4). By the highly precise calculations, anti-ferromagnetic (AFM) and ferromagnetic (FM) spin arrangements have a rather small energy difference (~0.01%; details in table 4.2(b)). Which phase comes to exist should depend on the synthetic chemical environments. Their essential properties will be discussed thoroughly in this work. Very interestingly, the crystal structures are greatly distorted during the decrease of lithium atoms from $x = 1.0$ to 0.25, i.e., a large distortion, which is closely to the outmost 2S-orbital in Li–O bond, is revealed as a meta-stable state during the variation of Li-concentration. Among four kinds of crystals, $Li_{x=1}FePO_4$ displays the lowest ground state energy, and it grows from the decline of x (table 4.2(b)). The prominent x-dependence will lead to the dramatic transformation of the spatial charge density distribution due to the active atomic

Table 4.2. (a) The space-group symmetries and (b) optimal geometric parameters of Li_xFePO_4 compounds (x = 1.0, 0.75, 0.5; 0.25).

(a) Space group	
	Crystal system: orthorhombic Hermann Mauguin: Pnma [62] Hall: -P 2ac 2n Point Group: mmm

(b)	Cell Volume (Å^3)	Li–O (Å)	Fe–O (Å)	Po-O (Å)	# of Li–O	# of Fe–O	# of P–O	Magnetic Moment (μB)	Ground state energy (eV)
$Li_{1.00}FePO_4$ (AFM)	298.868	2.094–2.202	2.086–2.219	1.536–1.573	24	24	16	0 (AFM)	−191.811
$Li_{1.00}FePO_4$ (FM)	298.910	2.098–2.203	2.087–2.219	1.536–1.572	24	24	16	15.736	−191.803
$Li_{0.75}FePO_4$	295.889	2.112–2.242	1.991–2.268	1.522–1.587	18	24	16	16.731	−185.601
$Li_{0.50}FePO_4$	291.822	2.132–2.207	2.005–2.198	1.531–1.590	12	24	16	17.755	−179.774
$Li_{0.25}FePO_4$	287.549	2.079–2.270	1.903–2.218	1.524–1.590	6	24	16	18.703	−174.592

	Lattice cont. (Å)	# of Li	# of Fe	# of P	# of O	# of Li–O	# of Fe–O	# of P–O per P
$Li_{1.00}FePO_4$ (AFM)	4.74/6.06/10.41	4	4	4	16	6	6	4
$Li_{1.00}FePO_4$ (FM)	4.74/6.06/10.41	4	4	4	16	6	6	4
$Li_{0.75}FePO_4$	4.77/6.02/10.28	3	4	4	16	6	6	4
$Li_{0.50}FePO_4$	4.80/5.97/10.17	2	4	4	16	6	6	4
$Li_{0.25}FePO_4$	4.85/5.93/10.01	1	4	4	16	6	6	4

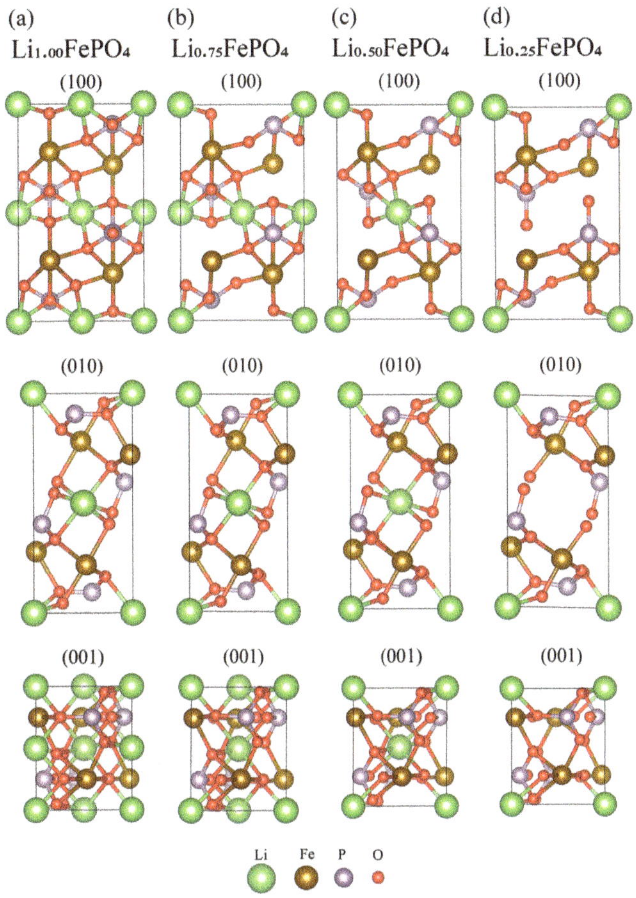

Figure 4.1. The optimal crystal structures of quaternary Li_xFePO_4 compounds: (a) $x = 1.0$, (b) 0.75, (c) 0.5 and (d) 025 projection along the [100], [010] and [001] planes.

orbitals in chemical bonds and thus determines the diversified quasiparticle properties. Whether the other x-related intermediate states could present similar lattice symmetries [34], and how to fully explore the dynamic transformation among them deserve systematic examinations [58–61].

The specific crystal structures of Li_xFePO_4 compounds ($x = 1$, 0.75, 0.5 and 0.25), corresponding to charging and discharging processes in Li^+-base batteries have been verified by the x-ray diffraction patterns (details in chapter 3 section 3.1; [7]). The high-resolution measurements have provided the constructive diffraction peaks, strongly depending on the incident wave length and angles [7]. The reciprocal lattice vectors (the optimal crystals) are directly identified from the observed x-ray spectrum. The effective elastic x-ray scatterings are dominated by the whole charge density in a large unit cell due to the position-dependent active orbitals. Whether this most important information could be estimated from the available data, as done for NaCl [62], is worthy of further experimental analyses. On the theoretical side, the

VASP simulations are capable of presenting reliable carrier distributions. How to link with the well-known x-ray scattering theory seems to be absent to date. Of course, the tight-binding model might be available in predicting the featured diffraction peaks about their intensity, angle, form and number, while the various hopping integrals and magnetic interactions are suitable within a moiré superlattice (discussed later). In short, the Li^+-ion extractions and insertions through the chemical reactions clearly illustrate the diversified crystal symmetries. Both quaternary and ternary transition-metal lithium oxides should be outstanding candidates for framework development of theoretical models and updated experimental progress.

4.2 Rich electronic properties

Very interestingly, quaternary Li_xFePO_4 compounds ($x = 1.0$, 0.75, 0.5 and 025), as shown in figures 4.2(a)–(d), display the unusual electronic energy spectra and wave functions. Their main features cover strong wave vector dependences along the high-symmetry points of the highly reduced first Brillouin zone [63–65], a lot of valence and conduction energy subbands due to Li/Fe/P/O atoms in a specific unit cell and their active orbitals (figures 4.1(a)–(d)), the highly asymmetry of valence hole and conduction electron band structure about the Fermi level ($E_F = 0$) under the

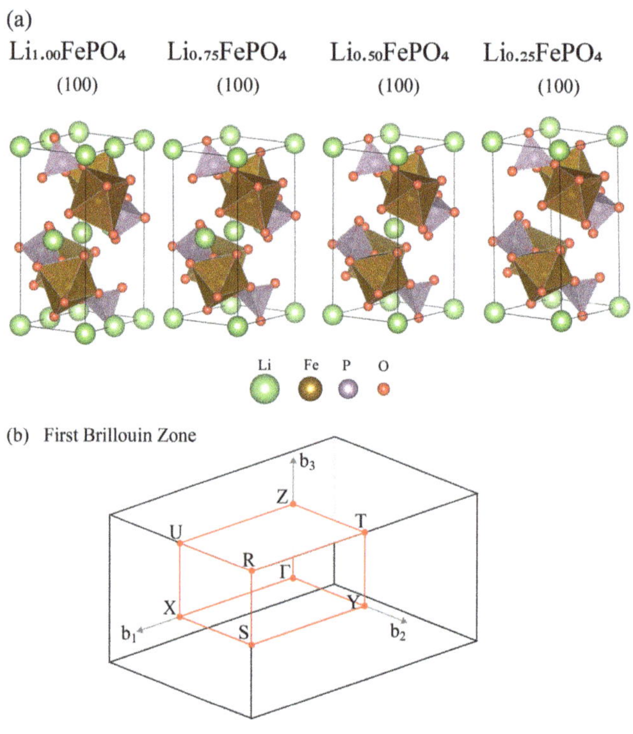

Figure 4.2. (a) The 3D orthorhombic structure of Li_xFePO_4 ($x = 1.0$, 0.75, 0.5 and 025) compounds and (b) its greatly reduced first Brillouin zone.

complicated intrinsic interactions, distinct atom dominances at separate energy ranges (indication by the radii of colored circles), spin-split/spin-degenerate ferromagnetic/anti-ferromagnetic electronic states [53], high anisotropy and non-monotonous energy dispersions closely related to the large non-uniformity of chemical environment, plenty of band-edge states (many critical points in energy-wave-vector space), and metallic/semiconducting behavior in the absence/presence of band gap. However, direct experimental examinations by high-resolution ARPES (details in chapter 3 section 3.3; [66]) are absent to date. The main reason lies in the complicated energy dispersions, many of them, and the important barrier in well characterizing k_z along the z-direction. All the diversified features will be delicately analyzed to get the most important quasiparticle information.

The close cooperation/competition between quasiparticle charge distributions and spin configurations can diversify the physical/chemical/material phenomena (the latter will be emphasized in the next section). Band structures and wave functions are easily modulated by the lithium concentrations, as clearly illustrated in figures 4.3(a)–(e). The number of occupied energy subbands below the Fermi level is equal to that of the total active orbitals in Li–/Fe–/P–O bonds (figure 4.4). Under the $x = 1$ case, there exist anti-ferromagnetic or ferromagnetic band structure, with almost the same ground state energies (table 4.2(b)). The former and the latter, respectively, exhibit spin-degenerate (figure 4.3(a)) and spin-spit (figure 4.3(b)) electronic states. Furthermore, the occupied charge density is dominated by the spin-up configurations, that is, there are more occupied spin-up valence bands (the blue curves). This is consistent with magnetic moments in table 4.3 and spin density (discussed later). Both of them belong to the wide-gap semiconductors. The AFM system presents the highest occupied state/the lowest unoccupied one (HOS/LUS) at the Γ/R point with an indirect gap of $E_g^i = 4.358$ eV. But for the FM one, HOS and LUS are, respectively, associated with the spin-down states at the Γ and R points, leading to $E_g^i = 4.318$ eV. It should be noticed that the wave-vector dependence is very weak for the valence and conduction subbands nearest to the Fermi level (consistent with delta-function-like density of states in figures 4.5(a) and 4.5(f)). The spin-created features are dramatically transformed during the variation of x. Most importantly, the low-lying valence and conduction bands are totally changed in the presence of the unusual semiconductor–metal transition. Only the metallic behavior comes to exist, in which it is characterized by certain single-split conduction subbands across the Fermi level. Some valence charges become the free conduction carriers. As to $x = 0.75$, 0.50 and 0.25 cases (figure 4.3(e)), the number of spin-up conduction subbands, crossing E_F, is 2, 4 and 2, respectively. Moreover, there exists an obvious energy spacing above/below E_F, covering $0<E^c<2.0$ eV / -0.5 eV$<E^v<0$, $0<E^c<2.5$ eV / -1.25 eV$<E^v<0$ eV, and $0<E^c<1.30$ eV / -1.30 eV$<E^v<0$ respectively, for $x = 0.75$, 0.50, and 0.25. As a result, the asymmetry of hole and electron spectra are very sensitive to the change of x. These clearly indicate the strong cooperation/competition due to the orbital hybridization and the spin-dependent many-particle interactions [1].

Electronic wave functions, which are characterized by wave-vector-dependent probability distributions, are able to provide information about atom dominances in the different energy ranges and thus will be very useful for supporting active orbital

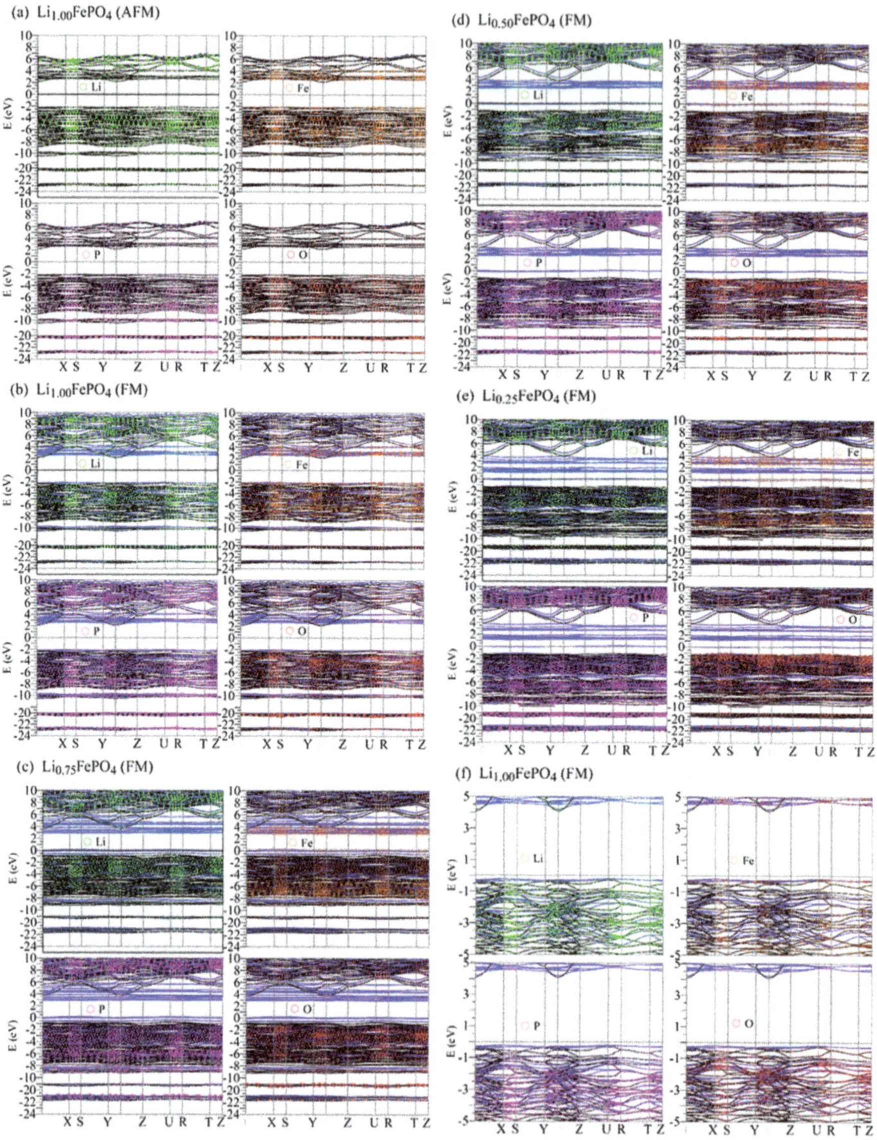

Figure 4.3. Electronic energy spectra for quaternary Li_xFePO_4 compounds: (a) $x = 1$ and AFM, (b) $x = 1$ and FM, (c) $0.75 = 1$ and FM, (d) $x = 0.5$ and FM, and (e) $x = 0.25$ and FM. (f) shows the low-lying bands of FM metals. Spin-up and spin-down states are, respectively, represented by the blue and solid curves. Furthermore, four kinds of atom dominances are revealed by the colored circles in each energy subband.

hybridizations of chemical bonds. The contribution percentage due to Li, Fe, P and O atoms are, respectively, represented by the radii of green, orange, pink and red balls. It is observed that lithium atoms only make a minor contribution for the whole energy range, but play a very significant role (the enhanced green balls). This is closely related to the only active Li-2s orbitals (orbital-projected density of states

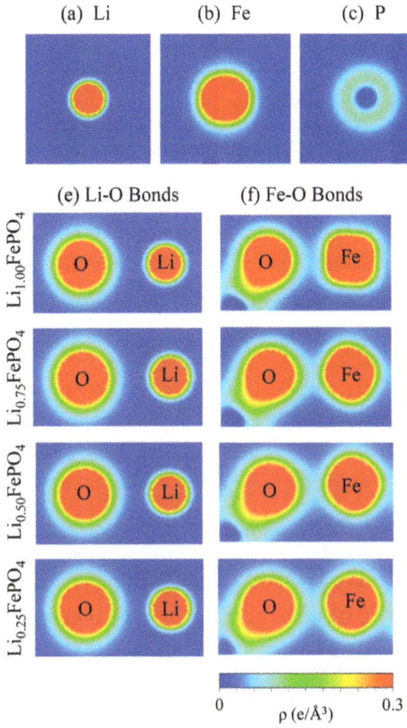

Figure 4.4. The spatial charge density distributions near the isolated (a) Li, (b) Fe, (c) P and (d) O atoms and the chemical bonds of (e) Li–O, (f) Fe–O and (g) P–O, in which the latter also cover the $x = 1$, 0.75, 0.5 and 0.25 cases.

(DOS) in figure 4.5). The easy modulations of their concentrations, which occur during charging and discharging processes, drastically change the chemical environment and then greatly diversify the various phenomena. Very interestingly, most of the contributions mainly arise from O, Fe and P atoms, especially for the two former. The evaluation of energy-dependent atom dominances is based on band structure (figure 4.3) and atom-decomposed DOS (figure 4.5), simultaneously. Roughly speaking, there are four/five specific energy ranges in semiconducting/metallic systems. As to $x = 1$ cases, both AFM and FM present the similar four regions: (I) -10 eV$<E^v<-6$ eV, (II) -6 eV$<E^v<-2.15$ eV, (III) 2.15 eV$<E^c<4.5$ eV and (IV) 4.5 eV$<E^c<7$ eV. In general, such categories are, respectively, characterized by the dominating O/comparable Fe and P, the dominant O/observable Fe/very small P, the dominating Fe/minor O/negligible P, and the same order for three kinds of atoms. Specifically, the highest/lowest II-/III-regions, corresponding to the partially flat valence/conduction subbands are predominated by Fe atoms (figure 4.3(a) and 4.3(b)). Obviously, the lithium concentration only slightly modifies the atom-projected contributions (figures 4.3(c)–(f)). Most importantly, the decrease of x leads to the Fe-dominated conduction and valence near the Fermi level; that is,

Table 4.3. The atom- and orbital-decomposed magnetic moments in quaternary Li_xFePO_4 compounds: (a) $x = 1$ and AM, (b) $x = 1$ and AFM, (c) 0.75 = 1 and FM, (d) $x = 0.5$ and FM, and (e) $x = 0.25$ and FM.

(a)		$Li_{1.00}FePO_4$ (FM)		
Atom	s	p	d	Total
Li1	0.000	0.000	0.000	0.000
Li2	0.000	0.000	0.000	0.000
Li3	0.000	0.000	0.000	0.000
Li4	0.000	0.000	0.000	0.000
Fe1	0.021	0.022	3.779	3.822
Fe2	0.021	0.022	3.779	3.822
Fe3	0.021	0.022	3.779	3.822
Fe4	0.021	0.022	3.779	3.822
P1	0.002	0.006	0.000	0.008
P2	0.002	0.006	0.000	0.008
P3	0.002	0.006	0.000	0.008
P4	0.002	0.006	0.000	0.008
O1	0.003	0.013	0.000	0.016
O2	0.005	0.029	0.000	0.034
O3	0.005	0.029	0.000	0.034
O4	0.003	0.013	0.000	0.016
O5	0.003	0.017	0.000	0.020
O6	0.003	0.013	0.000	0.016
O7	0.005	0.029	0.000	0.034
O8	0.005	0.029	0.000	0.034
O9	0.005	0.029	0.000	0.034
O10	0.005	0.029	0.000	0.035
O11	0.005	0.029	0.000	0.034
O12	0.003	0.017	0.000	0.020
O13	0.003	0.017	0.000	0.020
O14	0.003	0.013	0.000	0.016
O15	0.005	0.029	0.000	0.035
O16	0.003	0.017	0.000	0.020
Total	0.155	0.464	15.117	15.736

(b)		$Li_{1.00}FePO_4$ (AFM)		
Atom	s	p	d	Total
Li1	0.000	0.000	0.000	0.000
Li2	0.000	0.000	0.000	0.000
Li3	0.000	0.000	0.000	0.000
Li4	0.000	0.000	0.000	0.000
Fe1	0.021	0.021	3.778	3.820
Fe2	0.021	0.021	3.778	3.820

(*Continued*)

Table 4.3. (*Continued*)

(b)		$Li_{1.00}FePO_4$ (AFM)		
Fe3	−0.021	−0.021	−3.778	−3.820
Fe4	−0.021	−0.021	−3.778	−3.820
P1	−0.001	−0.004	0.000	−0.005
P2	−0.001	−0.004	0.000	−0.006
P3	0.001	0.004	0.000	0.006
P4	0.001	0.004	0.000	0.005
O1	−0.003	−0.012	0.000	−0.015
O2	0.005	0.027	0.000	0.032
O3	0.005	0.027	0.000	0.032
O4	0.003	0.012	0.000	0.015
O5	−0.003	−0.008	0.000	−0.011
O6	−0.003	−0.012	0.000	−0.015
O7	−0.005	−0.027	0.000	−0.032
O8	0.005	0.027	0.000	0.032
O9	−0.005	−0.027	0.000	−0.032
O10	−0.005	−0.027	0.000	−0.032
O11	−0.005	−0.027	0.000	−0.032
O12	−0.003	−0.009	0.000	−0.011
O13	0.003	0.009	0.000	0.011
O14	0.003	0.012	0.000	0.015
O15	0.005	0.027	0.000	0.032
O16	0.003	0.008	0.000	0.011
Total	0.000	0.000	0.000	0.000

(c)		$Li_{0.75}FePO_4$ (FM)		
Atom	s	p	d	Total
Li1	x	x	x	x
Li2	0.000	0.000	0.000	0.001
Li3	0.000	0.000	0.000	0.000
Li4	0.000	0.000	0.000	0.000
Fe1	0.022	0.025	4.060	4.108
Fe2	0.021	0.022	3.802	3.844
Fe3	0.021	0.022	3.802	3.844
Fe4	0.022	0.025	4.060	4.108
P1	0.002	0.007	0.000	0.009
P2	0.002	0.006	0.000	0.008
P3	0.002	0.006	0.000	0.008
P4	0.002	0.007	0.000	0.009
O1	0.003	0.022	0.000	0.025
O2	0.008	0.084	0.000	0.092
O3	0.007	0.055	0.000	0.061
O4	0.003	0.026	0.000	0.030

O5	0.005	0.072	0.000	0.077
O6	0.003	0.026	0.000	0.030
O7	0.007	0.055	0.000	0.061
O8	0.006	0.041	0.000	0.046
O9	0.006	0.041	0.000	0.046
O10	0.006	0.039	0.000	0.044
O11	0.008	0.084	0.000	0.092
O12	0.003	0.017	0.000	0.020
O13	0.003	0.017	0.000	0.020
O14	0.003	0.022	0.000	0.025
O15	0.006	0.039	0.000	0.044
O16	0.005	0.072	0.000	0.077
Total	0.174	0.833	15.723	16.730

(d)		$Li_{0.50}FePO_4$ (FM)		
Atom	s	p	d	Total
Li1	x	x	x	x
Li2	x	x	x	x
Li3	0.000	0.000	0.000	0.001
Li4	0.000	0.000	0.000	0.001
Fe1	0.021	0.024	4.108	4.153
Fe2	0.021	0.024	4.113	4.159
Fe3	0.021	0.024	4.113	4.159
Fe4	0.021	0.024	4.108	4.153
P1	0.002	0.008	0.000	0.011
P2	0.002	0.008	0.000	0.011
P3	0.002	0.008	0.000	0.011
P4	0.002	0.008	0.000	0.011
O1	0.004	0.040	0.000	0.044
O2	0.008	0.082	0.000	0.090
O3	0.007	0.072	0.000	0.079
O4	0.004	0.040	0.000	0.044
O5	0.005	0.054	0.000	0.059
O6	0.004	0.040	0.000	0.044
O7	0.007	0.072	0.000	0.079
O8	0.007	0.072	0.000	0.079
O9	0.007	0.072	0.000	0.079
O10	0.008	0.082	0.000	0.090
O11	0.008	0.082	0.000	0.090
O12	0.005	0.054	0.000	0.059
O13	0.005	0.054	0.000	0.059
O14	0.004	0.039	0.000	0.044
O15	0.008	0.082	0.000	0.090
O16	0.005	0.054	0.000	0.059
Total	0.193	1.119	16.443	17.755

(e)		$Li_{0.25}FePO_4$ (FM)		
Atom	s	p	d	Total
Li1	x	x	x	x
Li2	x	x	x	x
Li3	x	x	x	x
Li4	0.000	0.000	0.000	0.001
Fe1	0.022	0.026	4.126	4.173
Fe2	0.021	0.027	4.348	4.396
Fe3	0.021	0.027	4.348	4.396
Fe4	0.022	0.026	4.126	4.173
P1	0.002	0.008	0.000	0.010
P2	0.004	0.013	0.000	0.016
P3	0.004	0.013	0.000	0.016
P4	0.002	0.008	0.000	0.010
O1	0.005	0.060	0.000	0.065
O2	0.009	0.101	0.000	0.110
O3	0.008	0.093	0.000	0.101
O4	0.005	0.061	0.000	0.066
O5	0.004	0.049	0.000	0.053
O6	0.005	0.061	0.000	0.066
O7	0.008	0.093	0.000	0.101
O8	0.010	0.113	0.000	0.123
O9	0.010	0.113	0.000	0.123
O10	0.009	0.110	0.000	0.119
O11	0.009	0.101	0.000	0.110
O12	0.007	0.111	0.000	0.118
O13	0.007	0.111	0.000	0.118
O14	0.005	0.060	0.000	0.065
O15	0.009	0.110	0.000	0.119
O16	0.004	0.049	0.000	0.053
Total	0.215	1.541	16.947	18.703

it can create the fifth region: (V) -2.1 eV$<E^v/E^c<$0.2 eV. Also, oxygen atoms make certain contributions to the metallic behavior.

A whole real-space distribution of charge density, as done for the current lithium oxides, is very useful in comprehending the critical orbital hybridizations in various oxidation bonds [2]. The VASP-based simulations can provide the total charge densities of Li_xFePO_4 (figures 4.4(e)–(g)) and those of isolated Li/Fe/P/O atoms (figures 4.4(a)–(d)). The dramatic transformations of electron clouds come to exist everywhere, especially for the neighboring regions. The calculated results are

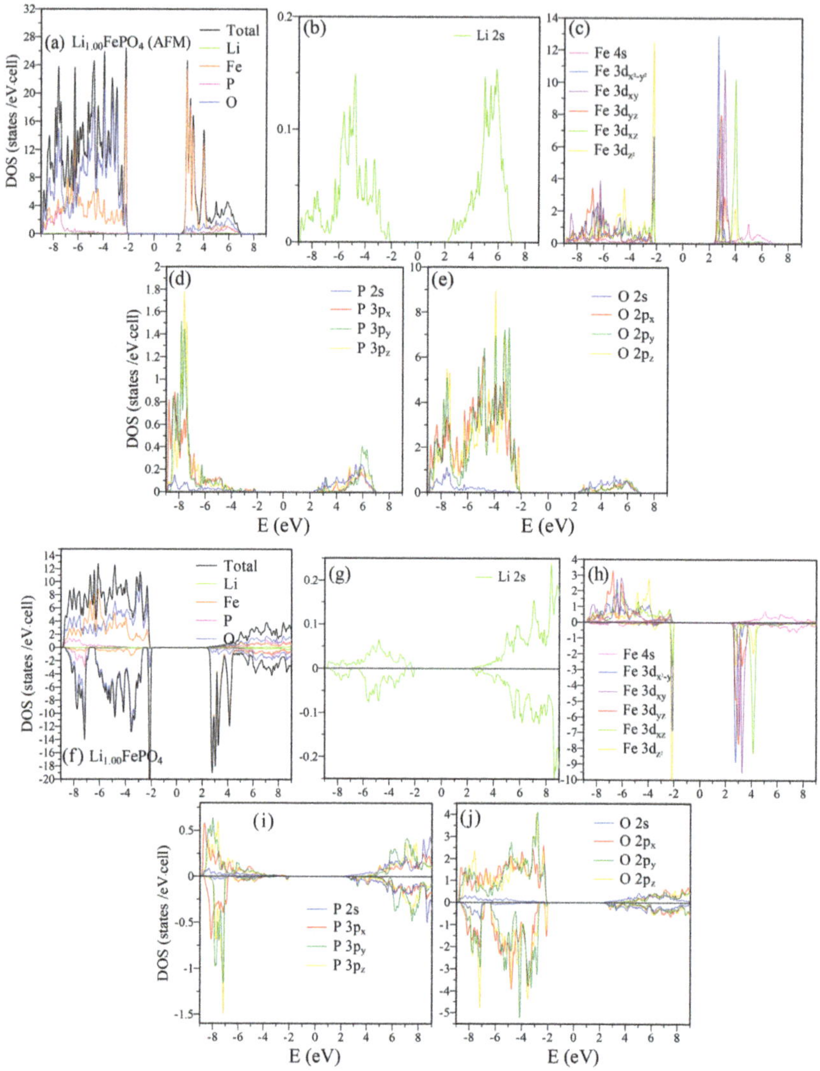

Figure 4.5. The atom-/orbital-/spin-decomposed density of states under the various cases: (a)/(b)/(c)/(d)/(e) AFM and $x = 1$, (f)/(g)/(h)/(i)/(j) FM and $x = 1$ and, (k)/(l)/(m)/(n)/(o) FM and $x = 0.75$, (p)/(q)/(r)/(s)/(t) FM and $x = 0.5$, and (u)/(v)/(w)/(x)/(y) FM and $x = 0.25$.

thoroughly analyzed to extract the significant orbital hybridizations through the delicate comparisons between the real and isolated cases. As a result, there exist a lot of multi-orbital mixings, namely 2s-[2s, 2p*x*, 2p*y*, 2p*z*], [4s, $3d_{x2\text{-}y2}$,$3d_{z2}$, $3d_{xy}$, $3d_{yz}$, $3d_{xz}$]-[2s, 2p*x*, 2p*y*, 2p*z*], and [3s, 3p*x*, 3p*y*, 3p*z*]-[2s, 2p*x*, 2p*y*, 2p*z*] in Li–, Fe– and P–O bonds, respectively. These results are estimated from the obvious changes of different-color regions, since the spatial charge density distributions are isotropic and anisotropic forms. Furthermore, they principally cover the green, green/red, green/red, green/red regions near the lithium, iron, phosphorus, and oxygen atoms,

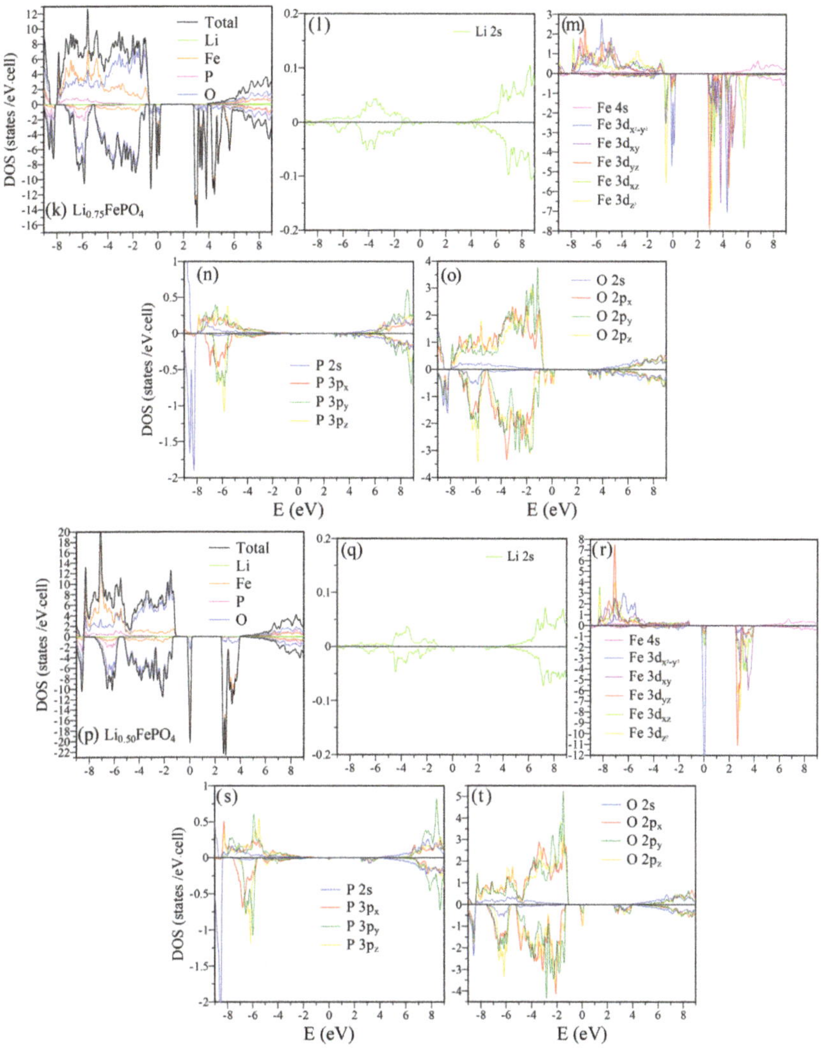

Figure 4.5. (Continued.)

repectively. It should be noticed that three-p/five-d orbital in [P, O]/Fe atoms cannot be distinguished from one another, i.e., they might have the same contributions. Consequently, the π and σ bonding, respectively, due to the p_z and [2s, 2px, 2py, 2pz] orbitals totally disappear in Li_xFePO_4 When the active multi-orbital hybridizations are combined with the orbital-decomposed van Hove singularities, their detailed energy dependences could be clarified in further analyses (figure 4.5). How to utilize the total charge density within a unit cell in further verifying a specific crystal structures, or by the x-ray diffraction peaks in obtaining the spatial carrier distribution has become an open issue. In addition, the predicted charge

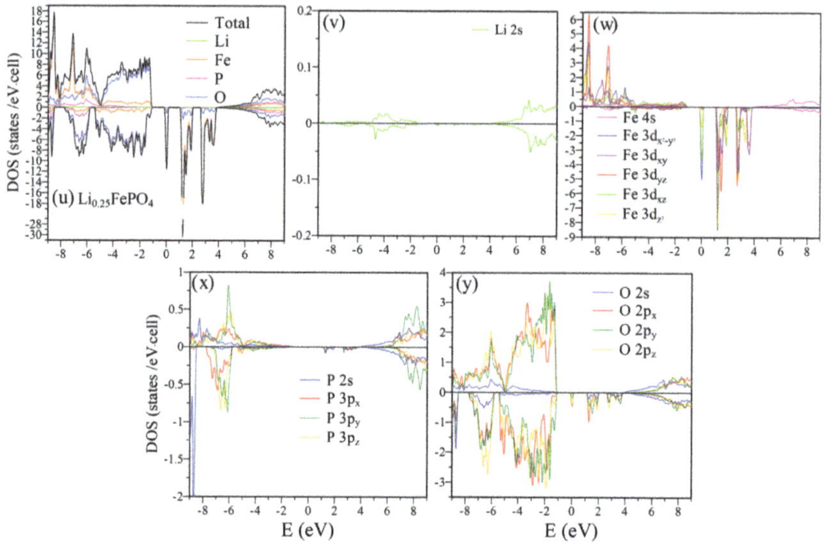

Figure 4.5. (Continued.)

distributions in a unit cell could be verified by the further analyses of the x-ray measurements [37, 38].

DOS, which is characterized as the number of electronic states between an infinite small energy range of $E \rightarrow E+dE$, is able to comprehend the E-dependent features across a whole band structure. It covers the occupied and unoccupied energy spectrum across the Fermi level ($E<0$ and $E>0$, details in chapter 4 section 4.2). However, the ARPES examinations are only suitable for the former (section 4.3). According to the standard calculations [67], DOS just corresponds to the inverse integration of group velocity on a specific constant-energy space, e.g., the closed surface/loop/segment/discrete point for 3D/2D/1D/0D condensed-matter systems. It creates the unusual van Hove singularities under the vanishing/singular first derivative of state energy versus wave vector. Apparently, their form, number, energy and intensity are mainly determined by the various band-edge states and distinct dimensionalities. Very interestingly, such special states belong to the critical ones in the energy-wave-vector space, in which they include the local maxima/minima, saddle points, constant-energy loops, and partially flat energy dispersions. As a result, there exist diverse structures of van Hove singularities. For example, as to 2D few-layer graphene systems, these special critical points are, respectively, the broadening shoulders, the logarithmically symmetric peaks, the asymmetric peaks in the square-root divergent form, and the delta-function-like peaks. Most importantly, the STS examinations (details in chapter 3 section 3.3), as done for 2D, 1D and 0D graphene-related systems [68–71], are responsible for the simultaneous verifications of the lattice symmetries and dominating electronic properties. That is, the high-resolution detections can determine the close relations between nanoscaled geometries and electronic energy spectra. However, this technique cannot measure the very weak tunneling quantum currents with the bulk features because of too thick samples.

The high-precision DOS calculations, which are mainly conducted on the atom-, orbital- and spin-decomposed contributions, are able to provide detailed information about the critical mechanisms of quasiparticle charges and spins (the concise physical/chemical/material pictures). How to unify the complicated projections in clearly identifying the significant orbital hybridizations and spin configurations within the specific energy ranges is the study focus. The featured van Hove singularities, being fully supported by the position-dependent chemical bonds (figures 4.1(a)–(d)), the band-edge states and atom dominances (figures 4.3(a)–(f)), and the spatial charge density distributions (figures 4.4(a)–(g)), are the most important direct evidence. Their prominent intensities are thoroughly examined from the dominating contributions of multi-orbitals in Li–/Fe–/P–O bonds. Furthermore, the emerged peaks due to few orbitals might survive within certain energy ranges. As a result, one can get the active orbitals in chemical bondings and their strengths.

The separate contributions due to Li, Fe, P, and O atoms are clearly illustrated in figure 4.5(a)/figure 4.5(f)/figure 4.5(k)/figure 4.5(p)/figure 4.5(u) under the cases of AFM and $x = 1$/FM and $x = 1$/FM and $x = 0.75$/ FM and $x = 0.5$/FM and $x = 0.25$. The various atom dominances could be roughly divided into four, four, five, five, five categories of energy ranges. This is consistent with the atom-dominated band structures in figures 4.3(a)–(f). The orbital- and spin-projected DOSs can further identify the significant orbital hybridizations. As to the ns orbitals, only those of lithium atoms make significant, but weak contributions in the whole energy ranges under the various cases. This is clearly indicated by the green curves in figure 4.5(b) of AFM and $x = 1$, figure 4.5(g) of FM and $x = 1$, figure 4.5(l) of FM and $x = 0.75$ figure 4.5(q) of $x = 0.5$ and figure 4.5(u) of $x = 0.25$. However, 4s/3s/2s of Fe/P/O atoms reveal their important weights at the deeper and higher energies, being roughly within/near (I) and IV regions through the blue curves in (c)/(d)/(e) AFM and $x = 1$, (h)/(i)/(j) FM and $x = 1$ and, (m)/(n)/(o) FM and $x = 0.75$, (r)/(s)/(t) FM and $x = 0.5$, and (v)/(w)/(x)/(y) of $x = 0{,}25$. Very interestingly, five/three kinds of d/p orbitals due to Fe/[P, O] atoms (clearly illustrated by the different colors) make the comparable contributions to the intensity and form of the same van Hove singularities. Such results indicate they possess the almost identical chemical bondings, being also confirmed from the charge density distributions (figures 4.4(a)–(g)). For example, II-, III- and V-categories are dominated by the multi-orbital hybridizations [$3d_{x2\text{-}y2}$,$3d_{z2}$, $3d_{xy}$, $3d_{yz}$, $3d_{xz}$]- [2s, $2px$, $2py$, $2pz$] in Fe–O bonds However, the other two categories become more complicated by adding 4s and 2s orbitals from Fe and O atoms, respectively. In short, the above-mentioned features can determine the energy-dependent orbital hybridizations,

4.3 Unusual magnetic configurations

Another quasiparticle characteristic, spin configuration, is thoroughly examined from the spin-degenerate or spin-split energy bands (figures 4.3(a)–(f)), the spin-decomposed van Hove singularities (figures 4.5(a)–(y)), the atom- and orbital-induced magnetic moments (tables 4.3(a)–(f)), and the spin density distributions (figures 4.6(a)–(e)). The first and second properties clearly illustrate that the

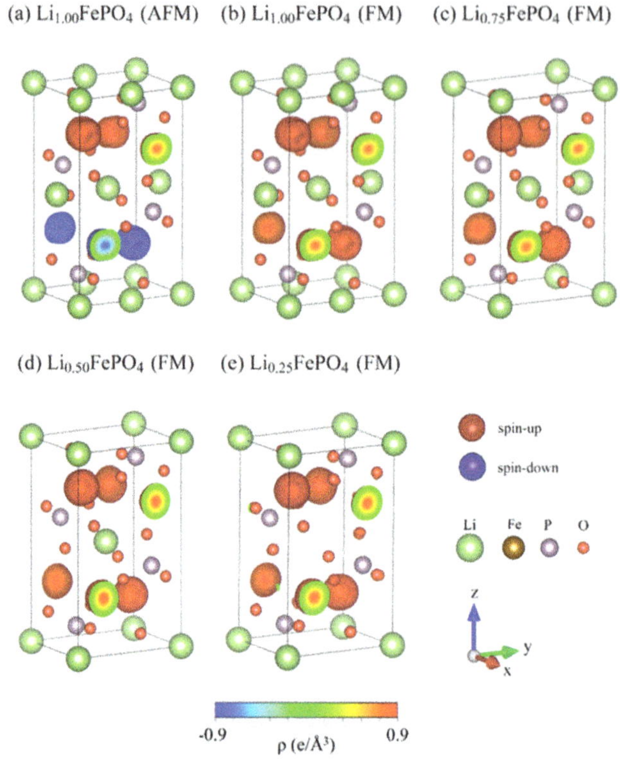

Figure 4.6. The spin density arrangements of quaternary Li_xFePO_4 compounds: (a) $x = 1$ and AFM, (b) $x = 1$ and FM, (c) 0.75 = 1 and FM, (d) $x = 0.5$ and FM, and (e) $x = 0.25$ and FM. Also shown are those of the enhanced results near P and O atoms.

spin-related magnetic properties are directly reflected in a whole energy range. Only lithium atoms under any x cases do not present evidence of spin arrangement, being determined from the cancellation of spin-up and spin-down DOS (the green curves in figure 4.3(b), 4.3(g), 4.3(l), 4.3(q) and 4.3(v)) and the fully vanishing magnetic moment (the first four rows in tables 4.3(a)–(f)). Among the active orbitals of Fe, P and O atoms, the five kinds of 3d orbitals make the most important contributions to the net magnetic moments (about 3.339 μB for each Fe atom). This clearly indicates their dominances on the anti-ferromagnetic or ferromagnetic configurations. The other s and p orbitals reveal the weak weights in the spin-induced magnetic properties (roughly 1% of that 3d orbitals). Such featured magnetic moments in table 4.3 are fully consistent with the spatial spin arrangements. The $x = 1$ AFM displays two giant spin-up (red colors in figure 4.6(a)) and spin-down (blue colors) distributions localized about four Fe atoms, while the other FM systems only present the dominating spin-up configurations (figures 4.6(b)–(e)).

The atom-, orbital- and spin-created diverse phenomena is worthy of closer examination. AFM of $x = 1$ displays the fully vanishing magnetic moments (table 4.3(a)), since Fe, P and O atoms at the upper and lower planes just make the opposite contributions in the spin density distributions (figure 4.6(a)). This reflects in the spin-degenerate band structure (figure 4.3(a)) and density of states (figures 4.5(a)–(e)). But for the FM systems under various x, all the active orbitals of Fe/P/O atoms create the positive magnetic moments (tables 4.3(b)–(e)), being consistent with the spin-up-dominated density arrangements (figures 4.6(b)–(e)). Furthermore, both spin-up and spin-down DOSs due to Fe/P/O atoms and their significant orbitals cover the quite different area for the occupied/unoccupied region of $E<0/E>0$ (figures 4.6(h)–(j), (m)–(o), (r)–(t), and (w)–(y)), in which the former dominates the valence energy spectrum. It should be noticed that the spin-down van Hove singularities (the strong orange peaks of Fe in the lower plane) dominate the low-lying near or across the Fermi level. That is, the metallic or semiconducting transport behavior is related to the spin-down states, while the magnetic configuration mainly arises from the occupied spin-up ones. This unusual phenomenon is expected to create the unique magnetic properties.

In summary, the detailed analyses of the different physical and chemical quantities can clearly identify the energy-dependent orbital hybridizations and spin configurations. Li-, Fe- and P-bonds, respectively, possess the active multi-orbital hybridizations of 2s-[2s, 2px, 2py, 2pz], [4s, $3d_{x2\text{-}y2}$,$3d_{z2}$, $3d_{xy}$, $3d_{yz}$, $3d_{xz}$]- [2s, 2px, 2py, 2pz], and [3s, 3px, 3py, 3pz]-[2s, 2px, 2py, 2pz]. The various p and d orbitals dominate [II], [III], and [V] regions, corresponding to the valence and conduction energy spectra near or across the Fermi level. Furthermore, Fe/[P, O] atoms make the most important/minor contributions to the ferromagnetic configurations, since five kinds of d orbitals in [I] and [II] regions present the spin-up dominances. However, the metallic behavior is characterized by the spin-down energy bands crossing E_F.

References

[1] Lin S-Y, Tran N T T, Chang S-L, Su W-P and Lin M-F 2018 *Structure-and Adatom-enriched Essential Properties of Graphene Nanoribbons* (Boca Raton, FL: CRC Press)

[2] Tran N T T, Lin S Y, Lin C Y and Lin M F 2017 *Geometric and Electronic Properties of Graphene-Related Systems: Chemical Bondings* (Boca Raton, FL: CRC Press)

[3] Lin C-Y, Chen R-B, Ho Y-H and Lin M-F 2017 *Electronic and Optical Properties of Graphite-Related Systems* (Boca Raton, FL: CRC Press)

[4] Lin C Y, Do T N, Huang Y K and Lin M F 2017 *Optical Properties of Graphene in Magnetic and Electric Fields* (Bristol: IOP Publishing) e-book

[5] Kang B and Ceder G 2009 Battery materials for ultrafast charging and discharging *Nature* **458** 190–3

[6] Malik R, Zhou F and Ceder G 2011 Kinetics of non-equilibrium lithium incorporation in $LiFePO_4$ *Nat. Mater.* **10** 587–90

[7] Andersson A S, Kalska B, Häggström L and Thomas J O 2000 Lithium extraction/insertion in $LiFePO_4$: an x-ray diffraction and Moessbauer spectroscopy study *Solid State Ionics* **130** 41–52

[8] Kanno R, Shirane T, Kawamoto Y, Takeda Y, Takano M, Ohashi M and Yamaguchi Y 1996 Synthesis, structure, and electrochemical properties of a new lithium iron oxide, $LiFeO_2$, with a corrugated layer structure *J. Electrochem. Soc.* **143** 2435

[9] Sakurai Y, Arai H, Okada S and Yamaki J I 1997 Low temperature synthesis and electrochemical characteristics of $LiFeO_2$ cathodes *J. Power Sources* **68** 711–5

[10] Ozawa K 1994 Lithium-ion rechargeable batteries with $LiCoO_2$ and carbon electrodes: the $LiCoO_2$/C system *Solid State Ionics* **69** 212–21

[11] Cho J, Kim Y J and Park B 2000 Novel $LiCoO_2$ cathode material with Al_2O_3 coating for a Li ion cell *Chem. Mater.* **12** 3788–91

[12] Ohzuku T, Ueda A and Nagayama M 1993 Electrochemistry and structural chemistry of $LiNiO_2$ (R3m) for 4 volt secondary lithium cells *J. Electrochem. Soc.* **140** 1862

[13] Dahn J R, von Sacken U, Juzkow M W and Al-Janaby H 1991 Rechargeable $LiNiO_2$/carbon cells *J. Electrochem. Soc.* **138** 2207

[14] Scrosati W A S B 2002 *Advances in Lithium-Ion Batteries* (Boston, MA: Springer)

[15] Chan C K, Peng H, Liu G, McIlwrath K, Zhang X F, Huggins R A and Cui Y 2008 High-performance lithium battery anodes using silicon nanowires *Nat. Nanotechnol.* **3** 31–5

[16] Wang L, He X, Sun W, Wang J, Li Y and Fan S 2012 Crystal orientation tuning of $LiFePO_4$ nanoplates for high rate lithium battery cathode materials *Nano Lett.* **12** 5632–6

[17] Lin C-Y, Ho C-H, Wu J-Y, Do T-N, Shih P-H, Lin S-Y and Lin M-F 2019 *Diverse Quantization Phenomena in Layered Materials* (Boca Raton, FL: CRC Press)

[18] Nakayama M, Yamada S, Jalem R and Kasuga T 2016 Density functional studies of olivine-type $LiFePO_4$ and $NaFePO_4$ as positive electrode materials for rechargeable lithium and sodium ion batteries *Solid State Ionics* **286** 40–4

[19] Ong S P, Wang L, Kang B and Ceder G 2008 Li-Fe-P-O_2 phase diagram from first principles calculations *Chem. Mater.* **20** 1798–807

[20] Ottakam Thotiyl M M, Freunberger S A, Peng Z, Chen Y, Liu Z and Bruce P G 2013 A stable cathode for the aprotic Li-O_2 battery *Nat. Mater.* **12** 1050–6

[21] Moreau P, Guyomard D, Gaubicher J and Boucher F 2010 Structure and stability of sodium intercalated phases in olivine $FePO_4$ *Chem. Mater.* **22** 4126–8

[22] Lara-García H A, Vera E, Mendoza-Nieto J A, Gómez-García J F, Duan Y and Pfeiffer H 2017 Bifunctional application of lithium ferrites (Li_5FeO_4 and $LiFeO_2$) during carbon monoxide (CO) oxidation and chemisorption processes. A catalytic, thermogravimetric and theoretical analysis *Chem. Eng. J.* **327** 783–91

[23] Ning F, Li S, Xu B and Ouyang C 2014 Strain tuned Li diffusion in $LiCoO_2$ material for Li ion batteries: a first principles study *Solid State Ionics* **263** 46–8

[24] Laubach S, Laubach S, Schmidt P C, Ensling D, Schmid S, Jaegermann W, Thißen A, Nikolowski K and Ehrenberg H 2009 Changes in the crystal and electronic structure of $LiCoO_2$ and $LiNiO_2$ upon Li intercalation and de-intercalation *Phys. Chem. Chem. Phys.* **11** 3278–89

[25] Chang H H, Chang C C, Wu H C, Yang M H, Sheu H S and Wu N L 2008 Study on dynamics of structural transformation during charge/discharge of $LiFePO_4$ cathode *Electrochem. Commun.* **10** 335–9

[26] O'Connor D T, Welland M J, Kam Liu W and Voorhees P W 2016 Phase transformation and fracture in single Li_xFePO_4 cathode particles: a phase-field approach to Li-ion intercalation and fracture *Model. Simul. Mater. Sci. Eng.* **24** 035020

[27] Xiao P and Henkelman G 2018 Kinetic Monte Carlo study of Li intercalation in $LiFePO_4$ *ACS Nano* **12** 844–51

[28] Ouyang C Y, Shi S Q, Wang Z X, Li H, Huang X J and Chen L Q 2004 The effect of Cr doping on Li ion diffusion in $LiFePO_4$ from first principles investigations and Monte Carlo simulations *J. Phys. Condens. Matter* **16** 2265

[29] Li J, Klee Barillas J, Guenther C and Danzer M A 2014 Sequential Monte Carlo filter for state estimation of $LiFePO_4$ batteries based on an online updated model *J. Power Sources* **247** 156–62

[30] Masrour R, Jabar A, Benyoussef A, Hamedoun M and Hlil E K 2015 Magnetic properties of $LiFePO_4$ compound: a Monte Carlo study *Chem. Phys. Lett.* **635** 268–72

[31] Harinipriya S, Diwakar V D and Subramanian V R 2008 Performance characteristics of cathode materials for Lithium-ion batteries: a Monte Carlo strategy *J. Electrochem. Soc.* **155** A875

[32] Zahid T, Qin F, Li W, Xu K and Zhou Y 2017 Sequential Monte Carlo based technique for SOC estimation of $LiFePO_4$ battery pack for electric vehicles 2016 *IEEE Int. Conf. on Information and Automation, IEEE ICIA 2016*

[33] Axmann P, Stinner C, Wohlfahrt-Mehrens M, Mauger A, Gendron F and Julien C M 2009 Nonstoichiometric $LiFePO_4$: Defects and related properties *Chem. Mater.* **21** 1636–44

[34] Islam M S, Driscoll D J, Fisher C A J and Slater P R 2005 Atomic-scale investigation of defects, dopants, and lithium transport in the $LiFePO_4$ olivine-type battery material *Chem. Mater.* **17** 5085–92

[35] Padhi A K, Nanjundaswamy K S and Goodenough J B 1997 Characterization of $LiFePO_4$ as the cathode material for rechargeable lithium batteries *J. Electrochem. Soc.* **144** 1188–94

[36] Janssen Y, Santhanagopalan D, Qian D, Chi M, Wang X, Hoffmann C, Meng Y S and Khalifah P G 2013 Reciprocal salt flux growth of $LiFePO_4$ single crystals with controlled defect concentrations *Chem. Mater.* **25** 4574–84

[37] Gregory N W 1957 Elements of x-ray diffraction *J. Am. Chem. Soc.* **79** 1773–4

[38] Cullity B D and Weymouth J W 1957 Elements of x-ray diffraction *Am. J. Phys.* **25** 394

[39] Yamada A, Chung S C and Hinokuma K 2001 Optimized $LiFePO_4$ for lithium battery cathodes *J. Electrochem. Soc.* **148** A224

[40] Lu C Z, Fey G T K and Kao H M 2009 Study of $LiFePO_4$ cathode materials coated with high surface area carbon *J. Power Sources* **189** 155–62

[41] Kim J K, Choi J W, Cheruvally G, Kim J U, Ahn J H, Cho G B, Kim K W and Ahn H J 2007 A modified mechanical activation synthesis for carbon-coated $LiFePO_4$ cathode in lithium batteries *Mater. Lett.* **61** 3822–5

[42] Kim J K, Cheruvally G, Choi J W, Kim J U, Ahn J H, Cho G B, Kim K W and Ahn H J 2007 Effect of mechanical activation process parameters on the properties of $LiFePO_4$ cathode material *J. Power Sources* **166** 211–8

[43] Rangappa D, Sone K, Kudo T and Honma I 2010 Directed growth of nanoarchitectured $LiFePO_4$ electrode by solvothermal synthesis and their cathode properties *J. Power Sources* **195** 6167–71

[44] Zheng Z, Pang W K, Tang X, Jia D, Huang Y and Guo Z 2015 Solvothermal synthesis and electrochemical performance of hollow $LiFePO_4$ nanoparticles *J. Alloys Compd.* **640** 95–100

[45] Teng F, Santhanagopalan S, Lemmens R, Geng X, Patel P and Meng D D 2010 *In situ* growth of $LiFePO_4$ nanorod arrays under hydrothermal condition *Solid State Sci.* **12** 952–5

[46] Li J, Zhang L, Zhang L, Hao W, Wang H, Qu Q and Zheng H 2014 In-situ growth of graphene decorations for high-performance $LiFePO_4$ cathode through solid-state reaction *J. Power Sources* **249** 311–9

[47] Orikasa Y *et al* 2013 Phase transition analysis between $LiFePO_4$ and $FePO_4$ by in-situ time-resolved x-ray absorption and x-ray diffraction *J. Electrochem. Soc.* **160** A3061
[48] Abd Mutalib M, Rahman M A, Othman M H D, Ismail A F and Jaafar J 2017 Scanning electron microscopy (SEM) and energy-dispersive x-ray (EDX) spectroscopy *Membrane Characterization* (Amsterdam: Elsevier)
[49] Gabrisch H, Wilcox J and Doeff M M 2008 TEM study of fracturing in spherical and plate-like $LiFePO_4$ particles *Electrochem. Solid-State Lett.* **11** A25
[50] Chen G, Song X and Richardson T J 2006 Electron microscopy study of the $LiFePO_4$ to $FePO_4$ phase transition *Electrochem. Solid-State Lett.* **9** A295
[51] Liang G, Croft M and Zhong Z 2013 *In Situ* energy dispersive x-ray diffraction study of prototype $LiMnO_4$- and $LiFePO_4$-based coin cell batteries *ECS Trans.* **50** 293
[52] Chowdhury S, Majumdar A and Jana D 2017 Search for magnetism in transition metal atoms doped tetragonal graphene: a DFT approach *J. Magn. Magn. Mater.* **441** 523–30
[53] Castro L, Dedryvère R, El Khalifi M, Lippens P E, Bréger J, Tessier C and Gonbeau D 2010 The spin-polarized electronic structure of $LiFePO_4$ and $FePO_4$ evidenced by in-lab XPS *J. Phys. Chem.* C **114** 17995–8000
[54] Whittingham M S 2004 Lithium batteries and cathode materials *Chem. Rev.* **104** 4271–302
[55] Fergus J W 2010 Recent developments in cathode materials for lithium ion batteries *J. Power Sources* **195** 939–54
[56] Julien C M, Mauger A, Zaghib K and Groult H 2014 Comparative issues of cathode materials for Li-ion batteries *Inorganics* **2** 132–54
[57] Perdew J P, Burke K and Ernzerhof M 1996 Generalized gradient approximation made simple *Phys. Rev. Lett.* **77** 3865
[58] Liu H, Strobridge F C, Borkiewicz O J, Wiaderek K M, Chapman K W, Chupas P J and Grey C P 2014 Capturing metastable structures during high-rate cycling of $LiFePO_4$ nanoparticle electrodes *Science (80-.).* **344** 1252817
[59] Wang L, Lu D, Liu Q, Liu L and Zhao X 2019 State of charge estimation for $LiFePO_4$ battery via dual extended kalman filter and charging voltage curve *Electrochim. Acta* **296** 1009–17
[60] Yang L, You W, Zhao X, Guo H, Li X, Zhang J, Wang Y and Che R 2019 Dynamic visualization of the phase transformation path in $LiFePO_4$ during delithiation *Nanoscale* **11** 17557–62
[61] Wang D, Wang H, Yang J, Zhou J, Hu Y, Xiao Q, Fang H and Sham T K 2016 Dynamic study of sub-micro sized $LiFePO_4$ cathodes by *in situ* tender X-ray absorption near edge structure *J. Power Sources* **302** 223–32
[62] Pathak P D and Vasavada N G 1970 Thermal expansion of NaCl, KCl and CsBr by X-ray diffraction and the law of corresponding states *Acta Crystallogr., Sect.* A **A26** 655–8
[63] Chadi D J and Cohen M L 1973 Special points in the Brillouin zone *Phys. Rev.* B **8** 5747
[64] Baldereschi A 1973 Mean-value point in the Brillouin zone *Phys. Rev.* B **7** 5212
[65] Monkhorst H J and Pack J D 1976 Special points for Brillouin-zone integrations *Phys. Rev.* B **13** 5188
[66] Kevan S D 1992 *Angle-Resolved Photoemission: Theory and Current Applications* (Amsterdam: Elsevier)
[67] Wang F and Landau D P 2001 Efficient, multiple-range random walk algorithm to calculate the density of states *Phys. Rev. Lett.* **86** 2050

[68] Söde H, Talirz L, Gröning O, Pignedoli C A, Berger R, Feng X, Müllen K, Fasel R and Ruffieux P 2015 Electronic band dispersion of graphene nanoribbons via Fourier-transformed scanning tunneling spectroscopy *Phys. Rev.* B **91** 45429
[69] Lauffer P, Emtsev K V, Graupner R, Seyller T, Ley L, Reshanov S A and Weber H B 2008 Atomic and electronic structure of few-layer graphene on SiC (0001) studied with scanning tunneling microscopy and spectroscopy *Phys. Rev.* B **77** 155426
[70] Cherkez V, de Laissardière G T, Mallet P and Veuillen J-Y 2015 Van Hove singularities in doped twisted graphene bilayers studied by scanning tunneling spectroscopy *Phys. Rev.* B **91** 155428
[71] Li G, Luican A and Andrei E Y 2009 Scanning tunneling spectroscopy of graphene on graphite *Phys. Rev. Lett.* **102** 176804

IOP Publishing

First-Principles Calculations for Cathode, Electrolyte and Anode Battery Materials

Thi Dieu Hien Nguyen, Shih-Yang Lin, Hsien-Ching Chung, Ngoc Thanh Thuy Tran and Ming-Fa Lin

Chapter 5

Unusual phenomena in $AlCl_4^-$/$AlCl_4$-graphite intercalation compounds of aluminum-ion-based battery cathodes

Ming-Hsiu Tsai, Wei-Bang Li, Thi Dieu Hien Nguyen, Shih-Yang Lin, Hsien-Ching Chung, Ngoc Thanh Thuy Tran, Nguyen Thi Han, Hsin-Yi Liu, Hai Duong Pham and Ming-Fa Lin

5.1 Optimal crystal structures: a first-principles study

Pristine graphite and its intercalation/de-intercalation compounds with guest atoms/molecules/ions [1] displays rich geometric symmetries, mainly owing to the chemical modifications [2]. Bernal graphite consists of periodical carbon-honeycomb lattices along the z-direction through AB stacking, i.e., it is the so-called AB-stacked bulk graphite. Each graphitic sheet remains a planar structure, clearly illustrating the orthogonal features of the significant π and σ bondings. The former, which is due to the carbon-$2p_z$ orbital hybridizations, can survive in the intralayer [3] and interlayer (van Der Waals;[4]) atomic interactions. It is responsible for the low-energy electronic properties of a 3D graphite (figure 5.2(b); [5]). However, the strength of the latter remains the same even in the presence of interlayer couplings. As a result, it is easy to distinguish the σ and π electronic states. In general, this phenomenon stays unchanged during the chemical reaction processes, i.e., planar graphene layers exist under the very strong sp^2- σ bondings.

The sufficiently wide spacing of ~3.35 Å between two neighboring graphene layers, which are created by the weak, but significant van der Waals interactions (the interlayer $2p_z$ orbital hybridizations; [6, 7]), are available for easy intercalations of the guest $AlCl_4^-$ ions and $AlCl_4$ molecules intercalations. The interlayer distance (I_c = 3.35 Å under a pristine case; [8, 9]) is greatly enhanced under the various chemical environments, as clearly indicated in tables 5.1 and 5.2 for ion and

Table 5.1. Optimal geometric structures of $AlCl_4^-$ graphite intercalation compounds with various concentrations: (a) 1:18, (b) 1:24, (c) 1:32 and (d) 1: 54 for the ratio of $AlCl_4^-$ and C.

	Concentration ($AlCl_4$/C)	Layer-Distance (Å)	C–C Bond-length (Å)	Al–Cl Bond-length (Å)	Al–Cl Bond-angel (°)
Primitive		3.35	1.42	2.159	109.0
1:54	1.85%	10.65	1.424	2.165	108.01
1:32	3.12%	10.81	1.426	2.163	107.94
1:24	4.16%	11.29	1.427	2.162	107.46
1:18	5.55%	11.30	1.428	2.160	107.19

Table 5.2. Similar results to table 5.1, but illustrated for $AlCl_4$ graphite intercalation compounds. The blue shifts of the Fermi levels under various concentrations are also shown for the charge transfer effects.

	Concentration ($AlCl_4$/C)	Layer Distance(Å)	C–C Bond-length (Å)	Al–Cl Bond-length (Å)	Al–Cl Bond-angel (°)	Red shifts (eV)
Primitive		3.35	1.42	2.159	109.27	
1:54	1.85%	8.80	1.421	2.165	113.03	0.978
1:32	3.12%	8.81	1.422	2.163	112.00	0.911
1:24	4.16%	8.78	1.421	2.162	111.45	0.862
1:18	5.55%	8.77	1.422	2.160	111.7	0.753

molecule intercalations, respectively. It is very sensitive to the change of intercalation concentration and arrangement (figure 5.1). As to the ion/molecule cases, I_c^s are, respectively, 11.30, 11.29, 10.81 and 10.65 Å. 8.77, 8.78, 8.81 and 8.80 Å are for 1:18, 1:24, 1:36 and 1:54 concentrations in terms of ratio between $AlCl_4^-$/$AlCl_4$ and carbon. A very large I_c clearly indicates the thorough disappearance of the interlayer $2p_z$ orbital couplings. Very interestingly, the multi-orbital hybridizations in C– $AlCl_4^-$ or C–$AlCl_4$ bonds can account for the interlayer atomic interactions. As a result, there exist intralayer C–C bonds, intra-ion/intra-molecule bondings of $AlCl_4^-$/$AlCl_4$, inter-ion/inter-molecule ones, and carbon–intercalant interactions. The active orbital hybridizations in distinct chemical bondings need to be identified from the other physical quantities (figures 5.2–5.6; [10]). In addition, whether the van der Waals interactions could survive in large-I_c graphite intercalation compounds requires very detailed numerical examination.

The regular arrangements of large ions/molecules, which possess high projection symmetries, are chosen for a model study. Very interestingly, pristine graphite has a periodical AB stacking configuration along the *z*-direction, being dramatically transformed into the AA one during the strong chemical intercalations/de-intercalations. This will lead to dramatic changes of the other fundamental properties [11]. As for each intercalant an aluminum atom is just situated at the hollow site above a hexagon (the top view of the *x*–*y*-plane projection in figure 5.1); furthermore, it is accompanied by

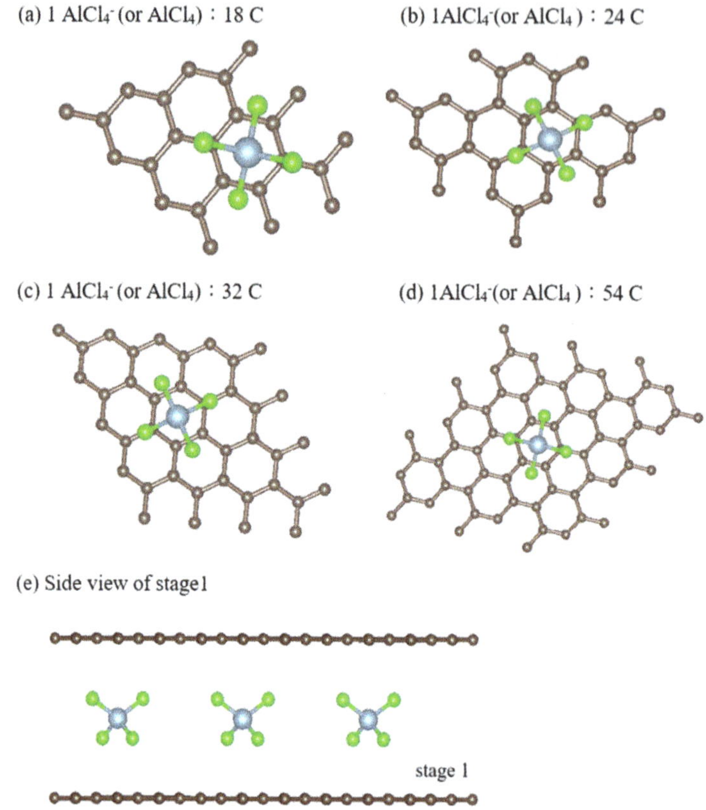

Figure 5.1. The optimal geometric structures of $AlCl_4^-$ ($AlCl_4$) graphite intercalation compounds with the various concentrations: (a) 1:18, (b) 1:24, (c) 1:32 and (d) 1: 54 for the ration of ($AlCl_4$) and C. Both top and side views (e) are shown simultaneously.

two chloride atoms along the dimer and bridge-middle directions simultaneously. The periodical ion/molecule distribution fully occupies the whole interlayer spacing so that the intercalant layer is formed after the chemical modification [12, 13]. This is the so-called stage-1 graphite intercalation compound, where there is only monolayer graphene between two neighboring intercalant layers. And then, the intercalant concentrations decline within the larger unit cells. There are larger moiré superlattices under the dilute cases [14]. These graphite de-intercalation systems are expected to have the lower lattice symmetries and thus more complicated calculations/phenomena [15], compared with alkali atoms [16]. Most importantly, the atomic configurations in the saturated ions and the unbalanced molecules (the excited angling bonds) [17] play a critical role about an obvious difference in the interlayer distances (~11Å in table 5.1 and ~8.8 Å), since the carbon–intercalant interactions are weaker under the former chemical environment. During the concentration variation of large ions/molecules, the sensitive dependences of their internal bonding angles indicate the significant contributions to the total ground state energies. Specifically, the ion–ion interaction might strongly modify the optimal interlayer distances.

The intercalant configurations deserve closer examination. The previous experimental [18] and theoretical [19] studies have proposed the characterizations of stage-n systems during the chemical modification processes, mainly owing to the unchanged σ bonding honeycomb lattices [20] and the drastic changes of free carrier densities [21]. As for $n \geqslant 2$ cases, intercalants exhibit a periodical distribution along the z-direction, but a non-uniform one. However, this unusual configuration might not agree with the natural ion/molecule diffusion phenomena under the external factors (e.g., pressure and thermal energy; [22]). It is mission impossible to create the critical mechanisms in forbidding their intercalations and de-intercalations inside any spacings of the nearest-neighbor graphitic layers. That is, the guest-intercalant transport, which obeys the thermal dynamical laws [23], will be revealed in the successfully synthesized compounds. Obviously, there are certain important differences between the stage-n graphite intercalation compounds and the stage-1 systems [24, 25] with the various concentrations, covering the active chemical bonds, the interlayer distances, the crystal symmetries of moiré superlattices, the bonding angles, the charge transfers, and the metallic or semiconducting behaviors. How to clarify which kind of stacking configuration is the optimal one after the experimental synthesis. This interesting issue could be settled through the method of molecular dynamics [26], in which the physical/chemical/materials environments are the necessary conditions for the delicate numerical simulations. Systematic investigations are required in near-future basic science research.

Very apparently, high-precise x-ray diffraction spectroscopy, as clearly illustrated in chapter 3, is reliable in fully exploring the optimal crystal structures of $AlCl_4^-$/$AlCl_4$ graphite intercalation compounds. The examined quantities cover the periodical distances along the z-direction and the lattice constants on the (x, y) plane. Whether this method could detect the intercalant-dependent bonding angles is worthy of further thorough investigations. Up-to-date x-ray patterns claim successful observations about stage-3 and stage-4 large-intercalant graphite intercalation compounds [18]. The former/latter is deduced to be relatively stable at lower/room temperatures [~250 K/300 K] during the charging and discharging processes in aluminum-ion-based batteries [18]. Apparently, temperature is one of the critical factors in determining the lattice symmetries. For example, the thermal excitation energies are expected to be comparable with the interlayer graphene–intercalant interactions. Theoretical predictions on the stage-1 systems with the various intercalant concentrations could be generalized to stage-n ones. The stacking configurations would strongly modify similar physical and chemical phenomena (great enhancement or reduction of similar quantities). This is under current study [27].

5.2 Rich and unique electronic properties

Electronic properties of $AlCl_4^-$/$AlCl_4$ graphite intercalation compounds, 3D band structures, charge density distributions and density of states, are fully explored by numerical Vienna *ab initio* simulation package (VASP) calculations and further generalized by the phenomenological models. The critical mechanisms, the intralayer π and σ bondings, the interlayer van Der Waals interactions, the carbon–intercalant orbital hybridizations, and intra- and inter-ion/inter-molecule interactions, are examined and identified from the delicate analyses. These will be revealed in the semi-metallic or metallic behaviors (density of free carriers due to the weak or strong charge transfer), well-characterized/unusual π and σ bands, C-, Al- and Cl-dominated energy spectra at different energy range, spatial orbital bondings after and before intercalations, and merged special structures of the energy-dependent van Hove singularities.

Bernal graphite and $AlCl_4^-$, $AlCl_4$ graphite intercalation compounds, as clearly shown in figures 5.2 and 5.3, present diverse electronic energy spectra and wave functions. Their first Brilloun zone in figure 5.2(a) possesses a hexagonal symmetry [28], in which the (kx, ky)-projection is similar to that of a layered graphene [29]. Band structure of a pristine graphitic system (figure 5.2(b)), being illustrated along the high-symmetry points (ΓKMΓAHLA; [30]), is rich and unique. This system is an unusual semimetal, while a monolayer graphene is a zero-gap semiconductor with a vanishing density of states at the Fermi level [31]. Such difference obviously indicates the important role of the interlayer van der Walls interactions. The intralayer and interlayer C-$2p_z$ orbital hybridizations, respectively, create the gapless Dirac-cone structure [32] and the weak, but significant overlaps of valence and conduction bands [33]. The latter are responsible for the asymmetric energy spectra of valence holes and conduction electrons about the Fermi level. The low-lying π electronic states are initiated from the K and H valleys (the inset of figure 5.2). Furthermore, their state energies are, respectively, lower and higher than the Fermi energy in terms of a weak energy dispersion, i.e., there exist 3D free carriers of the hole and electron pockets. Very interestingly, the other essential properties are easily further modulated by the external factors, such as the intercalation-/de-intercalation- [34], temperature-, pressure-, and magnetic-field-enriched phenomena [35–37]. In addition, systematic investigations on bulk graphite systems can be found in review articles and books [38]. Most importantly, the well-behaved π bands could be clearly identified for the energy spectra along KMΓ and HLA with the whole widths more than 7 eV. The wide π-band widths are attributed to a close cooperation of the intralayer and interlayer carbon-$2p_z$ orbital hybridizations. On the other side, the σ orbitals of [$2p_x$, $2p_y$, 2s] can build two degenerate bands and one band below at the Γ point below −3 and −10 eVs, respectively. These features are mainly determined by the σ-electronic hopping integrals and ionization energies [39].

$AlCl_4^-$-ion and $AlCl_4$ -molecule intercalations, which are, respectively, shown in figure 5.2(c)–5.2(f) and figure 5.3(a)–5.3(d), are able to create the diverse energy spectra and wave functions. The dramatic changes cover variation of high-symmetry points, creation of a lot energy subbands, greatly enhanced asymmetry of occupied and unoccupied spectra about the Fermi level, obvious reduction or enhancement of band overlaps (the diversified free carrier densities), almost isotropic/highly anisotropic features near/away from $E_F = 0$, various energy dispersions with the different critical points, frequently crossing and anti-crossing behaviors, non-well-behaved π-σ-band widths, carbon-, aluminum- and chloride-dominances at the different energy ranges (blue circles, red triangles and green squares, respectively). Moreover, the π and σ electronic states could be easily identified from the original valleys, but not their whole band widths. After the chemical modifications, the enlarged moiré superlattices possess many atoms/ions in primitive unit cells (figure 5.1), so that the hexagonal first Brillouin zone in figure 5.2(a) is diminished quickly, especially for low concentration cases. This leads to many valence and conductions with smaller wave-vector ranges [40]. Due to the zone-folding effects [41], the low-lying electronic states are initiated from the Γ and A valleys (the K and H ones) under the cases of 1:18, 1:24 and 1:54 [1:32]. As to the ionic chemical environments, a pair of anisotropic valence and conduction bands across the Fermi level, which appears in a pristine Bernal graphite (figure 5.2(b)), is changed into an isotropic Dirac-cone structure of monolayer graphene. Furthermore, the energy spectra are dispersed less along the ΓA or KH directions. This clearly illustrates the semiconducting behaviors with a zero-band gap and density of state at $E_F = 0$, simultaneously indicating the very weak carbon–intercalant-ion orbital hybridizations under the saturated atomic configurations. These graphite intercalation compounds are expected to present the lower electrical conductivities after the chemical reactions [42], However, they become outstanding merits in aluminum-ion transports and cathode intercalations/de-intercalations [43], Very interestingly, the π-electronic dominance in the energy range of $E^{c,v} \leqslant 1.0$ eV also comes to exist in $AlCl_4$-molecule graphite intercalations, The obvious red shifts of E_F's is revealed in any chemical cases. Electrons are largely transferred from carbon atoms to molecules, where the latter possess larger affinities [44]. The strong p-type doping effects should be attributed to the significant carbon–molecule orbital hybridizations (illustrated later in figures 5.4 and 5.5). In short, three types of band structures, semimetal, semiconductor and metal, respectively, arise from the interlayer van der Waals, carbon-saturated-ion and carbon–molecule

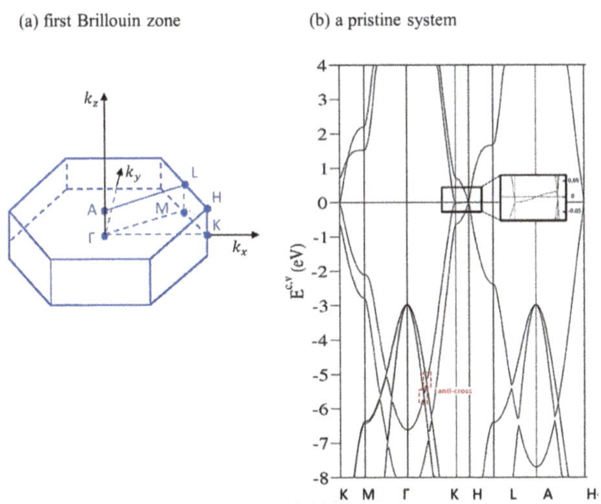

Figure 5.2. Band structures of $AlCl_4^-$-related graphite intercalation compounds (a) along the high-symmetry points within the first Brillouin zone under various cases: (b) a pristine system, (c) 1:18, (d) 1:24, (e) 1:32 and (f) 1: 54, in which the atom dominances are represented by the colored balls.

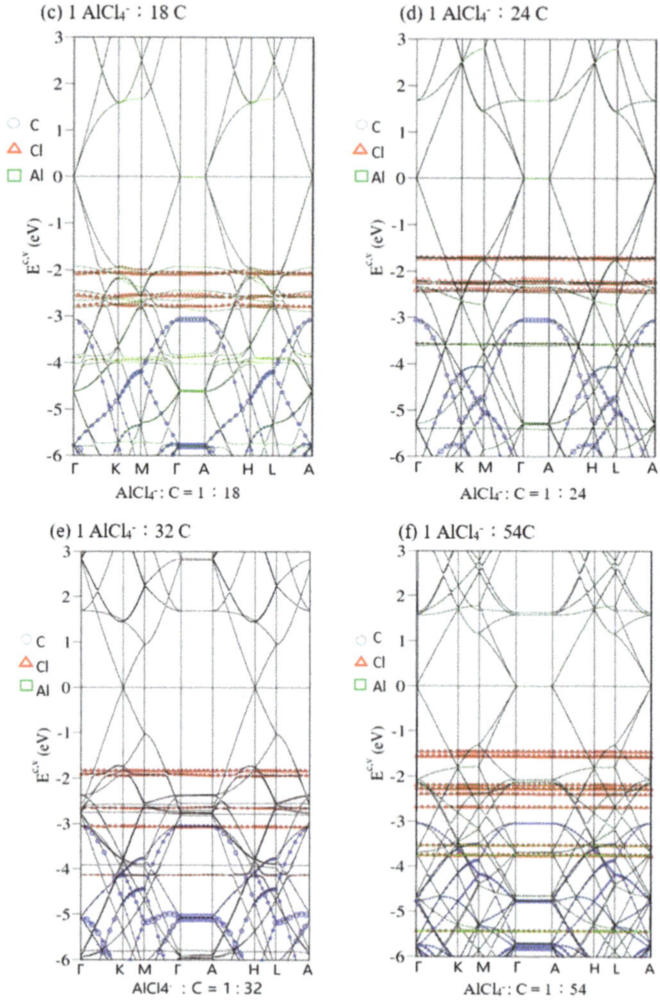

Figure 5.2. (Continued.)

interactions. Whether similar phenomena could be found in other graphite intercalation compounds deserves closer VASP simulation.

The atom dominance, which corresponds to the spatial distribution probability of each wave-vector state, is clearly revealed in the specific energy ranges. It is determined by the intrinsic orbital hybridizations of chemical bonds. Most electronic states in the whole energy spectrum are dominated by carbon atoms for the various $AlCl_4^-$-ion and $AlCl_4$-molecule intercalations (blue open circles in figure 5.2(c)–5.2(f) and 5.3(a)–5.3(d) for the π, σ and π* energy subbands). This is attributed to the dominating π, σ, carbon-intercalant bondings, since their initial state energies remain there and the frequent anti-crossing behaviors come to exist [45]. Specifically, aluminum atoms make observable contributions near E^v~−4 and −6 eVs

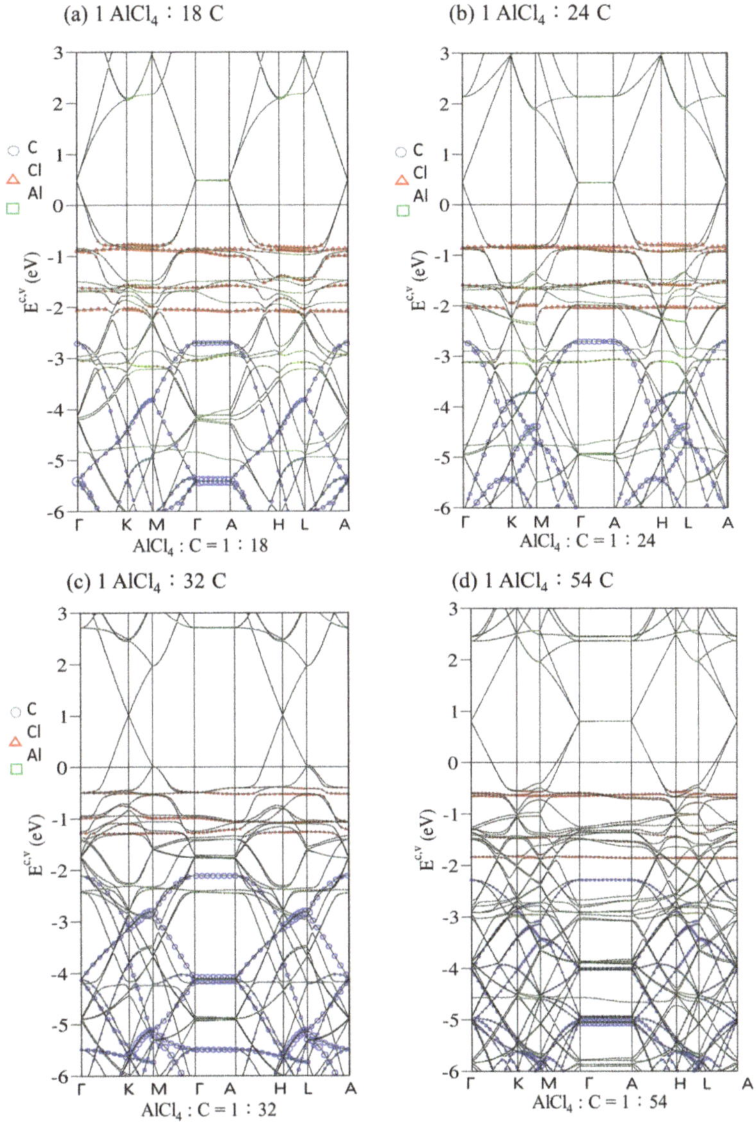

Figure 5.3. Electronic energy spectra in $AlCl_4$-related graphite intercalation compounds for the various guest-molecule concentrations: (a) 1:18, (b) 1:24, (c) 1:32 and (d) 1: 54.

(green squares), suggesting the linking roles through a large ion/molecule structure. As for chloride atoms, the unusual roles are revealed as the weakly dispersive valence bands, at least four of them, below the Fermi level more than 1 eV (red triangles). Their main features, energy, degeneracy, spacing and group velocity of valence subband, are very sensitive to the change of intercalant configuration. The partially flat subbands, with the zero velocities (the localized behaviors) frequently appear under the large-ion intercalations (figure 5.2(c)–5.2(f)). However, they might exhibit observable modifications under molecular cases (figure 5.3(a)–5.3(d)).

The larger carrier mobility clearly indicates more extensive charge distributions. The above-mentioned characteristics might be closely related to all the active chemical bondings, being supported by further discussions in charge density distributions (figures 5.4 and 5.5) and van Hove singularities (figures 5.6 and 5.7).

The theoretical predictions of occupied electronic states below the Fermi level could be examined by angle resolved photoemission spectroscopy (ARPES; [46]), as discussed in chapter 3 section 3.3 in detail. In general, it is very difficult to measure the k_z-dependent energy spectra because of the destruction of momentum conservation through the surface boundary. How to utilize the most important band features along KH, ML and ΓA would become a critical technique in identifying

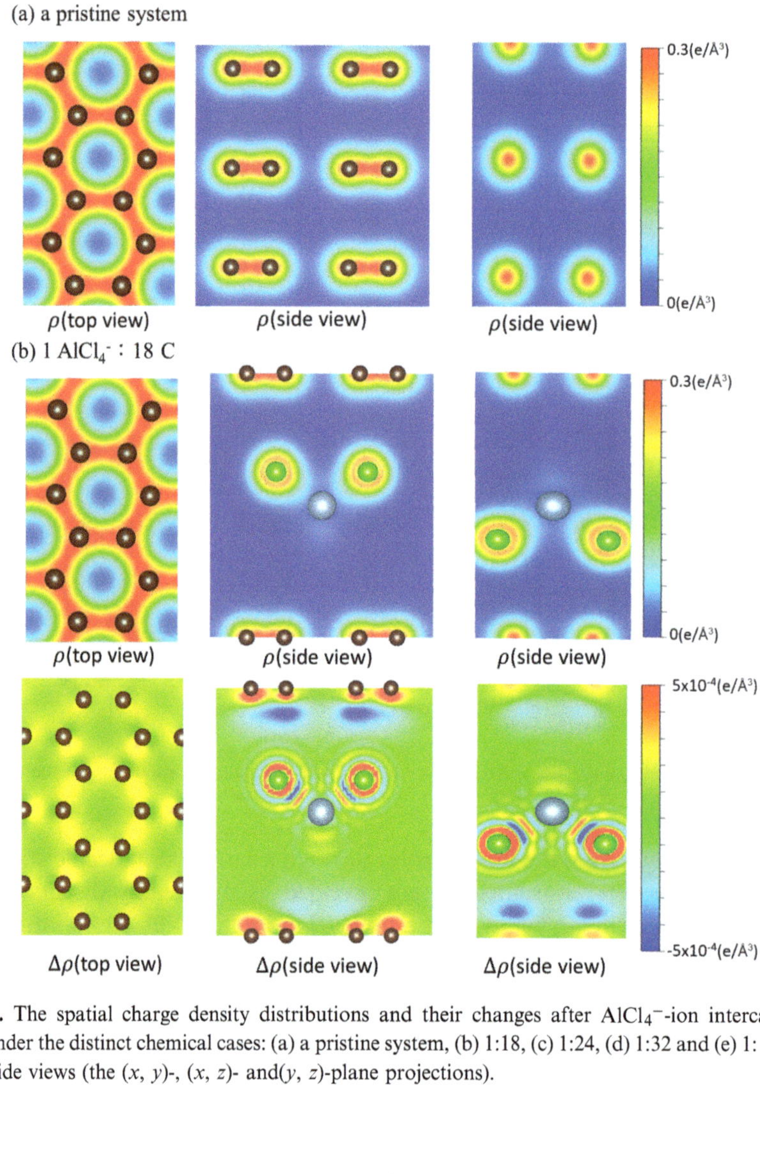

Figure 5.4. The spatial charge density distributions and their changes after $AlCl_4^-$-ion intercalation into graphite under the distinct chemical cases: (a) a pristine system, (b) 1:18, (c) 1:24, (d) 1:32 and (e) 1: 54, with the top- and side views (the (x, y)-, (x, z)- and(y, z)-plane projections).

(c) 1 $AlCl_4^-$: 24 C

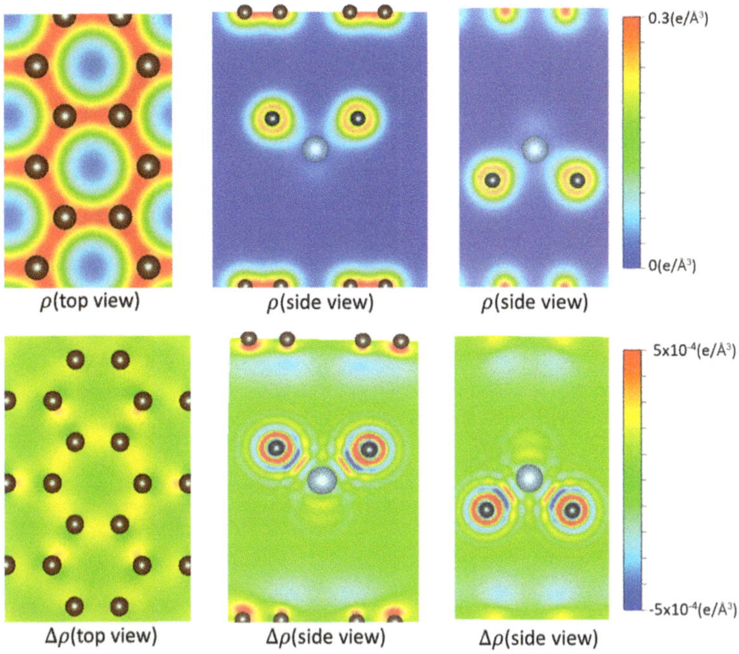

(d) 1 $AlCl_4^-$: 32 C

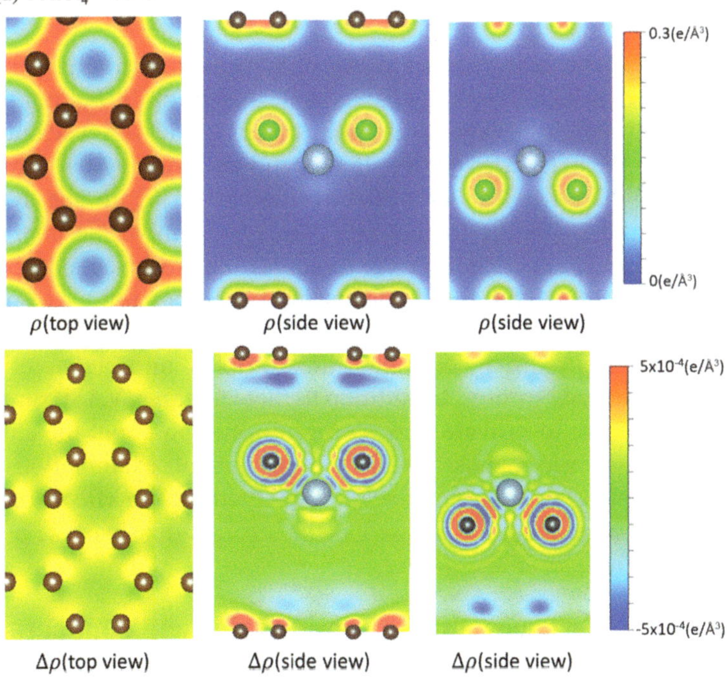

Figure 5.4. (Continued.)

(e) 1 $AlCl_4^-$: 54C

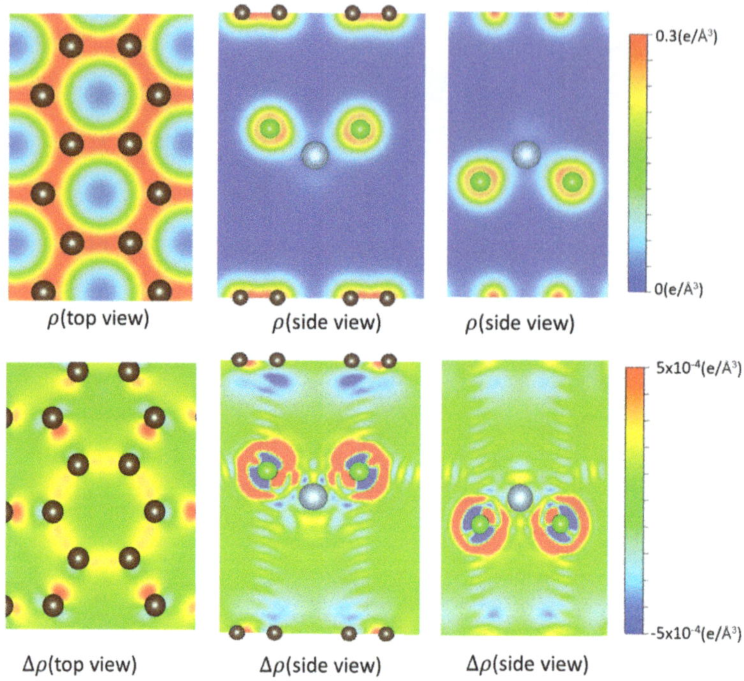

Figure 5.4. (Continued.)

energy dispersions. The high-resolution ARPES measurements have successfully conducted on the semi-metallic energy bands in Bernal graphite, but not those of the rhombohedral and simple hexagonal graphites (ABC- and AA-stacked ones; [47, 48]). According to the calculated results, the second and third systems, respectively, the lowest and highest free carrier densities, mainly owing to the symmetry of stacking configuration [49]. As to the $AlCl_4^-$-ion/$AlCl_4$-molecule intercalation of stage-3/stage-4 graphite intercalation compounds, the observed occupied energy spectra are expected to exhibit greatly diversified phenomena in terms of stacking-symmetry dependences, red shifts of the Fermi level, band overlaps (free carrier densities), strong energy dispersions, high anisotropies, and characterizations of π- and σ-electronic energy spectra. In addition to these features, VASP simulations on stage-1 systems are able to provide very useful information about the chloride- and aluminum-related valence bands. Further experimental examinations are very helpful to thoroughly clarify the intercalation/de-intercalation effects on electronic energy spectra and wave functions, as well as the intrinsic quasiparticle properties of orbital hybridizations. [50]

5.3 Important differences among semimetals, semiconductors and p-type metals

The spatial charge distributions [$\rho(r)s$] and their variations [$\Delta\rho(r)$s] after the chemical intercalations, as clearly shown in figures 5.4(a)–5.4(e) and 5.5(a)–5.5(d), are capable of providing very useful chemical pictures in fully comprehending the critical chemical bonds with the active orbital hybridizations. The [*x*, *y*]-top, [*x*, *z*]-side and [*y*, *z*]-side views fully illustrate the rich and unique intrinsic interactions: the C–C bonds in a honeycomb lattice, the interlayer C–Cl bonds, and the intra-ion/intra-molecule Al–Cl and Cl–Cl bonds. The significant chemical bondings agree with density of states (the merged van Hove singularities in figures 5.6 and 5.7) and band structures (atom dominances in figures 5.2 and 5.3). The strong evidence is thoroughly identified from $\rho(r)$s and $\Delta\rho(r)$s.

First, the prominent σ binding, which survives in a pristine Bernal is revealed as a very strong covalent between two neighboring carbon atoms (a super-high charge density by the red color on the [*x*, *y*] plane, the first plot of figure 5.4(a); [51]).

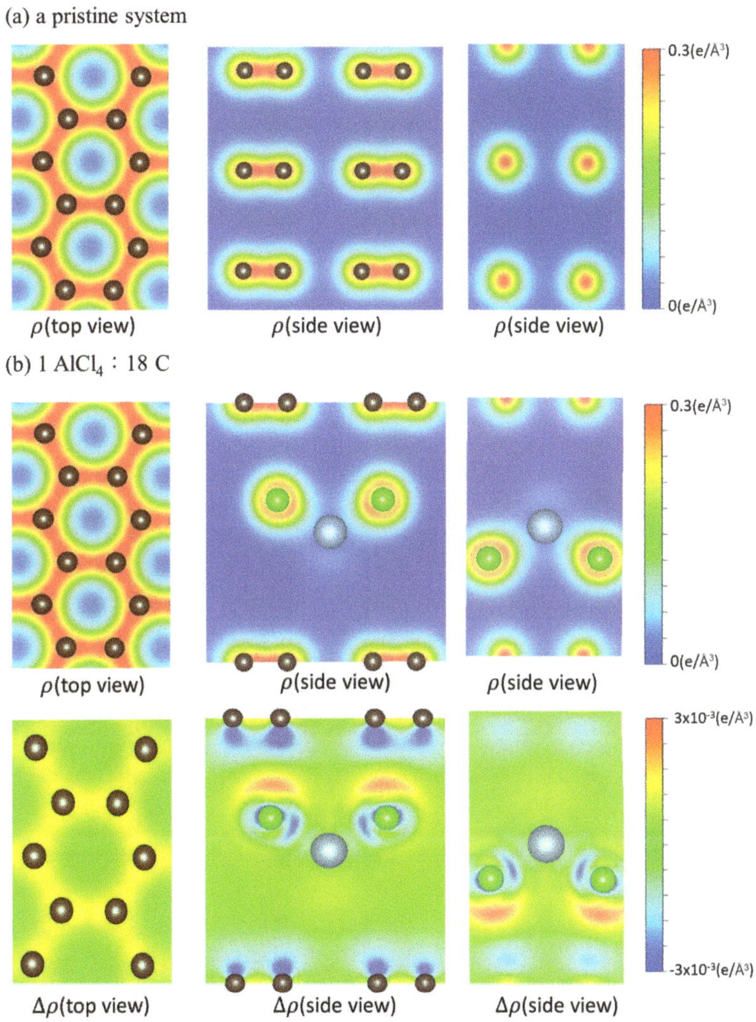

Figure 5.5. Similar plots to figure 5.4, but shown for $AlCl_4$-molecule intercalations in (a), (b), (c) and (d).

(C) 1 $AlCl_4$: 24 C

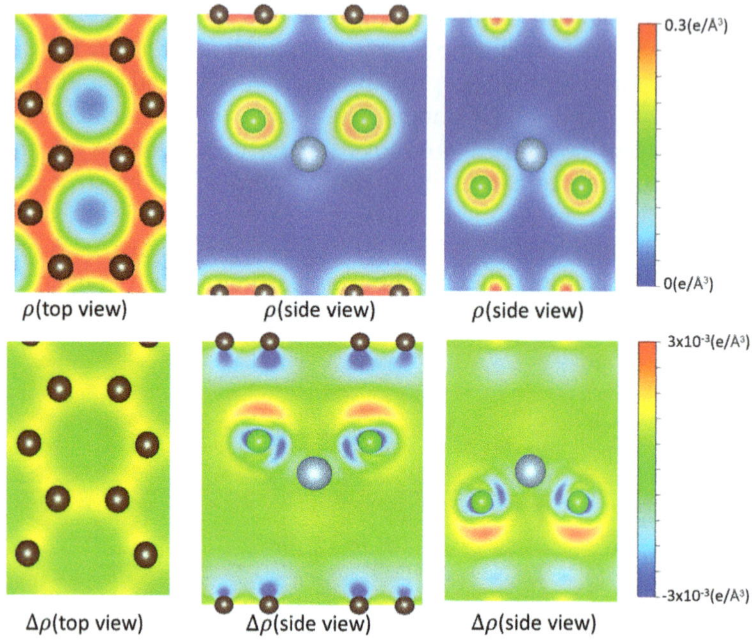

(d) 1 $AlCl_4$: 32 C

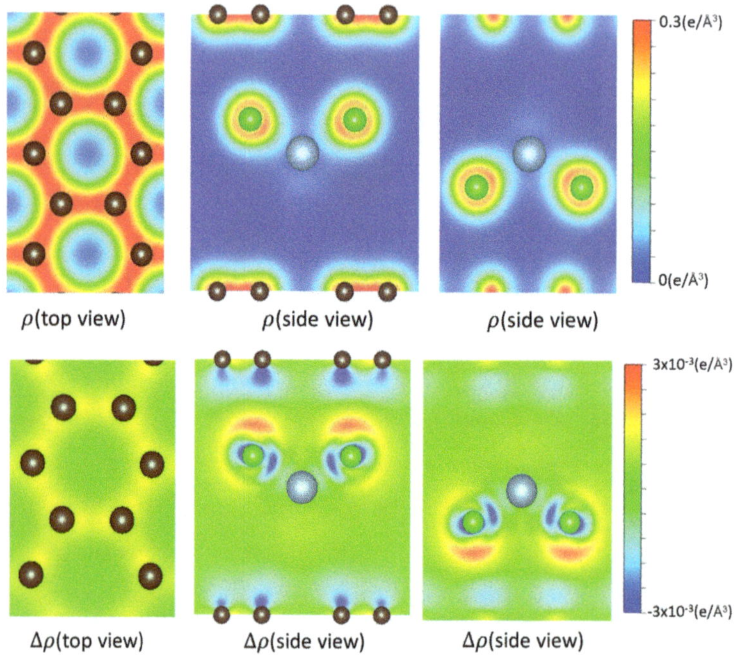

Figure 5.5. (Continued.)

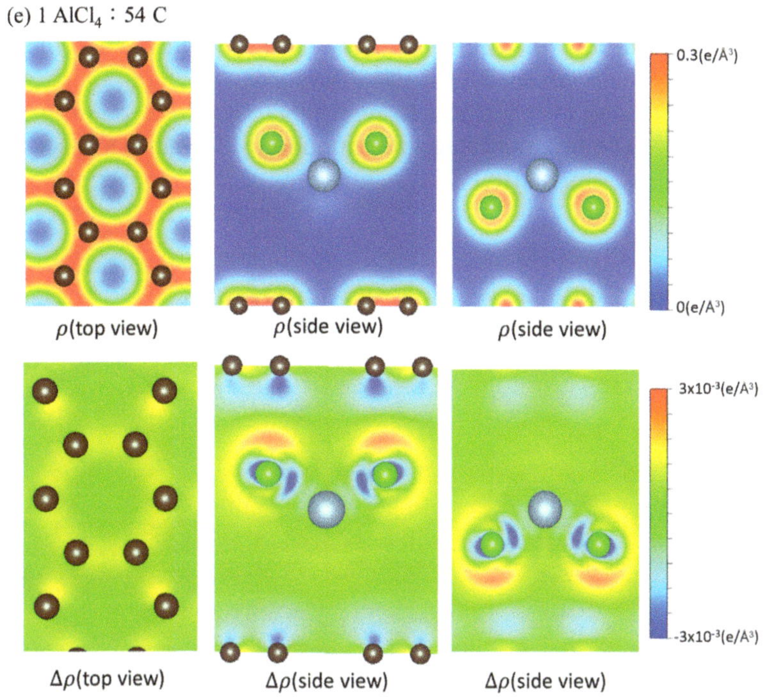

Figure 5.5. (Continued.)

Furthermore, the π bonding is characterized by wave-like charge distribution ρ(*r*)*s* and their variations Δρ(*r*)s due to the parallel significant $2p_z$-orbital hybridizations, as examined from ρ(*r*) on both [*x*, *z*] and [*y*, *z*] planes (the second and third plots in figure 5.4; [51]). Its distribution along the *z*-direction is somewhat extended by the interlayer van der Waals interactions [52]. Secondly, the significant carbon–intercalant couplings, corresponding to the large-ion cases (figure 5.4(b)–5.4(e)), are directly reflected in the strongly anisotropic charge distributions ρ(*r*)*s* in the first, second and third plots and their variations Δρ(*r*)s near chloride atoms and the drastic changes between them, especially for the [*x*, *z*]- and [*y*, *z*]-plane projections. On the other side, the large-molecule intercalations, as indicated in figures 5.5(a)–(d), are able to greatly enhance charge density distributions in C–Cl bonds, since their atomic configuration belongs to the non-closed-shell status. As for the Al–Cl and Cl–Cl bonds, they present prominent bondings through the obvious distorted charge densities near chloride and aluminum atoms. In addition, the calculated results make it very difficult to examine the existence of Al–C and Al–Al bonds. Furthermore, they cannot provide enough information in examining the effects of the inter-ion/inter-molecule intercations. Similar analyses could be generalized to other multi-component graphite intercalation compounds, e.g., the active chemical bonds in H_2SO_4- [53], HNO_3- [54], and $FeCl_3$- [55] related systems.

According to the good definition of density of states, D(*E*) is expressed as the integration of the inverses of group velocities on the constant-energy configuration.

For example, the 3D/2D/1D D's are greatly enriched by the various first derivatives of the gradient operations on electronic energy spectra and the specific-E integrations on the closed shells/circles/but the two discrete wave-vector points. A vanishing group velocity comes to exist while it corresponds to a critical point in the energy-wave-vector space. The singular integration function will lead to a special structure, namely, a van Hove singularity. The main features of singular structures, their forms, intensities, energies and numbers are very sensitive to the characteristics of

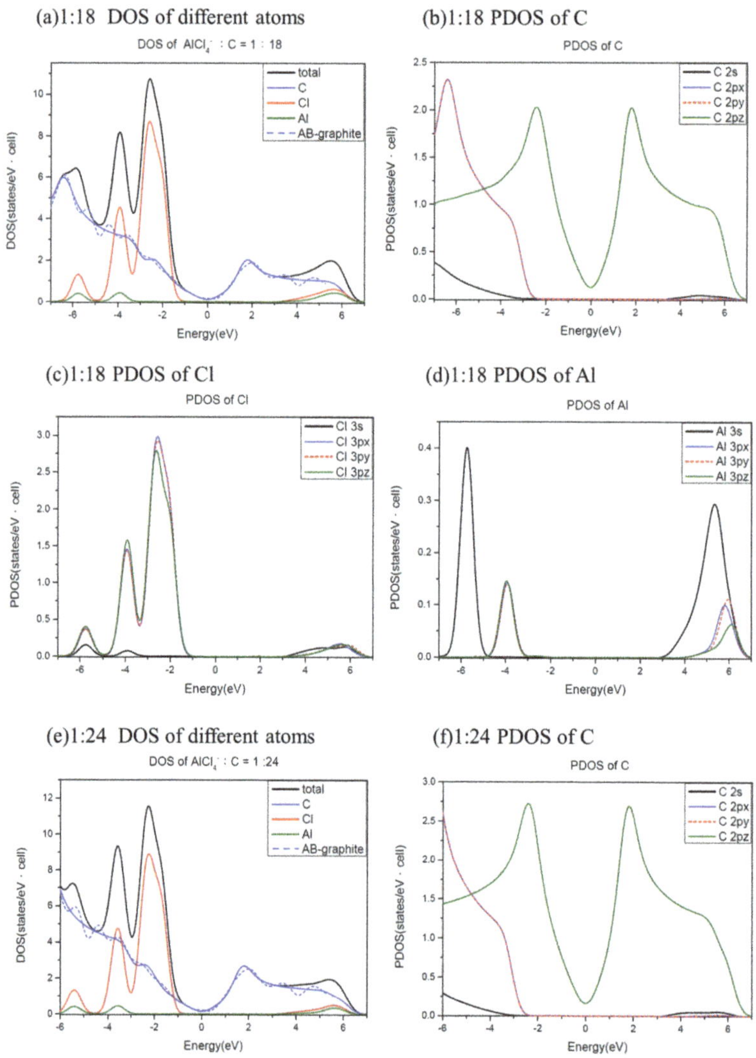

Figure 5.6. The atom- and orbital-projected density of states for the $AlCl_4^-$-intercalation graphite compounds through the molecular concentrations: (a)/(b)/(c)/(d) 1:18 (are illustrated by four plots (the contributions of different atoms and the orbital decompositions of three kinds of atoms) by the red curve is that of Bernal graphite). (e)/(f)/(g)/(h) 1:24, (i)/(j)/(k)/(l) 1:32 and (m)/(n)/(o)/(p) 1: 54.

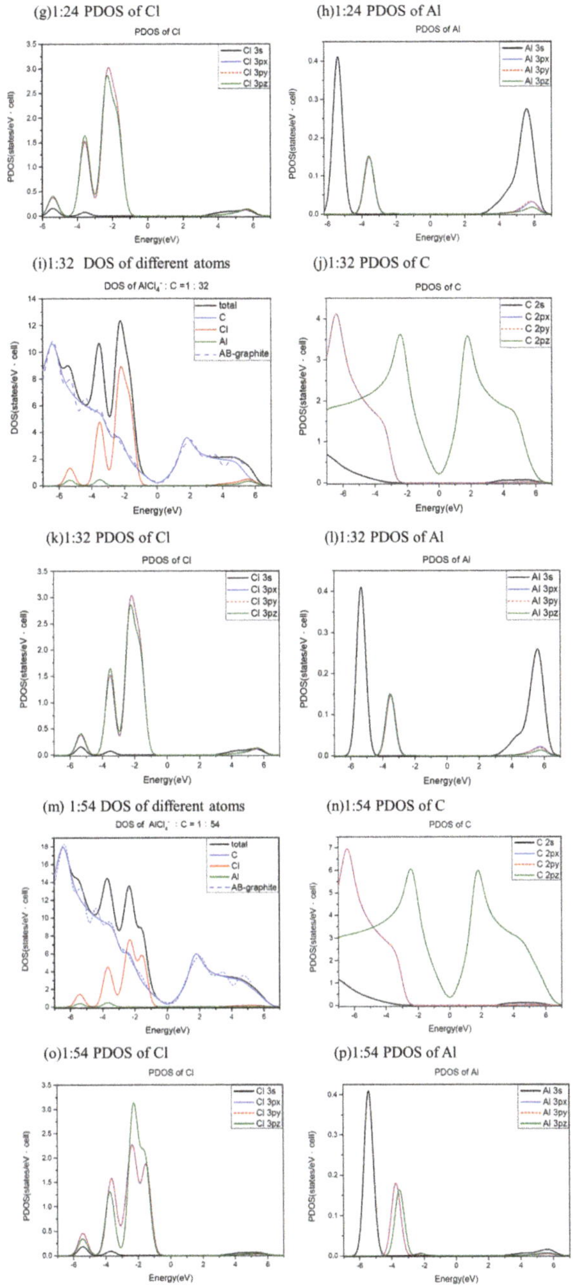

Figure 5.6. (Continued.)

distinct critical points and dimensionalities [56]. In general, the former are classified into the extreme, saddle and partially flat points, being clearly illustrated by the linear, parabolic, almost dispersionless and sombrero-shape energy dispersions of few-layer graphene systems [57]. When the orbital- and orbital-decomposed density of states are done for any condensed-matter systems, the various singular structures, with prominent intensities, are available in determining the active orbital hybridizations of different chemical bonds [58]. This is based on their great enhancements through the emerged van Hove singularities [59]. As for the large-intercalant graphite intercalation compounds, there are plenty of atom- and orbital-projected components. The very complicated results need to be analyzed in detail [60].

For each intercalation case of $AlCl_4^-$/$AlCl_4$, there are one atom- and three orbital-decomposed densities of states, being rather sufficient in providing useful information about the active multi-/single-orbital hybridizations of distinct chemical bonds [61]. Very apparently, figure 5.6(a)–5.6(p) and 5.7(a)–5.7(p) show the rich and unique van Hove singularities mainly due to C, Cl, Al atoms and their significant orbitals. The magnitude of D(E) at the Fermi level represents the characteristics of free carriers. Bernal graphite and ion intercalation systems (figure 5.6(a), 5.6(i) and 5.6(m)) have low values at $E_F = 0$, as well as band structures (figure 5.2(b)–5.2(f)), respectively, suggesting semi-metallic and semi-conducting behaviors. However, each large-molecule case exhibits a finite value there. Most importantly, the difference between the Fermi level and the featured energy with the smallest density of states could be regarded as its red shift (details in table 5.2). Furthermore, this covered area just corresponds to the total free carriers per unit cell after the obvious charge transfers from carbon to chloride (figure 5.5). The stronger affinity of the latter is responsible for the p-type doping effects (the free valence holes; [62]). This quantity is deduced to be proportional to the $AlCl_4$-molecule concentration. In addition, the Fermi-momentum states of electronic spectra (figure 5.3(a)–5.3(d)) are not reliable in evaluating the transferred valence hole density in the presence of complicated zone-folding effects.

The significant chemical bonds and their active multi-/single-orbital hybridizations are further achieved from the delicate analyses, covering all the separated and merged van Hove singularities (figures 5.6 and 5.7). The concise physical and chemical pictures are also supported by electronic energy spectra (figure 5.2 and 5.3) and spatial charge distributions (figures 5.4 and 5.5). Both $AlCl_4^-$ and $AlCl_4$ graphite intercalation compounds possess intralayer carbon–carbon, interlayer carbon–intercalant, and intra-/inter-intercalant interactions, respectively, leading to the C–C, C–Cl, and Al–Cl and Cl–Cl bonds. However, the observable evidence of merged van Hove singularities are absent for Al–C and Al–Al. The prominent chemical bondings are thoroughly illustrated as follows. Since $AlCl_4^-$ has a closed 0 shell atomic configuration, each graphitic sheet recovers to a pure honeycomb lattice. The π- and σ-electronic spectra are well separated from each other (figure 5.6(b)/figure 5.6(f)/figure 5.6(j)/figure 5.6(n)). Furthermore, the former (pink curves) and the latter (red, blue and black curves) are, respectively, characterized by the initial/prominent structures at ~ −2.0 eV and ~ −3.12/−6.15 eV. In addition, two strong peaks mainly arise from the saddle points of valence π and σ bands.

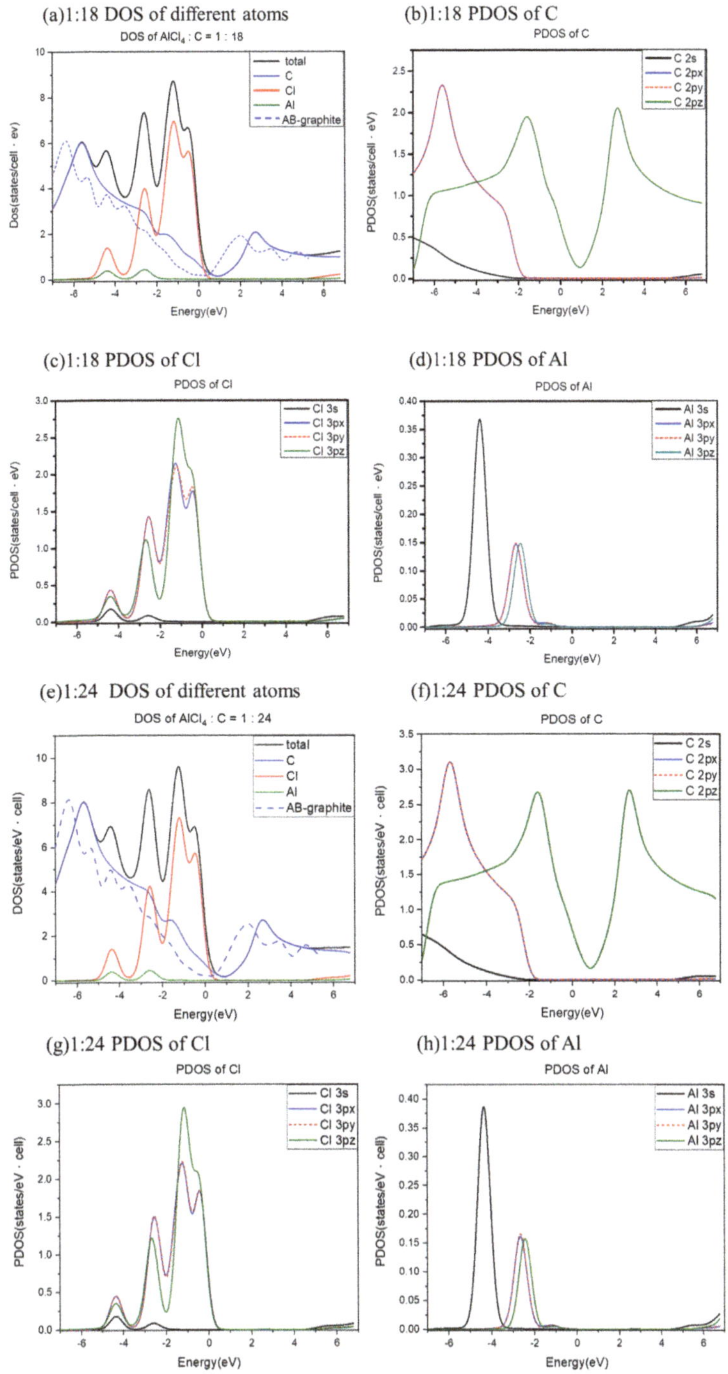

Figure 5.7. Similar DOSs to figure 5.6, but displayed for $AlCl_4$-molecule graphite intercalation compounds in (a)–(p).

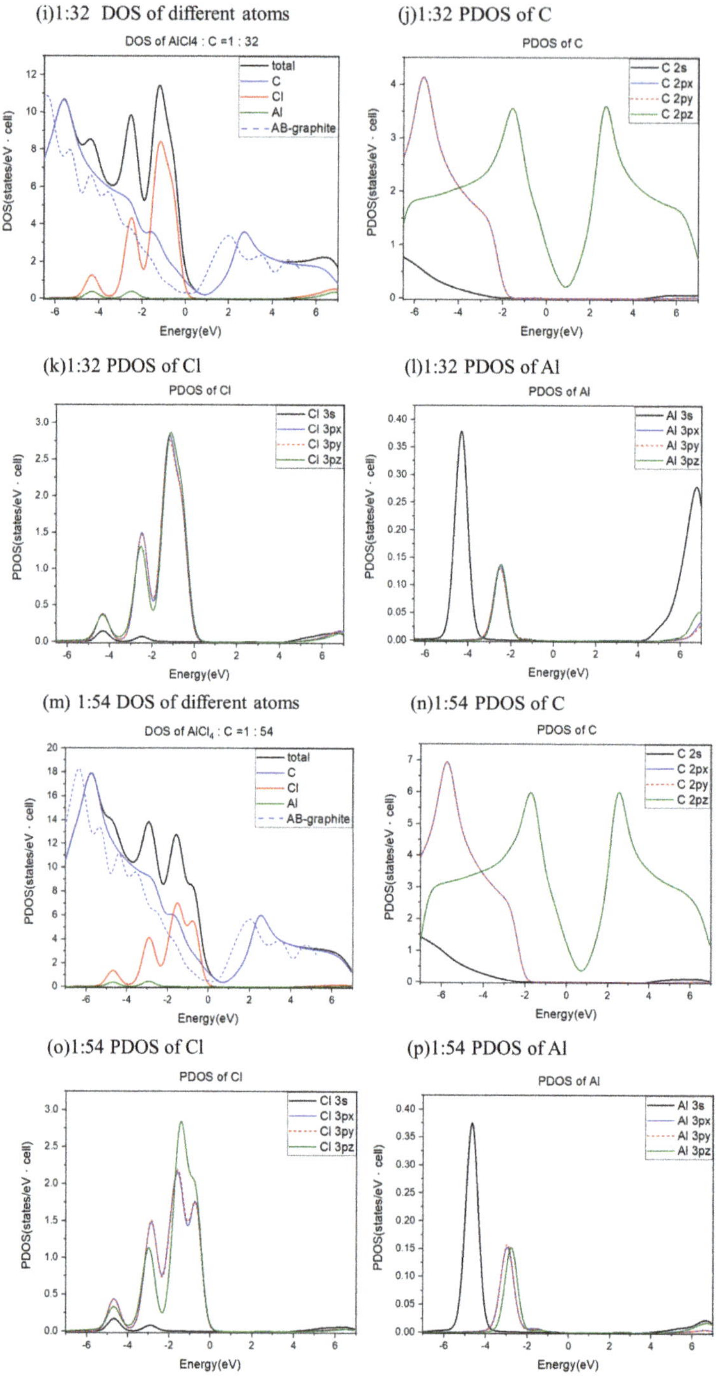

Figure 5.7. (Continued.)

Apparently, π and sp^2 bondings are orthogonal to each other and thus survive in C–C bonds. The interlayer C–Cl bonds are revealed as the multi-orbital hybridizations of [$2p_x$, $2p_y$, $2p_z$]-[$3p_x$, $3p_y$, $3p_z$] through the emerged structures within -6.1 eV$\leqslant E \leqslant -1.8$eV (figure 5.6(a)–5.6(c)/figure 5.6(e)–5.6(g)/figure 5.6(i)–5.6(l)). As for Al–Cl/Cl–Cl bonds, the obvious four-orbital hybridizations of [3s, $3p_x$, $3p_y$, $3p_z$]-[3s, $3p_x$, $3p_y$, $3p_z$]/[3s, $3p_x$, $3p_y$, $3p_z$]-[2s, $2p_x$, $2p_y$, $2p_z$] are clarified from the van Hove singularities at −4.0eV and −5.9eV. Such unusual results are due to the fact 3s and [$3p_x$, $3p_y$, $3p_z$]-decomposed, respectively, appear the same and different energies for Al and Cl. The similar features of density of states could be found in the molecular intercalation cases (figure 5.7(a)–5.7(p)), while the red-shift phenomena are created by the very strong p-type doping effects. That is, the important C–Cl bonds can enrich the valence van Hove singularities near the Fermi level (figure 5.7(a), 5.7(e), 5.7(i) and 5.7(m)). This is consistent with more charge variations between honeycomb lattice and intercalant layer. In addition, it is almost mission impossible to investigate the inter-ion and inter-molecule interactions from the van Hove singularities. [63]

The direct examinations of scanning tunneling spectroscopy (STS; details in chapter 3.2; [64]) are available only when a condensed-matter is sufficiently thin. The high-resolution STS measurements are very suitable for fully exploring the low-lying electronic properties of 2D/1D/0D few-layer group-IV and group-V systems [65]. For example, layered graphene systems, with/without chemical modifications, could present semi-conducting/semi-metallic/metallic behaviors through different van Hove singularities across the Fermi level, and stacking- and dimensionality-enriched singular structures [66, 67]. On the other hand, such observations are meaningless for thin-film $AlCl_4^-$-ion and $AlCl_4$-molecule graphite intercalation compounds, since the electronic properties are expected to be strongly modified by the z-direction quantum confinement. The theoretical predictions on the main features in density of states could be indirectly verified from the other fundamental properties, e.g., optical absorption peaks [68], energy loss spectra and electrical conductivities [69] related to the joint van Hove singularities.

In summary, Bernal graphite, stage-1 $AlCl_4^-$-ion and $AlCl_4$-molecule graphite intercalation compounds exhibit the diverse quasiparticle behaviors under the distinct orbital hybridizations of intralayer and interlayer chemical bonds. The effects, which are due to the significant van der Waals interactions, the rich intercalations of closed-shell ions, and the strong charge transfers, are responsible for the rich and unique properties. They are directly reflected in distinct crystal symmetries (periodical configurations perpendicular/on the (x, y) plane), largely enhanced moiré superlattices, the existence of many valence and conduction energy subbands (the zone-folding effect), variations about the initial high-symmetry valleys, largely enhanced asymmetry of hole and electron spectra about the Fermi level, vanishing band overlaps/p-type doping effect (zero/finite free carrier densities after ion/molecule interactions), almost isotropic/highly anisotropic features close/away from the Fermi level, various energy dependences at the distinct critical points, frequent band crossings and mixings (complicated orbital hybridizations), non-well-defined π- and σ- electronic band widths, carbon-, aluminum- and chloride-determined energy spectra at the distinct energy ranges. The active multi-/single-orbital hybridizations of C–C/C–Cl/Al–Cl/Cl–Cl chemical bonds are identified to be [2s, $2p_x$, $2p_y$]-[2s, $2p_x$, $2p_y$] and $2p_z$–$2p_z$/[2s, $2p_x$, $2p_y$, $2p_z$]-[3s, $3p_x$, $3p_y$, $3p_z$]/[3s, $3p_x$, $3p_y$, $3p_z$]-[3s, $3p_x$, $3p_y$, $3p_z$]/[3s, $3p_x$, $3p_y$, $3p_z$]-[3s, $3p_x$, $3p_y$, $3p_z$]. However, the observable evidence is very difficult to examine for C–Al and Al–Al bonds.

There exist important differences between alkali-atom and molecule-related graphite intercalation compounds in the main features of fundamental properties, such as, n- or p-type dopings [70], optical absorption spectra [71], and electron–hole and plasmon excitations [72]. Very interestingly, the intralayer/interlayer C–C orbital hybridizations, the interlayer C–intercalant interactions, and intra- and intercalant ones can provide useful information in establishing the reliable hopping integrals for the tight-binding model [73]. Up-to-date experiments only verify the observation of stage-3 and stage-4 cases [74, 75], in which the molecular dynamics could examine the existence of stage-1 and stage-2 configurations. Concerning the complicated intercalation and de-intercalation processes, their critical roles on the large-ion transports within high-performance batteries require systematic investigations, [75] especially for the development of theoretical frameworks.

References

[1] Sutter P, Sadowski J T and Sutter E A 2010 Chemistry under cover : tuning metal-graphene interaction by reactive intercalation *J. Am. Chem. Soc.* **132** 8175–79

[2] Dale A, Brownson C and Banks C E 2010 Graphene electrochemistry: an overview of potential applications *Analyst* **135** 2768–78

[3] Lin C Y, Wu J Y, Ou Y J, Chiu Y H and Lin M F 2015 Magneto-electronic properties of multilayer graphenes *Phys. Chem. Chem. Phys.* **17** 26008

[4] RozpLlocha F, Patyk J and Stankowski J 2007 Graphenes bonding forces in graphite *Acta Phys. Pol.* **112** 557–62

[5] Brownson D A C and Banks C E 2010 Graphene electrochemistry: an overview of potential applications *Analyst* **135** 2768–78

[6] Hu M, Dong X, Wu Y J, Liu L Y and Zhao Z S 2018 Low-energy 3D sp(2) carbons with versatileproperties beyond graphite and graphene *Dalton Trans.* **47** 6233–9

[7] Konschuh S, Gmitra M and Fabian J 2010 Tight-binding theory of the spin-orbit coupling in graphene *Phys. Rev.* B **82** 245412

[8] Huang J R, Lin J Y, Chen B H and Tsai M H 2008 Structural and electronic properties of few-layer graphenes from first-principles *Phys. Status Solid B-Basic Solid State Phys.* **245** 136–41

[9] de Andres P L, Ramirez R and Verges J A 2008 Strong covalent bonding between two graphene layers *Phys. Rev.* B **77** 045403

[10] Abergel D S L, Apalkov V, Berashevich J, Ziegler K and Chakraborty T 2010 Properties of graphene: a theoretical perspective *Adv. Phys.* **59** 261–482

[11] Lee S H, Chiu C W and Lin M F 2010 Deformation effects on electronic structures of bilayer graphenes *Phys.* E **42** 732–5

[12] Briggs N, Gebeyehu Z M, Vera A, Zhao T, Wang K, Duran A D, Bersch B, Bowen T, Knappenberger K L and Robinson J A 2019 Epitaxial graphene/silicon carbide intercalation: a minireview on graphene modulation and unique 2D materials *Nanoscale* **11** 15440–7

[13] Silva C C, Cai J Q, Jolie W, Dombrowski D, zum Hagen F H F, Martinez-Galera A J, Schlueter C, Lee T L and Busse C 2019 Lifting epitaxial graphene by intercalation of alkali metals *J. Phys. Chem.* C **123** 3712–13719
[14] Yoo H, Engelke R, Carr S, Fang S A, Zhang K, Cazeaux P, Sung S H, Hoyden R, Tsen A W and Taniguchi T 2019 Atomic and electronic reconstruction at the van der Waals interface in twisted bilayer graphene *Nat. Mater.* **18** 448
[15] Kaneko T and Saito R 2017 First-principles study on interlayer state in alkali and alkaline earth metal atoms intercalated bilayer graphene *Surf. Sci.* **665** 1–9
[16] Silva C C, Cai J Q, Jolie W, Dombrowski D, zum Hagen F H F, Martinez-Galera A J, Schlueter C, Lee T and Busse C 2019 Lifting epitaxial graphene by intercalation of alkali metals *J. Phys. Chem.* C **123** 13712–9
[17] Ho J H, Lu C L, Hwang C C, Chang C P and Lin M F 2006 Coulomb excitations in AA- and AB-stacked bilayer graphites *Phys. Rev.* B **74** 085406
[18] Pana C-J, Yuana C, Zhua G, Zhangc Q, Huangb C-J, Chang M, Angella L M, Hwangb B-J, Kaghazchic P and Daia H 2018 An operando x-ray diffraction study of chloroaluminate anion-graphite intercalation in aluminum batteries *PNAS* **115** 22
[19] Bhauriyal P, Mahata A and Pathak B 2017 The staging mechanism of $AlCl_4$ inter and calation in a graphite electrode for an aluminium-ion battery *Phys. Chem. Chem. Phys* **19** 7980
[20] Wua M S, Xua B, Chenb L Q and Ouyanga C Y 2016 Geometry and fast diffusion of $AlCl_4$ cluster intercalated in graphite *Electrochim. Acta* **195** 158–65
[21] Do T N, Chang C P, Shih P H, Wu J Y and Lin M F 2017 Stacking enriched magneto transport properties of few-layer graphenes *Phys. Chem. Chem. Phys.* **19** 29525–33
[22] Leon A and Pacheco M 2011 Electronic and dynamics properties of a molecular wire of graphane nanoclusters *Phys. Lett.* A **375** 4190–97
[23] Mounet N and Marzari N 2005 First-principles determination of the structural, vibrational and thermodynamic properties of diamond, graphite, and derivatives *Phys. Rev.* B **71** 205214
[24] Gao Y, Zhu C, Chen Z Z and Lu G 2017 Understanding ultrafast rechargeable aluminum-ion battery from first-principles *J. Phys. Chem.* C **121** 7131–8
[25] Lin M F, Huang C S and Chuu D S 1997 Plasmons in graphite and stage-1 graphite intercalation compounds *Phys. Rev.* B **55** 13961
[26] Alder B J and Wainwright T E 1959 Studies in molecular dynamics. 1. General method *J. Chem. Phys.* **31** 459–66
[27] Wang D-Y *et al* 2019 Insights into dynamic molecular intercalation mechanism for Al–C battery by operando synchrotron x-ray techniques *Carbon* **146** 528–34
[28] Lin C Y, Chen R B, Ho Y H and Lin M F 2018 *Electronic and Optical Properties of Graphite-Related Systems* (Boca Raton, FL: CRC Press)
[29] Malko D, Neiss C, Vines F and Gorling A 2012 Competition for graphene: graphynes with direction-dependent dirac cones *Phys. Rev. Lett.* **108** 086804
[30] Castro Neto A H, Guinea F, Peres N M R, Novoselov K S and Geim A K 2009 The electronic properties of graphene *Rev. Mod. Phys.* **81** 109–62
[31] Abergel D, Apalkov V, Berashevich J, Ziegler K and Chakraborty T 2010 Properties of graphene: a theoretical perspective *Adv. Phys.* **59** 261–482
[32] Wu C J and DasSarma S 2008 $p_{x,y}$-orbital counterpart of graphene: cold atoms in the honeycomb optical lattice *Phys. Rev.* B **77** 23

[33] Sprinkle M, Siegel D, Hu Y, Hicks J, Tejeda A, Taleb-Ibrahimi, Le Fevre P, Bertran F, Vizzini S and Enriquez H 2009 First direct observation of a nearly ideal graphene band structure *Phys. Rev. Lett.* **103** 226803

[34] Dresselhaus M and Dresselhaus G 2002 Intercalation compounds of graphite *Adv. Phys.* **51** 1–186

[35] Lin C Y, Wu J Y, Chiu Y H and Lin M F 2014 Stacking-dependent magneto-electronic properties in multilayer graphenes *Phys. Rev.* B **90** 205434

[36] Wu J Y, Chen S C, Roslyak O, Gumbs G and Lin M F 2011 Plasma excitations in graphene: their spectral intensity and temperature dependence in magnetic field *ACS Nano* **5** 1026

[37] Goerbig M O 2011 Electronic properties of graphene in a strong magnetic field *Rev. Mod. Phys.* **83** 1193–243

[38] Lin S-Y, Chang S-L, Shyu F-L, Lu J-M and Lin M-F 2015 Feature-rich electronic properties in graphene ripples *Carbon* **86** 207–16

[39] Shallcross S, Sharma S, Kandelaki E and Pankratov O A 2010 Electronic structure of turbostratic graphene *Phys. Rev.* B **81** 165105

[40] Reich S, Maultzsch J, Thomsen C and Ordejon P 2002 Tight-binding description of graphene *Phys. Rev.* B **66** 035412

[41] Lin Y, Chen G, Sadowski J T, Li Y Z, Tenney S A, Dadap J I, Hybertsen M S and Osgood R M 2019 Observation of intercalation-driven zone folding in quasi-free-standing graphene energy bands *Phys. Rev.* B **99** 035428

[42] Loh K P, Bao Q L, Ang P K and Yang J X 2010 The chemistry of graphene *J. Mater. Chem.* **20** 2277–89

[43] Jung S C, Kang Y J, Yoo D J, Choi J and Han Y K 2016 Flexible few-layered graphene for the ultrafast rechargeable aluminum-ion battery *J. Phys. Chem.* C **120** 13384–9

[44] Wang Q, Zheng D, He L and Ren X 2019 Cooperative effect in a graphite intercalation compound: enhanced mobility of $AlCl_4$ in the graphite cathode of aluminum-ion batteries *Phys. Rev. Appl.* **12** 044060

[45] Kiss T, Shimojima T, Ishizaka K, Chainani A, Togashi T, Kanai T, Wang X Y, Chen C T, Watanabe S and Shin S 2008 A versatile system for ultrahigh resolution, low temperature, and polarization dependent laser-angle-resolved photoemission spectroscopy *Rev. Sci. Instrum.* **79** 023106

[46] Lin C Y, Yang C H, Chiu C W, Chung H C, Lin S Y and Lin M F 2021 *Many-Particle Interactions in Carbon Nanotubes* (Bristol: IOP Publishing) in print

[47] Hwang E H and Das Sarma S 2008 Quasiparticle spectral function in doped graphene: electron–electron interaction effects in Arpes *Phys. Rev.* B **77** 081412

[48] Lin C Y, Lee M H and Lin M F 2018 Coulomb excitations in trilayer ABC-stacked graphene *Phys. Rev. B Rapid Commun.* **98** 041408

[49] Tasaki K 2014 Density functional theory study on structural and energetic characteristics of graphite intercalation compounds *J. Phys. Chem.* C **118** 1443–50

[50] Li W B, Lin S Y, Tran N T T, Lin M F and Lin K I 2020 Essential geometric and electronic properties in stage-n graphite alkali-metal-intercalation compounds *RSC Adv.* **10** 23573–81

[51] Geim A K and Mac Donald A H 2007 Graphene: exploring carbon flatland *Phys. Today* **60** 35–41

[52] Duong D L F, Yun S and Lee Y H 2017 van der Waals layered materials: opportunities and challenges *Acs Nano* **11** 11803–30

[53] Shioyama H and Fujii R 1987 Electrochemical reactions of stage 1 sulfuric acid—graphite intercalation compound *Carbon* **25** 771–4
[54] Sorokina N E, Maksimova N V, Nikitin A V, Shornikova O N and Avdeev V V 2001 Synthesis of intercalation compounds in the graphite–HNO_3–H_3PO_4 system *Inorg. Mater.* **37** 584–90
[55] Zhan D, Sun L, Ni Z H, Liu L, Fan X F, Wang Y y, Yu T, Lam Y M, Huang W and Shen Z X 2010 $FeCl_3$-based few-layer graphene intercalation compounds: single linear dispersion electronic band structure and strong charge transfer doping *Adv. Funct. Mater.* **20** 3405–9
[56] Ro senzweig P, Karakachian H, Marchenko D, Kuster K and Starke U 2020 Overdoping graphene beyond the van Hove singularity *Phys. Rev. Lett.* **125** 176403
[57] Bostwick A, Ohta T, Seyller T, Horn K and Rotenberg E 2007 Quasiparticle dynamics in graphene *Nat. Phys.* **3** 36–40
[58] Dresselhaus M S and Dresselhaus G 2002 Intercalation compounds of graphite *Adv. Phys.* **51** 1–186
[59] Ro senzweig P, Karakachian H, Marchenko D, Kuster K and Starke U 2020 Overdoping graphene beyond the van Hove singularity *Phys. Rev. Lett.* **125** 176403
[60] Dresselhaus M S and Dresselhaus G 2002 Intercalation compounds of graphite *Adv. Phys.* **51** 1–186
[61] Tasaki k 2014 Density functional theory study on structural and energetic characteristics of graphite intercalation compounds *J. Phys. Chem.* C **118** 1443–50
[62] Liu H T, Liu Y Q and Zhu D B 2011 Chemical doping of graphene *J. Mater. Chem.* **21** 3335–45
[63] Li G H, Luican A, dos Santos J M B L, Castro Neto, Reina A, Kong J and Andrei E Y 2010 Observation of Van Hove singularities in twisted graphene layers *Nat. Phys.* **6** 109–13
[64] Garcia N 1986 Theory of scanning tunneling microscopy and spectroscopy—resolution, image and field states, and thin oxide layers *IBM J. Res. Dev.* **30** 533–42
[65] Bando H, Tokumoto H, Mizutani W, Endo K, Wakiyama S, Watanabe K and Kajimura K 1987 Measurement of local density of states by scanning tunneling spectroscopy at low-temperature *Jpn. J. Appl. Phys.* **26** 627–8
[66] Luican A, Li G H, Reina A, Kong J, Nair R R, Novoselov K S, Geim A K and Andrei E Y 2011 Single-layer behavior and its breakdown in twisted graphene layers *Phys. Rev. Lett.* **106** 126802
[67] Brihuega I, Mallet P, Gonzalez-Herrero H, de Laissardiere G T, Ugeda M M, Magaud L, Gomez-Rodriguez J M, Yndurain F and Veuillen J Y 2012 Unraveling the intrinsic and robust nature of van Hove singularities in twisted bilayer graphene by scanning tunneling microscopy and theoretical analysis *Phys. Rev. Lett.* **109** 196802
[68] Kravets V G, Grigorenko A N, Nair R R, Blake P, Anissimova S, Novoselov K S and Geim A K 2010 Spectroscopic ellipsometry of graphene and an exciton-shifted van Hove peak in absorption *Phys. Rev.* B **81** 155413
[69] Nozieres P and Pines D 1959 Electron interaction in solids—characteristic energy loss spectrum *Phys. Rev.* **113** 1254–67
[70] Csanyi G, Littlewood P B, Nevidomskyy A H, Pickard C J and Simons B D 2005 The role of the interlayer state in the electronic structure of superconducting graphite intercalated compounds *Nat. Phys.* **1** 42–5
[71] Lin C Y, Chen R B, Ho Y H and Lin M F 2018 *Electronic and Optical Properties of Graphite-Related systems.* (Boca Raton, FL: CRC Press)

[72] Tassin P, Koschny T, Kafesaki M and Soukoulis C M 2012 A comparison of graphene, superconductors and metals as conductors for metamaterials and plasmonics *Nat. Photonics* **6** 259–64
[73] Charlier J, Michenaud J, Gonze X and Vigneron J 1991 Tight-binding model for the electronic properties of simple hexagonal graphite *Phys. Rev.* B **44** 13237
[74] Wang D Y, Huang S K, Liao H J, Chen Y M, Wang S W, Kao Y T, An J Y, Lee Y C, Chuang C H and Huang Y C 2019 Insights into dynamic molecular intercalation mechanism for Al–C battery by operando synchrotron x-ray techniques *Carbon* **146** 528–34
[75] Childress A S, Parajuli P, Zhu J Y, Podila R and Rao A M 2017 A Raman spectroscopic study of graphene cathodes in high-performance aluminum ion batteries *Biotechnol. Adv.* **29** 189–98

Chapter 6

$FeCl_3$ graphite intercalation compounds: iron-ion-based battery cathodes

Wei-Bang Li, Shih-Yang Lin, Thi Dieu Hien Nguyen, Hsien-Ching Chung, Ngoc Thanh Thuy Tran, Nguyen Thi Han, Hsin-Yi Liu, Hai Duong Pham and Ming-Fa Lin

Graphene is a single layer with carbon honeycomb lattices [1]. A bulk graphite [2–5], which is composed of multi-layer graphene systems, has attracted many theoretical and experimental researches, e.g. basic science, engineering and applications. 2D graphene layer and 3D graphite systems possess various types of stackings, i.e., AA [6]/AB [7]/ABC [8] stackings. All above stackings represent semimetals, and there are weak but significant Van der Waal interactions between the graphene layers. In general, such condensed-matter systems become *n*-type or *p*-type metals depending on the species of interacting matter. For instance, the intercalation of an alkali metal into graphite will turn this system into an *n*-type metal because of free conduction electrons, whereas the intercalation of $FeCl_3$ turns it into a *p*-type metal because of free conduction holes. The interesting features of geometric structures and electronic properties lead to the application technology, such as green energy, and lithium-ion batteries (LIBs).

LIBs are well-known rechargeable batteries and are promising for various devices [9], e.g. cell phones, vehicles and almost all portable equipment, because of the reversible intercalation/de-intercalation of Li^+ into/from the host lattices of either graphite-layered anodes or other layer-based cathodes. The intercalation/de-intercalation features on graphite are very important for the development of LIBs. Recently, many research papers mainly focused on the distinguished guest atoms/ions like Li, Na, K [10, 11] or ion clusters like $AlCl_4$, Al_2Cl_7 [12–14]; also, battery capacity was gained with ferric chloride graphite intercalation compounds ($FeCl_3$-GICs) [15, 16]. The $FeCl_3$-GIC is found to possess a high reversible capacity, cycling stability and an outstanding rate, demonstrating its potential as an electrode

doi:10.1088/978-0-7503-4685-6ch6

material in LIBs. $FeCl_3$ is known to extract electrons from graphite, thereby creating a *p*-type doping effect [17].

In experiments, researchers synthesized and exfoliated $FeCl_3$-GICs by mixing the $FeCl_3$ and graphite under proper temperature in dimethyl formamide (DMF) solvent [15]. These studies monitored the evolution of graphite in intercalation and exfoliation processes by x-ray diffraction (XRD) with Cu Kα radiation and a Renishaw inVia Raman microscope, and then analyzed the results on a Zeiss Supra 55 field-emission scanning electron microscope (SEM), FEI Tecnai G220 high-resolution transmission electron microscope (TEM), and Bruker multimode 8 atomic force microscope (AFM) under scan Asyst mode. Taking the stage-1 $FeCl_3$-GICs as an example, the pristine structures of graphite are broken with the expanded interlayer spacing. According to the results, the stage-1 structure of $FeCl_3$-GICs is explored by XRD. For pristine graphite, the periodic interlayer distance between each layer is about 0.33 nm; after intercalation with $FeCl_3$, the interlayer distance is significantly increased to 0.96 nm. The $FeCl_3$-GIC results in a larger capacity than pure graphite. Also, the $FeCl_3$-GIC exhibits a better cycling stability and rate capability compared with pure $FeCl_3$ [16]. Other experiments even show that few-layered $FeCl_3$-GIC has a better rate capacity than stage-1 $FeCl_3$-GIC, which may be due to the residual $FeCl_3$ in few-layered $FeCl_3$-GIC as the main active material; Li-ions rapidly insert themselves into or extract from the interlayered spacing due to the nanoscale thickness and porous structure.

As an electrode for Li-ion batteries, stage-1 $FeCl_3$-GIC possesses reversible capacities of 525 mAh g^{-1} after 100 charge–discharge cycles [16]. This is considerably higher than the capacity for expanded graphite, which is 113 mAh g^{-1}. Moreover, the in-plane conductivity at room temperature for stage-1 $FeCl_3$-GIC is $1.1 \times 10^5\ \Omega\ cm^{-1}$, which is higher than the $2.5 \times 10^4\ \Omega\ cm^{-1}$ in graphite. Not only that, but stage-2 $FeCl_3$-GIC also exhibits a much higher reversible capacity with a specific capacity of 719 mAh g^{-1} after 100 charge–discharge cycles, and even 910 mAh g^{-1} after 170 charge–discharge cycles. The special and important performance of $FeCl_3$-GICs undoubtedly results from the unique geometric structure of intercalations and the stage index; in other words, the interlayer spacings of graphite are enlarged because of the intercalations of $FeCl_3$-GICs, and the enlarged spacings provide larger channels for the electrochemical reaction while promoting fast Li-ion transport. Although the reversible cycle capacity of stage-1 $FeCl_3$-GIC is lower than that of stage-2 $FeCl_3$-GIC, it is still a significant improvement and worthy of further exploration. Our work is focused mainly on the study of stage-1 $FeCl_3$-GIC, and we will show details of the essential properties of electrons, e.g. geometric structure/band structure/density of state/spatial charge distribution/charge variation.

In theory, the quantum mechanical wave functions provide all information about the system, but it is hard to solve the Schrödinger equation in a many-particle condensed matter system because of the electron–electron interactions, electron–ion interactions and so on. Up to now, first-principles calculations [18] have been the dominant method for solving the complicated interactions through some approximations and simplifications to obtain the eigenvalues and eigenfunctions; going a step further, we can also model and explore the essential properties. The Vienna *ab initio* simulation package (VASP) [19] computes the approximate solutions within density functional theory (DFT) [20–23] to solve the Kohn–Sham equations. The exchange-correlation energy due to the electron–electron interactions is calculated from the Perdew–Burke–Ernzerhof functional (PBE) [24–26] under generalized gradient approximation (GGA); projector-augmented wave methods (PAW) [27, 28] are utilized to evaluate the electron–ion interactions.

6.1 Crystal symmetry in a large primitive unit cell

In this chapter, the optimal structures and electronic properties of 3D graphite-related systems are studied by utilizing VASP. These systems include monolayer, Bernal graphite, hexagonal graphite, and intercalation compounds. A primitive unit cell depends on the geometric distributions of host and guest atoms. In our calculations, the Brillouin zone is set to be $20 \times 20 \times 20$ and $50 \times 50 \times 50$ k point meshes for geometric structures and electronic distributions.

In pristine AA-stacking graphite, each layer is bound by the weak but significant van der Waals interactions due to the perpendicular P_z orbital between the neighboring layers. The strong σ bondings appear between carbon atoms in hexagonal structures on every plane, and the bond length is 1.42 Å. In AA-stacking configurations, the interlayer distance of each neighboring layer is roughly 3.55 Å [29, 30], which is large enough to be intercalated/de-intercalated by certain atoms, ions, or even layered materials. Up to now, the graphite intercalation compounds of alkali metal atoms, e.g. Li, Na, K and Rb, have been well-known, as are some molecular clusters, e.g. $AlCl_4$, N_2O_5, Na(ethylenediamine)C_5 [31]. The interlayer distances between neighboring carbon layers will depend on the species of intercalations, for instance, the distances are 3.76 Å and 5.65 Å for, respectively, Li atoms and Rb atoms [11]. Apparently, the interlayer distances are easily affected by intercalations. However the bond lengths between carbons in hexagonal structures, or in the same planes, are only slightly changed from 1.42 Å to 1.40 Å. The planar hexagonal structure is retained, so that the σ bondings due to the (2s, $2P_x$, $2P_y$) orbitals of the carbon atoms seldom take part in the orbital hybridizations between intercalated atoms/ions/clusters and carbon atoms.

Previous experimental and theoretical studies have been performed on the properties of stage-n systems of $FeCl_3$-graphite intercalation compounds [15, 16, 32, 33]. The σ-bonding remains the same, but the carriers are quite different when intercalation occurs. In experimental methods, some stage-n $FeCl_3$-graphite intercalation compounds are prepared from highly oriented pyrolytic graphite (HOPG), which is well-ordered in the c-axis direction. A standard two-temperature-zone technique was applied for intercalation. By regulating the temperature difference between graphite and $FeCl_3$, it is probable to derive different stage numbers n [34–38]. The configurations for all stages present periodical arrangements in the c-axis. According to some studies, the stage-2 $FeCl_3$-graphite intercalation compound is better than the stage-1 one compound in electrochemical performance, e.g. reversible cyclic capacity and stability. Even though stage-2 shows good performance, its configuration is not uniform. As mentioned in chapter 5, all guests must comply with the thermodynamic laws so that a stage-1 compound is more likely to be obtained in syntheses with diverse concentrations.

This work is mainly focused on stage-1 $FeCl_3$-graphite intercalation compounds. The stage-1 configuration means that every spacing of graphitic interlayers is fully occupied by the intercalant materials, which stands for layered $FeCl_3$ in this chapter; in other words, each intercalant material is sandwiched between two graphitic layers. It is also the highest concentration for layered intercalant compounds.

The optimized structure for the $FeCl_3$-graphite intercalation compound is shown in figure 6.1. This supercell with lattice constants $a = b = 12.12$ Å, which is constructed by combining 2×2 unit cells of $FeCl_3$ and 5×5 unit cells of graphene as mentioned in previous works [33, 35], covers 50 carbon atoms, 8 iron atoms and 24 chlorine atoms, due to the high symmetry, and is quite big. In addition, the $FeCl_3$ also forms layered structures between the carbon layers. The large spacings of interlayer distances between graphenes allow the guest $FeCl_3$ to be intercalated. Calculation results show that the optimized interlayer distances between the pristine carbon layers are enlarged from 3.55 Å to 9.200 Å. The calculated results are very similar to the x-ray diffraction experimental observations, which showed an interlayer distance of 9.36 Å [15, 33, 39–41]. Interestingly, the tremendous changes of interlayer distance reveal that the interactions between interlayered $2P_z$ orbitals become much weaker after intercalation. The pristine graphite has very stable AB-stacking configurations, but it turns into AA-stacking; this is also the case for other stage-1 intercalations, e.g. alkali metal atoms.

6.2 Free valence holes by molecular anions

Up to know, the usual stacking configurations of AAA, ABA and ABC have been synthesized in experiments. The AAA stacking is often presented in stage-1 graphite intercalation compounds because of the higher symmetry. However, natural graphite almost only takes on ABA and ABC structures. The three stackings of AAA, ABA and ABC are predicted to have unusual energy bands, respectively, and this indicates the important and essential properties of the special charge distributions and the ratio of free electrons/holes. It is well-known that the monolayer graphene is a semiconductor with zero gap, and there exist linearly intersecting valence and conduction bands at the Dirac-cone point. Thus the density of states (DOS) vanishes at the Fermi level. The linear Dirac-cone structures on the k-space (k_x, k_y) planes are clearly revealed in the AAA-stacking and ABC-stacking graphites. In general, the energy of the Dirac point is quite sensitive to the k_z component wave vector. Among

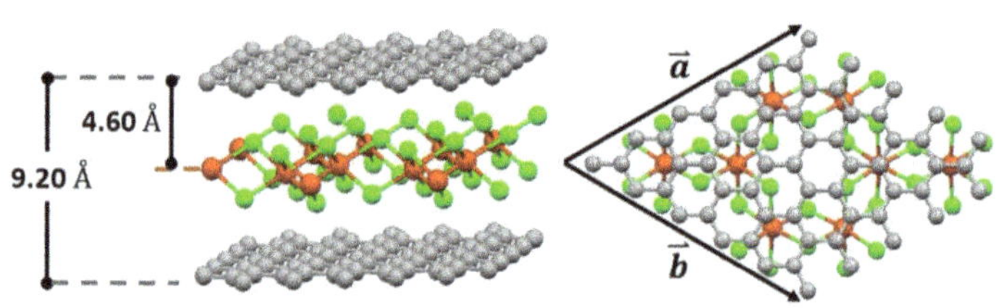

Figure 6.1. The geometric structure of $FeCl_3$-graphite intercalation compound.

the three kinds of stacking, the AAA-stacking graphite possesses a periodical arrangement of graphite layers in the z-direction and creates a band width as roughly 1 eV. Also, the density of state exists at the Fermi level. That is to say, the free electrons and holes appear simultaneously in the range of −0.5–0 eV and 0–0.5 eV, respectively. For ABA- and ABC-stacking graphites, the band widths in terms of the k_z-dependent π-electronic band are 0.2 and 0.03 eV, respectively, near the Fermi level along KH line in k_z-dependent band structure. As a result, the AAA-, ABA- and ABC-stacking graphites are semimetalic with the free electrons and holes appearing simultaneously, while the latter two only possess relatively few free carriers. The electronic structures exhibit severe changes after the intercalation. That is to say, the electronic structures strongly depend on the properties of the intercalant materials. In this chapter, we intercalate $FeCl_3$ as an anion into graphitic layers.

It is well-known that the electrons situated on the outermost orbitals of an atom or molecule, i.e. the valence electrons, will dominate the size, electron binding energy and other chemical reactivity. Moreover, the valence electron of anions tends to be removed due to its relatively weak binding energy. In theory, the Coulomb attractions are $(Z-1)e^2/r$ for anions, where Z is the magnitude of the negative charge; and $-Ze^2/r$ is for neutral molecules with $Z = 1$ and for cations with $Z = n+1$, where n is the magnitude of the positive charge. Specifically, an electron in the valence region of an anion possesses an almost zero net Coulomb attraction in its asymptotic regions, whereas the electrons in neutral molecules and cations experience non-zero Coulomb attraction. The physical and chemical properties of anions are very different from those of natural molecules or cations. The negative charges in anions drive them to interact with surrounding molecules and ions. Also, the anions polarize the electronic clouds of nearby molecules in the opposite direction that cations do. The anions have a polarizability and thus tend to strongly interact more with surrounding molecules than the less polarizable neutral molecules and cations. In addition, the valence electron binding energy in anions is often smaller than neutral molecules and cations. In many cases of intercalation, the intercalant species will critically govern the p- or n-type, as mentioned above. The system of graphite intercalation compounds exhibit n-type doping with the alkali-metal atom intercalants, for instance, the alkali-metal atoms will provide the electrons of the outmost orbitals, i.e., s-orbitals will lose electrons. And thus the graphite will behave like a metal due to the extra free electrons from the guest compound [11]. This result can be analyzed by the band structures and the DOS as well. In the band structure, the Fermi level shifts from the Dirac point to the conduction points. The DOS allows a complete understanding of the relations among the different chemical bondings. The lowest valley in DOS with respect to the energy presents the blue shift of the Fermi level in agreement with the shift in band structure. Furthermore, the DOS indicates the hybridization in host–guest, host–host and guest–guest atoms. By the Bader analysis in VASP calculations, we can predict the charge transfers of the host and guest atoms. After intercalation, the alkali-metal atoms behave like cations due to the loss of electrons.

In our work, we explore the band structures, DOS, charge distributions and the differences between charge distributions of stage-1 $FeCl_3$-graphite intercalation compounds on VASP calculations with proper parameters. The above-mentioned properties allow us to analyze and determine further essential mechanics, e.g. host–guest orbital hybridizations, the interlayer interactions, the intralayer interactions, and the intralayer σ- and π-bondings. For $FeCl_3$-graphite intercalation compounds, the first Brillouin zone in the reciprocal lattice space is a symmetric hexagon, as exhibited in graphene. The 3D band structures of the hexagon are described along the high symmetry points in reciprocal lattice [Γ–K–M–Γ–A–H–*L*–A], as shown in figure 6.2.

As shown in figure 6.3, the calculated band structures of $FeCl_3$-graphite intercalation compounds possess a Dirac cone with a blue shift of the Fermi level around 0.9 eV. The complicated horizontal bands mainly arise from the d-orbitals, i.e., the band structures of $FeCl_3$-graphite intercalation compounds are very similar to those of monolayer graphene, except for the d-orbitals. Zhan *et al* published a research paper about $FeCl_3$-based few layer graphene intercalation compounds, which contains 2-layer- to 4-layer-intercalation compounds [33]. The band structures in the low-lying energy region are in agreement with our results. The interlayer distances between carbon layers are approximately 9.4 Å and 9.2 Å in $FeCl_3$-few-layer graphene intercalation compounds and $FeCl_3$-graphite intercalation compounds, respectively; the lattice constants of graphene and $FeCl_3$ are 6.06 Å and 2.46 Å, respectively. The rather large interlayer distance indicates that the feature of low-lying energy of this system is in good agreement with the monolayer graphene due to the loss of coupling interactions between the neighboring graphene layers.

The band structures at about 0 and 1.35 eV present d-orbitals of the iron atoms. The band structures excluding complicated d-orbitals are very similar to graphene, which possesses the Dirac points at 0 eV, with 0.9 eV with a red shift of the Fermi level. The red shifts indicate that there exist free holes in valence bands below the Dirac-cone point. That is to say, the system of stage-1 $FeCl_3$-graphite intercalation compounds turns into a p-type system. In addition, the charge transfers for our systems are examined and analyzed by Bader analyses in the VASP calculations. The charge transfers per atom (electrons/atom) for C, Fe, Cl are −0.02 e/atom, −0.28 e/atom and +0.46 e/atom, respectively. In our stage-1 $FeCl_3$-graphite intercalation

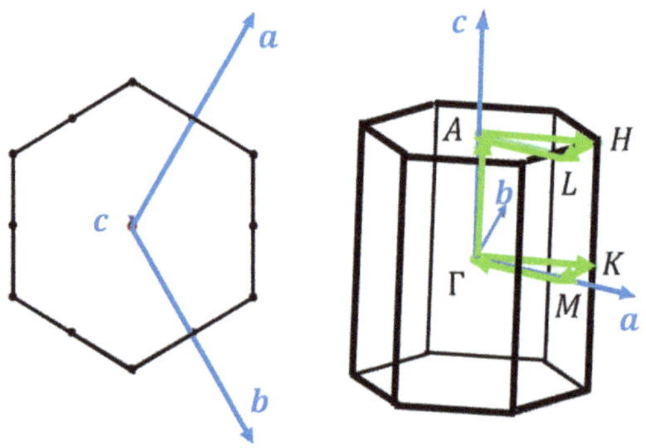

Figure 6.2. The first-Brillouin zone of $FeCl_3$-graphite intercalation compound.

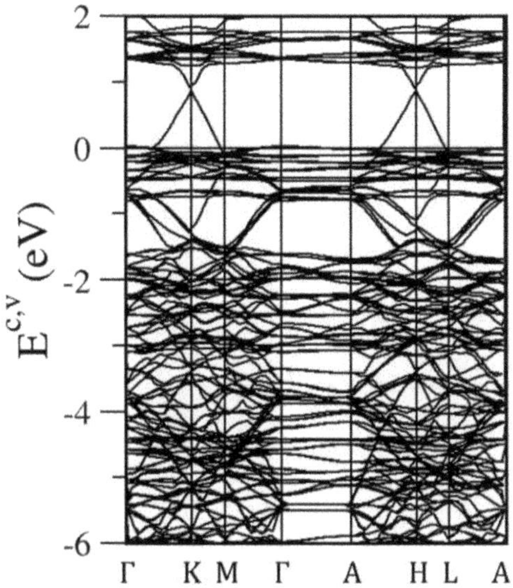

Figure 6.3. The band structures of $FeCl_3$-graphite intercalation compound.

compound unit cell, which contains 50 C atoms, 8 Fe atoms and 24 Cl atoms, the total transfers for C, Fe, Cl are −1 electrons, −2.24 electrons and +3.68 electrons, respectively. Thus, C atoms and Fe atoms present as positive ions, but the Cl atoms present as negative ions. Due to the above, the pristine Bernal graphite will possess many more free holes after intercalations, i.e., it exhibits p-type doping features.

6.3 Diverse dopings through different anion and cation intercalations/de-intercalations

Previous works showed that the graphite system can change into p- or n-type doping. During the intercalation, intercalants interact with the host material and transfer charges, which can be electrons or holes, to carbon atoms resulting in p- or n-type doping. As the diverse species of intercalants progress, the special orbital hybridizations come into play. The band structures, DOS, spatial charge distributions and the charge variations can give more information about the important chemical bondings, e.g., Fe–Cl bonds, Cl–Cl bonds, C–Cl bonds and C–C bonds.

In pristine graphite, the (2s, $2p_x$, $2p_y$) orbitals dominate the σ bondings in the planar honeycomb structure, which mainly form the geometric structures. The Van Hove singularities in 3D graphite-related systems are revealed as special structures. The atom- and orbital-decomposed DOS, as shown in figure 6.4, could provide full information and are useful in examining the host/guest effects after intercalations. In principle, these structures originate from the critical points in the energy-wave-vector space. In AB-stacking graphite, the parabolic dispersions have band-edge states at the M-point, in which they belong to the saddle points in the energy-wave-vector space. This indicates that π electronic states are accumulated near the M-point and will make important contributions to the essential physical properties.

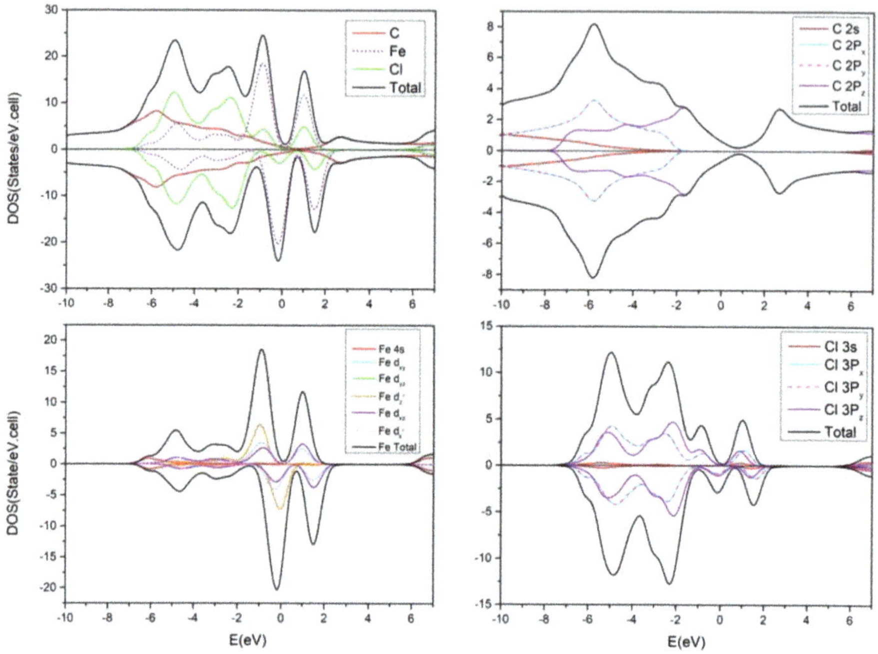

Figure 6.4. The total density of states and decomposed density of states for C, Fe, and Cl, respectively.

The saddle point is mainly owed to the interlayered interaction. In other word, the interactions between neighboring layer atoms will result in the asymmetry of valence and conduction bands. The low-lying $2p_z$ orbitals will be easily influenced after intercalation and even sensitive to the kinds of intercalants. On the other hand, the 2s, $2p_x$ and $2p_y$ orbitals can generate much deeper energy bands initiated by the Γ points at $E = -3$ eV.

In previous cases of diverse intercalations such as alkali-metals, e.g. Li/Na/K/Rb/Cs atoms, the weak but significant van der Waals interactions arise from the interlayer $2p_z$–$2p_z$ and $2p_z$-s orbital hybridizations in C–C and C–A bonds, respectively, where A represents alkali metals; these interactions also make the most important contribution to the low-lying π-electronic structures and thus dominate the essential physical properties. For instance, the number of stages will lead to quite different amounts of asymmetric free electrons and free holes; the distinct alkali metal atoms also cause much different interlayer distances because of their very different atom sizes.

In the case of $FeCl_3$ as an intercalant, this super big intercalant will have a charge distribution and charge variation in C–Cl bonds, as shown in figure 6.5. The charge variation reveals that the amount of difference between C and Cl atoms is not too much, and this agrees with the charge transfers mentioned above of only 0.02 electrons per atom from C to Cl atoms. This phenomenon probably indicates that the Cl atoms are close to Fe atoms and mainly form bonds with Fe atoms. Similarly, the charge distributions between intralayer neighboring C atoms are very strong due

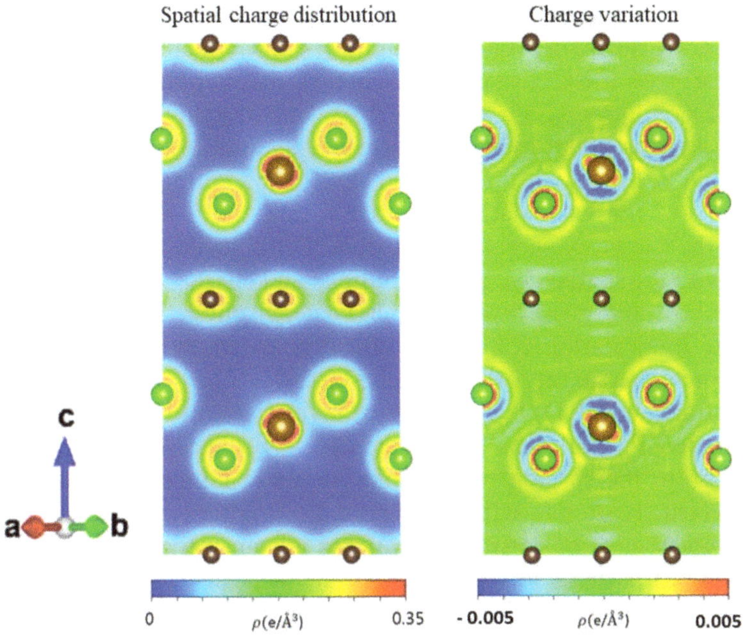

Figure 6.5. The spatial charge distribution and charge variation.

to the parallel $2p_z$ orbital hybridizations; after intercalations, the variation almost vanishes between the intralayer neighboring C atoms. That is to say, the C–C bonds in intralayer remain even intercalating. The anisotropic charge distributions between Fe atoms and Cl atoms reveal that there might exist interactions between them.

The van Hove singularities in graphite-related systems are very important and useful, being revealed as special structures and quite different from those in layered graphene systems. The atom- and orbital-decomposed DOS, as shown in figure 6.4, could provide full information for examining the intercalation effects. In principle, these structures originate from the critical points in the energy-wave-vector space. The band structures are very complicated due to the d-orbitals below 0 eV and above 1.3 eV. However the Dirac-cone point remains with a shift of Fermi level about 0.9 eV and is in good agreement with the lowest valley at $E = \sim 0.9$ eV in the DOS. In figure 6.4, the orbital-decomposed DOS for C atoms presents that the σ bands are initiated from −2 eV and the difference between the initiation and Fermi level is roughly −2.9 eV to −3 eV centered from Fermi level, and it almost remains the same as pristine AB-stacking graphite. The orbital-decomposed DOS for C atoms also shows that the π bondings are orthogonal to the sp^2 bondings, and the π bondings are situated at −1.8 eV without other singularities there. This means that there exist C–C bonds in the intralayer C atoms.

Based on the van Hove singularities in DOS, the Fe atoms and Cl atoms form the Fe–Cl bondings at −4.7 eV to −5 eV, −3.2 eV to −2.3 eV, −1.5 eV to −1.8 eV, −1 eV to 0 eV and 1 eV to 2 eV, presenting the multi-orbital hybridizations of [4s, $3d_{xy}$, $3d_{yz}$, $3d_{xz}$, $3d_{x^2-y^2}$, $3d_{z^2}$]-[$3p_x$, $3p_y$, $3p_z$]. Corresponding to the band structures, apparently, the complicated d orbitals of Fe atoms, which originate from $E < 0$ eV and $E > 1$ eV, are the main orbitals of bonding with Cl atoms, especially −1 eV–0 eV and 1 eV–2 eV. As for C–Cl and Cl–Cl bonds, the unique van Hove singularities exhibit their couplings in −6 eV to −2 eV and roughly −6.2 eV, respectively, revealing the multi-orbital hybridizations of [$2p_x$, $2p_y$, $2p_z$]-[$3p_x$, $3p_y$, $3p_z$] and [3s, $3p_x$, $3p_y$, $3p_z$]-[3s, $3p_x$, $3p_y$, $3p_z$]. The red shift of the Fermi level reveals the p-type doping effects of the intercalants. In other words, the C–Cl bonds will affect the diversity of the van Hove singularities around Fermi level. Also, these results are in agreement with the spatial charge distributions and charge variations in the spacing between the host and guest layers. The low-energy physical properties of graphite and graphite intercalation compounds are dominated by the multi-orbital interactions between intercalants and graphene layers.

X-ray diffraction (XRD) is available for investigation of crystalline materials or graphene-related intercalation compounds [42–50]. The used wavelength of the x-rays is close to the spacing between the atoms in lattice. The incident photons experience elastic scattering within the periodic structure of the lattice atoms and emerge as constructive or destructive patterns in some special directions. Graphite has typical XRD patterns in which the dominant peak relates to the basal planes (002). The scattering peak appears at an angle of $2\theta \approx 26.4°$ and corresponds to $d_{002} = 3.35$ Å. The intercalation of stage-1 $FeCl_3$ increases the interlayer distance and thus the peak is displaced to new scattering angles at 9.1°, 18.6°, 28.1°, 37.8°, 50.4°. The interlayer distance d of $FeCl_3$-graphite intercalation compounds is drastically increased to 9.6 Å, corresponding to the scanning electron microscope (SEM) measurements in previous studies. In addition, the characteristic Raman scattering is utilized for estimating the Fermi level of the graphite intercalation compounds.

Raman spectroscopy uses a monochromatic laser to interact with molecular vibrational modes and phonons in a sample. A shifting-down of the laser energy is called Stoke Raman scattering, while shifting-up is called Anti-Stoke Raman scattering through inelastic scattering. The peaks in Raman spectra exhibit some important information, e.g., the G peak (1580 cm^{-1}) comes from the C–C sp^2-orbital hybridizations, the D peak and the 2D peak present the first order scattering of the zone-boundary phonons and the second order scattering of the zone-boundary phonons, respectively. The G peak is also used to estimate the staging. In general, stage-1 graphite intercalation compounds possess a single G peak. For p-type graphite intercalation compounds:

$$E_T = E_L - \hbar\omega_{2D},$$

where E_L, E_T and ω_{2D} represent the excitation energy, the energy corresponding to the electron–hole recombination process and the 2D frequency, respectively [51]. This is due to the ratio of 2D peaks and G peaks. The E_T and E_L are situated at the valence band and conduction band, respectively. Apparently, the equation is only applicable if $E_L > 2E_F$. Applying a distinct wavelength of a laser allows us to detect the Fermi level. The previous experimental works put the Fermi level of 0.9 eV[39], which is in agreement with our calculations.

In summary, the stage-1 $FeCl_3$-graphite intercalation compounds present diversified electronic properties due to the intercalant. The first-principles calculations on VASP are utilized to analyze the essential properties, such as the geometric structures, spatial charge distributions, charge variations, band structures and density of states. The interlayer distances of pristine graphite are enlarged from 3.35 Å to 9.40 Å by the super big molecule $FeCl_3$, in which the unit cell consists of 50 C atoms, 8 Fe atoms and 24 Cl atoms. The density of states displays full information for an explanation of the hybridizations with the special structures van Hove singularities on it. The lowest valley in the DOS shifts from 0 to 0.9 eV. In other words, the Fermi level exhibits a red shift. Furthermore, the intercalant $FeCl_3$ results in a p-type doping. The interaction between interlayer C atoms becomes very weak but almost remains the same between the intralayer C atoms. The Fe–Cl bondings arise from multi-orbital hybridizations of [4s, $3d_{xy}$, $3d_{yz}$, $3d_{xz}$, $3d_{x^2-y^2}$, $3d_{z^2}$]-[$3p_x$, $3p_y$, $3p_z$]. C–Fe bondings do apparently not exist in our results. In addition, for the $FeCl_3$-graphite intercalation compound, the large size and thickness hinder the transport of lithium ions and electrons. On the other hand, the layered structures with enlarged interlayer spacings allow the transport of lithium ions and electrons and provide additional sites for the accommodation of lithium, and thus exhibit a better storage performance as an electrode in lithium-ion batteries. The differences of essential properties between distinct staging configurations are worth investigating as well. The stage-*n* systems with increasing *n* will possess decreasing concentrations and lower symmetries. Interestingly, for the $FeCl_3$-graphite intercalation compounds, the stage-1 case performs better than pure graphite in both theory and experiment, whereas the stage-2 case exhibits a higher reversible capacity and better stability than the stage-1 case [16]. The complicated mechanics of intercalation and de-intercalation in stage-*n* configurations, where *n* is greater than 2, are very important and worth investigating further in the future in order to construct a theoretical framework coinciding with the presently performed experiments.

References

[1] Tran N T T, Lin S Y, Lin C Y and Lin M F 2017 *Geometric and Electronic Properties of Graphene-Related Systems: Chemical Bonding Schemes* (Boca Raton, FL: CRC Press)

[2] Lee S M, Kang D S and Roh J S 2015 Bulk graphite: materials and manufacturing process *Carbon Lett.* **16** 135–46

[3] Wang Y, Panzik J E, Kiefer B and Lee K K 2012 Crystal structure of graphite under room-temperature compression and decompression *Sci. Rep.* **2** 520

[4] Ho J H, Chang C P and Lin M F 2006 Electronic excitations of the multilayered graphite *Phys. Lett.* A **352** 446–50

[5] Lin C Y, Chen R B, Ho Y H and Lin M F 2017 *Electronic and Optical Properties of Graphite-Related Systems* (Boca Raton, FL: CRC Press)

[6] Zhang Z, Huang H, Yang X and Zang L 2011 Tailoring electronic properties of graphene by π-π stacking with aromatic molecules *J. Phys. Chem. Lett.* **2** 2897–905

[7] Cheng Q *et al* 2017 Graphene-like-graphite as fast-chargeable and high-capacity anode materials for lithium ion batteries *Sci. Rep.* B **14782**

[8] Mecklenburg M *et al* 2011 Spin and the honeycomb lattice: lessons from graphene *Phys. Rev. Lett.* **106** 116803
[9] Cheng Q 2017 Graphene-like-graphite as fast-chargeable and high-capacity anode materials for lithium ion batteries *Sci. Rep.* **7** 14782
[10] Natori A, Ohno T and Oshiyama A 1985 Work function of alkali-atom adsorbed graphite *J. Phys. Soc. Jpn.* **54** 3042–50
[11] Li W B, Lin S Y, Tran N T T, Lin M F and Lin K I 2020 Essentail geometric and electronic properties in stage-n graphite alkali-metal-intercalation compounds *RSC Adv.* **10** 23573
[12] Bhauriyal P, Mahata A and Pathak B 2017 The staging mechanism of $AlCl_4$ intercalation in a graphite electrode for an aluminium-ion battery *Phys. Chem. Chem. Phys.* **19** 7980
[13] Wu M S, Xu B, Chen L Q and Ouyang C Y 2016 Geometry and fast diffusion of $AlCl_4$ cluster intercalated in graphite *Electronchem. Acta* **195** 158
[14] Wand D Y *et al* 2017 Advanced rechargeable aluminium ion battery with a high-quality natural graphite cathode *Nat. Commun.* **8** 14283
[15] Qi X *et al* 2015 $FeCl_3$ intercalated few-layer graphene for high lithium-ion storage performance *J. Mater. Chem.* A **3** 15498
[16] Wang L, Zhu Y C, Guo C, Zhu X, Liang J and Qian Y T 2014 Ferric chloride-graphite intercalation compounds as anode materials for Li-ion batteries *ChemSusChem* **7** 87
[17] Liu W, Kang J and Banerjee K 2016 Characterization of $FeCl_3$ intercalation doped CVD few-layer graphene *IEEE Electron Device Lett.* **37** 9
[18] Leenaerts O, Partoens B and Peeters F M 2008 Adsorption of H_2O, NH_3, CO, NO_2, and NO on graphene: a first-principles study *Phys. Rev.* B **77** 125416
[19] Medeiros P V, Mascarenhas A J, de Brito Mota F and de Castilho C M C 2010 A DFT study of halogen atoms adsorbed on graphene layers *Nanotech* **21** 485701
[20] Muhammad R *et al* 2018 Ab-initio investigations on physisorption of alkaline earth metal atoms on monolayer hexagonal boron nitride (h-BN) *J. Phys. Chem. Solids* **118** 114–25
[21] Muhammad R, Yong S, Tan H P and Muhammad H 2017 Theoretical perspective on structural, electronic and magnetic properties of 3d metal tetraoxide clusters embedded into single and di-vacancy graphene *Appl. Surf. Sci.* **408** 21–33
[22] Kresse G and Hafner J 1993 *Ab initio* molecular dynamics for liquid metals *Phys. Rev.* B **47** 558–61
[23] Kresse G and Hafner J 1994 *Ab initio* molecular-dynamics simulation of the liquid-metal–amorphous-semiconductor transition in germanium *Phys. Rev.* B **49** 14251–69
[24] Madsen G K K 2007 Functional form of the generalized gradient approximation for exchange: the PBE α functional *Phys. Rev.* B **75** 195108
[25] Lin M F, Chuang Y C and Wu J Y 2012 Electrically tunable plasma excitations in AA-stacked multilayer graphene *Phys. Rev.* B **86** 125434
[26] Wu J Y, Chen S C, Roslyak O, Gumbs G and Lin M F 2011 Plasma excitations in graphene: their spectral intensity and temperature dependence in magnetic field *ACS Nano* **5** 1026–32
[27] De Padova P *et al* 2010 Evidence of graphene-like electronic signature in silicene nanoribbons *Appl. Phys. Lett.* **96** 261905
[28] Rufieux P *et al* 2012 Electronic structure of atomically precise graphene nanoribbons *ACS Nano* **6** 6930–35
[29] Lee J K, Lee S C, Ahn J P, Kim S C, Wilson J I and John P 2008 The growth of AA graphite on (111) diamond *J. Chem. Phys.* **129** 234709

[30] Bhattacharyya S and Singh A K 2016 Lifshitz transition and modulation of electronic and transport properties of bilayer graphene by sliding and applied normal compressive strain *Carbon* **99** 432438
[31] Quan L *et al* 2020 The electromagnetic absorption of a Na-ethylenediamine graphite intercalation compound *ACS Appl. Mater. Interfaces* **12** 16841
[32] Wang Z D and Inagaki M 1991 Intercalaction of $FeCl_3$ into film in molten salts *Synth. Met.* **44** 165–76
[33] Zhan D *et al* 2010 $FeCl_3$-based few-layer graphene intercalation compounds: single linear dispersion electronic band structure and strong charge transfer doping *Adv. Funct. Mater.* **20** 3504–9
[34] Ritsko J J and Mele E J 1980 Experimental determination of the momentum-dependent dielectric response of stage-1 $FeCl_3$-intercalated graphite *Phys. Rev.* B **21** 2
[35] Prietsch M, Wortmann G and Kaindl G 1986 Mossbauer study of stage-2 $FeCl_3$-graphite *Phys. Rev.* B **33** 11
[36] Abe T *et al* 1995 Bi-intercalation of ICl into a stage-5 $FeCl_3$ graphite intercalation compound *Phys. Rev.* B **52** 19
[37] El-shazly O and Abou-Aly A I 1991 Transport properties of stage 6 of $FeCl_3$ graphite intercalation compounds *J. Phys. Chem. Solids* **52** 643–46
[38] Zhang Y *et al* 2008 XPS and XRD study of $FeCl_3$–graphite intercalation compounds prepared by arc discharge in aqueous solution Spectrochimica *Acta Part* A **70** 1060–64
[39] Dresselhaus M S and Dresswlhaus G 2002 Intercalation compounds of graphite *Adv. Phys.* **51** 1–186
[40] Caswell N and Solin S A 1978 Vibrational excitations of pure $FeCl_3$ and graphite intercalation with ferric chloride *Solid State Commun.* **27** 961
[41] Underhill C, Leung S Y, Dresselhaus G and Dresselhaus M S 1979 Infrad and Raman spectroscopy of graphite-ferric chloride *Solid State Commun.* **29** 769
[42] Aurbach D and Ein-Li Y 1995 The study of Li-graphite intercalation processes in several electrolyte systems using *in situ* x-ray diffraction *J. Electrochem. Soc.* **142** 1746
[43] Wertheim G K *et al* 1980 Electronic structure of lithium graphite *Solid State Commun.* **33** 1127–30
[44] Kganyago K R and Ngoepe P E 2003 Structural and electronic properties of lithium intercalated graphite LiC_6 *Phys. Rev.* B **68** 205111
[45] Pfluger P *et al* 1980 Electronic properties of graphite intercalation compounds *Physica* B **99** 395–400
[46] Schulk W *et al* 1991 Interlayer states of Lic_6 investigated by inelastic x-ray scattering spectroscopy (IXSS) using li-core excitation *Solid State Commun.* **79** 657–60
[47] Gaocedo F M G *et al* 2020 Synthesis of graphene oxide from graphite by ball milling *Diam. Relat. Mater.* **109** 108064
[48] Winter M *et al* 1998 Insertion electrode materials for rechargeable lithium batteries *Adv. Mater.* **10** 725–63
[49] Cancado L G *et al* 2011 Quantifying defects in graphene via Raman spectroscopy at different excitation energies *Nano Lett.* **11** 3190–96
[50] Emtsev K V, Speck F, Seyller T and Ley L 2008 Interaction, growth, and ordering of epitaxial graphene on SiC{0001} surfaces: a comparative photoelectron spectroscopy study *Phys. Rev.* B **77** 155303
[51] Zhao W J *et al* 2020 Intercalation of few-layer graphite flakes with $FeCl_3$: Raman determination of fermi level, layer by layer decoupling, and stability *J. Am. Chem. Soc.* **133** 5941–46

Chapter 7

Unique phenomena of vanadium pentoxide in iron-ion-based battery cathodes

Thi Dieu Hien Nguyen, Shih-Yang Lin, Hsien-Ching Chung, Wei-Bang Li, Ngoc Thanh Thuy Tran, Hsin-Yi Liu, Hai Duong Pham and Ming-Fa Lin

Nowadays, lithium-ion batteries make subsequent roles in industrial products based on their advantages such as high energy density [1–3], low self-discharge [1, 2], quick charging, high open-circuit voltage, and long service life. However, this rechargeable battery also has limitations in aging effect [4], safety concerns, fast degradation due to high temperature [4], and the most important drawback being its high expense. Therefore, many new types of rechargeable batteries are simultaneously being developed, especially multivalent ion, e.g. iron-ion batteries. These secondary batteries provide significant merits such as cheap price and high energy densities compared with Li^+-ion based batteries. From the experimental point of view, the developed batteries have been successfully achieved with a specific capacity of 207 $mAhg^{-1}$ at 30 $mAhg^{-1}$ [5].

Interestingly, there are three main components in the iron-ion battery, including anode, cathode and electrolyte, which are composed of mild steel, layered vanadium pentoxide (V_2O_5; [5]), and tetraethylene glycol dimethylether (TEGDME; [5]), respectively. In general, 99.33 % in weight of the first component comes from iron, while carbon (0.15 %) and manganese (0.52 %) are represented by cathode and electrolyte. The essential properties of the above strongly depend on the majority of materials. It is noted that the three-dimensional ternary iron compound has a moiré supper lattice cell due to the low substituted atom. The specific chemical bonds with the bond-length modulations have been calculated in theoretical analysis, which caused the non-uniform chemical environment in an identical unit cell [5, 6].

There exist many potential cathode materials for ion-based batteries. Among them, the V_2O_5 compound with the pristine system belongs to a well-behaved crystal symmetry [7–9]. The typical V_2O_5 structure is classified as a fascinating ferroelectric

doi:10.1088/978-0-7503-4685-6ch7

semiconductor [8]. Moreover, many merits are shown by V_2O_5 such as abundance, low price, high theoretical capacity, and environmental safety. The successful synthesis of the binary compound using the solvothermal method [10, 11], is a great opportunity to investigate in detail the physical and chemical properties. In this chapter, we focus on the essential properties in terms of hybridization, which is absent in previous research. The chemical bonding of oxygen with other atoms in the geometric structures will be determined based on the multi-orbital hybridizations, which can be fully understood in geometric structures, bands, atom dominances, atom, orbital-decomposed density of states, charge density, complex dielectric functions, and the various optical characteristics. Furthermore, in this chapter, the linking among many physical qualities such as the rich electronic, excitation optical properties, that provide the crucial essential behaviors are described by the numerical analysis. The study of dielectric functions and other optical properties under the phonon–electron interactions will be comprehended in the composite orbital pictures. This theoretical framework can apply to cathode, anode, and electrolyte materials of Fe^{3+}-ion based battery. The research of the charging and discharging process, mainly coming from intercalating and de-intercalating Fe^{2+} ions in or out of the cathode material would be fascinating, particularly the existence of meta-stable and intermediate configurations and the diversified electronic properties. The performance of iron-ion batteries will depend on the diversified materials which play critical roles in electrode and electrolyte, requiring a systematic experimental investigation [68].

The binary V_2O_5 compound is completely explored by first-principles calculations based on density functional theory (DFT; [12]). The main simulation uses the Vienna *ab initio* Simulation Package (VASP) [13]. The generalized gradient approximation in Perdew–Burke–Ernzerhof (PBE; [14]) exchange-correlation applied for the electron–electron and many-particle interaction. The electron–ion interactions are assessed by the projector-augmented wave (PAW; [15, 16]). Analytic form for the single- and many-particle Hamiltonian is difficult due to the existence of inexact formulae for the above intrinsic interactions [17]. In common, the kinetic energy cutoff is set at 520 eV to settle the number of plane waves basis sets [13]. It is exceptionally dependable and adequate for evaluating Bloch wave functions and electronic energy spectra. The convergence condition for the ground-state energy is 10^{-6} eV for continuous calculations steps. The most extreme Hellmann–Feynman drive acting on each particle is less than 0.01 eV $Å^{-1}$. The layer structure of vanadium pentoxide requires adding the weak van der Waals interaction in the simulation calculation. In this work, we use semi-empirical quantum chemical DFT-D3 correlation, developed by Grimme *et al.* [18–20]. K-point meshes are $4 \times 16 \times 8$ and $10 \times 40 \times 20$ in the Monkhorst–Pack scheme of the first Brillouin zone, which is used to calculate geometric and electronic properties, respectively. During the simulation calculation, a potential of 3.25 eV will be included due to the 3d orbital of vanadium.

To calculate the excitonic effect calculations, the quasiparticle wave function will use the GW approximation under the quasi-particle Green's function G and the screened Coulomb interaction W with the Wannier system. The significant analysis is calculated for the self-energy covering electron many-body systems. We include 50 bands for calculating the valence hole unoccupied and electron conduction occupied spectrum with an energy cutoff of 350 eV. The general computational analysis uses 4 × 16 × 8 k-point meshes of the Monkhorst–Pack sample. The involved excitonic effects will solve the standard Bethe–Salpeter equation (BSE) within the Tamm–Dancoff approximation, mainly related to election–hole interactions that typically appear in semiconductor materials. Other parameters of energy cutoff, and the number of subbands are the same numbers in GW calculations. The excitonic phenomenon will require the 20 occupied hole valence bands and also 15 unoccupied electron conduction bands. In addition, the broadening parameter Γ sets at 0.1, which originated from various de-excitation mechanisms to build precise calculation results.

The numerical VASP calculation could provide certain significant physical properties, including the band structures along with the high symmetric points and their atom-dominances in subbands, the charge densities of isolated and V–O bonds, the atom- and orbital-decomposed densities of states, the spin distribution configurations, the spin-split or spin-degenerate energy bands, and also the net magnetic moments. The concise characteristics could explain by the multi-, single-orbital hybridizations with atom chemical bonds and the spin configurations. As a result, a full understanding of physical, chemical, and materials properties could be achieved. Previous researches showed that the grand theoretical framework can be unified in the systematic examinations of the rich geometric, electronic, and magnetic properties about 1D graphene nanoribbons [21], few-layer 2D graphene systems [22], and 2D silicene-related compounds [23]. Therefore, this can be generalized thoroughly for the various properties of Li^+-based anode, cathode, electrolyte materials in batteries [24, 25]. In general, the combination between numerical analysis and phenomenological models is one of the key factors to comprehend the essential properties of condensed matter systems. For example, few-layer graphene systems could be fully investigated from first-principles calculations and the tight-binding model, e.g. well-fitted through the electronic energy spectra across the uncomplicated Fermi level. It is noted that the tight-binding model is concerned with the parameterized Hamiltonian with diversified hopping integrals of critical orbital hybridizations under the diverse magnetic quantization due to any external fields [26]. For current V_2O_5 materials, the phenomenological model might face difficulty in finding a suitable hopping integral with the diversified chemical bonds and nonuniform environment.

7.1 Rich geometric symmetries of V_2O_5 compound

In this chapter, V_2O_5 is chosen as one of the metastable configurations, which possesses unusual geometric structure. Figure 7.1 provides optimal calculations, in which the lowest energy is achieved at −116.9 eV and the volume is 262.2 Å^3. Figure 7.2(a) presents the general structure of V_2O_5 corresponding to orthorhombic symmetry. There are 14 atoms (4 vanadium atoms and 10 oxygen atoms) in *a* primitive cell. The binary compound possess the lattice constant parameter of

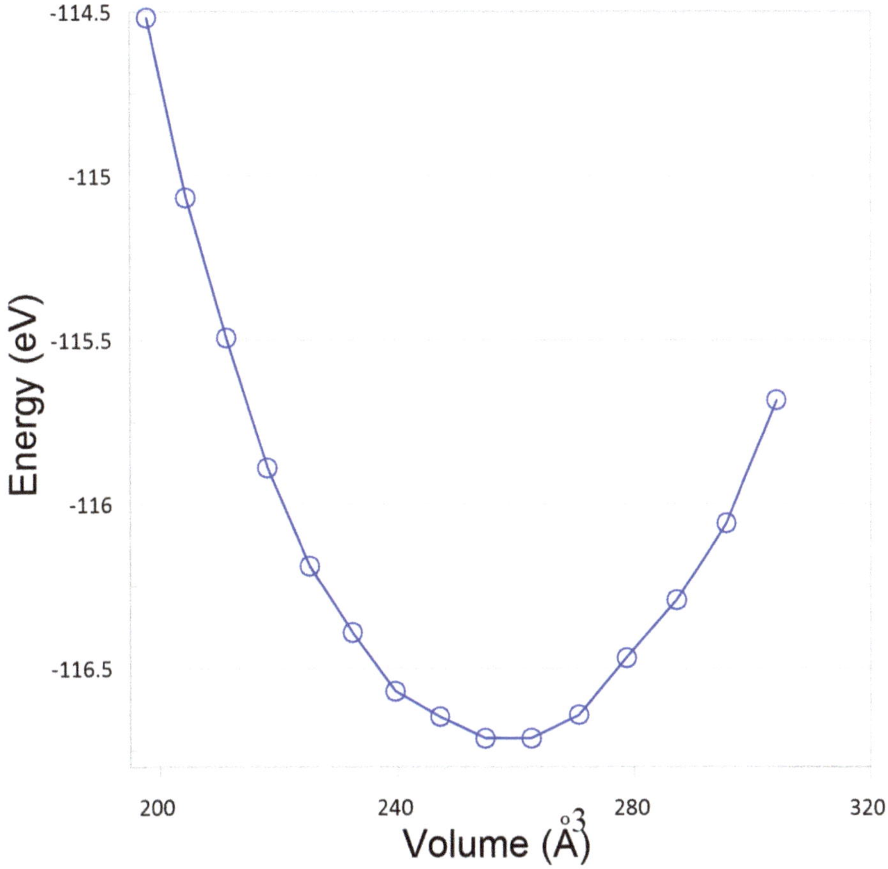

Figure 7.1. The calculation of lowest energy at −116.9 eV with the volume 262.2 Å^3.

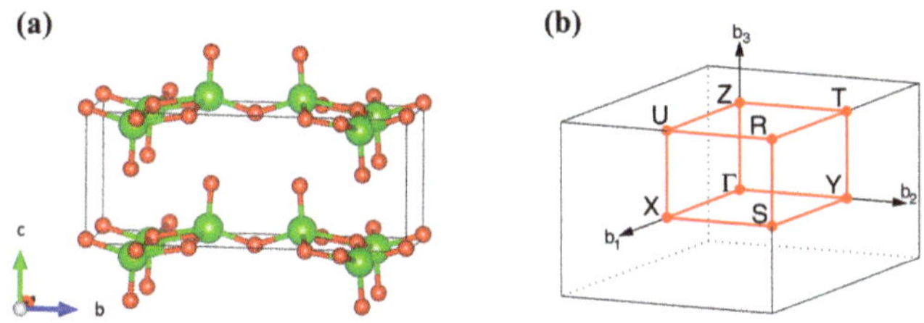

Figure 7.2. (a) The geometric structure of V_2O_5 in the unit cell, where O and V are presented by red and green balls, respectively. (b) The wave vector dependent high symmetric points in first Brillouin zone.

a = 11.697 Å, b = 3.570 Å and c = 4.726 Å, compared with experimental measurements of a = 11.510 Å, b = 3.653 Å and c = 4.369 Å [27] and a = 11.515 Å, b = 3.566 Å and c = 4.376 Å [7] (listed in table 7.1), and angles of x–y, y–z and x–z are $\alpha = \beta = \gamma = 90°$, respectively. The structure of V_2O_5 is formed by some layers perpendicular to the c-axis direction. Between layers, the weak van der Waals interaction the significant p_z orbital interactions come into existence. Vanadium

Table 7.1. The comparison of the lattice parameters in this calculation and the experimental measurements.

This work	$a = 11.697$ Å, $b = 3.569$ Å, $c = 4.726$ Å
Experiment	$a = 11.510$ Å, $b = 3.563$ Å, c = 4.369 Å [11]
	$a = 11.515$ Å, $b = 3.566$ Å, $c = 4.376$ 78(19) Å [12]

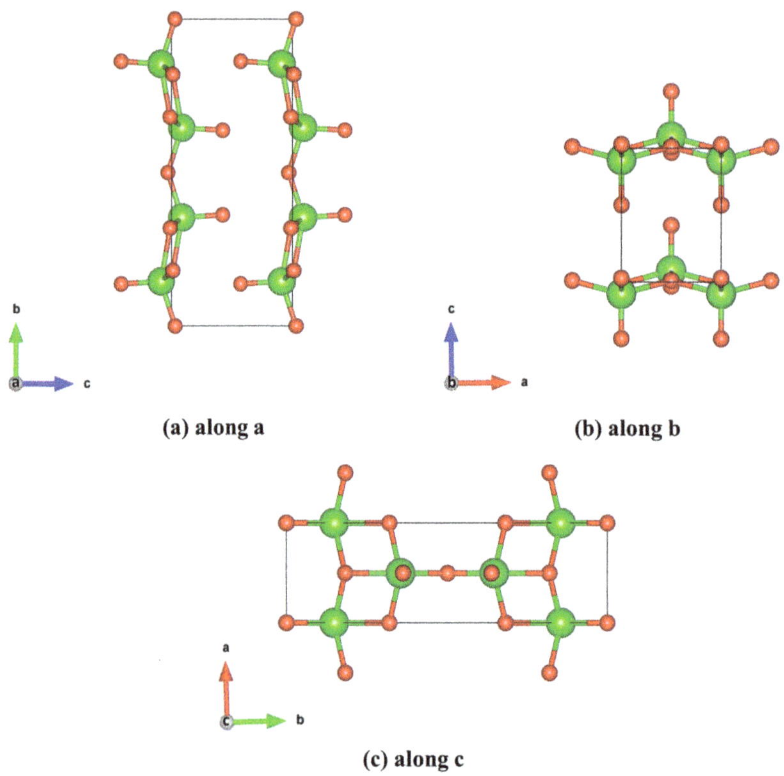

Figure 7.3. The projection of V_2O_5: (a) along a, (b) along b, and (c) along c, which are presented by green and red atoms, respectively.

is encompassed by five oxygen atoms creating square pyramids, where the edge-sharing and the corner-sharing pyramids are inverse alignments along the c direction. It might be known that the layer distance is measured with 3.130 Å. The general basic structure of V_2O_5 is classified as Hermann–Mauguin Pmmn space group, P 2 2ab 1ab Hall, mmm point group. The geometric atom arrangements can be observed in various projected directions, for example for those of (a) a, (b) b, and (c) c directions in figures 7.3(a)–(c), where vanadium and oxygen are shown by blue and red balls, respectively. The investigations band structure properties can be done by the wave vector-independent space, which directly converses from the real-space lattice with the identical Γ-X-S-Y-Γ-Z-U-R-T-Z|Y-T|U-X|S-R high symmetry paths in figure 7.2(b).

The featured properties can be in part supported by investigations of the unique chemical bonds in the geometric structure. The bond length modulation is about 27.957 %, staying in a large range from 1.5997 Å to 2.0466 Å (shown in table 7.2). The binary compound shows an anisotropic property and also the chemical bonds induce a greatly nonuniform chemical environment. This characteristic can be straightforwardly confirmed under the form of the spatial charge density (examined below in figure 7.4). Therefore, the process of intercalation and deintercalation of iron-ion will be easier and might lead to the transfer to different metastability, basically based on the large modulation of the chemical bonds. These phenomena can be explained under a grand theoretical framework of multi- or single-orbital hybridizations, which is provided in detail in figure 7.5, of the orbital-decomposed density of states. Besides the numerical method, phenological analysis, i.e., the tight-binding model can also provide information of diverse orbital hybridizations through the hopping integrals parameters.

V_2O_5 is successfully synthesized in the experimental examinations, which use many analysis tools in investigating physical behaviors. The crystalline phase or lattice symmetry can be estimated by x-ray diffraction (XRD) [7, 8]. The recording XRD pattern along 2θ angles can be used to compare with the data of each material. The morphology with the real layer structure [7] was characterized by a scanning electron microscope (SEM) [8] and transmission electron microscopy (TEM) [28, 29]. The high-resolution experimental analysis can be utilized to clearly investigate the chemical V–O bonds, which is an important step to comprehending the complicated multi-orbital hybridizations associated with the different chemical bond environments.

Table 7.2. The various bond lengths of V_2O_5 from 1.5997 Å to 2.0466 Å.

Number	Atom1	Atom2	Length (Å)
1	V1	O3	1.5997
2	V1	O7	2.0466
3	V1	O6	1.8890
4	V1	O9	1.7970
5	V1	O6	1.8890
6	V2	O4	1.5997
7	V2	O5	2.0466
8	V2	O8	1.8890
9	V2	O10	1.7970
10	V3	O1	1.5997
11	V3	O6	2.0466
12	V3	O10	1.7970
13	V3	O7	1.8890
14	V4	O9	1.7970
15	V4	O2	1.5997
16	V4	O8	2.0466
17	V4	O5	1.8890

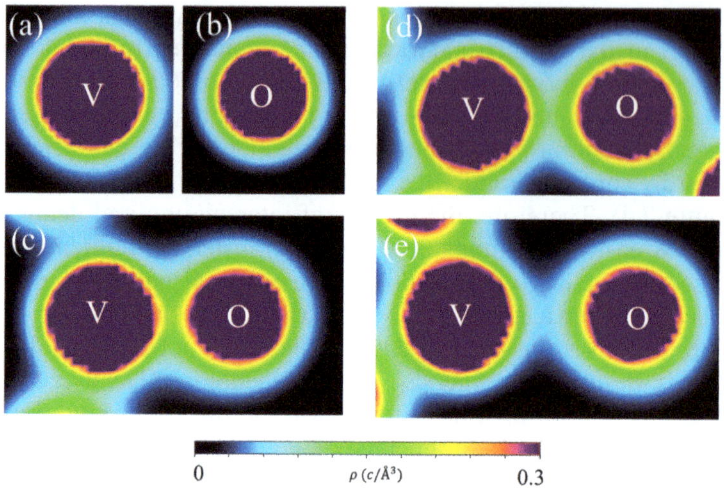

Figure 7.4. The spatial charge distributions associated with (a) isolated V atom, (b) isolated O atom, and the various distances of V–O lengths with (c)shortest, (d) middle and (e) longest ones.

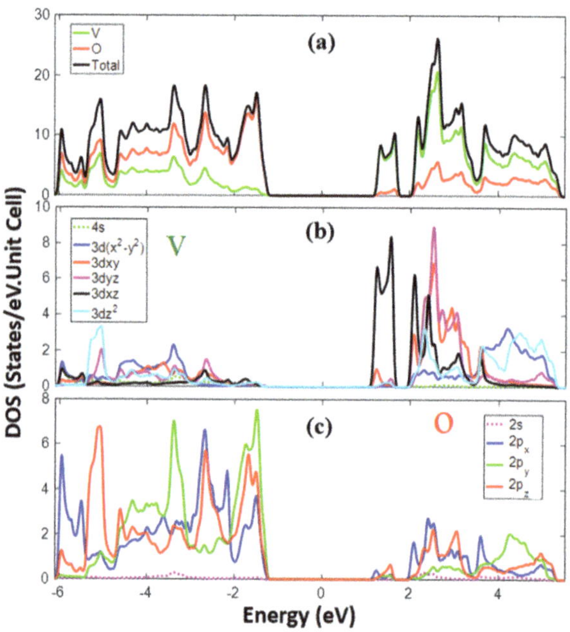

Figure 7.5. (a) The total and atom density of states and the orbital-projected DOSs of (b) vanadium and (c) oxygen.

7.2 Specific electronic properties of vanadium pentoxide

Rich and unusual electronic properties of V_2O_5 under first-principles calculations will be exhibited by the energy band spectrum, the atom dominances along the subbands, the spatial electron distributions, the atom- and orbital-projected density

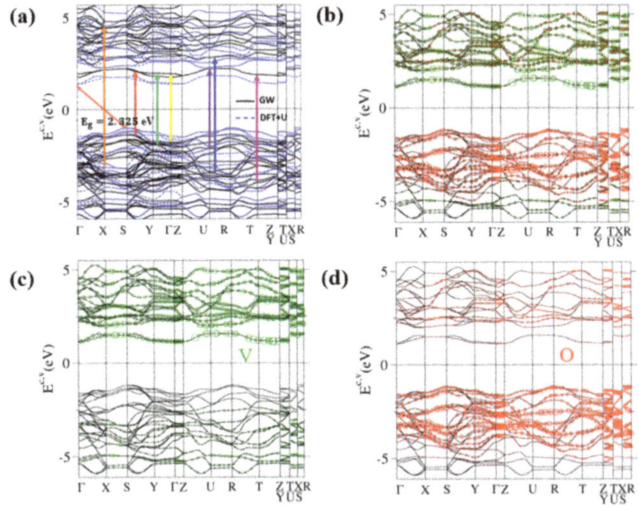

Figure 7.6. (a) The wave vector dependent band structures using GW and DFT+U for the specific atom-dominated (b) vanadium and oxygen (c) vanadium (green circles) (d) oxygen (red circles).

of states. The featured band structure is shown in figure 7.6(a), which consists of various subbands spreading the valence occupied state to conduction an unoccupied one. It is noted that the Fermi level will be located in the middle of the valance and conduction bands. The band spectrum changes along with the different energy ranges and depends on various wave-vectors high symmetric points. The appearance of asymmetric properties about the fermi level ($E_{\mathrm{F}} = 0$) can be explained by the orbital hybridizations in the high variation chemical bonds of vanadium and oxygen, which is known as the non-uniform environment. The indirect gap comes into existence in different propagation constants from S to Γ point of $E^{\mathrm{i}}_{\mathrm{g}} \sim 2.325$ eV of the DFT+U calculation (indicated by the red arrow in figure 7.6(a)) and $E^{\mathrm{i}}_{\mathrm{g}}{\sim}3.049$ eV for GW, which belongs to the semiconductor. This can be identified through the maximum occupied valence-band spectrum and the minimum unoccupied conduction one. Theoretically, the indirect gap is made by the phonon creation or absorption due to the different positions of wave vector values, which is sensitive to the temperature. In general, the indirect gap is smaller than the optical gap or threshold frequency. The measurement of $E^{\mathrm{i}}_{\mathrm{g}}$ can be done by several methods such as optical absorption or photoluminescence spectroscopy. Almost the magnetic properties of V_2O_5 can be neglected due to the difficulty of forming the spin-split electronic states and the spin-degenerate valence and conduction bands below and above E_{F}, respectively. The calculated band gap is similar to those with the simulation values of 2.32 eV [30] and 2.344 eV [31] reported by Szymanski *et al.* and Jain *et al.*, respectively; and the experimental measurement of 2.30 eV [32]. The atoms on the unit cell and orbitals of each element will mainly determine the number of subbands. Obviously, V_2O_5 also exists in a large number of energies subbands, in which it is difficult to distinguish π and σ bands. The number of subbands that appear in the whole band spectrum [33], depends on the number of atoms and the outer shell orbitals. The band structures show parabolic, oscillatory, linear, partially flat energy dispersions, which exist in various band-edge states in various subbands. Many specific phenomena can be classified in the structure such as anisotropic behaviors, crossing, non-crossing, and anti-crossing properties, which have featured energy width of 0.303–0.786 eV. The appearance of band-edge states, associated with zero group velocity might form specific van Hove singularities. This phenomenon can be used to explain the creation of absorption peak or shoulder channels.

There are two identical ranges of band energy, including the occupied valence subbands and unoccupied conduction one. Oxygen will dominate in the first group from −5.67 eV to 0.97 eV (red circles) and the inversely true for vanadium atom from 1.05 to 3.99 eV (green circles), which is shown in figures 7.6(c) and (d). The contributions of vanadium and oxygen will spread in the whole energy spectrum, which provides the significant roles of the atoms in creating the chemical bondings of V–O. In comparison with orbital-decomposed density of states (DOS), they are totally agreed with each other. The atom dominances of vanadium and oxygen will take part in the material properties based on the V-($3d_{x^2-y^2}$, $3d_{xy}$, $3d_{yz}$, $3d_{xz}$, $3d_{z^2}$) and the (2s, $2p_x$, $2p_y$, $2p_z$)-orbital energies

The complicated electronic properties due to the energy or wave vector dependence such as band structures, i.e. the Fermi level, number of energy subbands, energy dispersions, crossing, non-crossing, or anti-crossing behaviors, and free carrier density, the numerical simulations, might present a struggle in calculating by phenomenological models. Remarkably, the general electronic properties could be explained from the orbital hybridizations because of the non-uniform environment in chemical bondings of vanadium and oxygen. From the phenomenological point of view, the Hamiltonians are necessary to calculate the number of hopping integrals based on the orbital site energy energies. This might be very difficult for calculating the precise parameters in a large unit cell.

The numerical simulation results, which are conducted on the wave-vector-dependent energy spectra below the Fermi level, especially for the occupied electronic states, can be tested by angle-resolved photoemission spectroscopy (ARPES) [34, 35]. This technique is very successful in observing the diversified band structures such as the diverse energy dispersions in the sp^2-bonding carbon systems, 3D transition metal oxide compounds, and group-IV and group-V layered materials. Similar experimental measurements are available in examining the main features of the V_2O_5 compound, such as the wave vector dependence, the large band gap, the highly asymmetric unoccupied and occupied states, the parabolic or partial flat or weak energy dispersions, and the frequent non-crossing, crossing, and anti-crossing behaviors. The experimental analysis plays a substantial role in comprehending the critical orbital hybridizations of various chemical bonds.

The spatial charge density is investigated with isolated atoms, the shortest, middle, and longest V–O bonds from figures 7.4(a)–(e). The various colors dispositions show the significant contributions of s, p, and d orbital states, mainly classified into deep purple, light red, and yellow-green areas. This characteristic will be apart supposed by orbital van Hove singularities (figure 7.5). Figures 7.4(a) and (b) present the isolated V and O atoms, in which the deep represents s-orbital in the atomic structures. This s-contribution will not affect the essential properties of the Fermi level because it is dominantly in the deep energy. The light red, yellow–green regions will directly take part in the chemical interactions under the multi-hybridizations theory. Due to the different chemical bonds of vanadium and oxygen (V–O (5)), the great non-uniform environment will determine the various charge arrangements in the orthorhombic structures. The overlap will be appeared stronger in the shortest distance (1.6003 Å in figure 7.4(c)) of the V–O bond and decrease with the middle (1.824 Å in figure 7.4(d)) and longest (2.0477 Å in figure 7.4(e)). The atomic number of vanadium is 23 and oxygen is 8, therefore, the charge density around vanadium will be higher compared with oxygen, identified from the orbitals ($3d_{x^2-y^2}$, $3d_{xy}$, $3d_{yz}$, $3d_{xz}$, $3d_{z^2}$) and ($2p_x$, $2p_y$, $2p_z$), respectively. From the center core of V, the directly participating range is 0.71 for the case of the short bond. The charge will transfer from vanadium to oxygen to satisfy the deformed electron distributions between these atoms.

The DOSs show in atom and orbital along with different atoms in figures 7.5(a)–(c). The specific properties are associated with the amount of van Hove singularities due to the structures of band-edge states. There might exist merged areas in diversified orbitals so one is unable to identify π or σ bonding of p_z or (p_x, p_y) interactions. With the bandgap range, DOS become vanishing, and the DOS appearance will start outside of E^{i}_{g} ~ 3.049 eV. Interestingly, the intensity of vanadium will be higher than the oxygen in the unoccupied states and inversely in the occupied band spread in the energy range of -5.785 eV $\leqslant E_v \leqslant -1.154$ eV and 1.154 eV $\leqslant E_c \leqslant$ 5.058 eV with green and red lines. Figure 7.5(b) shows the highest peak of the value of 2.25 eV of $3d_{yz}$, while oxygen orbital-projected DOS possess the prominent intensity at $2p_y$ in figure 7.5(c). Therefore, the chemical bondings of V–O in V_2O_5 compounds are identified from the critical hybridizations of V-($3d_{x^2-y^2}$, $3d_{xy}$, $3d_{yz}$, $3d_{xz}$, $3d_{z^2}$) and O-($2p_x$, $2p_y$, $2p_z$). The covalent interactions are created from most of the orbitals in figures 7.5(b) and (c), from ($3d_{x^2-y^2}$, $3d_{xy}$, $3d_{yz}$, $3d_{xz}$, $3d_{z^2}$) and ($2p_x$, $2p_y$, $2p_z$), except 4s and 2s with very small intensity components. The explanation might depend on the full occupied in the orbital configuration with spin up and spin down.

The discussed contents above show that scanning tunneling spectroscopy (STS) can be used to achieve the information of density of states in various energy ranges, in particular the diversified van Hove singularities, because the band-edge states are worth studying. In this method, the metal tip is placed close to the material surface but they always avoid touching each other, which is based on the tunneling quantum current measurements. First of all, electrons versus voltage (I–V) curves will be measured at high precision. And then, the calculations of normalized differential conductance of $(dI/dV)/(I/V)$ will directly approximate the DOS. This examination can apply to the typical sample, i.e., few-layer graphene systems by measuring the coupling effects. For example, STS is successful in achieving an almost symmetric V-shape structure with a dip at the Fermi level for zero-gap semiconductor monolayer, twisted bilayer graphene with the critical logarithmically symmetric peaks, bilayer AB and tri-layer ABC stacking with a gate-voltage-induced energy gap, a delta-function-like symmetric peak occurs at the Fermi level in tri-layer ABC stackings. This high-resolution measurement is successful in presenting the features of the electronic DOS at each atomic site [9]. Current imaging tunneling spectroscopy (CITS) is achieved, which clarifies the atomic locations in the unoccupied state image coinciding with those in the occupied-state image observed with a small sample bias voltage V_s. This experimental technique is useful to measure the large bandgap of V_2O_5 $E^i_g \sim 2.308$ eV. It is noted that this system possesses a lot of asymmetric and symmetric channels and widening shoulders with various line widths. The combination between the experimental results and the theoretical predictions of van Hove singularities is very important to understand the complicated multi-orbital hybridizations of V–O. Also, the linking with the phenomenological investigation, e.g. tight-binding model (how to obtain the hopping integral values) could become an open issue for future research. This is a challenge due to the existence of different chemical bonds and the complicated valence and conduction subbands. All of the theoretical, phenomenological, experimental pictures can be extended to the other Fe^{3+}-based anode [5], cathode [5], and electrolyte materials [5].

7.3 Optical properties

In addition to the electronic properties of V_2O_5, the optical excitation properties are investigated to provide more understanding in terms of classical and quantum pictures. Electromagnetic waves propagating into the materials will produce a special screening caused by the charge and time and position-dependent wave interactions under the dispersion equation $c^2k^2 = \varepsilon\omega^2$, where c is the wave velocity. The dielectric function $\varepsilon(\omega)$ will illustrate the vertical excitation properties, presenting the bare function under the long-wavelength limit (wave vector will not be taken into account). The theoretical calculations of electric-dipole channels can be identified using gradient approximations. Further investigations on many-body effects will be included in this material, which are effectively attractive Coulomb interactions between excited holes and electrons under the quantum pictures points of view. It is noted that the excitonic effects are influenced directly by the optical absorption peaks of semiconductor materials. The whole picture of this effect would be combined with orbital hybridization investigations simultaneously.

The absorption spectra depend on the joint density of states of the occupied hole and unoccupied electron conduction bands and the square of electric dipole moments. Other factors are the numbers and intensities of the specific optical absorptions. In addition, the conservation of energy and the Pauli principle must be satisfied in calculating the bare response function $Im(\epsilon\,(\omega\,))$

$$Im[\epsilon(\omega)] = \frac{4\pi e^2}{m^2\omega^2}\int_{BZ}\frac{2}{(2\pi)^3}|\langle f|\hat{e}\cdot\vec{p}\,|i\rangle|^2\left(f(E^V)-f(E^C)\right)\delta\left(E_f-E_i-\omega\right)d^3k$$

The imaginary part can be calculated from the numerical analysis, in which the prominent absorption peaks have a close relation to orbital hybridization from the initial and final state of specific van Hove singularities. Therefore, the replacement of electron–phonon interaction by the orbital-photon composites is significant. The linking of absorption peaks with the band structure inter-transitions (figure 7.6), spatial charge density (figure 7.4), and orbital-decomposed DOS (figures 7.5(b) and (c)) need to be identified in detail. Figures 7.7(a) and (b) show the real and imaginary parts of dielectric functions in three directions (x, y, z), which is calculated by Kramers–Kronig relations. The real part of the dielectric function presents in normal form almost in the whole spectrum and there also exists anomalous dispersion. The former needs to except the part close to the resonant frequency. The negative component of the real dielectric function shows the transmission completely. The identification of static dielectric constant is 5.022 eV in the x direction in figure 7.7(a).

The imaginary part dielectric function in figure 7.7(b) provides the values of optical threshold frequency of 2.320 eV in the x-direction, 2.335 eV in the y-direction, and 2.255 eV in the z-one, which is given by black, red, and blue lines, respectively. This value will be increased to 2.624 eV in the case of single-particle phenomena (the black lines in figure 7.8(a)). Figures 7.8(a) and (b) exhibit the absorption channels without and with excitonic effects. It is remarkable that there exist two extra peaks P_1 and P_2 due to the combination of excited electron–hole pair interactions. The various absorptions channels are given in table 7.3 by red, yellow, green, purple, pink, blue, and orange, respectively. These identical values can be investigated in linking with intertransitions in band structures with specific band-edge states of zero group velocity and the orbital van Hove singularities.

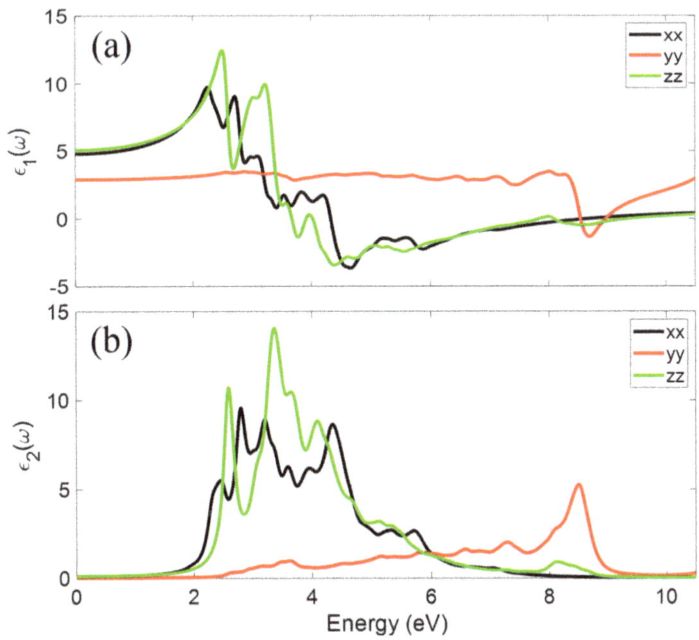

Figure 7.7. The complex dielectric functions (a) the imaginary and (b) real-part under the excitonic effects in xx, yy, zz electric-polarization directions (the black, red and green curves). Reproduced from [41], copyright (2021) with permission from Elsevier.

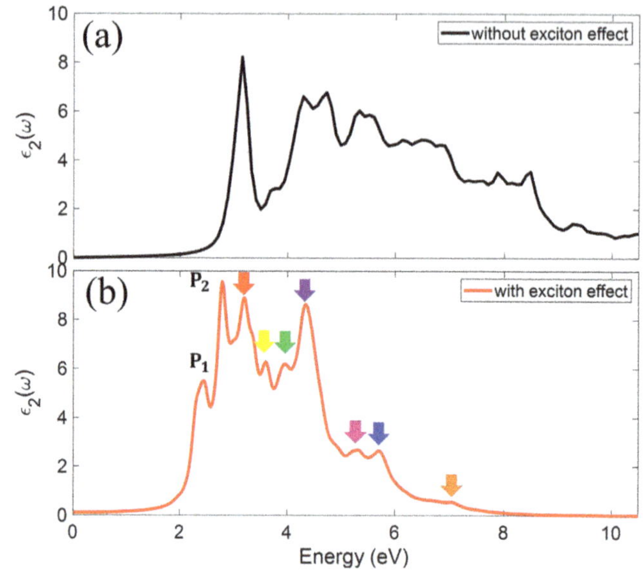

Figure 7.8. The imaginary parts of dielectric function when considered (a) without excitonic effects and with (b) excitonic effects in *xx* polarized direction.

Table 7.3. The absorption spectrum of imaginary part dielectric function linking with orbital hybridizations.

Peak/ Color	Exciton effect	Without exciton effect	Interband transitions	Orbital channels
P_1	2.457 eV	_		—
P_2	2.760 eV	_		—
Red	3.191 eV	3.149 eV	X-Y	O(2s,$2p_x$,$2p_y$,$2p_z$)–V($3d_{xy}$,$3d_{xz}$,$3d_{z^2}$)
Yellow	3.599 eV	3.673 eV	Γ	O(2s,$2p_x$,$2p_y$,$2p_z$)–V($3d_{xy}$,$3d_{xz}$,$3d_{z^2}$)
Green	3.972 eV	3.935 eV	Y-Γ	O(2s,$2p_x$,$2p_y$,$2p_z$)–V($3d_{xy}$,$3d_{xz}$,$3d_{z^2}$)
Purple	4.355 eV	4.023 eV	U	O(2s,$2p_x$,$2p_y$,$2p_z$)–V($3d_{yz}$,$3d_{xz}$,$3d_{z^2}$)
Pink	5.296 eV	5.335 eV	T	O(2s,$2p_x$,$2p_y$,$2p_z$)–V (4s,$3d_{x^2-y^2}$,$3d_{xy}$,$3d_{yz}$,$3d_{xz}$,$3d_{z^2}$)
Blue	5.727 eV	5.510 eV	U-R	O(2s,$2p_x$,$2p_y$,$2p_z$)–V (4s,$3d_{x^2-y^2}$,$3d_{xy}$,$3d_{yz}$,$3d_{xz}$,$3d_{z^2}$)
Orange	7.051 eV	6.812 eV	X	O(2s,$2p_x$,$2p_y$,$2p_z$)–V($3d_{x^2-y^2}$,$3d_{xy}$,$3d_{z^2}$)

The strong Coulomb interactions in the many-particles picture might create some variation peak positions. The first peak is presented by the red color with the energy of 3.191 eV, resulting from the excited state from the valence hole states to the conduction electron states in a specific band-edge spectrum. The vertical excited transition is placed in the X and Y of wave vector-dependent high symmetric points. According to the partial orbital DOS, the corresponding position from

O(2s,2p$_x$,2p$_y$,2p$_z$) to V(3d$_{xy}$,3d$_{xz}$,3d$_{z^2}$). When we consider these peak and shoulder responses along y and z polarization, the intensities might be changed but not the energy positions based on the identical dipole matrix appearance with the atom configurations. Another 3.599 eV dielectric function peak shows the excited channel at Γ point, which identify from O(2s,2p$_x$,2p$_y$,2p$_z$) to V(3d$_{xy}$,3d$_{xz}$,3d$_{z^2}$) van Hove singularities. A similar method can be applied to fully comprehend other structures, particularly the electrode and electrolyte compounds of ion-based batteries.

Besides the discussion of the complex dielectric function, other optical properties might be worthy of investigation, such as the energy loss function (ELF), reflectance, absorption, and refraction. $Im\left[1/\varepsilon\right] = \frac{\varepsilon_2(\omega)}{\varepsilon_1^2(\omega) + \varepsilon_2^2(\omega)}$ is the energy loss function, which is known as the screened response ability by the effect of the Coulomb field of vanadium and oxygen. For consideration in the long-wavelength limit, the coherent charge oscillations can be created under the screening of electromagnetic waves. The plasmon modes or collective excitations come into existence in the ELF. In x, y, and z polarization directions, the peaks appear at 8.678 eV, 7.8010 eV, and 9.428 eV (figure 7.9(b)). The multi-hybridizations of orbitals of O-[2p$_x$, 2p$_y$, 2p$_z$] and V-[4s, $3d_{x^2-y^2}$, 3d$_{xy}$, 3d$_{yz}$, 3d$_{xz}$, $3d_{z^2}$] might determine the featured properties of 3D V_2O_5 compounds. It is noted that the real part $\epsilon_1(\omega)$ tending to zero will form the plasmon peaks.

Figure 7.9(c) shows the measured reflectance of vanadium pentoxide in three directions in the photon energy function. The reflectivity starts from the small value of less than 20% and s up to near 80% along the x- and z-axis, while this value is 40% in the y-one. Connecting with the complex dielectric function, the formula of reflectivity is given by $\left|\frac{\sqrt{\epsilon(\omega)} - 1}{\sqrt{\epsilon(\omega)} + 1}\right|^2$ [17]. The origin of this phenomenon comes from the

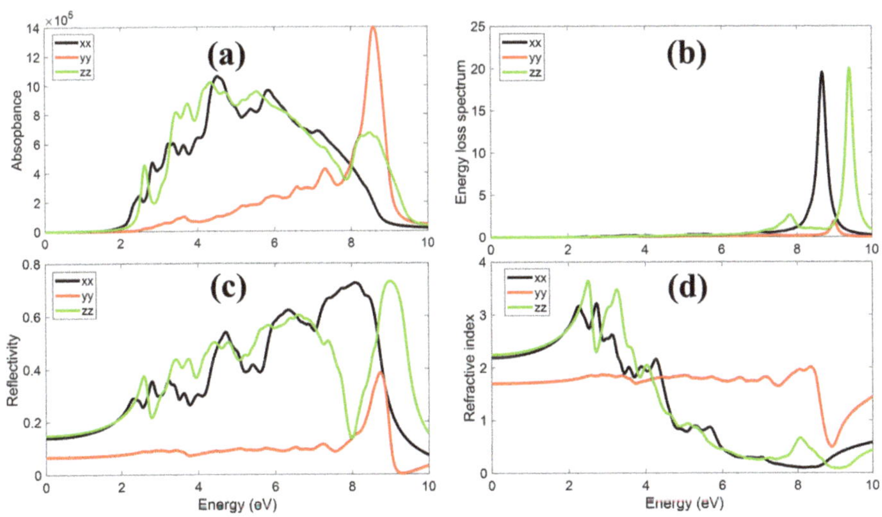

Figure 7.9. The various optical characteristics (a) absorption structures, (b) energy loss spectrum and (c) reflectivity, and (d) refractive index.

electromagnetic wave propagating in the medium with a small velocity compared with the free space. The absorption spectrum also provides useful information which is directly connected with interband transitions and the partial orbital DOS, reflecting the chemical, physical and material properties of V_2O_5. Reflectance $R(\omega)$ and the absorption coefficient $\alpha(\omega)$ are inversely proportional to the propagation decay length [36] and play an important role in determining the single-particle and collective excitation properties. The reflection will not be influenced by the exciton effect at the small frequency, which leads to strong transmission. The variations of reflectance due to the interband transitions will appear when the frequency increases. Compared with the plasmon mode at the resonant frequency, the changing of energy loss spectrum with reflectance is 65% in the electric field along the x-direction. As the frequency is smaller than 2 eV, the absorption peaks disappear by directly corresponding to the real dielectric function $\epsilon_1(\omega)$. The weak Landau damping that exists at $\epsilon_1(\omega)$ tends to zero, also, $\epsilon_2(\omega)$ will have low value, which presents at 8 eV in x, z electric polarization and 8.2 eV in the y-direction. Another coefficient also included in figure 7.9(d) is the refractive index, which can determine the change of electromagnetic waves inside materials. This value depends on the complex dielectric function by the formula $n(\omega) = \left[\left(\sqrt{\varepsilon_1^2(\omega) + \varepsilon_2^2(\omega)} + \varepsilon_1(\omega)\right)/2\right]^{1/2}$. The 3D compound shows the positive value of the refractive index, which exhibits the largest bend appearing in the low frequency of x and z directions, while the y direction presents an unusual characteristic where $n(\omega)$ remains unchanged at the low frequency.

Many experimental measurements have been done about geometric structure (morphology of side view or top view, lattice parameters), electronic properties (band structures, van Hove singularities), and optical spectrum (energy loss function, reflectivity, absorption). The popular method to study the lattice symmetry is x-ray diffractions pattern [37, 38] and low-energy electron diffraction (LEED) [39]. The most important chemical bonds V–O in the current structure can be examined in the above techniques and will help to comprehend the complicated orbital hybridization theory. The angle-resolved photoemission spectroscopy (ARPES) [34, 40] is a common technique to measure the band structure in the occupied hole states at the diversified energy in the first Brillouin zone. The atom numbers and orbitals of vanadium and oxygen will determine the contributions of wave-vector dependent subbands. However, the moiré superlattice in the unit cell might create barriers in the identification of the consisted bands in ARPES measurements. The optical spectroscopies are available to observe theoretical predictions such as energy loss reflectivity absorption and refraction. Currently, the first excitonic peak is calculated at $\sim$ 2.16 eV, while the experimental examination shows at $\sim$ 2.2 eV by one photoluminescence.

The featured properties of the V_2O_5 compound are fully investigated under first-principles calculations. The specific geometric, electronic, and optical properties are successfully linked to the multi-orbital hybridizations in chemical bonds from V-($3d_{x^2-y^2}$, $3d_{xy}$, $3d_{yz}$, $3d_{xz}$, $3d_{z^2}$) and O-($2p_x$, $2p_y$, $2p_z$). The diversified chemical nonuniform geometric structures, rich atom-dominated energy bands, atom, orbital-projected DOS, the specific spatial charge density, and the excitonic effects in optical properties are revealed through a unified theoretical framework. Other optical characteristics such as energy loss function, reflectance, and refraction support the whole essential phenomena of vanadium pentoxide compounds. The intermediate states of the cathode materials need to be identified and might be an interesting topic for further studies. Using the same strategy, the various cathode, anode, and electrolyte materials of ion-related batteries could be applied in terms of theoretical quasiparticle points of view.

References

[1] Nitta N, Wu F, Lee J T and Yushin G 2015 Li-ion battery materials: present and future *Mater. Today* **18** 252–64

[2] Goodenough J B and Park K-S 2013 The Li-ion rechargeable battery: a perspective *J. Am. Chem. Soc.* **135** 1167–76

[3] Tarascon J-M and Armand M 2011 *Materials for Sustainable Energy: a Collection of Peer-reviewed Research and Review Articles from Nature Publishing Group* (Singapore: World Scientific) 171–9

[4] Leng F, Tan C M and Pecht M 2015 Effect of temperature on the aging rate of Li ion battery operating above room temperature *Sci. Rep.* **5** 12967

[5] Saroja A, Samantaray S S and Sundara R 2019 A room temperature multivalent rechargeable iron ion battery with an ether based electrolyte: a new type of post-lithium ion battery *Chem. Commun.* **55** 10416–9

[6] Greenwood N N and Earnshaw A 2012 *Chemistry of the Elements* (Amsterdam: Elsevier)

[7] Shevchuk V, Usatenko Y N, Demchenko P Y, Antonyak O and Serkiz R Y 2011 Nano- and micro-size V_2O_5 structures *Chem. Metals Alloys* **4** 67–71

[8] Fiermans L, Clauws P, Lambrecht W, Vandenbroucke L and Vennik J 1980 Single crystal V_2O_5 and lower oxides. A survey of their electronic, optical, structural, and surface properties *Phys. Status Solidi (a)* **59** 485–504

[9] Sakai Y and Ehara S 1999 Surface structure and electronic states of V_2O_5 surface studied by scanning tunneling microscopy/spectroscopy *Jpn. J. Appl. Phys.* **38** 2944

[10] Wang H g, Ma D l, Huang Y and Zhang X b 2012 Electrospun V_2O_5 nanostructures with controllable morphology as high-performance cathode materials for lithium-ion batteries *Chem. Eur. J.* **18** 8987–93

[11] Pan A, Wu H B, Yu L and Lou X W 2013 Template-free synthesis of VO_2 hollow microspheres with various interiors and their conversion into V_2O_5 for lithium-ion batteries *Angew. Chem.* **125** 2282–6

[12] Hafner J 2008 Ab-initio simulations of materials using VASP: Density-functional theory and beyond *J. Comput. Chem.* **29** 2044–78

[13] Kresse G and Furthmüller J 1996 Efficiency of ab-initio total energy calculations for metals and semiconductors using a plane-wave basis set *Comput. Mater. Sci.* **6** 15–50

[14] Perdew J P, Burke K and Ernzerhof M 1996 Generalized gradient approximation made simple *Phys. Rev. Lett.* **77** 3865

[15] Blöchl P E 1994 Projector augmented-wave method *Phys. Rev.* B **50** 17953

[16] Rangel T, Caliste D, Genovese L and Torrent M 2016 A wavelet-based projector augmented-wave (PAW) method: Reaching frozen-core all-electron precision with a systematic, adaptive and localized wavelet basis set *Comput. Phys. Commun.* **208** 1–8

[17] Jackson J D 2007 *Classical Electrodynamics* (New York: Wiley)

[18] Grimme S, Antony J, Ehrlich S and Krieg H 2010 DFT-D3—a dispersion correction for density functionals, Hartree–Fock and semi-empirical quantum chemical methods DFT-D3 *J. Chem. Phys.* **132** 154104

[19] Grimme S, Antony J, Ehrlich S and Krieg H 2010 A consistent and accurate *ab initio* parametrization of density functional dispersion correction (DFT-D) for the 94 elements H-Pu *J. Chem. Phys.* **132** 154104

[20] Grimme S, Ehrlich S and Goerigk L 2011 Effect of the damping function in dispersion corrected density functional theory *J. Comput. Chem.* **32** 1456–65

[21] Lin S Y, Tran N T T, Chang S L, Su W P and Lin M F 2018 *Structure- and Adatom-Enriched Essential Properties of Graphene Nanoribbons* (Boca Raton, FL: CRC Press)

[22] Tran N T T, Lin S Y, Lin C Y and Lin M F 2017 *Geometric and Electronic Properties of Graphene-related Systems: Chemical Bonding Schemes* (Boca Raton, FL: CRC Press)

[23] Shih-Yang Lin H-Y L, Nguyen D K, Thuy Tran N T, Pham H D, Chang S-L, Lin C-Y and Lin M-F 2020 *Silicene-Based Layered Materials: Essential Properties* (Bristol: IOP Publishing)

[24] Nguyen T D H, Pham H D, Lin S Y and Lin M F 2020 Featured properties of Li+-based battery anode: $Li_4Ti_5O_{12}$ *RSC Adv.* **10** 14071–9

[25] Khuong Dien V, Thi Han N, Nguyen T D H, Huynh T M D, Pham H D and Lin M-F 2020 Geometric and electronic properties of Li_2GeO_3 *Front. Mater.* **7** 288

[26] Lin C-Y, Wu J-Y, Chiu C-W and Lin M-F 2019 *Coulomb Excitations and Decays in Graphene-related Systems* (Boca Raton, FL: CRC Press)

[27] Shklover V, Haibach T, Ried F, Nesper R and Novak P 1996 Crystal structure of the product of Mg2+ insertion into V2O5 single crystals *J. Solid State Chem.* **123** 317–23

[28] Egerton R, Li P and Malac M 2004 Radiation damage in the TEM and SEM *Micron* **35** 399–409

[29] Gotić M, Popović S, Ivanda M and Musić S 2003 Sol–gel synthesis and characterization of V_2O_5 powders *Mater. Lett.* **57** 3186–92

[30] Szymanski N, Liu Z, Alderson T, Podraza N, Sarin P and Khare S 2018 Electronic and optical properties of vanadium oxides from first principles *Comput. Mater. Sci.* **146** 310–8

[31] Jain A, Ong S P, Hautier G, Chen W, Richards W D, Dacek S, Cholia S, Gunter D, Skinner D and Ceder G 2013 Commentary: the materials project: a materials genome approach to accelerating materials innovation *APL Mater.* **1** 011002

[32] Li Y, Kuang J-L, Lu Y and Cao W-B 2017 Facile synthesis, characterization of flower-like vanadium pentoxide powders and their photocatalytic behavior *Acta Metall. Sin. (Engl. Lett.)* **30** 1017–26

[33] Gao J F, Zhang J F, Liu H S, Zhang Q F and Zhao J J 2013 Structures, mobilities, electronic and magnetic properties of point defects in silicene *Nanoscale* **5** 9785–92

[34] Lv B, Qian T and Ding H 2019 Angle-resolved photoemission spectroscopy and its application to topological materials *Nat. Rev. Phys.* **1** 609–26

[35] Bussolotti F, Chi D, Goh K J, Huang Y L and Wee A T 2020 *2D Semiconductor Materials and Devices* (Amsterdam: Elsevier) 199–220

[36] Fox M 2002 *Optical Properties of Solids* (College Park, MD: American Association of Physics Teachers)

[37] Li J and Sun J L 2017 Application of x-ray diffraction and electron crystallography for solving complex structure problems *Acc. Chem. Res.* **50** 2737–45

[38] Speakman S A 2013 Introduction to x-ray powder diffraction data analysis *Center for Materials Science and Engineering at MIT*

[39] Van Hove M A, Weinberg W H and Chan C-M 2012 *Low-Energy Electron Diffraction: Experiment, Theory and Surface Structure Determination* vol 6 (Berlin: Springer)

[40] Tepper B, Richter B, Dupuis A-C, Kuhlenbeck H, Hucho C, Schilbe P, Bin Yarmo M and Freund H-J 2002 Adsorption of molecular and atomic hydrogen on vacuum-cleaved V_2O_5 (001) *Surf. Sci.* **496** 64–72

[41] Nguyen T D H, Vo K D, Pham H D, Huynh T M D and Lin M-F 2021 Electronic and optical excitation properties of vanadium pentoxide V_2O_5 *Comput. Mater. Sci.* **198** 110675

Chapter 8

Comprehensive understanding of electronic and optical properties of Li_2SiO_3 compound

Nguyen Thi Han, Vo Khuong Dien, Shih-Yang Lin, Hsien-Ching Chung, Wei-Bang Li, Ngoc Thanh Thuy Tran, Hsin-Yi Liu, Hai Duong Pham and Ming-Fa Lin

This chapter has been adapted from [1] with permission from Springer.

8.1 Introduction

Lithium-ion based batteries (LIBs) are one of the mainstream materials in basic science and play important roles in life and technology since they have many outstanding properties, namely low cost, light weight, long lifetime, rapid charging/discharging, and work with a voltage of the order of 4 V with specific energy ranging between 100 Wh kg^{-1} and 150 Wh kg^{-1} [2–6]. However, recent commercial LIBs with liquid electrolytes contain potential security risks concerning volatilization, flammability, and explosion. Currently, experimental and theoretical studies evidence that secondary batteries using solid electrolytes would overcome these safety issues. Additionally, the solid states' battery cell design would be more simplified, providing a wider electrochemical window than liquid ones [7–11].

Some potential candidates for solid-state electrolyte materials can be listed, such as Li_3ClO, Li_2SiO3, Li_2PS_3 [12–14], and so on. Among them, the Li_2SiO_3 compound has gained considerable attention because it satisfies many applicable standards of an electrolyte material in LIBs. For example, the ternary Li_2SiO_3 compound with low-cost synthesis exhibits a reliable ionic conductivity (2.5×10^{-8} S cm^{-1} at 25 °C) [15] and high mechanic strength, which make it suitable for a fast and flexible thin-film ionic conductor [16], and a CO_2 fine absorption effect evidences that this material can be utilized for gas captures [17]. Furthermore, many previous studies also indicated that the Li_2SiO_3 compound also has been actively explored as a luminescent material [18, 19] due to its exceptional optical properties.

doi:10.1088/978-0-7503-4685-6ch8

The orthorhombic structure's polar orthorhombic symmetry suggests that this material can be suitable for piezoelectric, pyroelectric, and electro-optic applications [20–23].

Up to now, Li_2SiO_3 compound crystals have been successfully synthesized by various methods [18, 19, 24–26]. Its essential physical/chemical properties, such as optimized geometry structures and surface monologies, are usually examined by x-ray diffraction [18, 19, 27], scanning electronic microscope (SEM) [28] and tunneling electronic microscope (TEM) [29, 30]. The energy spectrum depending on frequencies and wave vectors at the valence states and the van Hove singularities in the density of states are not achieved because of a lack of angle-resolved photoemission spectroscopy (ARPES) [31] and scanning tunneling spectroscopy (STS) measurements [32], respectively. Optical responses, photoluminescence, absorption, and transmission measurement have been conducted [18, 19, 33]. However, they have only been investigated for doped Li_2SiO_3 or measured at high temperatures, in which optical properties are strongly affected by surface defects.

On the other hand, the first-principles calculations, are frequently utilized to understand the geometric, electronic and optical properties of the cathode/electrolyte/anode materials of LIBs. Apparently, despite the previous theoretical investigations that have investigated the electronic properties, transport mechanisms and optical properties of the Li_2SiO_3 compound [34–36], most studies do not propose a concise physical/chemical picture that closely relates to the performance of the energy storage system. Our recent studies have linked the orbital hybridizations with the geometric, electronic properties of this material [37]. However, systematic combination of the geometric, electronic, and excitonic effects on the optical response with energy-dependent orbital hybridizations has not been achieved so far.

Therefore, the presented theoretical framework, which is established on the significant orbital hybridizations, is utilized to examine the crucial features in electrolyte materials of Li^+ related batteries. This strategy is based on first-principles simulation on the optimized geometric structure with position-dependent chemical bonding, atom-dominated energy spectrum at various energy ranges, spatial charge distribution due to different orbitals, and atom- and orbital-projected density of states associated with the overlap of orbitals. The specific orbital hybridizations will be used to interpret the onset of the optical frequency, stable excitonic states, a lot of prominent absorption structures, a very strong plasmon mode in terms of the dielectric functions, energy loss functions, absorption coefficients, and reflectance spectra under the distinct electric polarization. The predictions in this chapter require highly resolved experimental measurements.

8.2 Computational details

We used the density functional theory (DFT) method and many-body perturbation method (GW) via the Vienna *ab initio* simulation opackage (VASP) [38] to perform the optimization of the structure and the electronic and optical properties. The Perdew–Burke–Ernzerhof (PBE) generalized gradient approximation was used for the exchange-correlation functional [39]. The projector augmented wave (PAW) method was utilized for characterization of the ion and valence electron interactions [40]. The cutoff energy for the expansion of the plane wave basis was set to 550 eV. The Brillouin zone was integrated with a special k-point mesh of 8×8×8 in the Γ-centered sampling technique [41] for the structural optimization. The convergence condition of the ground-state is set to be 10^{-8} eV between two consecutive simulation steps, and all atoms could be allowed to fully relax during the geometric optimization until the Hellmann–Feynman force acting on each atom was smaller than 0.01 eV.

In this chapter, the optical properties of Li_2SiO_3 compound have been investigated by adopting the method of Rohlfing and Louie [42]. Following the DFT results, the quasiparticle energies are obtained within GW approximation for the self-energy [43] (where G is the Green' function and W the screened coulomb potential), the cutoff energy for the response function was set to 220 eV, 8×8×8 Γ-centered k-points sampling was used to represent reciprocal space. The Wannier interpolation procedure performed in the WANNIER90 code [44] was used to plot the quasiparticle band structure. Regarding the optical properties involved in the excitonic effect, the standard Bethe-Salpeter equation (BSE) in the Tamm–Dancoff approximation was employed to take the electron–hole interactions into account [45]. The k-point sampling, energy cutoff, number of bands, were set to the same values as in the GW calculation. In addition, the eight lowest unoccupied conduction bands (CBs) and the 20 highest occupied valence bands (VBs) are included as a basis for the excitonic states with a photon energy region from 0 eV to 26 eV. In addition, the Lorentzian broadening γ, which arises from various de-excitation mechanisms, was set at 0.1 eV for all optical calculations. Other optical properties, such as the energy loss function, absorption coefficient, and reflectivity can be obtained from the dielectric function by the following relations:

$$L(\omega) = \frac{\varepsilon_2(\omega)}{\varepsilon_1^2(\omega) + \varepsilon_2^2(\omega)},\ R(\omega) = \left| \frac{\sqrt{\varepsilon_1(\omega) + \varepsilon_2(\omega)} - 1}{\sqrt{\varepsilon_1(\omega) + \varepsilon_2(\omega)} + 1} \right|^2 \text{ and } \alpha(\omega)$$
$$= \sqrt{2}\,\omega\left[\sqrt{\varepsilon_1^2(\omega) + \varepsilon_2^2(\omega)} - \varepsilon_1(\omega)\right]^{1/2}$$

8.3 Results and discussions

8.3.1 Rich geometric and electronic properties

In this work, we have chosen to investigate the electronic and optical properties of the Li_2SiO_3 compound. This solid-state electrolyte crystallizes in an orthorhombic structure with space group [Cmc21] (figure 8.1(a)). The calculated lattice constants of this compound are 9.467 Å, 5.440 Å and 4.719 Å for *x*-, *y*- and *z*-directions, respectively, which are close to the value in recent experimental [46] and theoretical investigations [47]. The conventional Li_2SiO_3 cell contains 24 atoms (8-Li, 4-Si and 12-O atoms), where each Li/Si atom is tetrahedrally coordinated and bonded to four

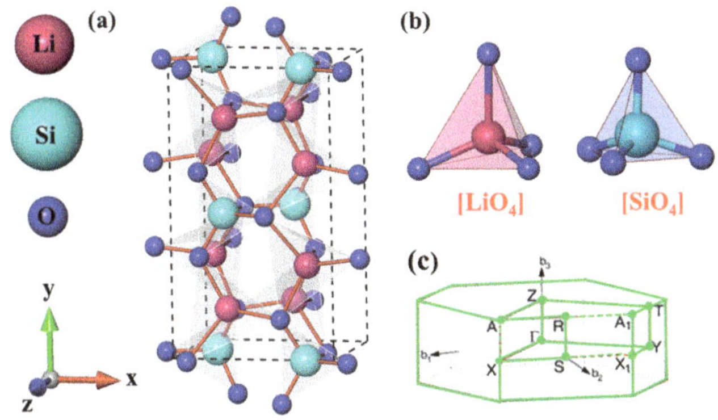

Figure 8.1. (a) The optimal geometric structure of Li_2SiO_3 compound. The dashed black line is unit cell; (b) oxygen atoms around each Li/ Si atom; and (c) the first brillouin zone along the high symmetry points.

oxygen atoms (figure 8.1(b)). The basic structural unit is comprised of 32 Li–O and 16 Si–O chemical bonds, in which the bond length fluctuation ranges are ~1.94 Å–2.20 Å and ~1.61 Å–1.70 Å for the former and the latter, respectively. Obviously, the extremely non-uniform chemical/physical environment is indicated to be very useful for supporting the Li^+ migration in electrolyte materials [48]. Furthermore, this behavior is also responsible for the anisotropic optical properties. The unusual phenomena presented here evidence the complex nature of the orbital-hybridizations in the chemical bonding.

The unusual physical/chemical environment is generated by the complex orbital hybridizations and is directly reflected in the electronic structure. To examine this property, the energy spectrum along high symmetry points (figure 8.1(c)) of this material is calculated and shown in figure 8.2(a). As can be seen from the diagram, the Li_2SiO_3 compound displays rich and unique electronic properties, a high asymmetry of the valence hole and the conduction electron states were observed within a wide range of energy (−28 eV to 13 eV). The presence of many energy sub-bands with complicated energy dispersion, such as parabolic, linear, oscillatory or almost flat energy bands, are a result of the contribution of many atoms and orbitals in the large moiré cell. Additionally, band edge-states which are the extreme or saddle points in the energy-wave-vector, are situated at the highly symmetric points and the other wave vector between them. These critical points in the energy spectrum will create strong van Hove singularities and therefore are responsible for the prominent peaks in the optical absorption spectrum. Very interestingly, the highest occupied state and the lowest unoccupied state, respectively, are located at the Z and Γ points, leading to a wide indirect insulator gap of 8.1 eV. This indirect band gap is located very close to the HSE06 calculated band gap [49]. The mentioned results are associated with the complex orbital hybridizations in the chemical bonds, the close combinations of which with the charge density distribution, the atom and orbital project-density of states, the specific orbital hybridizations

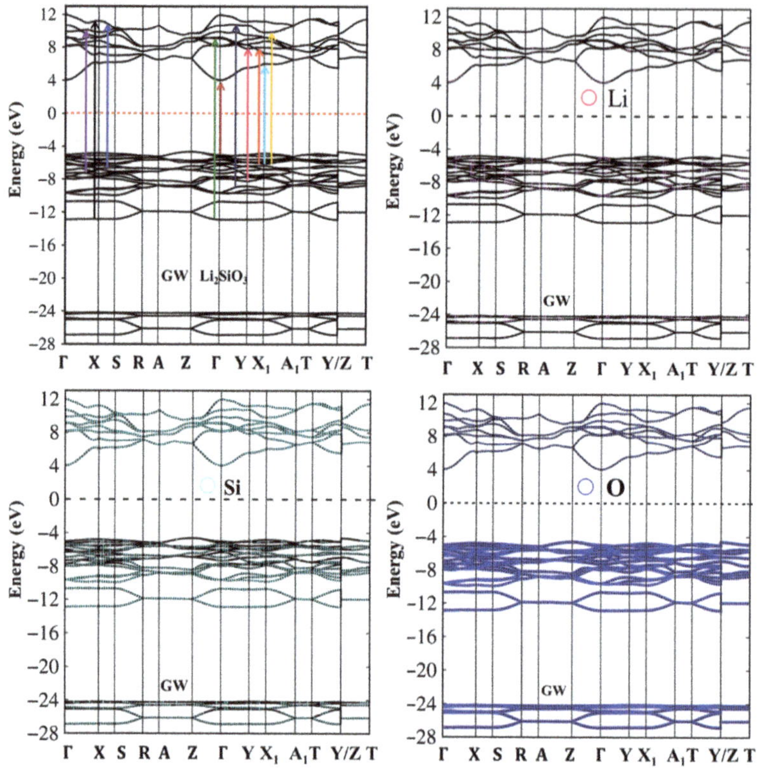

Figure 8.2. (a) Quasi-particle band structure along the high-symmetry points in the wave-vector space, with (b) Li-, (c) Si- and (d) O-atom dominances (pink, cyan and blue circles, respectively).

in chemical bonds, are related to the prominent absorption structures, could be successfully achieved.

In addition to the band structures, the atom dominated-energy-spectrum also provides partly useful information about orbital hybridizations. For the entire band structure, the Li atoms' contribution is negligible. Yet they cannot be ignored since in the absence of the Li–O chemical bonds, some noticeable features will disappear. According to the domination of the various atoms and orbitals, as shown in figure 8.2(b)–(d), the electronic band structure could be systematically divided into four sub-groups: (I) 4.2 eV < E^c < 13 eV, (II) −10 eV < E^v < −4 eV, (III) −13.5 eV < E^v < −10 eV, and (IV) −28 eV < E^v < −23 eV. Obviously, the IV group below – 23 eV is separated by a huge gap from the IV group. Based on the obvious features, each energy sub-group could be revealed as the dominance of (I) (Si, O) atoms, (II) (O) atoms, (III) (Si, O) atoms and (IV) (Si, O) atoms. Through the combination of these dominant parts with the charge density distribution, the atom and orbital-projected density of states, the concise chemical/physical pictures in Li–O, Si–O chemical bonds could be identified.

In order to generalize the orbital hybridization between the constituents, the charge density distribution of the Li, Si and O atoms before/after modification was performed. As presented in figure 8.3(a), the localization of the charge on isolated Li atom (the blue and green parts), correspond to the 1s and 2s orbitals. The similar but wider charge accommodation is associated with (1s, 2s) and ($2p_x$, $2p_y$, $2p_z$) orbitals of the O atoms (figure 8.3(c)). After combination to form the chemical bonds, the inner/outer regions of the Li and O atoms of the Li–O chemical bonds along three electric-polarizations (figure 8.3(d)–(f)) present the small but finite/obvious deformations, especially for the shortest ones. This reflects the presence of non-negligible single (2s-2s)/remarkable multi-2s-($2p_x$, $2p_y$, $2p_z$) orbital interactions between these atoms. As for the Si–O chemical bonds, the highest charge density distribution residing around Si atom arises from the (1s, 2s, $2p_x$, $2p_y$, $2p_z$, 3s) and ($3p_x$, $3p_y$, $3p_z$) orbitals (figure 8.3(b)). Similar but stronger than Li–O, the Si–O bonds are presented by the significant changes of the blue-green part combined with the obvious deformation of the green region (figures 8.3(g)–(i)) of the Si and O atoms, these propose the significant hybridization of the (3s, $3p_x$, $3p_y$, $3p_z$) – (2s, $2p_x$, $2p_y$, $2p_z$) orbitals. The orbital character of the Li–O and Si–O chemical bonds, which dominate the essential properties, could be sufficiently elucidated from the atom- and orbital-projected density of states. The main characteristic of their special

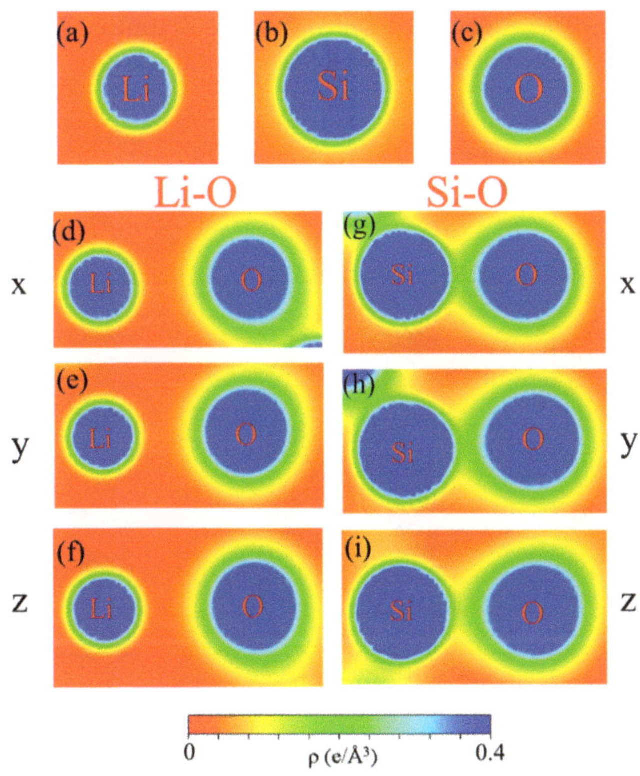

Figure 8.3. Comparison of the isolated Li/Si/O atoms, respectively, (a)/(b)/(c). The charge density distributions related to the significant orbital hybridizations in Li–O and Si–O bonds, respectively, (d)/(e)/(f) and (g)/(h)/(i) along the *x*-/*y*-/*z*-directions for the shortest bonds.

structures is governed by the energy dispersions in the energy band structure and the dimensionality of the material. As for the three-dimensional Li_2SiO_3 compound, the Van Hove singularities mainly come from the extreme or saddle points and the dispersionless energy sub-bands. This leads to the formation of the shoulder structure and pronounced asymmetric peaks (figure 8.4(a)). Furthermore, the considerable mixed density of states of difference atoms/orbitals (figures 8.4 (a)–(d)) clearly evidences the very strong coupling of specific orbital hybridizations. Depending on their co-existence, we can divide four domination parts in term: (I) Li (2s)–O (2s, $2p_x$, $2p_y$, $2p_z$) and Si (3s, $3p_x$, $3p_y$, $3p_z$)–O (2s, $2p_x$, $2p_y$, $2p_z$), (II) Li (2s)–O ($2p_x$, $2p_y$, $2p_z$) and Si (3s, $3p_x$, $3p_y$, $3p_z$)–O ($2p_x$, $2p_y$, $2p_z$), (III) Li (2s)–O ($2p_x$, $2p_y$, $2p_z$) and Si (3s, $3p_x$, $3p_y$, $3p_z$)–O ($2p_x$, $2p_y$, $2p_z$), and (IV) Li (2s)–O (2s) and Si (3s, $3p_x$, $3p_y$, $3p_z$)–O (2s), where each part, respectively, is examined to be associated with each identified sub-group of the energy band structure. This delicate analysis of the orbital characteristics illuminates the excitation mechanism of various strong optical responses.

8.3.2 Rich optical properties

Following the absorption of a photon with suitable energy, an electron will be vertically excited from occupied to unoccupied states. The single-particle excitation spectrum can be described by using the linear Kubo formula [50]:

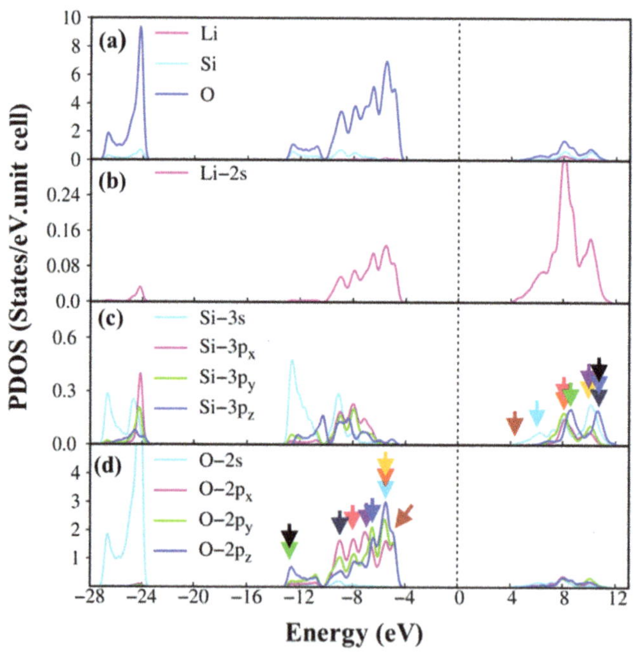

Figure 8.4. Density of states under the different components: (a) the total magnitude with lithium, silicon and oxygen atom contributions, (b) lithium and 2s-decomposed results, (c) silicon and (3s, $3p_x$, $3p_y$, $3p_z$)-projected calculations, and (d) oxygen and (2s, $2p_x$, $2p_y$, $2p_z$)-related ones.

$\epsilon_2(\omega) = \frac{8\pi^2 e^2}{\omega^2} \sum_{v c \mathbf{k}} | e\langle v\mathbf{k}|\mathbf{v}|c\mathbf{k}\rangle |^2 \delta(\omega - (E_{c\mathbf{k}} - E_{v\mathbf{k}}))$, where the first part, $| e\langle v\mathbf{k}|\mathbf{v}|c\mathbf{k}\rangle |^2$ is the square of the electric dipole moment, which is responsible for the strength of the excitation peaks, and the second part, $\delta(\omega - (E_{c\mathbf{k}} - E_{v\mathbf{k}}))$, is the joining of the density of states, which corresponds to the available excitation transition channels. In addition to the independent single-particle excitations, the bound state of the electron–hole pair, or so-called exciton is formed through the mutual Coulomb interaction between excited electrons and holes. Under a sufficiently strong attractive interaction, this quasi-particle may be stable enough against collisions with a phonon. The tightly bound exciton is usually observed in large gap materials, strongly affecting the main feature of the optical excitation, and closely related to the critical orbital hybridizations. The physics of the selective optical absorption and the effect of the robust exciton are the main study focus of the current work.

As a direct consequence of the electromagnetic (EM) wave absorption, the dielectric function is complex and provides very important information for comprehending the optical transition mechanisms, in which the prominent responses are due to the electronic excitation and therefore, strongly related to the orbital hybridization. As clearly presented in figure 8.5(a), the single-particle spectrum is described by the imaginary part $\varepsilon_2(\omega)$ of the dielectric function. It presents a serious change under the three different polarizations as a consequence of the non-uniform physical/chemical environment. The threshold frequency, also called the optical gap,

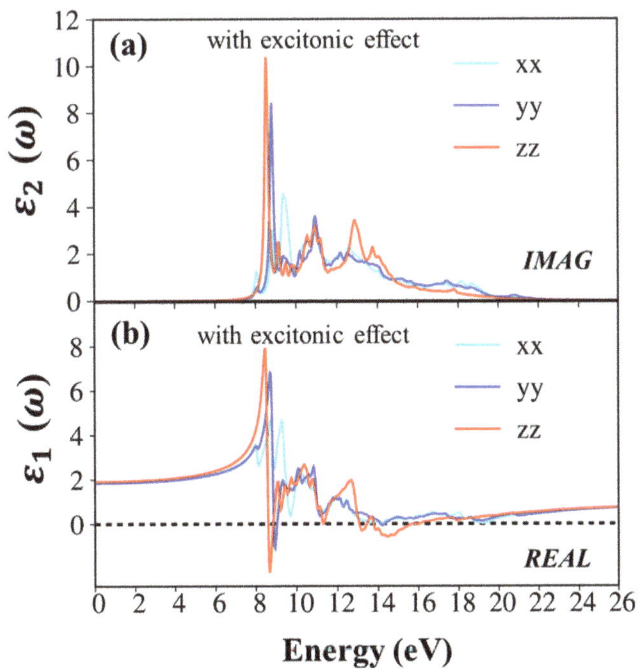

Figure 8.5. (a) The imaginary-part, and (b) real-part dielectric functions with the excitonic effects under *x-/y-/z*-polarization directions (the cyan, blue and red curves).

around 8.8 eV (figure 8.6(a)–(c)) for independent particle excitations may be a little larger than the fundamental indirect gap $E_g^i = 8.1$ eV due to the conservation of the momentum. However, when the Coulomb interactions are considered (figure 8.5(a)), its value is reduced by 1.9/1.8/1.85 under x-/y-/z-directions. The great redshift indicates very strong electron–hole couplings, and therefore, the excitonic states may be stable at room temperature.

In addition to the threshold frequency, there also exist an extra nine non-vanishing shoulders and/or pronounced peaks (the distinct colored arrows). These are associated with the vertical promotion of available valence electrons from the strong van Hove singularities/extreme points in the density of states/band structures and the atom-dominated energy spectrum. The presence of an extra two peaks below the band gap is due to the strong mutual electron–hole interactions at the extreme band edge states. According to the state-of-the-art analysis of the concise physical and chemical pictures in the atom-dominated band structure, atom-/orbital-van Hove singularities, and strong optical responses, the excitation mechanism was achieved and described by the color arrowheads in the energy band structure (figure 8.2(a)). The corresponding chemical characters are denoted by the distinct colors of circles in the orbital-projected density of states (figure 8.4(b)–(d)) and the orbital hybridizations related to these optical responses are successfully identified and clearly shown in table 8.1. Generally, these peaks of $\varepsilon_2(\omega)$ mainly arise from the electron transition between the occupied O-($2p_x$,$2p_y$,$2p_z$) states and the unoccupied Si-(3s, $3p_x$, $3p_y$, $3p_z$) ones. For example, the transition from the O-$2p_z$ to the Si-3s at the Gamma high symmetry point leads to the optical gap at 8.8 eV (brown arrow) (figure 8.6(a)–(c)). Based on this procedure, the orbital hybridizations associated with the specific optical transitions can be thoroughly understood. This viewpoint could be developed for other anode or cathode compounds of LIBs as well.

The real part $\varepsilon_1(\omega)$ of the dielectric function is related to the imaginary part $\varepsilon_2(\omega)$ through the Kramers–Kronig relationship [51]. Under this relation, the pair pronounced peaks of the real part exist simultaneously with the prominent structures

Table 8.1. Prominent absorption structures: frequencies, colored indicators and identified orbital hybridizations without electron–hole interaction in Si–O bonds.

Energy (eV)	Colored arrows	Specific orbital hybridizations in Si–O bonds
8.8	Brown (optical gap)	Si(3s)–O ($2p_x$, $2p_y$, $2p_z$)
11.5	Cyan	Si(3s)–O ($2p_x$, $2p_y$, $2p_z$)
13.5	Red	Si (3s, $3p_x$, $3p_y$)–O ($2p_x$, $2p_y$, $2p_z$)
15.5	Orange	Si (3s, $3p_x$, $3p_y$)–O ($2p_x$, $2p_y$, $2p_z$)
16	Pink	Si (3s, $3p_x$, $3p_y$)–O ($2p_x$, $2p_y$, $2p_z$)
17	Purple	Si (3s, $3p_x$, $3p_y$)–O ($2p_x$)
17.5	Blue	Si ($3p_z$)–O ($2p_y$, $2p_z$)
20	Dark blue	Si ($3p_z$)–O (2s, $2p_x$, $2p_y$, $2p_z$)
22	Green	Si ($3p_z$)–O (2s, $2p_x$,$2p_y$,$2p_z$)
23.5	Black	Si ($3p_z$)–O (2s, $2p_x$,$2p_y$,$2p_z$)

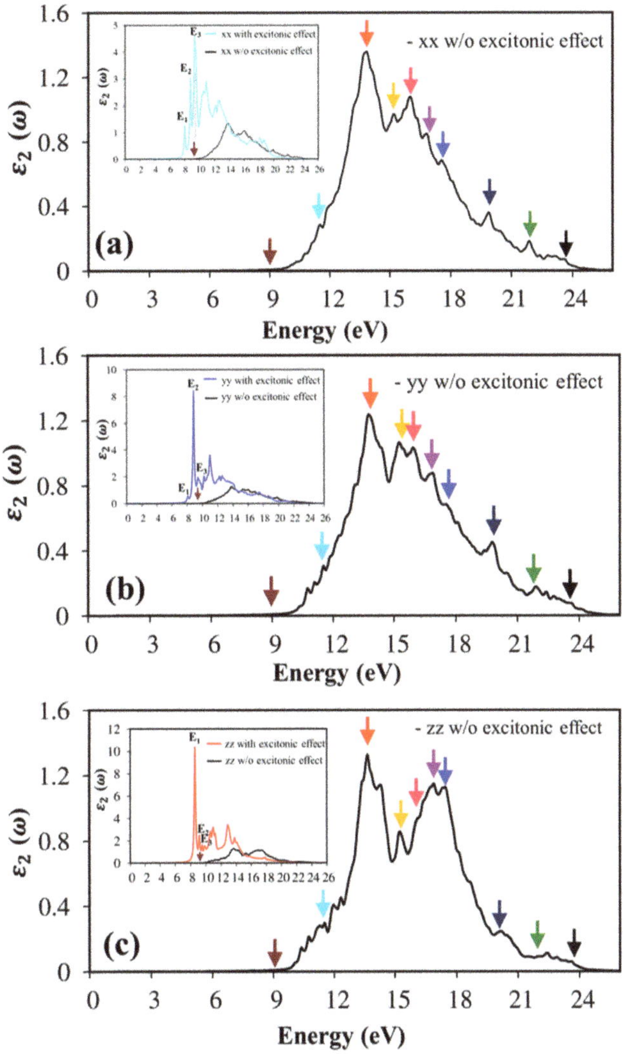

Figure 8.6. Comparison of the imaginary part of dielectric function (a)/(b)/(c) with and w/o excitonic effect (electron–hole interactions) under *xx-*/*yy-*/*zz-*polarization directions, respectively.

of the imaginary one, as presented in figure 8.5(b). Within the inactive region, the real part is weakly dependent on the energy. As an example, the background dielectric constants, $\varepsilon_1(0)$, are about 1.9–2.0 for the x-, y- and z-directions and almost stay dispersionless for frequencies less than or equal to the optical gap through which the low energy reflectance spectrum $R(\omega) = \left| \frac{\sqrt{\varepsilon_1(0)} - 1}{\sqrt{\varepsilon_1(0)} + 1} \right|^2$ and the vanishing range of absorption coefficients could be identified. Contrarily, during the excitation events, the real part is very sensitive to a change of photon energy. Very interestingly, at some frequencies, the real part almost vanishes. The simultaneous

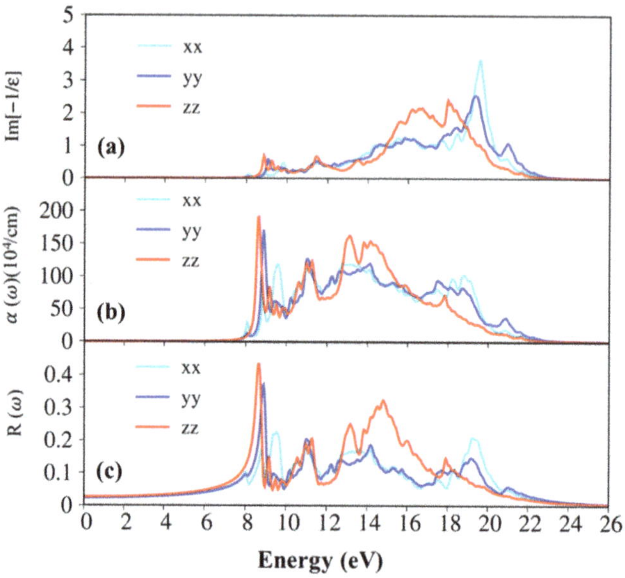

Figure 8.7. The various optical properties: (a) energy loss functions, (b) absorption coefficients and (c) reflectance spectra under the three electric-polarization directions (the cyan, blue and red curves).

existence of the zero points of the real part with very weak single-particle transitions indicated corresponds to the weak damping of the plasmon resonances (shown in detail in figure 8.7(a)). For example, the insignificant Landau damping by the single partial excitation at the real part is 19 eV, 18.5 eV and 17 eV for *x*-, *y*- and *z*-directions, respectively. On the other hand, the zero points for the *z*-direction at 8.1 eV combine with a very strong interband transition peak, and therefore, correspond to the serious Landau damping.

The energy loss function (ELF), being defined as $\mathrm{Im} = \left| \frac{-1}{\varepsilon(\omega)} \right|$ [52] is the screened response function due to the various valence charges of Li, Si and O atoms. Each spectral peak in the ELF corresponds to the excitation of a plasmon mode (a collective excitation). As we mentioned before, the zero points in the real part, $\varepsilon_1(\omega)$, is also one way for explaining the coherent behavior of the free electrons. However, due to the combination with the remarkable imaginary $\varepsilon_2(\omega)$ part at certain zero points, the prominent peaks in the ELF may be better in this scenario. As clearly indicated in figure 8.7(a), the strongest peak is located at frequency ω_p = 19.0/18.5/17.0 eV for the *x*-/*y*-/*z*-directions of electric polarizations due to almost un-damping of the plasmon resonances. These most prominent/important peaks are contributed from the significant orbitals of Li-2s, Si- (3s, $3p_x$, $3p_y$, $3p_z$) and O- ($2p_x$, $2p_y$, $2p_z$). Because the contributions appear mostly at very low energy below −23.5 eV and hence, the contribution of the O-2s atom to this mode is not considered. In addition to the most pronounced peak, a few plasmon modes with relatively weak intensity exist owing to the considerable Landau damping.

When a light beam incident on the Li_2SiO_3 compound medium along the normal direction, some of the light will be attenuated by the valence electrons and be reflected from the front surface, propagating through a finite-width of the sample. The absorption coefficient and the reflectance of solid-state materials are two general optical phenomena and clearly reflect the principal characteristics of single and coherent excitations. As evident in figure 8.7(b), (c), when photon energy is smaller than the onset frequency, due to the lack of the electronic excitation contribution, the absorption coefficient $\alpha(\omega)$ vanishes, while the reflectance $R(\omega)$ is weakly dependent on the frequency. The reflective index is roughly examined by $\left|\frac{\sqrt{\varepsilon_1(0)}-1}{\sqrt{\varepsilon_1(0)}+1}\right|^2$ as we mentioned before. However, beyond the threshold frequency, both $R(\omega)$ and $\alpha(\omega)$ are changing significantly and sensitively to the excitation events. As for the absorption coefficient, the rapid increase is contributed by the various interband transitions, especially, the most obvious change at the plasmon resonance frequency due to the participation of the collective excitations. The inverse values of the absorption coefficient are in the range of 60–400 Å, indicating that the EM waves propagating in the medium are easily absorbed by the rich electronic excitations. On the other hand, the reflectance is significantly enhanced and displays large fluctuation. The abrupt prominence corresponding to the plasmon mode appears at ω_p. For example, the reflectance has a 22% variation under the *E//xx* case (the cyan curve) at the resonance frequency.

In addition to the geometric structures, the measurement of other features, such as electronic and optical properties requires high-resolution techniques. For example, the density of states near the Fermi level can be tested by STS. Unfortunately, this method might only be suitable for a thin-film sample because of the weak quantum current. ARPES is utilized for the examination of the occupied energy band structure. Apparently, the complicated orbital hybridizations in Li_2SiO_3 would create a daunting challenge in the examination of the energy sub-bands. Very interestingly, the rich optical properties of the Li_2SiO_3 compound, such as the onset of excitonic peaks, a lot of prominent absorption structures, and the strongest coherent excitation mode at $\omega_p > 16.5$ eV could be obtained by various optical measurements. For example, by combining the absorption, reflectance or transmission measurements with the Kramers–Kronig relationship, the frequency-dependent optical excitations could be verified.

8.4 Conclusions

In this chapter, first-principles calculations were used to investigate the geometric, electronic, and optical properties of the Li_2SiO_3 compound. Under this state-of-the-art analysis, a concise physical/chemical picture and the crucial characteristics of the electrolyte in Li^+ related materials are achieved, through which the diversified essential properties of other cathode/electrolyte/anode materials in ion-related materials in general could be fully understood.

The orthorhombic Li_2SiO_3 compound exhibits special properties, a large moiré cell with extremely non-uniform physical/chemical environments, wide-indirect insulator gap, a significant spatial charge distribution is generated by the orbital hybridizations, and atom-/orbital-projected density of states. Consequently, the distinct orbital hybridization related to the optical response could be identified. The outstanding optical properties involve the un-isotropic optical behaviors under three electronic dipole directions, the robust excitonic states with a largely reduced optical gap ($E_g^O = 6.95$eV) much smaller than the band gap ($E_g^i = 8.1$eV), strongly modified single-particle interband transitions, the reflectance/high transparency of electromagnetic waves for $\omega < E_g^o$, a high absorption coefficient, nine prominent single-particle excitation peaks, and many pronounced peaks in ELF at the plasmon resonance frequency ω_p, 17–19 eV. The main optical features of the Li_2SiO_3 compound could be verified by experimental optical spectroscopy measurements. The presented theoretical framework could be developed for further studies of other anode/cathode/electrolyte materials for LIBs.

References

[1] Han N T, Dien V K and Lin M-F 2021 Excitonic effects in the optical spectra of Li_2SiO_3 compound *Sci. Rep.* **11** 7683

[2] Boydston A J, Cao B, Nelson A, Ono R J, Saha A, Schwartz J J and Thrasher C J 2018 Additive manufacturing with stimuli-responsive materials *J. Mater. Chem.* A **6** 20621

[3] Liu Y, Zhu Y and Cui Y 2019 Challenges and opportunities towards fast-charging battery materials *Nat. Energy* **4** 540–50

[4] Scrosati B and Garche J 2010 Lithium batteries: status, prospects and future *J. Power Sources* **9** 2419–30

[5] Lisbona D and Snee T 2011 A review of hazards associated with primary lithium and lithium-ion batteries *Process Saf. Environ. Protect.* **6** 434–42

[6] Scrosati B, Hassoun J and Sun Y-K 2011 Lithium-ion batteries: A look into the future *Energy Environ. Sci.* **9** 3287

[7] Wang A, Kadam S and Li H *et al* 2018 Review on modeling of the anode solid electrolyte interphase (SEI) for lithium-ion batteries *NPJ Comput. Mater.* **4** 15

[8] Heiska J, Nisula M and Karppinen M 2019 Organic electrode materials with solid-state battery technology *J. Mater. Chem.* A **7** 18735–58

[9] Chen R, Qu W, Guo X, Li L and Wu F 2016 The pursuit of solid-state electrolytes for lithium batteries: from comprehensive insight to emerging horizons *Mater. Horizon.* **6** 487–516

[10] Zhao Q, Stalin S and Zhao C *et al* 2020 Designing solid-state electrolytes for safe, energy-dense batteries *Nat. Rev. Mater.* **5** 229–52

[11] Manthiram A, Yu X and Wang S 2017 Lithium battery chemistries enabled by solid-state electrolytes *Nat. Rev. Mater.* **2** 16103

[12] Lü X, Howard J W, Chen A, Zhu J, Li S, Wu G and Jia Q 2016 Antiperovskite Li_3OCl superionic conductor films for solid-state Li-ion batteries *Adv. Sci.* **3** 1500359

[13] Zhu Y, He X and Mo Y 2016 First principles study on electrochemical and chemical stability of solid electrolyte–electrode interfaces in all-solid-state Li-ion batteries *J. Mater. Chem.* A **9** 3253–66

[14] Lian P, Zhao B, Zhang L, Xu N, Wu M and Gao X 2019 The inorganic sulfide solid electrolytes for all-solid-state lithium secondary batteries *J. Mater. Chem.* A **7** 20540–57

[15] Nakagawa A, Kuwata N, Matsuda Y and Kawamura J 2010 Characterization of stable solid electrolyte lithium silicate for thin film lithium battery *J. Phys. Soc. Jpn.* **79** 98–101

[16] Shi L, Zhu T and Gao G *et al* 2018 Highly stretchable and transparent ionic conducting elastomers *Nat. Commun.* **9** 2630

[17] Watanabe M *et al* 2017 Synthesis of Li_2SiO_3 using novel water-assisted solid-state reaction method *J. Ceram. Soc. Jpn.* **6** 472–75

[18] Alemi A, Khademinia S and Sertkol M 2015 Part III: lithium metasilicate (Li_2SiO_3) – mild condition hydrothermal synthesis, characterization and optical properties *Int. Nano Lett.* **2** 77–83

[19] Alemi A and Khademinia S 2014 Part I: lithium metasilicate (Li_2SiO_3) – mild condition hydrothermal synthesis, characterization, and optical properties *Int. Nano Lett.* **1** 15–20

[20] Sakiyama H, Yamasaki M, Suzuki T, Watanabe Y, Ito R, Ohnishi A and Nishida Y 2006 Formation of polar crystals of a chiral dinuclear nickel (II) complex and their pyroelectric effect and electro-optic effect *Inorg. Chem. Commun.* **9** 18–20

[21] Recker K and Wallrafen F 1984 Crystal growth and dielectric, piezoelectric, electrooptic, elastic and thermoelastic properties of orthorhombic lithium hydrogen selenite, $LiHSeO_3$ *Z. Kristallogr. Cryst. Mater..* **169** 249–56

[22] Schwartz R W, Ballato J and Haertling G H 2004 Piezoelectric and electro-optic caramics *Ceramic Materials for Electronics* (Boca Raton, FL: CRC Press) ch 4

[23] Halliyal A, Bhalla A S and Newnham R E 2013 Glass-ceramics for piezoelectric and pyroelectric devices *Glasses and Glass-Ceramics* (Dordrecht: Springer) pp 272–315

[24] Habasaki J and Okada I 1992 Molecular dynamics study of Li_2SiO_3 in the liquid and glassy states *Mol. Simul.* **8** 179–95

[25] Tang T, Chen P, Luo W, Luo D and Wang Y 2012 Crystalline and electronic structures of lithium silicates: a density functional theory study *J. Nucl. Mater.* **420** 31–8

[26] Zulueta Y A, Froeyen M and Nguyen M T 2017 Structural properties and mechanical stability of lithium-ion based materials. A theoretical study *Comput. Mater. Sci.* **136** 271–79

[27] Richet P, Mysen B O and Andrault D 1996 Melting and premelting of silicates: Raman spectroscopy and X-ray diffraction of Li_2SiO_3 and Na_2SiO_3 *Phys Chem Mineral.* **23** 157–72

[28] Ortiz-Landeros J, Contreras-García M E, Gómez-Yáñez C and Pfeiffer H 2011 Surfactant-assisted hydrothermal crystallization of nanostructured lithium metasilicate (Li_2SiO_3) hollow spheres: (I) Synthesis, structural and microstructural characterization *J. Solid-State Chem.* **5** 1304–11

[29] Li X and Yang H 2014 Morphology-controllable Li_2SiO_3 nanostructures *CrystEngComm.* **21** 4501–07

[30] Furusawa S, Tsurui T and Shimizu K 2013 TEM observation and ionic conductivity study of Li_2SiO_3 thin-film on sapphire substrate *Key Eng. Mater.* **596** 15–20

[31] Damascelli A 2004 Probing the electronic structure of complex systems by ARPES *Phys. Scr.* **109** 61

[32] Feenstra R M 1994 Scanning tunneling spectroscopy *Surf. Sci.* **299** 965–79

[33] Haussühl S, Liebertz J and Stähr S 1982 Single crystal growth and pyroelectric, dielectric, piezoelectric, elastic, and thermoelastic properties of orthorhombic Li_2SiO_3, Li_2GeO_3, and Na_2 GeO_3 *Cryst. Res. Technol.* **4** 521–26

[34] Ching W Y, Li Y P, Veal B W and Lam D 1985 Electronic structures of lithium metasilicate and lithium disilicate *J. Phy. Rev. B.* **32** 1203

[35] Yohandys A. Zulueta, Minh Tho Nguyen, and James A. Dawson 2020 Na- and K-doped Li_2SiO_3 as an alternative solid electrolyte for solid-state lithium batteries *J. Phys. Chem. C.* **9** 4982–88

[36] Alemi A, Khademinia S and Joo S W *et al* 2013 Lithium metasilicate and lithium disilicate nanomaterials: optical properties and density functional theory calculations *Int. Nano Lett.* **3** 14
[37] Han N T, Dien V K, Tran N T T, Nguyen D K, Su W-P and Lin M-F 2020 First-principles studies of electronic properties in lithium metasilicate (Li_2SiO_3) *RSC Adv.* **41** 24721–29
[38] Kresse G and Hafner J 1993 *Ab initio* molecular dynamics for liquid metals *Phys. Rev.* B **47** 558
[39] Perdew J P, Burke K and Ernzerhof M 1997 Generalized gradient approximation made simple *Phys. Rev. Lett.* **7** 1396
[40] Kresse G and Joubert D 1999 From ultrasoft pseudopotentials to the projector augmented-wave method *Phys. Rev.* B **59** 1758–75
[41] Wisesa P, McGill K A and Mueller T 2016 Efficient generation of generalized Monkhorst–Pack grids through the use of informatics *Phys. Rev.* B **93** 155109
[42] Giustino F, Cohen M L and Louie S G 2010 GW method with the self-consistent Sternheimer equation *Phys. Rev.* B **81** 115105
[43] Shishkin M, Marsman M and Kresse G 2007 Accurate quasiparticle spectra from self-consistent GW calculations with vertex corrections *Phys. Rev. Lett.* **24** 246403
[44] Pizzi G, Vitale V, Arita R, Bluegel S, Freimuth F, Géranton G and Yates J R 2019 Wannier90 as a community code: new features and applications *J. Phys. Condens. Matter* **32** 16
[45] Paier J, Marsman M and Kresse G 2008 Dielectric properties and excitons for extended systems from hybrid functionals *Phys. Rev.* B **78** 121201
[46] Völlenkle H 1981 Refinement of the crystal structures of Li_2SiO_3 and Li_2GeO_3 *Z. Kristallogr. Cryst. Mater..* **154** 77–81
[47] Tang T, Chen P, Luo W, Luo D and Wang Y 2012 Crystalline and electronic structures of lithium silicates: a density functional theory study *J. Nucl. Mater.* **420** 31–8
[48] Huan Y, Fan Y, Li Y, Yin B, Hu X, Dong D and Wei T 2019 Factors influencing Li^+ migration in garnet-type ceramic electrolytes *J. Materiomi.* **2** 214–20
[49] Eladgham E H, Demchenko D O, Nakagawara T A, Ozgur U and Arachchige I U 2019 Facile synthesis of highly luminescent lithium silicate nanocrystals with varying crystal structure and morphology *CrystEngComm* **21** 1974–83
[50] Bischoff J-M and Jeckelmann E 2017 Density-matrix renormalization group method for the conductance of one-dimensional correlated systems using the Kubo formula *Phys. Rev.* B **96** 195111
[51] Hutchings D C, Sheik-Bahae M and Hagan D J *et al* 1992 Kramers–Krönig relations in nonlinear optics *Opt. Quant. Electron.* **24** 1–30
[52] Browning N D, Yuan J and Brown L M 1993 Theoretical determination of angularly-integrated energy loss functions for anisotropic materials *Philos. Mag.* A **1** 261–71

IOP Publishing

Chapter 9

Screening electrolyte additives for lithium-ion batteries

Ngoc Thanh Thuy Tran, Phung My Loan Le, Wen-Dung Hsu, Thi Dieu Hien Nguyen, Shih-Yang Lin, Hsien-Ching Chung, Wei-Bang Li, Nguyen Thi Han, Hsin-Yi Liu, Hai Duong Pham and Ming-Fa Lin

9.1 Introduction

Many research interests have focused on the electrodes and electrolyte materials. Apart from them, identifying suitable electrolyte additives also plays a significant role to enhance the performance of lithium ion batteries (LIBs). After the first cycle of charge/discharge, the solid state electrolyte interface (SEI) is formed, which consists of organic and inorganic decomposition products of electrolyte. This SEI layer separates the cathode and anode but still allows ions to pass through. To keep this layer thin and stable, electrolyte additives are often added [1–10]. A good functional additive can help stabilizing the electrolytes from degradation to enhance the cycle life of Li-ion and maintain a stable SEI film on the anode without side effects of the electrolyte reduction on the anode and oxidation on the cathode. The selection of the best additive among phosphides, sulfonate esters and lactam, etc, which perfectly fits the electrodes, can accelerate and improve the developments in LIBs. Up to now, there have been several commonly used electrolyte additives, namely vinylene carbonate (VC), vinyl ethylene carbonate (VEC), ally ethyl carbonate (AEC), vinyl acetate(VA), ethylene sulfite (ES), 1,3-propane sultone (PS), fluoroethylene carbonate (FEC), γ-butyrolactone (GBL), and lithium difluorophosphate (LiDFP) [11–20]. However, there is still a huge number of chemicals that scientists have not yet tried to verify and select the best candidates. Therefore, identifying suitable electrolyte additives is necessary to improve high performance LIBs.

doi:10.1088/978-0-7503-4685-6ch9

Electron affinity and ionization energy are the two important parameters needed to be taken into account used for first screening of additive selection. However, with a huge number of molecules waiting to be tested, calculations or experimental measurements are not an effective way to deal with them. To save much time on research work, the powerful method of so-called 'machine learning' or 'artificial intelligence' has have been established. The key merit of the machine learning approach is that it can learn by example in the way human brains do. It has the ability to learn and model non-linear and complex relationships from a given database and its targets.

In this work, an artificial neural network-based model is developed to predict the electron affinity and ionization energy, which are important parameters for determining good electrolyte and electrolyte additives. Base on training a given database (electrolytes additives molecules with calculated electron affinity and ionization energy values), machine learning techniques could help scientists predict the results for a large number of new electrolyte additive molecules they want instead of doing experiment or simulation to obtain them. The aforementioned studies will provide important insights into theoretical and practical issues in the enhancement of LIBs.

9.2 Electron affinity and ionization energy

During the charging process, the electrons flow from cathode to anode through the circuit, while some of the Li^+ moves out from the cathode and partially intercalates into the anode to maintain electron neutrality. The difference between electrons moving into the anode with Li^+ intercalating through electrolyte–electrode film causes the short non-electroneutrality. Therefore, there exist extra electrons and these electrons need to be accepted. Moreover, it is easier for electrons to move out from the cathode than Li^+ exfoliating from the cathode. Hence, electrons need to be donated to this oxidative environment to maintain the elecroneutrality.

Comparing the tendency of releasing or gaining extra electrons of additives with those of the electrolyte can conclusively determine whether the additives can prevent the electrolyte decomposing or not, which is one of the critical factors for electrolyte additives selection. For liquid electrolyte and electrolyte additive molecules, the highest occupied molecular orbital (HOMO) and the lowest unoccupied molecular orbital (LUMO) values act as key factors in determining the operating voltage of a battery [21, 22], as shown in figure 9.1. In general, a good electrolyte molecule should have HOMO value lower than chemical potential of cathode to prevent the oxidation and LUMO value higher than chemical potential of anode to prevent the reduction. Vice versa will be valid for the case of electrolyte additives with higher HOMO and lower LUMO value compared to the chemical potentials of cathode and anode, respectively. It is worth to know that the energy of the HOMO value approximates to the negative of the ionization energy (IE) according to Koopmans' theorem. Similarly, the LUMO is identified with the negative value of electron affinity (EA). Considering which of the environmental components is going to either

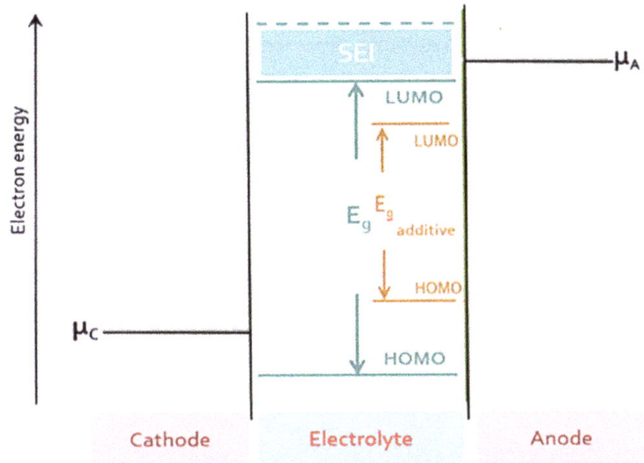

Figure 9.1. HOMO and LUMO levels for electrolyte and electrolyte additives.

release or accepte extra electrons, two of the most important parameters of an electrolyte component are EA and IE [23]. The former is the change in energy when a neutral atom attracts an electron to become a negative ion (the energy to reduce a molecule). On the other hand, the latter is the required energy to remove an electron from a neutral atom (energy required to oxidize a molecule). The formulas applied for EA and IE are as follows:

$$EA = -\frac{\Delta G_{\text{red}}(\text{sol})}{nF}$$

$$IE = -\frac{\Delta G_{\text{oxi}}(\text{sol})}{nF}$$

where F is the Faraday constant, and $\Delta G_{\text{red}}(\text{sol})$ and $\Delta G_{\text{oxi}}(\text{sol})$ are the Gibbs free energy variations as a result of reduction and oxidation in the solution phase, respectively. According to the thermodynamic diagram in figure 9.2, $\Delta G_{\text{red}}(\text{sol})$ and $\Delta G_{\text{oxi}}(\text{sol})$ can be calculated from the Gibbs free energy variations of gas phase as below:

$$\Delta G_{\text{red}}(\text{sol}) = \Delta G_{\text{red}}(\text{gas}) + \Delta G_{\text{solv}}(R^-) - \Delta G_{\text{solv}}(R)$$

$$\Delta G_{\text{oxi}}(\text{sol}) = \Delta G_{\text{oxi}}(\text{gas}) + \Delta G_{\text{solv}}(R^+) - \Delta G_{\text{solv}}(R)$$

where ΔG_{solv} is the solvation energy to transfer a molecule from vacuum to solution phase. Each of the additive molecules should undergo gas state and solvent state optimization, and charged system optimization in both states as well. However, these calculations are very time-consuming. To address this issue, another new approach is predicting the EA and IE values by means of machine learning.

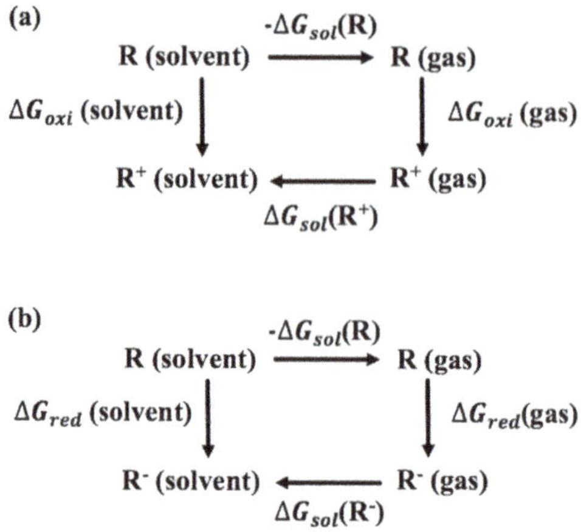

Figure 9.2. EA and IE calculations.

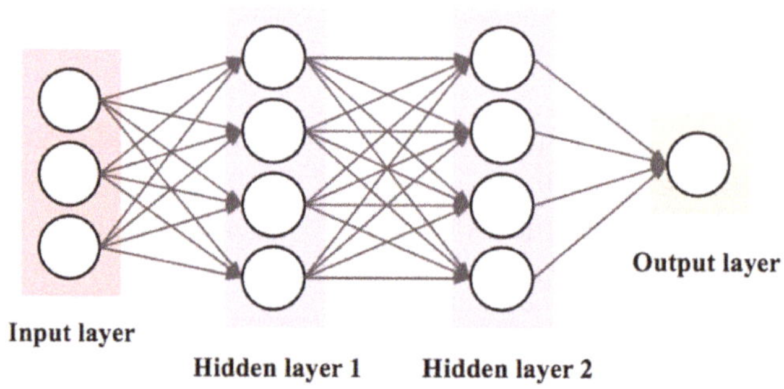

Figure 9.3. Artificial neural network scheme.

9.3 Artificial neural networks

Neural networks have been utilized in multiple fields, including banking, finance, securities, insurance, aerospace, automotive, entertainment, manufacturing, transportation, medical, and environment. 'Neural' here implies the brain-inspired systems which can replicate the way that humans learn. Of all the machine learning models, artificial neural networks (ANN) have been widely used, such as for system modeling [24], anomaly detection [25], and batteries also [26–29]. It is a computational intelligence tool comprising many nonlinear processing elements connected to each other. These elements of ANN are called neurons. By applying algorithms that imitate the real neurons processes, ANN can learn to solve many types of problems.

For neural network layers, there are input and output layers, as well as one or few hidden layers with nodes to imitate the nonlinearity between the system input so that the network learns and trains to adapt to convergence (figure 9.3). It can adjust any nonlinear function with sufficient layers and neurons per layer to make them suitable for a complex model. They are excellent tools for finding patterns which are far too complex or numerous for a human programmer to extract and teach the machine to recognize.

The input data of a neural network is typically divided into three sets: training, validation and testing dataset. The data in the training dataset are trained during the learning process, and the corresponded output data are utilized for regression analysis with the corresponded target data. After this, they fine-tune it using a validation data set. The data in the validation dataset are trained and monitored for regularization to prevent over-fitting Finally, the testing dataset is not involved in the training process but for testing the learning ability of the built model. It will be used to see if it can successfully turn the input into the desired output. To estimate the performance of ANN, the testing data must be those which have not been used to train the network.

In this work, the two important parameters for characterizing electrolytes, EA and IE, are chosen as targets. The input layer consists of the molecular structure factors based on SYBYL atom type [30, 31] shown in table 9.1. The output layer will be the EA and IE values. After testing several models, the neural network is optimally run in two hidden layers with 25 nodes per layer. In addition, the training, testing, and validation data are divided into three sets with ratios of 0.7, 0.15, 0.15.

9.4 Training and predicting results

One of the abilities of ANN is to adapt continuously to new data and trace the changes of input signals. The model in this study was established by Matlab [32, 33], and the details of the model architecture are described as follows. The schematic method of constructing an ANN model is demonstrated in figure 9.4. ANN might face an over-fitting problem when the network has too many parameters to be learned from the limited database. In order to solve this over-fitting problem, a big database from The Material Projects (about 20 000 molecules) is obtained for machine learning ANN. Python code is used to obtain the database from The Materials Project. Furthermore, Python codes are employed to analyse the data for the input file and target in the machine learning step. It was used to automatically convert SMILES [34, 35] to mol2 format and finally the SYBYL notation for features selection. An example of SYBYL notation for 1,3-propane sultone electrolyte additive is illustrated in figure 9.5.

The big challenge with ANN is that neural networks are 'black boxes', in which the user feeds in data and receives answers. They can fine-tune the answers, but they do not have access to the exact decision making process. The training from the big database of 21 954 molecules does not show an accuracy result, as shown in figure 9.6. After analyzing the database (figure 9.7), we found that the target values (EA and IE) of some molecules are too different from others. By checking the chemical structure of input molecules, it is interesting to know that those with ring-(17 014 molecules) and linear-structures (836 molecules) lead to good training results with a great prediction, in which the regression is around 0.97 (approximate to the perfect value $R = 1$) as observed in figure 9.8. Moreover, the error bars of EA and IE show about 95% confidence interval ±1.1 eV and 95% confidence interval ±0.7 eV, respectively, as shown in figure 9.9. Based on the error bars, the error standard deviations of EA and IE are about ±0.41 and ±0.27,

Table 9.1. SYBYL atom type description for database from The Materials Project.

Atom	C				
SYBYL	**C.1**	**C.2**	**C.3**	**C.ar**	**C.cat**
Atom type	Carbon sp	Carbon sp2	Carbon sp3	Carbon aromatic	cabocation (C+) used in guadinium group
Atom	N				
SYBYL	**N.1**	**N.2**	**N.3**	**N.4**	
Atom type	Nitrogen sp	Nitrogen sp2	Nitrogen sp3	Nitrogen sp3 positively charged	
Atom	N			P	
SYBYL	**N.am**	**N.ar**	**N.Pl3**	**P.3**	
Atom type	Nitrogen amide	Nitrogen aromatic	Nitrogen trigonal planar	Phosphorus sp3	
Atom	S				
SYBYL	**S.2**	**S.3**	**S.O**	**S.O2**	
Atom type	Sulfur sp2	Sulfur sp3	sulfoxide sulfur	Sulphone sulfur	
Atom	O			Others	
SYBYL	**O.2**	**O.3**	**O.co2**	**H/Li/Na/Ca/Mg/Al/B/Si/Zn/ F/Cl/Br/**	
Atom type	Oxygen sp2	Oxygen sp3	oxygen in carboxylate and phosphate groups	H/Li/Na/Ca/Mg/Al/B/Si/Zn/ Halogen	

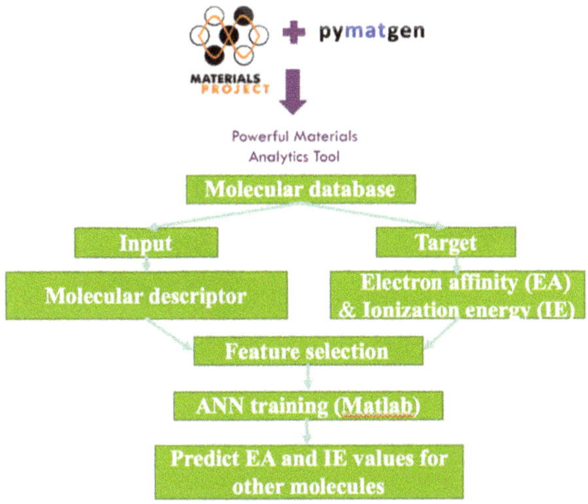

Figure 9.4. Artificial neural network study roadmap for electrolyte additives screening.

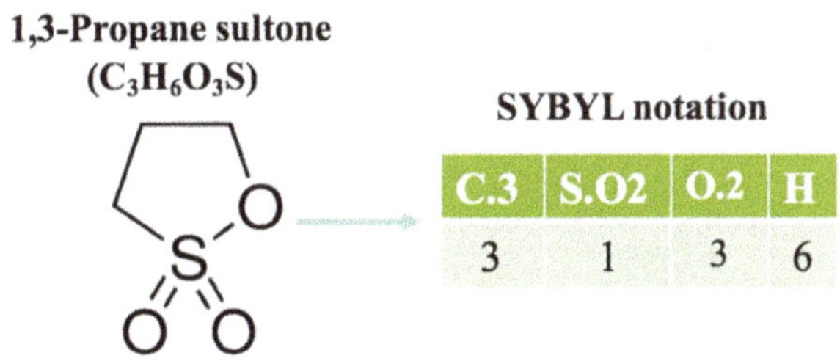

Figure 9.5. Example of SYBYL notation for 1,3-propane sultone (PS) electrolyte additive.

respectively, which indicates the ANN training model reaches high accuracy. After the training, the prediction EA and IE models can be used for testing other molecular databases.

In order to figure out if the results of prediction are convincing or not, it is necessary to pick up some of the molecules in the database for validation by calculating their EA and IE directly. Therefore, ethylene carbonate (EC), propane sultone (PS), diethyl carbonate (DEC), dimethyl carbonate (DMC) and γ-butyrolactone (GBL), the most commonly used electrolyte and additive molecules, are chosen for calculation of validation. After solvation and potential alignment. The calculation result can be really close to database and prediction results. Hence, this is the way to check the new modified molecules prediction. Furthermore, a functional new electrolyte and additive which can be stable at high working voltage instead of decomposing is imminently required before higher working voltage and higher energy density LIBs are developed. Therefore, the application of ANN by predicting a new modified molecule's EA and IE can omit the complicated and costly experimentation.

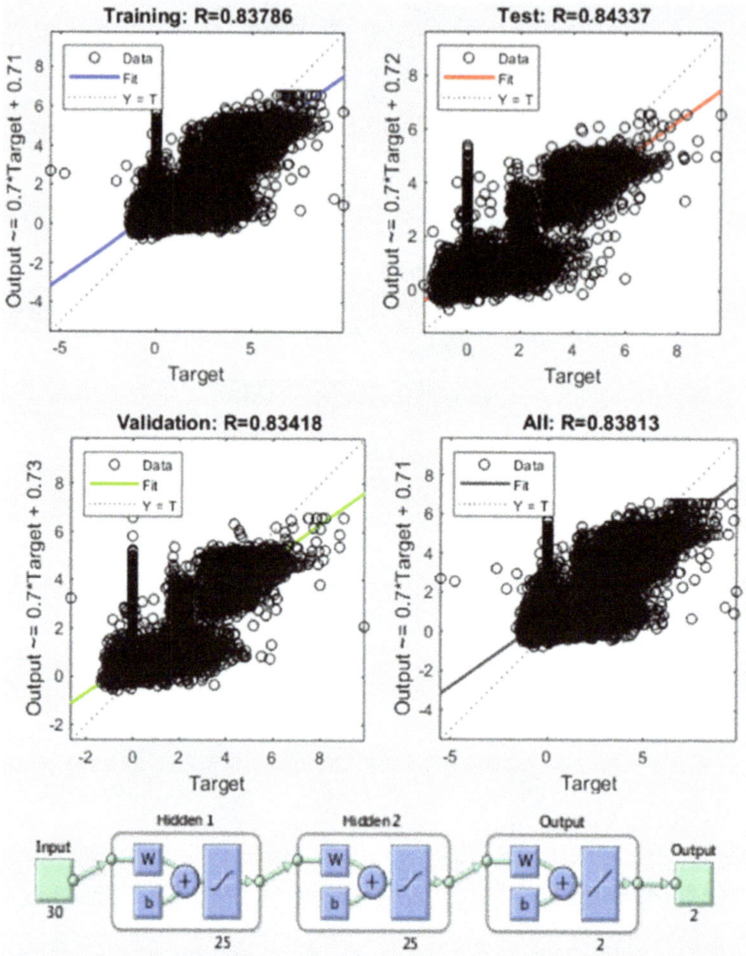

Figure 9.6. ANN training, testing, and validation results for EA and IE (21 954 molecules).

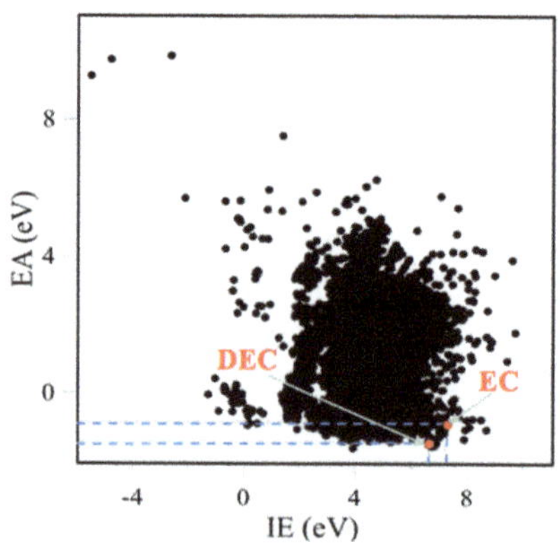

Figure 9.7. The Material Project data distribution versus EA and IE values (21 954 molecules).

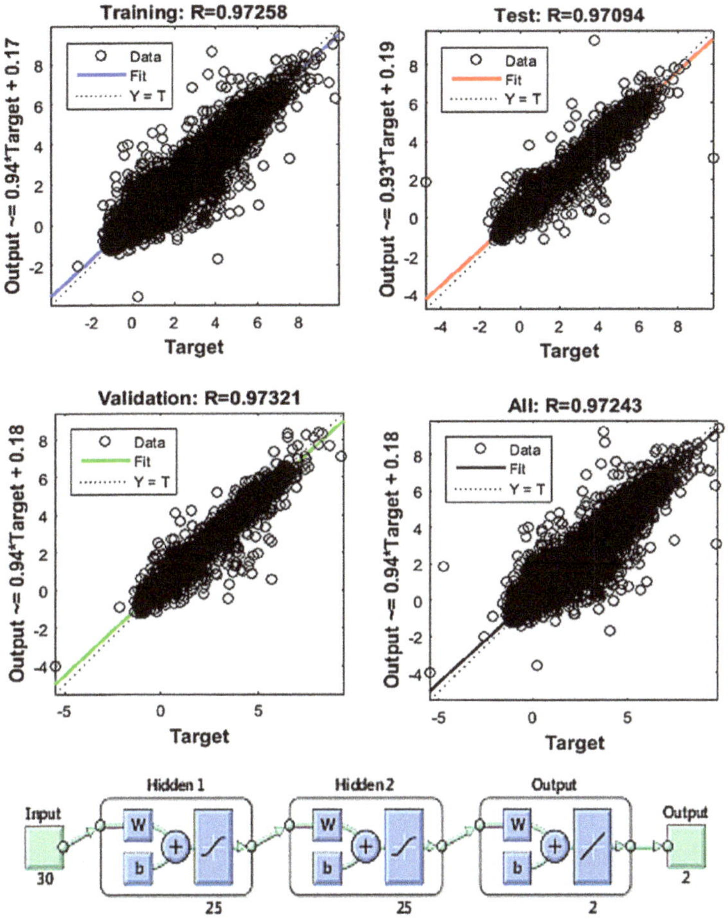

Figure 9.8. ANN training, testing, and validation results for EA and IE (17 850 molecules).

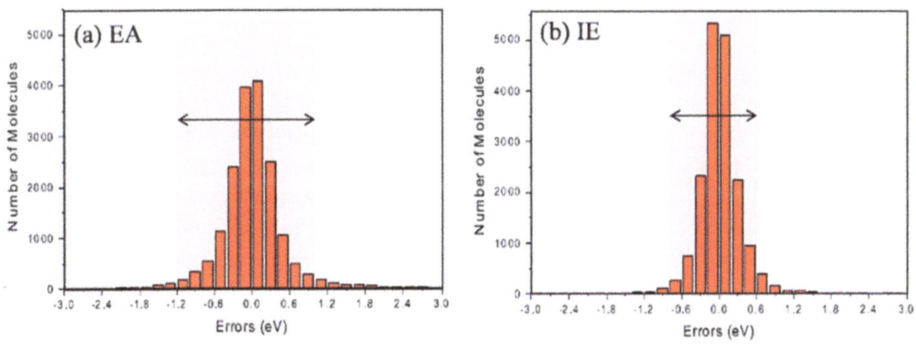

Figure 9.9. Error bar results for (a) EA and (b) IE (17 850 molecules).

9.5 Conclusions

The machine learning approach leads to a new way to create brand new functional molecules in the most economically efficient way. Based on this learning it can perform prediction for new data sets without doing complicated calculations and costly experiments. In this work, an ANN–based model is developed to predict the EA and IE values for determining good electrolyte additives by means of Matlab and Python codes. A large scale of 17 850 molecules database from The Materials Project was used for training and testing, and then from the validation results to estimate the expected values by the self-developed code. The results demonstrated that systematic machine learning of molecular structure factors is capable of describing the EA and IE parameters, in which the regression is around 0.97 (approximate to the perfect value $R = 1$) and the error standard deviation of EA and IE are about ±0.41 and ±0.27, respectively. Furthermore, the developed model can also be used for estimating the EA and IE of newly designed electrolyte additives from the structure factors with high accuracy.

Artificial intelligence by using machine learning is really amazing, however, it is still a new and difficult method in LIBs-related research. Further work on feature selection of the input data before the training will be implemented to enhance the accuracy. In addition to ANN, different metrics for molecular selection, training and predicting models will be considered to find out an optimal model. The generative model could serve as a first step design towards further investigation into other necessary properties not only for electrolytes additives but also for electrolyte molecules This could serve as a first step toward a better design of materials for LIBs and useful for experimental fabrications.

Acknowledgements

This work was financially supported by the Hierarchical Green-Energy Materials (Hi-GEM) Research Center, from The Featured Areas Research Center Program within the framework of the Higher Education Sprout Project by the Ministry of Education (MOE) and the Ministry of Science and Technology (MOST 110–2634-F-006–017) in Taiwan.

References

[1] Gireaud L, Grugeon S, Laruelle S, Pilard S and Tarascon J M 2005 Identification of Li battery electrolyte degradation products through direct synthesis and characterization of alkyl carbonate salts *J. Electrochem. Soc.* **152** A850

[2] Leggesse E G and Jiang J C 2012 Theoretical study of the reductive decomposition of 1, 3-propane sultone: SEI forming additive in lithium-ion batteries *RSC Adv.* **2** 5439–46

[3] Lin F W, Tran N T T and Hsu W D 2020 Effect of 1, 3-propane sultone on the formation of solid electrolyte interphase at Li-ion battery anode surface: a first-principles study *ACS Omega* **5** 13541–547

[4] Liu X, Zhou J, Xu Z and Wang Y 2020 Atomic thermodynamics and microkinetics of the reduction mechanism of electrolyte additives to facilitate the formation of solid electrolyte interphases in lithium-ion batteries *RSC Adv.* **10** 16302–12

[5] Zhang S S 2006 A review on electrolyte additives for lithium-ion batteries *J. Power Sources* **162** 1379–94
[6] Li S, Luo Z, Li L, Hu J, Zou G, Hou H and Ji X 2020 Recent progress on electrolyte additives for stable lithium metal anode *Ener. Storage Mater.* **32** 306–19
[7] Eshetu G G and Figgemeier E 2019 Confronting the challenges of next-generation silicon anode-based lithium-ion batteries: role of designer electrolyte additives and polymeric binders *ChemSusChem* **12** 2515–39
[8] Gao H, Maglia F, Lamp P, Amine K and Chen Z 2017 Mechanistic study of electrolyte additives to stabilize high-voltage cathode–electrolyte interface in lithium-ion batteries *ACS Appl. Mater. Inter.* **9** 44542–49
[9] Bar-Tow D, Peled E and Burstein L 1999 A study of highly oriented pyrolytic graphite as a model for the graphite anode in Li-ion batteries *J. Electrochem. Soc.* **146** 824
[10] Peled E, Tow D B, Merson A, Gladkich A, Burstein L and Golodnitsky D 2001 Composition, depth profiles and lateral distribution of materials in the SEI built on HOPG-TOF SIMS and XPS studies *J. Power Sources* **97** 52–7
[11] Xu K, Zhang S and Jow T R 2003 Formation of the graphite/electrolyte interface by lithium bis (oxalato) borate *Electrochem. Solid State Lett.* **6** A117
[12] Contestabile M, Morselli M, Paraventi R and Neat R J 2003 A comparative study on the effect of electrolyte/additives on the performance of ICP383562 Li-ion polymer (soft-pack) cells *J. Power Sources* **119** 943–47
[13] Aurbach D, Gnanaraj J S, Geissler W and Schmidt M 2003 Vinylene carbonate and Li salicylatoborate as additives in $LiPF_3$ $(CF_2CF_3)_3$ solutions for rechargeable Li-ion batteries *J. Electrochem. Soc.* **151** A23
[14] Chen G, Zhuang G V, Richardson T J, Liu G and Ross P N Jr 2005 Anodic polymerization of vinyl ethylene carbonate in Li-ion battery electrolyte *Electrochem. Solid State Lett.* **8** A344
[15] Sasaki T, Abe T, Iriyama Y, Inaba M and Ogumi Z 2005 Suppression of an alkyl dicarbonate formation in Li-ion cells *J. Electrochem. Soc.* **152** A2046
[16] Mogi R, Inaba M, Jeong S K, Iriyama Y, Abe T and Ogumi Z 2002 Effects of some organic additives on lithium deposition in propylene carbonate *J. Electrochem. Soc.* **149** A1578
[17] McMillan R, Slegr H, Shu Z X and Wang W 1999 Fluoroethylene carbonate electrolyte and its use in lithium ion batteries with graphite anodes *J. Power Sources* **81** 20–6
[18] Arakawa M and Yamaki J I 1995 Anodic oxidation of propylene carbonate and ethylene carbonate on graphite electrodes *J. Power Sources* **54** 250–54
[19] Heine J, Hilbig P, Qi X, Niehoff P, Winter M and Bieker P 2015 Fluoroethylene carbonate as electrolyte additive in tetraethylene glycol dimethyl ether based electrolytes for application in lithium ion and lithium metal batteries *J. Electrochem. Soc.* **162** A1094
[20] Xu Y, Wan L, Liu J, Zeng L and Yang Z 2017 γ-butyrolactone and glutaronitrile as 5 V electrolyte additive and its electrochemical performance for $LiNi_{0.5}Mn_{1.5}O_4$ *J. Alloys. Compounds* **698** 207–14
[21] Cheng L *et al* 2015 Accelerating electrolyte discovery for energy storage with high-throughput screening *J. Phys. Chem. Lett.* **6** 283–91
[22] Scheer A M, Gallup G A and Burrow P D 2008 Unoccupied orbital energies of 1, 4-benzenedithiol and the HOMO–LUMO gap *Chem. Phys. Lett.* **466** 131–35
[23] Qu X *et al* 2015 The Electrolyte Genome project: a big data approach in battery materials discovery *Comput. Mater. Sci.* **103** 56–67

[24] Chen C W 2009 Modeling and control for nonlinear structural systems via a NN-based approach *Expert Syst. Appl.* **36** 4765–72

[25] Schlechtingen M and Santos I F 2011 Comparative analysis of neural network and regression based condition monitoring approaches for wind turbine fault detection *Mech. Syst. Signal Process.* **25** 1849–75

[26] He W, Williard N, Chen C and Pecht M 2014 State of charge estimation for Li-ion batteries using neural network modeling and unscented Kalman filter-based error cancellation *Int. J. Elec.* **62** 783–91

[27] Halls M D and Tasaki K 2010 High-throughput quantum chemistry and virtual screening for lithium ion battery electrolyte additives *J. Power Sources* **195** 1472–78

[28] Korth M 2014 Large-scale virtual high-throughput screening for the identification of new battery electrolyte solvents: evaluation of electronic structure theory methods *Phys. Chem. Chem. Phys.* **16** 7919–26

[29] Wu B, Han S, Shin K G and Lu W 2018 Application of artificial neural networks in design of lithium-ion batteries *J. Power Sources* **395** 128–36

[30] Ash S, Cline M A, Homer R W, Hurst T and Smith G B 1997 SYBYL line notation (SLN): a versatile language for chemical structure representation *J. Chem. Inf. Comput. Sci.* **37** 71–9

[31] Homer R W, Swanson J, Jilek R J, Hurst T and Clark R D 2008 SYBYL line notation (SLN): a single notation to represent chemical structures, queries, reactions, and virtual libraries *J. Chem. Inf. Model.* **48** 2294–307

[32] Sivanandam S N and Deepa S N 2006 *Introduction to Neural Networks using Matlab 6.0* (New York: Tata McGraw-Hill Education)

[33] Ballabio D and Vasighi M 2012 A MATLAB toolbox for Self Organizing Maps and supervised neural network learning strategies *Chemometr. Intell. Lab. Syst.* **118** 24–32

[34] Weininger D, Weininger A and Weininger J L 1989 SMILES. 2. Algorithm for generation of unique SMILES notation *J. Chem. Inf. Comput. Sci.* **29** 97–101

[35] Weininger D 1988 SMILES, a chemical language and information system. 1. Introduction to methodology and encoding rules *J. Chem. Inf. Comput. Sci.* **28** 31–6

Chapter 10

Graphite alkali-atom-/ion-intercalated compounds: Li^+-related-based battery anodes

Wei-Bang Li, Shih-Yang Lin, Thi Dieu Hien Nguyen, Hsien-Ching Chung, Ngoc Thanh Thuy Tran, Nguyen Thi Han, Hsin-Yi Liu, Hai Duong Pham and Ming-Fa Lin

Recently, clean energy has become one of the most popular worldwide issues. Limitation of fossil fuels is the main reason, and they will be consumed eventually. In addition, concurrent environmental pollution is also unavoidable. Hence, it brings about the need for efficient storage of electrical energy [1]. On the other side, solar power is the most abundant energy for humans among all resources on the Earth [2]. As time passed, in order to solve the problems of energetic storage, rechargeable batteries emerged [3]. Such batteries possess anodes, cathodes, and electrolytes [4–7]. Among these, the anodes are composed of carbon and non-carbon materials [8–12]. Previous studies, covering theoretical and experimental aspects focused on lithium batteries and lithium-ion batteries because of their long cycle life, low cost and high safety [13–16]. The common anode material is graphite [9, 17–19]. Such anodes perform as graphite-intercalation compounds (GICs) [20–27]. In other words, the lithium atoms are intercalated into each graphite layer.

The layered structures of graphite makes it easier to be intercalated by alkali-metal atoms. There exist week but significant van der Waals (vdW) interactions between the hexagonal symmetric lattice [28–34], in which the layered system is purely composed of the carbon atoms. The low-lying energy band structures of graphite of GICs are modified by the vdW interactions, and so are the essential physical properties. The electronic structures are strongly related to the stacking configurations of the carbon layers. To be specific, AA (simple hexagonal), AB (Bernal), and ABC (rhombohedral) are the three well-known stacking configurations [35–39]. Previous experimentation and theoretical studies show that lithium-GIC (L-GIC) and Alkali-GIC (A-GIC; A stands for alkali metal, A = Na, K, Rb, Cs)

doi:10.1088/978-0-7503-4685-6ch10

present stage-1 structures in different forms, which are LiC_6 and AC_8, respectively [20, 29, 40–47]. Furthermore, among the three kinds of stacking configurations, AA-stacking is the most stable one for Li-GIC on stage-1 type [48, 49]. However, AB-stacking configuration has become the most energetic stable one for the other stage-n types (n = 2,3,4…). Li-GIC and A-GIC have been widely discussed in the past years for their similarity and differences, while experimental researches on Li-ion-GIC and alkali-ion-GIC are few and hence discussion is more often theoretical.

In this chapter, the orbital hybridizations in lithium and lithium-ion GICs are the main focus, presenting the essential structural and electronic properties. The different concentrations on uniform situation of intercalations are considered, that is, the stage-1 type for different concentrations are covered in this chapter. Moreover, the hollow site of hexagonal carbon layers is the optimal adsorption position [50, 51]. The total ground state energies, optimal geometric structures, energy band structures, density of states, and the spatial charge distributions are investigated by first-principles calculations. Specially, the intercalation-induced free electron densities could be predicted. The feature-rich phenomena in Li-GIC and Li^+-GIC are predicted to have significant differences, obtaining from the symmetric comparison with each other.

We use first-principles calculation and the density functional theory (DFT) to study the electronic and geometric properties on stage-1 Li- and Li^+-graphite-intercalation compounds (GIC) under various concentrations, that is, LiC_x/Li^+C_x with x = 6, 12, 18, 24, 32 and 36. The main focus of this chapter includes the following essential properties, e.g. geometric structures, spatial charge distributions, and band structures, which are determined by the hybridization of orbitals. The Li-GICs exhibit blue shift of fermi energy of band structures/density of states, being a similar feature with metals. However, the electronic structure of Li^+-GIC is still preserved as original graphite or so-called semimetal, possessing the same free electrons and holes at the Fermi level. Based on these properties, we find that there exist certain van der Waals interactions between graphitic layers, and weak 2s-2p_z hybridization between nearest Li and C atoms. In addition, strong interactions are found between Li^+ and C. The important interaction between the Li and C atoms should be 2s-2p_z orbital–orbital interaction, while the orbital–orbital couple is relatively less significant in Li^+ and C systems.

10.1 The strong dependence of essential properties on quest-atom/ion concentrations and arrangements

DFT [52–54] has been utilized for solving the many-electron problem. It is widely used in materials science, chemical physics, and condensed matter physics due to the computational efficiency. In general, the Vienna *ab initio* simulation package (VASP) [55] provides a great solution using DFT by solving the Kohn–Sham equations [56]. We apply first-principles calculations with VASP in this chapter. In order to deal with the problems of bulks with periodic boundary condition, VASP uses the Bloch's theorem for a suitable solution. Perdew–Burke–Ernzerhof function under the generalized gradient approximation (GGA) is utilized for the exchange-correlation energy [57–59]. For geometric and electronic optimization, the first Brillouin zone is sampled in a Gamma scheme by $10 \times 10 \times 10$ and $30 \times 30 \times 30$ k-points meshes, respectively. All the atoms are set to be relaxed to obtain the optimized structures, and hence the lowest total ground energy. By using the numerical self-consistent scheme, spatial charge densities can be obtained. Moreover, the above results can be used to study other fundamental physical properties. The energy convergence condition is set to be 10^{-5}eV. The Hellmann–Feynman force on each atom is less than 0.01 eV $Å^{-1}$.

After the chemical intercalation, graphite has its geometric symmetries to be diversified. The week but significant van der Waals interactions make the Li/Li^+ easily intercalated into the spacing between graphitic layers. There exist three common types of absorption sites for intercalant atoms: hollow site, top site, and bridge site. The lowest ground energies come to exist at the hollow-site adsorption in Li-GI. In other words, the hollow-site position is the most stable geometric configuration. The optimized structures of LiC_x and Li^+C_x possesses the same stacking configuration, i.e. stage-1. However, their concentrations are distinct, as listed in table 10.1. The variation of interlayer distances of Li-GIC/Li^+-GIC are quite different, which will lead to the distinct electronic properties (discussed later).

For lithium atom intercalating cases, the interlayer distances of LiC_6, LiC_{12}, LiC_{18}, LiC_{24}, LiC_{32} and LiC_{36} are 3.815 Å, 3.863 Å, 3.924 Å, 3.954 Å, 3.987 Å and 3.964 Å, respectively. This shows a good coincidence with the experimental results. As for lithium ion intercalating cases, the interlayer distances of Li^+C_6, Li^+C_{12}, Li^+C_{18}, Li^+C_{24}, Li^+C_{32} and Li^+C_{36} are 3.082 Å, 3.283 Å, 3.374 Å, 3.451 Å, 3.510 Å and 3.555 Å, respectively, being a significantly short distance from the atom ones, as seen in table 10.1.

The heights of LiC_x and Li^+C_x both decrease under higher concentrations. The height of the former is close to that of pristine graphite when the concentration rises, and that of pristine graphite lattice with experimental interlayer distance is

Table 10.1. Optimized interlayer distance of stage-1 Li-GICs and Li^+-GICs.

System	Interlayer distance (Å)
LiC_6	3.815
LiC_{12}	3.863
LiC_{18}	3.924
LiC_{24}	3.954
LiC_{32}	3.978
LiC_{36}	3.964
$LiC_6{}^+$	3.082
$LiC_{12}{}^+$	3.283
$LiC_{18}{}^+$	3.374
$LiC_{24}{}^+$	3.451
$LiC_{32}{}^+$	3.555
$LiC_{36}{}^+$	3.550

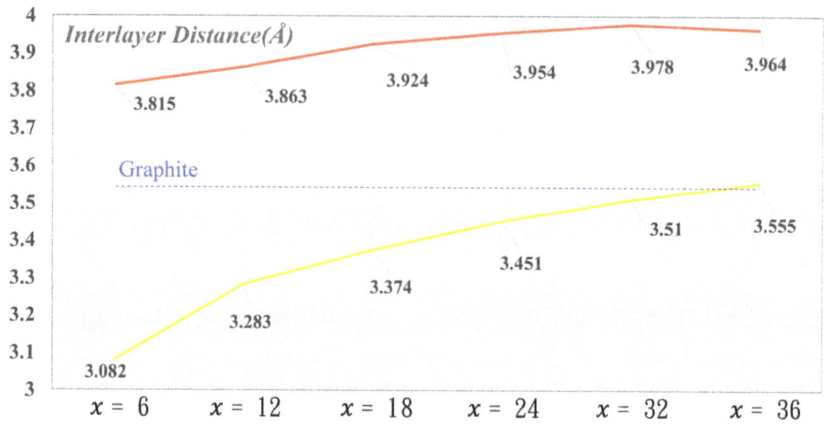

Figure 10.1. The variation of interlayer graphitic distance in Li-GICs and Li^+-GICs.

confirmed to be 3.550 Å [60]. On the other hand, the height of Li^+C_x is much closer to that of pristine graphite as concentration declines (figure 10.1).

10.2 The important differences between stage-1 and stage-*n* cases under the same alkali-atom concentrations

The simple hexagonal graphite with AA-stacking presents a special electronic structure, covering the equal number of free electrons and free holes within the energy range of −0.5 eV–0.5 eV, as shown by the band structure and density of state in figure 10.2(a) and (b). The AA-stacking pristine graphite belongs to semimetals since an obvious band overlap is exhibited in the k_z-dependent energy dispersion along ΓA. The energy bands near the Fermi level (also known as π bands) are fully dominated by the C-2p$_z$ orbitals within the energy range −2.0 eV to 4.0 eV. The C-(2s, 2p$_x$, 2p$_y$) orbitals forming the σ bands, which emerge at $E \leqslant -3.0$eV, being strongly contributed by the planar C–C bonds. Nevertheless, the above-mentioned band structures become very different after the intercalation of Li and Li^+.

After the intercalations of Li atom, the electron and hole bands become asymmetrical. The corresponding low-lying bands present a significant shift toward lower energy compared to the pristine graphite. Such bands do not intersect with any valence bands, as indicated in figure 10.3. The free conduction electrons are mainly aroused by the Li atoms. The energy shift of low-lying bands is determined by the minimum point of density of state. Those minimum points are calculated to be 1.800 eV, 1.326 eV, 1.138 eV, 0.991 eV, 0.882 eV and 0.810 eV, respectively, as shown in figure 10.4 for $x = 6$, 12, 18, 24, 32 (not shown) and 36 (not shown). The results indicate that C-2p$_z$ orbitals are easily affected by the intercalated Li atoms and are sensitive to the lithium concentrations.

The intercalations of Li^+ ions reveal quite different electronic properties to the Li case. Its energy shift of band structures is absent, and the corresponding minimum point of density states is 0.088 eV, 0.042 eV, 0.036 eV, 0.026 eV, 0.018 eV, and 0.026 eV, respectively, as indicated in figure 10.4 for $x = 6$, 12, 18, 24, 32 (not shown) and

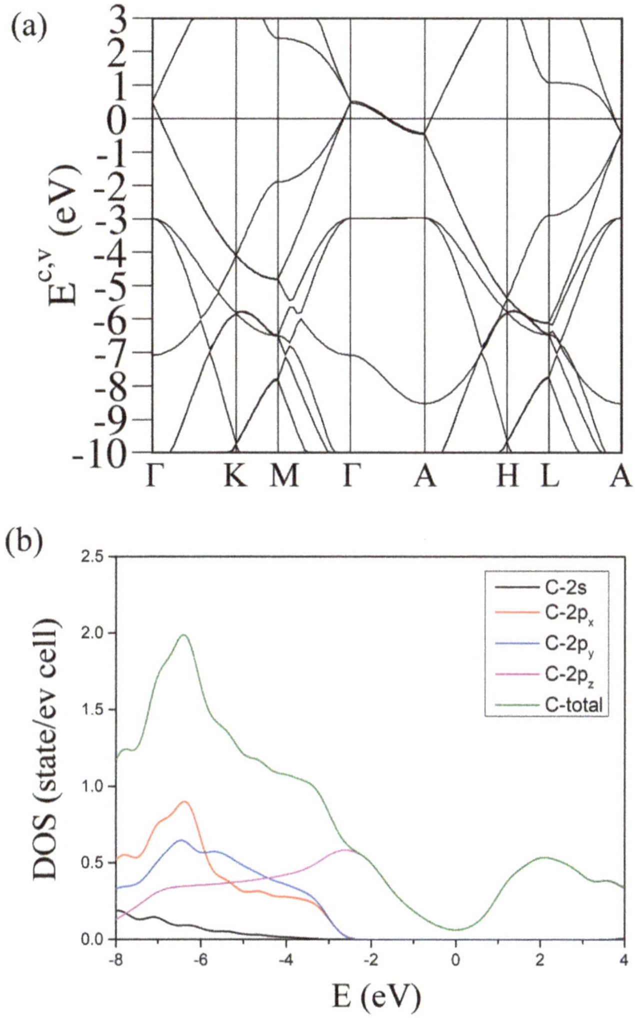

Figure 10.2. Energy band and density of states of pristine graphite.

36 (not shown) cases. The Fermi level is similar to the pristine graphite. Furthermore, shoulder structures are displayed in the σ bands, being exhibited by the orbit-projected density of states near −2.6 eV~−2.9 eV at Γ point. This can be treated as a blue shift compared to the Li atoms intercalation cases. All in all, the σ bands are presented at deeper energy and more stable in the Li intercalations case.

The spatial charge distributions (ρ) and variations ($\Delta\rho$) of Li- and Li^+-GICs can provide useful information to understand the charge bonding, as shown in figure 10.5. They are exhibited along x-z-plane for investigating the bonding changes after intercalations. The ρ and $\Delta\rho$ suggest that significant hybridizations exist in Li–C bonds for Li-atom intercalations, as indicated by red and yellow colors in figure 10.5 (dash rectangle). Such hybridizations are related to the Li concentrations. The variations of charge density become much higher as the Li concentration

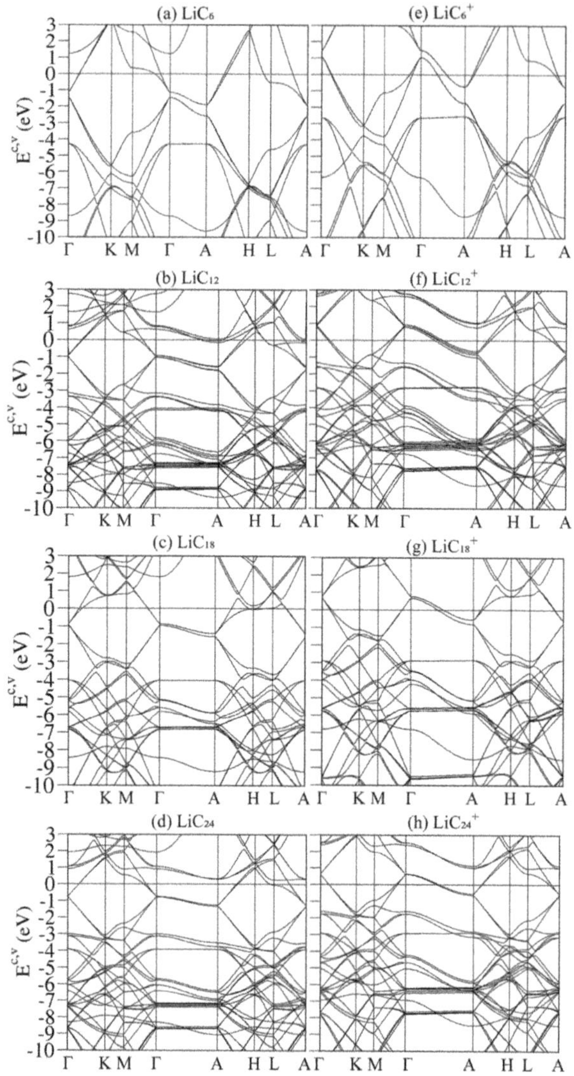

Figure 10.3. Band structures of (a) LiC_6, (b) LiC_{12}, (c) LiC_{18}, (d) LiC_{24}, (e) LiC_6^+, (f) LiC_{12}^+, (g) LiC_{18}^+, and (h) LiC_{24}^+.

increases. This indicates that the charge transfer in LiC_6 is much stronger than that of LiC_{36}. Furthermore, charge transfer mainly originates from the region near to C atoms, which accounts for the charge transfers from Li atom to C atom. In addition, the above-mentioned phenomenon is only exhibited by the C atoms neighboring Li atoms.

The Li^+-ion intercalations reveal quite different characteristics of charge variations. $\Delta\rho$ between Li^+ ions and C atoms are relatively weak but broader than the Li-atom cases, as shown by the light green and yellow colors in figure 10.6 (blue dash rectangle). The charge variation between the neighboring C atoms becomes

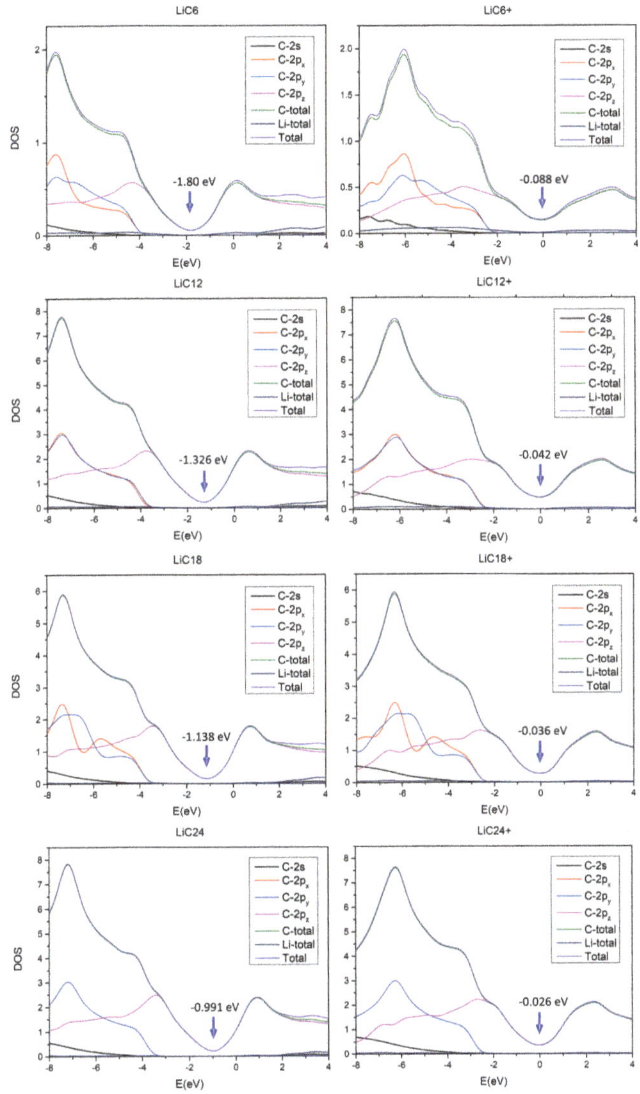

Figure 10.4. Density of states of LiC_6, LiC_{12}, LiC_{18}, LiC_{24}, LiC_6^+, LiC_{12}^+, LiC_{18}^+, and LiC_{24}^+.

significant as the Li^+ concentration increases, as indicated by light blue colors in figure 10.6 (purple dash rectangle). However, the variations of Li^+–C bonds are slightly close to the Li^+ ion, being a main difference from the Li atoms cases. Hence, the direction of the charge transfers is from C atoms to Li^+ ions. In addition, charge variation of the non-neighboring carbons atom to the Li^+ ions is hardly affected, which is a similar feature to that of Li atoms intercalation systems.

The important differences between stage-1 and stage-*n* cases under the same alkali-atom concentrations are also investigated, as shown in figure 10.7. The above-mentioned important feature, energy shift of the local minimum structure in density

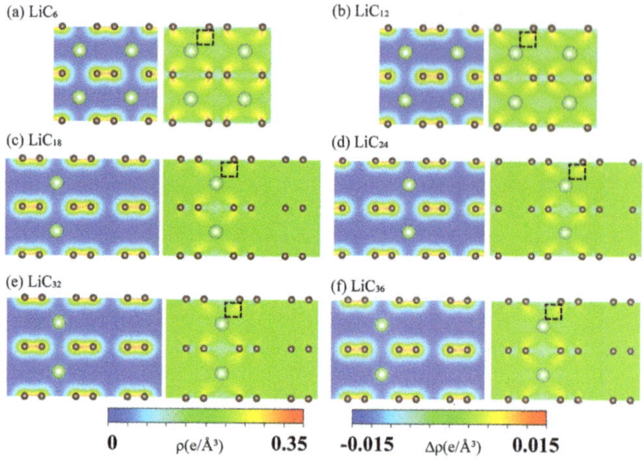

Figure 10.5. Spatial charge density distributions and variations of (a) LiC_6, (b) LiC_{12}, (c) LiC_{18}, (d) LiC_{24}, (e) LiC_{32}, and (f) LiC_{36}.

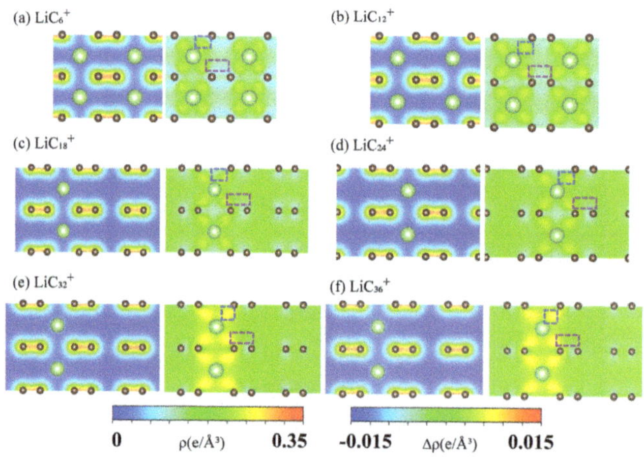

Figure 10.6. Spatial charge density distributions and variations of (a) LiC_6^+, (b) LiC_{12}^+, (c) LiC_{18}^+, (d) LiC_{24}^+, (e) LiC_{32}^+, and (f) LiC_{36}^+.

of states is slightly different in stage-*n* cases for LiC_{12}, LiC_{18}, and LiC_{24}. The cases with lower concentrations reveal a shift toward higher energy, while the LiC_{12} system presents an opposite shift. On the other hand, the orbital dominance of π bands and σ bands is hardly affected under this configuration modification. These special characteristics are expected to be further verified by STS measurement in the near future.

In summary, graphite Li-atom-/ion-intercalated compounds are studied by first-principles calculations within GGA method. The previous studies mainly focused on Li-atom intercalation graphites, while those on Li-ion intercalation graphites are relatively few. The Li-atom/Li-ion intercalation systems are thoroughly studied through the investigation of the orbital hybridizations. A comprehensive

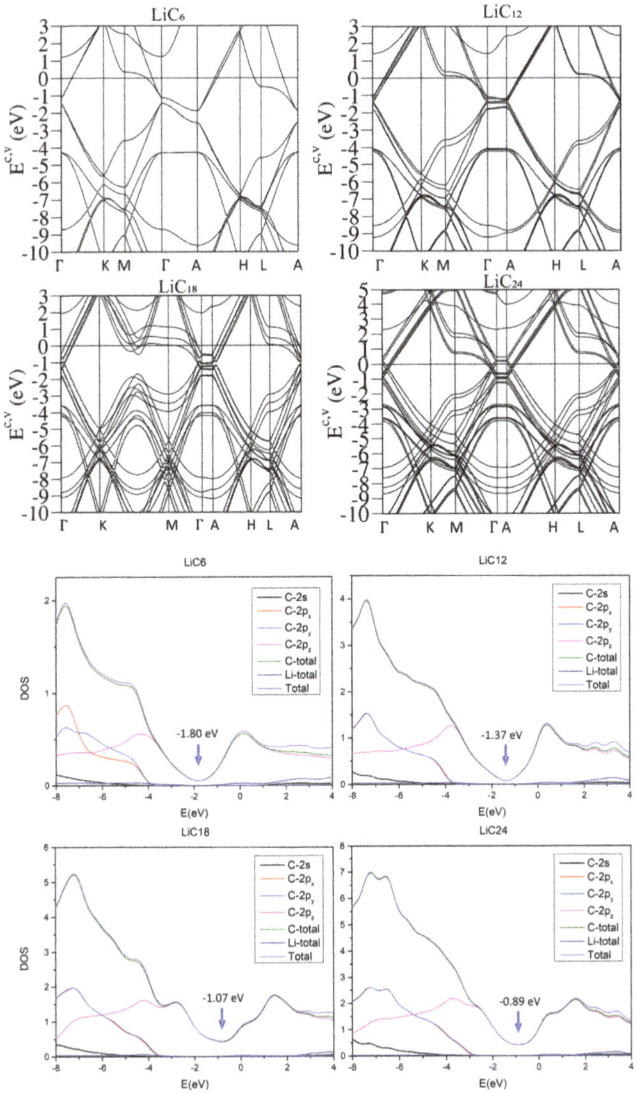

Figure 10.7. Energy bands of stage-*n* LiC_6, LiC_{12}, LiC_{18}, and LiC_{24}, and their density of states.

comparison between Li and Li^+ intercalation graphite is also made. The changes of essential physical properties are obtained and are very different between Li- and Li^+-GICs. The latter systems exhibit shortening of interlayer distances, red shift of energy bands and density of states, and smaller variation of spatial charge distributions. Those results clearly reveal that the Li–C bonds are formed by the 2s-$2p_z$ orbital hybridizations, leading to significant charge transfers from Li to C about 0.86 e–0.88 e. On the other hand, the variations are relatively week to Li^+ from C with the range of 0.163 e–0.191 e. The orbital–orbital interactions dominate essential properties between Li and C, while the main effects might be the dipole–

dipole interactions between Li^+ and C. The electronic configurations of Li^+ ions are saturated, which is similar to the inert gas helium, being the main reason for the low contributions of free carriers.

References

[1] Dunn B, Kamath H and Tarascon J M 2011 Electrical energy storage for the grid: a battery of choices *Science (80-.).* **334** 928–35

[2] Singh G K 2013 Solar power generation by PV (photovoltaic) technology: a review *Energy* **53** 1–13

[3] Tarascon J M and Armand M 2001 Issues and challenges facing rechargeable lithium batteries *Nature* **414** 359–67

[4] Xu W, Wang J, Ding F, Chen X, Nasybulin E, Zhang Y and Zhang J G 2014 Lithium metal anodes for rechargeable batteries *Energy Environ. Sci.* **7** 513–37

[5] Zhang X, Wang X G, Xie Z and Zhou Z 2016 Recent progress in rechargeable alkali metal–air batteries *Green Energy Environ.* **1** 4–17

[6] Muldoon J, Bucur C B, Oliver A G, Sugimoto T, Matsui M, Kim H S, Allred G D, Zajicek J and Kotani Y 2012 Electrolyte roadblocks to a magnesium rechargeable battery *Energy Environ. Sci.* **5** 5941–50

[7] Nakahara K, Iwasa S, Satoh M, Morioka Y, Iriyama J, Suguro M and Hasegawa E 2002 Rechargeable batteries with organic radical cathodes *Chem. Phys. Lett.* **359** 351–4

[8] Kanno R, Shirane T, Kawamoto Y, Takeda Y, Takano M, Ohashi M and Yamaguchi Y 1996 Synthesis, structure, and electrochemical properties of a new lithium iron oxide, $LiFeO_2$, with a corrugated layer structure *J. Electrochem. Soc.* **143** 2435–42

[9] Aurbach D and Ein-Eli Y 1995 The study of Li-graphite intercalation processes in several electrolyte systems using *in situ* x-ray diffraction *J. Electrochem. Soc.* **142** 1746–52

[10] Ozawa K 1994 Lithium-ion rechargeable batteries with $LiCoO_2$ and carbon electrodes: the $LiCoO_2$/C system *Solid State Ionics* **69** 212–21

[11] Walter S and Bruni S 2002 *Advances in Lithium-Ion Batteries* (New York: Kluwer Academic)

[12] Li N, Chen Z, Ren W, Li F and Cheng H M 2012 Flexible graphene-based lithium ion batteries with ultrafast charge and discharge rates *Proc. Natl. Acad. Sci. USA* **109** 17360–5

[13] Liu H, Strobridge F C, Borkiewicz O J, Wiaderek K M, Chapman K W, Chupas P J and Grey C P 2014 Capturing metastable structures during high-rate cycling of $LiFePO_4$ nanoparticle electrodes *Science (80–.).* **344**

[14] Yang L, You W, Zhao X, Guo H, Li X, Zhang J, Wang Y and Che R 2019 Dynamic visualization of the phase transformation path in $LiFePO_4$ during delithiation *Nanoscale* **11** 17557–62

[15] Sakurai Y, Arai H, Okada S and Yamaki J I 1997 Low temperature synthesis and electrochemical characteristics of $LiFeO_2$ cathodes *J. Power Sources* **68** 711–5

[16] Julien C M, Mauger A, Zaghib K and Groult H 2014 Comparative issues of cathode materials for Li-ion batteries *Inorganics* **2** 132–54

[17] Lin M C *et al* 2015 An ultrafast rechargeable aluminium-ion battery *Nature* **520** 325–8

[18] Chattopadhyay S, Lipson A L, Karmel H J, Emery J D, Fister T T, Fenter P A, Hersam M C and Bedzyk M J 2012 *In situ* x-ray study of the solid electrolyte interphase (SEI) formation on graphene as a model Li-ion battery anode *Chem. Mater.* **24** 3038–43

[19] Chan C K, Peng H, Liu G, McIlwrath K, Zhang X F, Huggins R A and Cui Y 2008 High-performance lithium battery anodes using silicon nanowires *Nat. Nanotechnol.* **3** 31–5

[20] Lee R S, Kim H J, Fischer J E, Thess A and Smalley R E 1997 Conductivity enhancement in single-walled carbon nanotube bundles doped with K and Br *Nature* **388** 255–7
[21] Yoo E J, Kim J, Hosono E, Zhou H S, Kudo T and Honma I 2008 Large reversible Li storage of graphene nanosheet families for use in rechargeable lithium ion batteries *Nano Lett.* **8** 2277–82
[22] Mele E J and Ritsko J J 1981 Low frequency plasmon spectra in stage I and stage II $FeCl_3$ intercalated graphite *Synth. Met.* **3** 89–98
[23] Lin M, Huang C and Chuu D 1997 Plasmons in graphite and stage-1 graphite intercalation compounds *Phys. Rev.* B **55** 13961–71
[24] Dresselhaus M S and Dresselhaus G 2002 Intercalation compounds of graphite *Adv. Phys.* **51** 1–186
[25] Nobuhara K, Nakayama H, Nose M, Nakanishi S and Iba H 2013 First-principles study of alkali metal-graphite intercalation compounds *J. Power Sources* **243** 585–7
[26] Siebentritt S, Pues R, Rieder K H and Shikin A M 1997 Surface phonon dispersion in graphite and in a lanthanum graphite intercalation compound *Phys. Rev.* B **55** 7927–34
[27] Kelty S P, Lu Z and Lieber C M 1991 Scanning-tunneling-microscopy investigations of ternary graphite intercalation compounds *Phys. Rev.* B **44** 4064–7
[28] Bhimanapati G R *et al* 2015 Recent advances in two-dimensional materials beyond graphene *ACS Nano* **9** 11509–39
[29] Petrović M *et al* 2013 The mechanism of caesium intercalation of graphene *Nat. Commun.* **4** 1–8
[30] Gerard Lavin J, Subramoney S, Ruoff R S, Berber S and Tománek D 2002 Scrolls and nested tubes in multiwall carbon nanotubes *Carbon N. Y.* **40** 1123–30
[31] McAllister M J *et al* 2007 Single sheet functionalized graphene by oxidation and thermal expansion of graphite *Chem. Mater.* **19** 4396–404
[32] Birowska M, Milowska K and Majewski J A 2011 Van der Waals density functionals for graphene layers and graphite *Acta Phys. Pol.* A **120** 845–8
[33] Lee H, Lee N, Seo Y, Eom J and Lee S 2009 Comparison of frictional forces on graphene and graphite *Nanotechnology* **20**
[34] McAllister M J *et al* 2007 Single sheet functionalized graphene by oxidation and thermal expansion of graphite *Chem. Mater.* **19** 4396–404
[35] Ho J H, Lu C L, Hwang C C, Chang C P and Lin M F 2006 Coulomb excitations in AA- and AB-stacked bilayer graphites *Phys. Rev.* B **74** 85406
[36] Jeong H K, Yun P L, Lahaye R J W E, Park M H, Kay H A, Ick J K, Yang C W, Chong Y P, Ruoff R S and Young H L 2008 Evidence of graphitic AB stacking order of graphite oxides *J. Am. Chem. Soc.* **130** 1362–6
[37] Rong Z Y and Kuiper P 1993 Electronic effects in scanning tunneling microscopy: moiré pattern on a graphite surface *Phys. Rev.* B **48** 17427–31
[38] Lee J K, Lee S C, Ahn J P, Kim S C, Wilson J I B and John P 2008 The growth of AA graphite on (111) diamond *J. Chem. Phys.* **129** 234709
[39] Yang Y *et al* 2019 Stacking order in graphite films controlled by van der Waals technology *Nano Lett.* **19** 8526–32
[40] Bower C, Suzuki S, Tanigaki K and Zhou O 1998 Synthesis and structure of pristine and alkali-metal-intercalated single-walled carbon nanotubes *Appl. Phys. A Mater. Sci. Process.* **67** 47–52

[41] Kelty S P and Lieber C M 1989 Atomic-resolution scanning-tunneling-microscopy investigations of alkali-metalgraphite intercalation compounds *Phys. Rev.* B **40** 5856–9

[42] Wang Z, Ratvik A P, Grande T and Selbach S M 2015 Diffusion of alkali metals in the first stage graphite intercalation compounds by vdW-DFT calculations *RSC Adv.* **5** 15985–92

[43] Li Y, Lu Y, Adelhelm P, Titirici M M and Hu Y S 2019 Intercalation chemistry of graphite: alkali metal ions and beyond *Chem. Soc. Rev.* **48** 4655–87

[44] Guerard D and Herold A 1975 Intercalation of lithium into graphite and other carbons *Carbon N. Y.* **13** 337–45

[45] Petrović M *et al* 2013 The mechanism of caesium intercalation of graphene *Nat. Commun.* **4** 2772

[46] Algdal J, Balasubramanian T, Breitholtz M, Kihlgren T and Walldén L 2007 Thin graphite overlayers: graphene and alkali metal intercalation *Surf. Sci.* **601** 1167–75

[47] Yoon G, Kim H, Park I and Kang K 2017 Conditions for reversible Na intercalation in graphite: theoretical studies on the interplay among guest ions, solvent, and graphite host *Adv. Energy Mater.* **7**

[48] Fiori S, Murata Y, Veronesi S, Rossi A, Coletti C and Heun S 2017 Li-intercalated graphene on SiC(0001): an STM study *Phys. Rev.* B **96**

[49] Virojanadara C, Watcharinyanon S, Zakharov A A and Johansson L I 2010 Epitaxial graphene on 6*H*-SiC and Li intercalation *Phys. Rev.* B **82** 205402

[50] Lin S Y, Tran N T T and Fa-Lin M 2018 *Structure- and Adatom-Enriched Essential Properties of Graphene Nanoribbons* (Boca Raton, FL: CRC Press)

[51] Tran N T T, Lin S Y, Lin C Y and Lin M F 2017 *Geometric and Electronic Properties of Graphene-Related Systems: Chemical Bondings* (Boca Raton, FL: CRC Press)

[52] Argaman N and Makov G 2000 Density functional theory: an introduction *Am. J. Phys.* **68** 69–79

[53] Sun J, Furness J W and Zhang Y 2018 Density functional theory *Mathematical Physics in Theoretical Chemistry* (Amsterdam: Elsevier) pp 119–59

[54] Zhang S, Wan S, Zeng Y, Jiang S, Gong X and You J 2019 *In situ* Raman spectroscopy and DFT Studies of the Li_2GeO_3 melt structure *Inorg. Chem.* **58** 5025–30

[55] Kresse G and Furthmüller J 1996 Efficient iterative schemes for *ab initio* total-energy calculations using a plane-wave basis set *Phys. Rev.* B **54** 11169–86

[56] Kohn W and Sham L J 1965 Self-consistent equations including exchange and correlation effects *Phys. Rev.* **140**

[57] Perdew J P, Burke K and Ernzerhof M 1996 Generalized gradient approximation made simple *Phys. Rev. Lett.* **77** 3865–8

[58] Perdew J P, Ernzerhof M and Burke K 1996 Rationale for mixing exact exchange with density functional approximations *J. Chem. Phys.* **105** 9982–5

[59] Perdew J P, Burke K and Ernzerhof M 1998 Perdew, Burke, and Ernzerhof reply *Phys. Rev. Lett.* **80** 891

[60] Lee J K, Lee S C, Ahn J P, Kim S C, Wilson J I B and John P 2008 The growth of AA graphite on (111) diamond *J. Chem. Phys.* **129** 234709

Chapter 11

Diversified phenomena in sodium-, potassium- and magnesium-related graphite intercalation compounds

Wei-Bang Li, Shih-Yang Lin, Thi Dieu Hien Nguyen, Hsien-Ching Chung, Ngoc Thanh Thuy Tran, Nguyen Thi Han, Hsin-Yi Liu, Hai Duong Pham and Ming-Fa Lin

The chemical elements in the periodical table [1, 2] have significant atomic configurations [1, 3, 4], induce active chemical environments [5–7], form various nanoclusters/composites/crystals/compounds ([8–10]/[7, 11–13]/[14–16]/[17, 18]), generate distinct functionalities [19, 20], and produce highly potential applications [21, 22]. By using up-to-date physical and chemical methods ([23, 24]; [25–27]), a lot of condensed-matter systems have been successfully synthesized in experimental laboratories, such as chemically modified systems (e.g., graphite intercalation compounds with n- and p-type dopings; [28–30]), emergent materials (graphene/silicene/germanene/tinene/plumbene; [31–33]/[34–36]/[37–40]/[37, 41, 42]/[43–46]), single-/multi-element semiconductors (binary III–V/III–VI/II–V/II–VI 2D and 3D compounds; [47–50]/[51–55]/[56]/[57–61]), as well as, green energy materials (batteries/solar cells/hydrogen energies; [62–71]/[72–80]/[81–86]). As a result of the diversified crystal symmetries [87–89], many interesting research topics come to exist as the mainstream issues in observing the rich and unique phenomena. The highly related systems are very suitable for systematic investigations in both progress about theoretical models and experimental measurements. For example, lithium-, phosphorus- and sulfur-based compounds/lithium- and oxygen-associated ones [90, 91]/[92, 93] are able to provide plenty of allotrope crystals, being fully supported by the multi-core components. Their different combinations are thoroughly tested and verified to be helpful in greatly enhancing the operation efficiencies of lithium-ion-based batteries [92, 94–98]. The theoretical quasiparticle viewpoints will be

doi:10.1088/978-0-7503-4685-6ch11

utilized to comprehend the fundamental properties, in which the close relations among the chemical environments, the significant orbital hybridizations in various bondings, the atom and orbital-decomposed spin configurations, and the featured electronic/optical properties will be established under the next book proposal.

In general, the chemical bondings in condensed-matter systems are classified into van der Waals [99–103], covalent [104–108], metallic [109–113], and ionic ones [114–118], in which the composite features might appear simultaneously in complicated multi-component compounds. For example, the stage-*n* alkali-metal-atom/$AlCl_4^-$ anions graphite intercalation compounds [30, 119]/[120] exhibit metallic behaviors, being in great contrast with the pristine semi-metallic transport properties. These systems clearly indicate the coexistence of very strong σ bondings on the carbon-honeycomb lattices and the Fermi sea due to free conduction electrons/valence holes, mainly owing to the significant atomic interactions between the neighboring intercalant and graphitic layers [121]. Their many merits, which are thoroughly examined, identified and verified from a series of basic and applied searches both experimentally [122] and theoretically [123], are responsible for high-performance commercial anode materials [124, 125]. The frequent intercalations and de-intercalations into the graphene spacings are easily driven by the charging/discharging processes of ion-based transports [126, 127]. There exist almost continuous intermediate crystal structures because of the rapid variations of intercalant concentrations and distribution configurations. All the crystal structures belong to the metal-stable configurations, except that only one or two of them are the most stable one. How to observe the time-dependent crystal symmetries from the simulation models and the high-precision experimental measurements has become one of the non-negligible barriers. In the current chapter, a few guest-atom intercalation configurations are chosen for a model study, in which the diverse multi-orbital hybridizations, not only chemical bondings, will be created by them, as well as three kinds of adatoms (sodium, potassium and magnesium; [128, 129]). This is consistent with the previous development of quasiparticle framework.

Graphite, which consists of the periodical stacking of carbon-honeycomb lattices through the weak, but significant van der Waals interlayer interactions [130, 131], has been extensively studied both theoretically [132] and experimentally [131] since one hundred years ago. This unusual material is a 3D semimetal in which there are small valence and conduction overlaps near the corners of the hexagonal first Brillouin zone. In up-to-date predictions and experiments, 3D bulk graphite systems can exhibit AA- [133–137], AB- [134, 138–141], ABC-stacked [142–146], and turbostratic configurations [147–151]. It should be noted that a simply hexagonal graphite is absent in Nature, and it is very difficult to produce in experimental syntheses. However, the opposite is true for the others. Semimetal–metal transitions come to exist by the intercalations of metal atoms and small/large molecules. Furthermore, the former and the latter clearly show strong evidence of free conduction electrons and valence holes, respectively. Such n- or p-type materials possess excellent electrical conductivities, as revealed in copper [152]. Specifically, they belong to the traditional superconductors with T_c~1–5 K. From theoretical viewpoints, the quasiparticle framework, being based on phenomenological models and first-principles simulations, could be developed to greatly promote their progress. By using the multi-/single-orbital tight-binding model, the random-phase approximation (RPA; [153–155]) and the Kubo formula [156–158], three kinds of well-stacked layered graphites are clearly identified to display the diverse low-frequency longitudinal Coulomb excitations/de-excitations [155, 159] and transverse optical absorption spectra [160, 161]. The calculated results are consistent with the experimental measurements of electron energy loss spectra (EELS; [162–164])/ distribution widths angle-resolved photoemission spectra [165, 166] and reflectance/absorption/transmission spectra. Furthermore, the essential properties are totally changed in the partners of intercalation compounds, such as the free-carrier-enriched (momentum, frequency)-excitation phase diagrams. In addition, Vienna *ab initio* simulation package (VASP) simulations are able to examine whether the excitonic effects survive in the semi-metallic graphite systems and the metallic graphite intercalation compounds. Also, how to link the essential properties of sodium, potassium and magnesium graphite intercalation compounds and the phenomenological models is one focus for study in the near future.

By a lot of experimental observations [167–169], both graphite and their intercalation compounds display rich and unique quasi-particle phenomena, being very useful in providing the specific functionalities of the commercial products [170], such as the featured mechanical properties of carbon fibers, and the graphitic spacings significant for adatom/molecule intercalations and de-intercalations [126]. A pristine Bernal graphite, with AB stacking, is examined to exhibit the polarization-, momentum- and temperature-dependent plasmon modes through the high-resolution measurements of reflection electron energy loss spectroscopy. Such quasiparticles mainly arise from the coherent oscillations of semi-metallic free electrons and holes. They are identified to dramatically transform into the intraband plasmons by the n- or p-type dopings, in which their frequencies are greatly enhanced from ~0.1 to 1.0 eV. The vertical optical excitations, with the vanishing momentum transfers behave so in terms of threshold absorption frequencies. This directly reflects the blue or red shift of the Fermi under the intercalation/ de-intercalation reactions. As for the well-known applications [128], the graphitic anode/cathode materials of ion-related batteries, with very high efficiencies [125], have become one of them, such as, the anode electrode of metal-atom graphite intercalation compounds (lithium-/sodium-/potassium-/magnesium-based systems; [98, 167, 171]/[123, 128, 172]/[173–175]/[176–178]), and the cathode electrode of $AlCl_4^-/FeCl_3^-$-related ones [120]/[179]. Up-to-date studies clearly illustrate that there exist five kinds of batteries, according to the featured charge transports, covering the common lithium- [70], lithium-/phosphorus-/sulfur- [91], sodium-, potassium- or magnesium- [79], aluminum- [180], and iron-associated ions [181]. Their merits and drawbacks need to be thoroughly analyzed under current and near-future investigations. Establishment of the important similarities and differences among these mainstream materials has become the focus of study.

The featured crystal symmetries and electronic properties of sodium-, potassium- and magnesium-metal-atom graphite intercalation compounds are fully investigated by first-principles simulations of the total group state energies at zero temperature [182, 183], stacking configuration of graphitic layers, the carbon-honeycomb and intercalant layers, periodical length along the *z*-direction, carbon–carbon, carbon–adatom and adatom–adatom bond lengths, atom-dominated electronic spectra within a whole energy range, spatial charge density distributions before and after intercalations [184–187], as well as atom- and orbital-decomposed density of states corresponding to the specific energy ranges. The relevant physical and chemical quantities are expected to be very sensitive to changes of guest-atom concentrations and distribution configurations [187–190]. The important differences among them are clearly illustrated by the crystal stabilities, the Fermi levels/the charge transfers, and the active carbon–adatom orbital hybridizations. Their merits and drawbacks in better applications will be fully analyzed from the viewpoint of basic science. The available and useless experimental measurements would be thoroughly discussed in this work, such as, the x-ray diffraction patterns on the stage structures of layered 3D graphite-related compounds, and the almost vanishing high-resolution measurements on the metallic compounds from angle-resolved photoemission spectroscopy (ARPES; [165, 166, 191]) and scanning tunneling spectroscopy (STS; [192–195]). Interestingly, the phenomenological models can be further developed because of the concise multi-/single-orbital hybridizations in intralayer and interlayer atomic interactions [196–198]. Finally, the above-mentioned features are unified under a quasiparticle framework. This is consistent with the previous, current and near-future developments [198, 199].

11.1 Crystal symmetries of sodium-, potassium- and magnesium-related graphite intercalation compounds

Graphitic layers are attracted together by the very weak, but important van der Waals interactions [146, 174], leading to the easy modulations by the adatom [200] or large-molecule intercalations/de-intercalations [201]. The intercalants between graphitic layers are assumed to present quasi-layer distributions through the various concentrations and configurations. The current work is only focused on the cases that two intercalant layers have monolayer graphene [169], and each one presents the adatom-distribution variations. The different doping cases are clearly illustrated in table 11.1 and figure 11.1(a)–(l). After the full relaxations of periodical crystal structures in the VASP simulations [202], the sensitive dependence of the total ground state energy on the intercalant-layer distance (not shown) is available in determining the stability of an optimal 3D structure from the lowest one with a vanishing slope. The concise unit cells would gradually become moiré superlattices with many intercalant and carbon atoms during the decrease of adatom concentrations. The giant numerical calculations come to exist in greatly suppressing the basic science research. Detailed comparisons between theoretical predictions and x-ray diffraction peaks are also discussed later.

In general, graphite intercalation compounds exhibit planar carbon-honeycomb symmetries and well-characterized intercalant distributions, especially for metal atom cases (even for large molecule ones; [120]). This clearly indicates the strong

Table 11.1. The calculated results of the groundstate energy, blue shift of Fermi level, and geometric structures for Na-, K-, Mg-GICs.

	Ground state energy (eV)	AA/AB stacking	Bond length (Å) Intralayer CC	Na–C/ K–C/ Mg–C	Na–Na/ K–K/ Mg–Mg	Periodical length along the z-direction (Å)	Interlayer distance between metal and honeycomb layer (Å)	Blue shift of the E_F (eV)
NaC_8	−75.017	AA	1.434	2.716	4.985	4.600	2.300	1.48
NaC_{18}	−169.527	AA	1.429	2.668	7.431	4.511	2.261	1.17
NaC_{24}	−225.407	AA	1.427	2.651	8.572	4.452	2.230	1.02
NaC_{32}	−299.823	AA	1.427	2.627	9.891	4.411	2.211	0.92
KC_8	−75.133	AA	1.442	3.027	4.985	5.321	2.661	1.48
KC_{18}	−169.426	AA	1.430	3.038	7.432	5.360	2.680	1.17
KC_{24}	−225.239	AA	1.429	3.039	8.572	5.350	2.670	1.03
KC_{32}	−299.578	AA	1.427	3.036	9.892	5.361	2.681	0.92
MgC_8	−74.425	AA	1.428	2.675	4.965	4.510	2.510	1.48
MgC_{18}	−167.323	AA	1.428	2.721	7.426	4.624	2.312	1.17
MgC_{24}	−223.149	AA	1.427	2.715	8.569	4.592	2.496	1.05
MgC_{32}	−297.520	AA	1.427	2.654	9.888	4.475	2.238	0.93

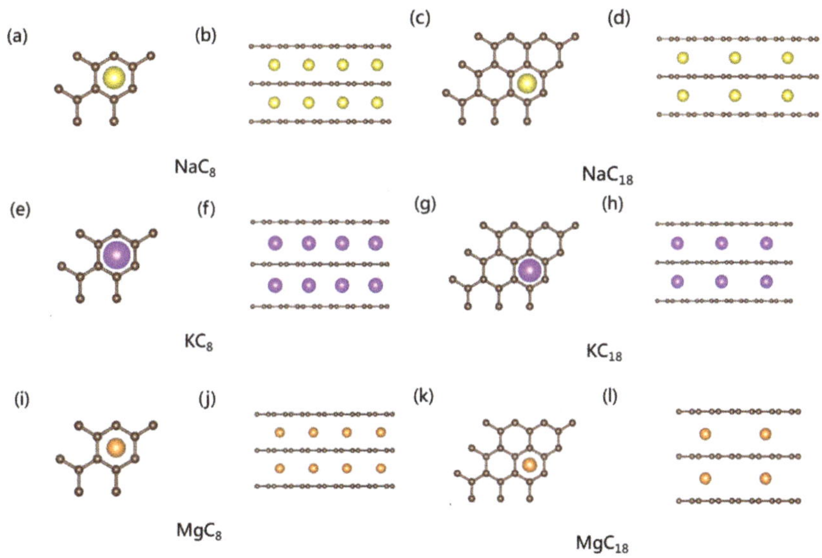

Figure 11.1. The optimal crystal structures of sodium-, potassium-and magnesium-based graphite intercalation compounds with the top/side view (the left-/right-hands); (a)/(b) NaC_8, (c)/(d) NaC_{18}, (e)/(f) KC_8, (g)/(h) KC_{18}, (i)/(j) MgC_8 as well as (k)/(l) MC_{18}.

σ bondings of graphitic layers, as well as the non-covalent interlayer and intra-intercalant interactions. From the delicate analyses of VASP simulations (table 11.1), Na-, K, and Mg-intercalations could be classified into stable and quasi-stable configurations according to their ground state energies (E_{gs}). For example, the latter, which have E_{gs} s of ~ −0.2 eV per unit cell, cover NaC_6, NaC_8, KC_6, KC_8, MgC_6, and MgC_8. Such 3D systems, with higher intercalant concentrations, are absent in up-to-date experimental syntheses [176, 177]. This issue might be solved by different synthesis methods.

The rich crystal symmetries of 3D graphite intercalation compounds, which are clearly revealed in the graphitic and intercalant layers, as well as their observable spacing, could be verified by high-resolution measurements of x-ray diffraction peaks [173, 203, 204], reflection electron diffraction patterns (top views; [205, 206]) and tunneling electronic microscopic spectra (side views; [207]), but not those by scanning tunneling microscopy (STM; [208, 209]). The former is frequently utilized in these layered materials, e.g., the strong evidence of stage-*n* alkali graphite intercalation compounds LiC_{6n} and MC_{8n}. However, the quasi-stable crystal structures of NaC_6, NaC_8, KC_6, KC_8, MgC_6, and MgC_8, corresponding to very high adatom concentrations, are required to be generalized under specific chemical environments. This enlarged capacity will be helpful for the performance of ion transports in batteries. Specifically, the electron beams are available in fully exploring the elastic scattering events with the periodical carrier density on layered structures, in which the diffraction patterns are detected for the reflection or transmission ones. Under the latter case, the standard samples should be sufficiently thin so that the tunneling electron beams are clearly observed under the constructive way. Such measurements are reliable, since the geometric properties of layered graphite intercalation compounds hardly depend on the layer number. In addition, both STM and TEM are able to thoroughly identify the surface morphologies for few-layer graphene systems with/without chemical absorptions, e.g., successful examinations of alkali adatoms on graphene surfaces [210]. They play a critical role in finding low-dimensional emergent materials.

11.2 Featured band structures and wave functions

It is well known that a 2D pristine graphene, a single carbon-honeycomb crystal, is a zero-gap semiconductor with zero density of states at the Fermi level [31]. A pair of linear and isotropic valence and conduction subbands intersect there so that a well-defined Dirac-cone structure is initiated from the K/K′ valley. This feature is modified by the perpendicular stacking configurations through the weak, but important van der Walls interactions. The linear and/or parabolic energy dispersions come to exist at the K and H valleys, depending on the stacking symmetries. Most important, the significant band overlaps, corresponding to the semi-metallic behaviors, are purely induced by the interlayer C-2p_z-orbital hybridizations. The AA-/ABC-stacked graphite has the highest/lowest densities of free conduction electrons/valence holes. The semiconductor–semimetal transition would become semimetal–metal after the n- or p-type dopings.

Band structures of sodium-, potassium- and magnesium-graphite intercalation compounds, as clearly displayed in figures 11.2(a)–(d), 11.3(a)–(d) and 11.4(a)–(d) within a sufficiently wide energy range of -15 eV $\leqslant E^{c,v} \leqslant 7$ eV, respectively, are fully supported by the carbon- and intercalant-atom dominances (the wave-function probability by the radii of blue/red balls for carbon/intercalant atoms; [198]). Their main features cover the number of valence subbands (the active atoms and orbitals in a unit cell; [198, 211]), the enhanced asymmetry of valence hole and conduction electron energy spectra about the Fermi level (the intercalant-carbon interlayer interactions; [212]), the blue shift of E_F (the semimetal–metal transitions due to the charge transfer of metal adatoms; [213]), the dependence of E_F on the chemical intercalations (the densities of free conduction electrons under the various intercalation cases), the slight/major modifications of carbon $2p_z$-π and [2s, $2p_x$, $2p_y$]-σ bands along the different wave-vector paths, the subband non crossing/crossing/anti-crossing behaviors [197], the diverse energy dispersions near the high-symmetry points [214], and their band-edge states (the critical points in energy-wave-vector spaces with zero group velocities [215]). Whether the high-precision ARPES

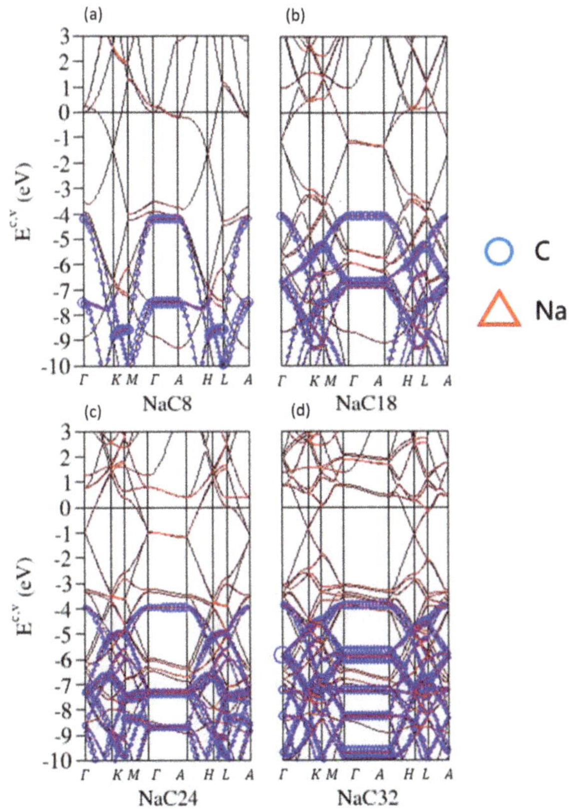

Figure 11.2. Band structures of Na-based graphite intercalation compounds within the active valence energy spectra, with atom dominances (the different colored circles): NaC_8, (b) NaC_{18}, (c) NaC_{24}, and (d) NaC_{32}. The radius of each blue/red circle corresponds to the relative probability by carbon/sodium atoms.

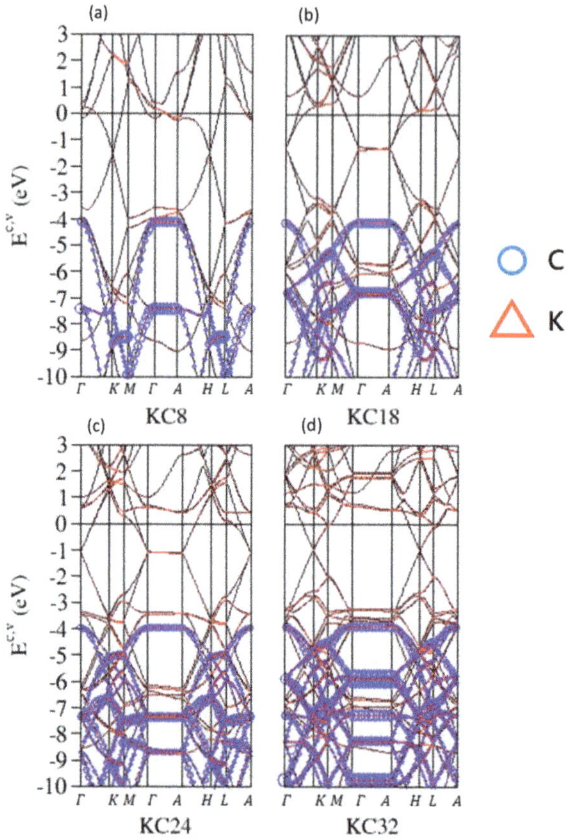

Figure 11.3. Similar plots as in figure 11.2(a)–(d), but shown for (a) KC_8, (b) KC_{18}, (c) KC_{24}, and (d) KC_{32}.

measurements are meaningful in examining the main features of occupied electronic states below the Fermi level will be discussed later.

The wave-vector dependences of occupied band structures in graphite intercalation compounds could only be examined by the direct measurements of ARPES [165, 191, 216]. ARPESs are very powerful in the few-layer graphene-related 2D systems (e.g., the diverse energy dispersions due to stacking configurations and layer numbers; [38, 191, 216, 217]), but not the 3D partners. As a result of the surface confinement, the ARPES measurements, being done a bulk material, will induce much trouble about the conservation/non-conservation issues of momentum transfers along the normal direction. Apparently, the accurate k_z-dependent energy dispersions are very difficult to achieve from the estimated electronic energy spectra along the high-symmetry paths. However, such issues have been well settled in a pristine Bernal graphite by using its main features near the K and H valleys (the low-density electrons and holes initiated from the corners of the first Brillouin zone, with k_z=0 and π, respectively; [218]). As for graphite intercalation compounds, the low-energy electronic states below the Fermi level might be also be available in determining their sensitive k_x, k_y, and k_z dependences, especially for those of

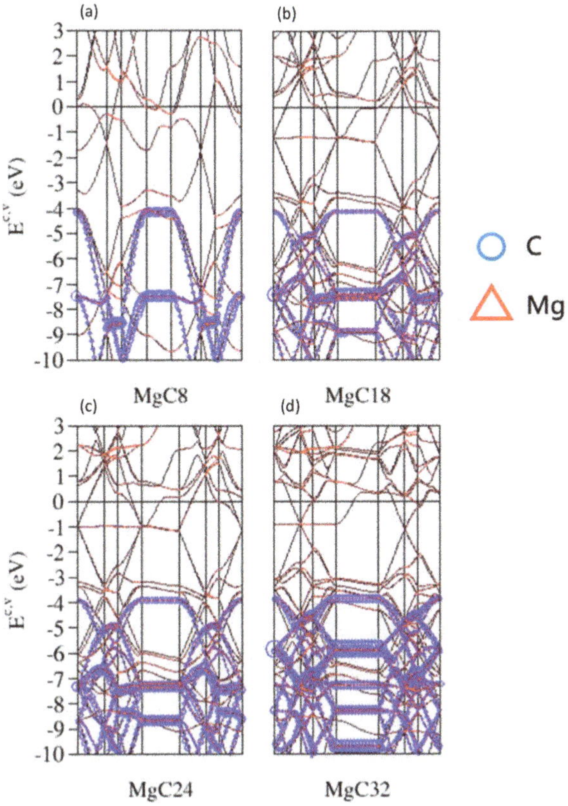

Figure 11.4. Similar plots to figure 11.2(a)–(d), while they are displayed for (a) MgC_8, (b) MgC_{18}, (c) MgC_{24}, and (d) MgC_{32}.

metal-atom n-type cases. The strong co-operations of experimental examinations and theoretical predictions could solve these open issues. When the featured Fermi level, π subbands and σ ones can be examined from ARPES measurements, these can partially support the specific orbital-hybridization in each chemical bond. On the other hand, the 3D complicated compounds, with moiré superlattices [219, 220], present unusual band structures, such as those related to ternary and quaternary lithium oxides in cathode/electrolyte/anode materials of ion-based batteries (details in other book chapters: [221]). Any ARPES measurements are absent to date, mainly owing to valence subbands in the presence of irregular energy dispersions. That is, the quasiparticle energy spectra and their Coulomb decay rates might become meaningless in these 3D compounds, according to the absence of experimental progress.

The Fermi level (the chemical potential: [222]), as clearly indicated in figures 11.2(a)–11.4(d), is mainly determined by the concentration, distribution configuration and charge transfer of intercalant metal atoms. Its characteristics of the Fermi–Dirac distribution at low temperatures strongly affect the other essential properties, especially for single- and many-particle optical absorption spectra and

Coulomb excitations [212, 223]. The intercalation-created free conduction electrons have been verified by the high-precision measurements of optical spectroscopies, and electron energy loss spectroscopy (EELS; [224, 225]). For example, these π-electronic carriers are capable of greatly enhancing the threshold absorption frequency (the optical gap; [226]) after the chemical intercalations (similar for the p-type dopings). Furthermore, the low-frequency plasmons, which belong to optical modes of coherent free carrier oscillations, exhibit the prominent EELS peaks below the critical momentum. Interestingly, the intraband collective excitations come to exist in the absence of Landau dampings at small momentum transfers, while they are in coexistence with interband electron–hole excitations for others. Apparently, the close cooperation between theoretical predictions and experimental examinations is available in determining the blue shift of the Fermi level. Up-to-date research shows an absence of model simulations about many-body excitonic effects [227]. The VASP results are under systematic investigation, being useful in clarifying certain differences of n- and p-type dopings. Whether the many-particle optical scatterings could be developed for graphite intercalation compounds is worth further evaluation. This also belongs to one part of the quasiparticle framework in this book [197].

11.3 Concise orbital hybridizations from charge density distributions and merged van Hove singularities

The spatial charge density [$\rho(\mathbf{r})$] and its variation [$\Delta\rho(\mathbf{r})$] after intercalation (figures 11.5–11.7) are clearly illustrated in a unit cell through the top- and side views under different chemical environments. The calculated results could be further utilized to test the x-ray diffraction peak structures [228–230]. By the delicate observations, the critical chemical bondings could be examined from the carbon-honeycomb lattices, the intercalant layers, and their significant spacings. The multi-/single-orbital hybridizations in different chemical bonds are determined by unifying the relevant quantities; that is, the critical quasiparticle pictures are reached through the full cooperation of the atom-dominated band structures (illustrated earlier in igures 11.2–11.4), the charge density distributions, and the orbital-decomposed van Hove singularities (illustrated later in figures 11.8–11.10). These will be very useful in establishing the concise phenomenological models, e.g., the tight-binding model [231–233]/the generalized tight-biding model [234, 235] in the absence/presence of a perpendicular magnetic field for the rich magnetic quantization phenomena [214, 233, 235].

From carrier distributions about each carbon-honeycomb lattice, the very strong σ bondings and the significant π ones, are respectively, observed in (x, y) and $(x, z)/(y, z)$ planes. For example, ($\rho(\mathbf{r})$) and ($\Delta\rho(\mathbf{r})$) of NaC_8 present high charge density between neighboring carbon atoms (the red color in figure 11.4(a)) and a peanut-like profile (the light yellow-green color). The former is hardly affected by the sodium-atom intercalations, being assisted by the vanishing ($\Delta\rho(\mathbf{r})$) in figure 11.4(c), However, the latter is modified by the interlayer C–Na interactions, as clearly illustrated by the obvious changes of carrier density after interactions (the red color near atoms in figure 11.4(d)). Detailed analyses further show the specific orbital hybridization of

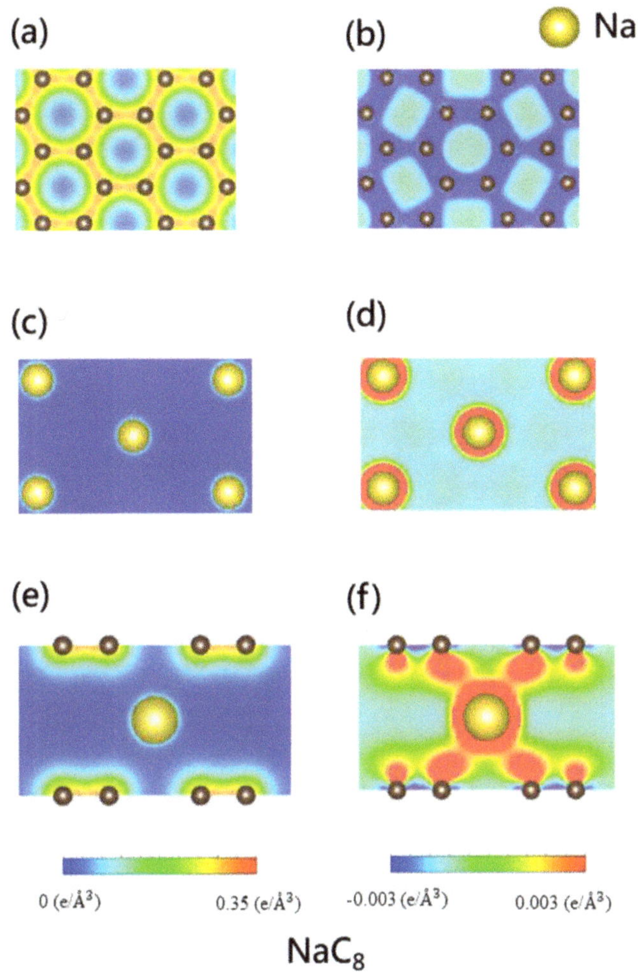

Figure 11.5. The spatial charge density distributions (ρ) and their variations before and after the chemical intercalations ($\Delta\rho$), with the top/side views (the left/right hands corresponding to (x, y) on graphene plane and Na plane/(x, z) planes): (a)/(c)/(e) and (b)/(d)/(f) and top view of intercalant layer NaC_8, and (g)/(i)/(k) and (h)/(j)/(l) NaC_{18}.

$2p_z$–3s orbitals in C–Na bond. Moreover, 3s-3s mixings in Na–Na bonds are also examined from the intercalant layers, e.g., the significant charge density deformation of 3s-orbital at the height of z = 3.50 Å. Apparently, the similar C–C, C–intercalant and intercalant–intercalant bondings are revealed in the different intercalation/intercalant figures 11.5(g)–(l)/figures 11.6(a)–(l) and figures 11.7(a)– (l) cases.

On the experimental examinations, the spatial charge density distributions could be directly deduced from the close co-operation of measured x-ray patters [228–230, 236] and elastic scattering predictions associated with atoms and active orbitals in a unit cell [229, 230]. For example, a sodium chloride crystal [237], with very strong ionic

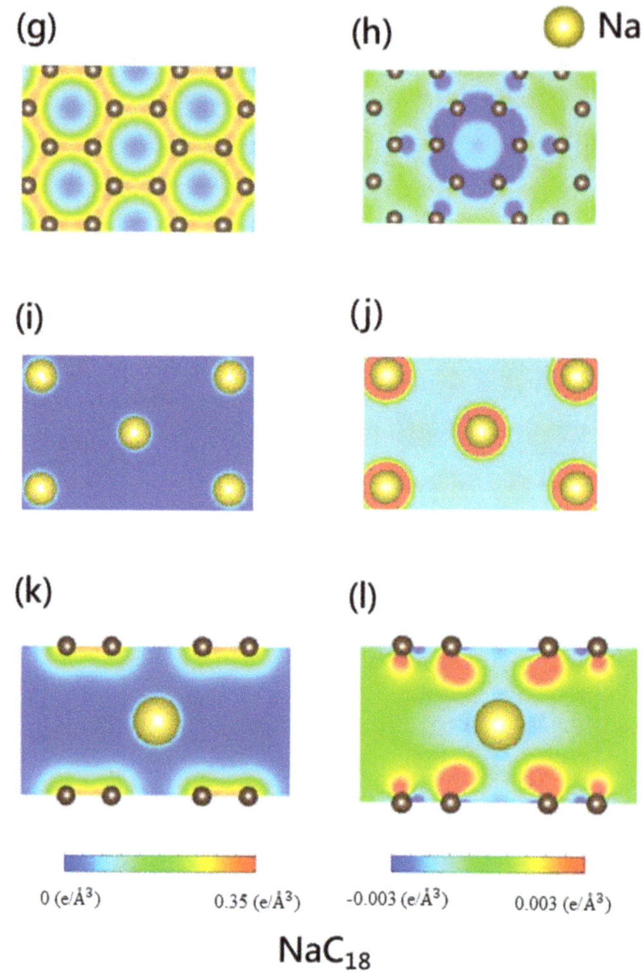

Figure 11.5. (Continued.)

bonds, is thoroughly investigated and presents an almost isotropic charge distribution near both Na^+-cations and Cl^--anions. Similar analyses are available in exploring those of metal-atom graphite intercalation compounds with concise orbital hybridizations, e.g., the x-ray tests on the theoretical predictions in figures 11.5–11.7 for Na-, K- and Mg-graphite intercalation compounds. These investigations will determine the active orbital hybridizations from the intralayer [119, 238] and interlayer [119, 238] atomic interactions Such works are very difficult to be done for large-molecule cases. Since the x-ray diffraction spectrum only depends on the whole distribution of charge density, the VASP results should be useful in determining it. How to link them is an open calculation issue.

The density of states is defined as the total electronic states in the closed constant-energy surfaces for 3D condensed-matter systems [239]. Since it is characterized by the integration about the inverse of group velocity on the wave-vector space, its van Hove singularities mainly arise from the band-edge states with zero Vg. The main features of the energy-dependent $D(E)$ is directly reflected in the other essential properties, such as, Coulomb-field excitations [214, 235], optical absorption structures [240], temperature dependences of heat capacities [239], transport conductivity/resistivity [241], and magnetic properties [197, 242, 243]. The atom- and orbital-/spin-projected density of states is very powerful in determining the concise orbital hybridization in various chemical bonds. A lot of van Hove singularities, which are

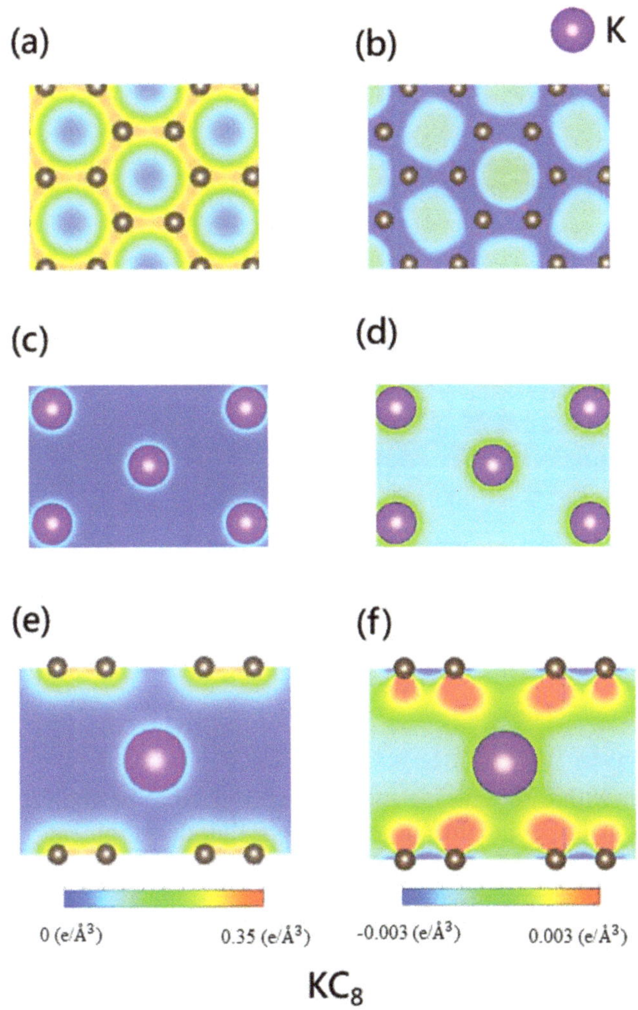

Figure 11.6. Similar plots to figure 11.5(a)–(l), but corresponding to (a)/(c)/(e) and (b)/(d)/(f) and top view of intercalant layer KC_8, and (g)/(i)/(k) and (h)/(j)/(l) KC_{18}.

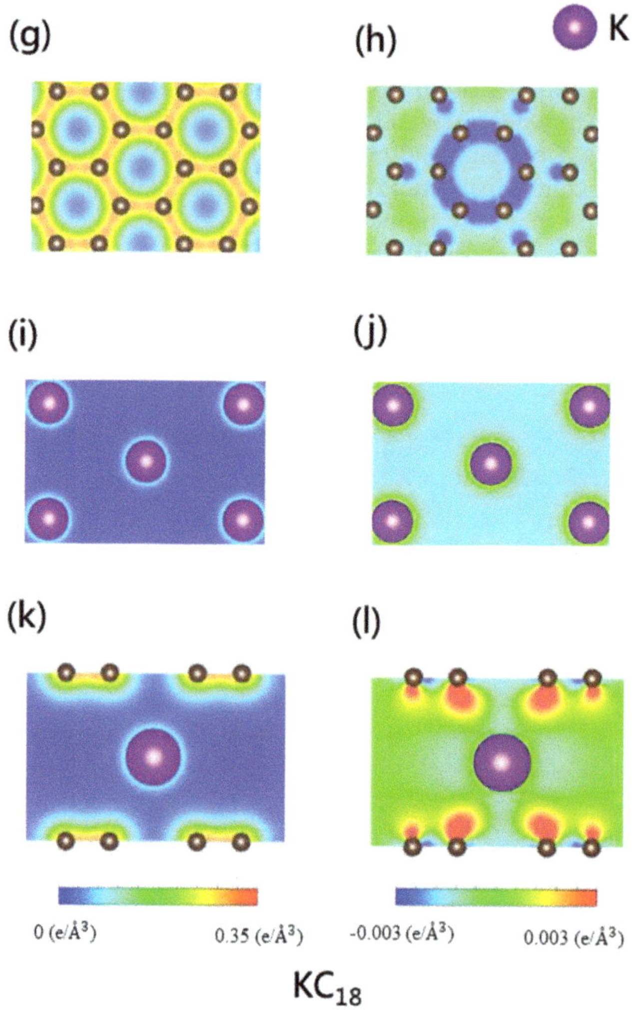

Figure 11.6. (Continued.)

merged together within the specific energy ranges, need to be analyzed in detail. The exact analyses will be helpful in enhancing the development of quasiparticle framework [198, 199, 215]. Direct STS examinations are almost absent to date, and the intrinsic limits are discussed later [244–246].

The atom- and orbital-decomposed van Hove singularities are clearly illustrated in figure 11.8(a)–(m), figure 11.9(a)–(m), and figure 11.10(a)–(m), respectively. These results are very suitable in determining the active orbital hybridizations of carbon–carbon, intercalant–intercalant, and carbon–intercalant bonds by using the merged van Hove singularities [119, 238]. In general, a moiré superlattice has a lot of carbon atoms with four significant orbitals, compared with a single metal atom with

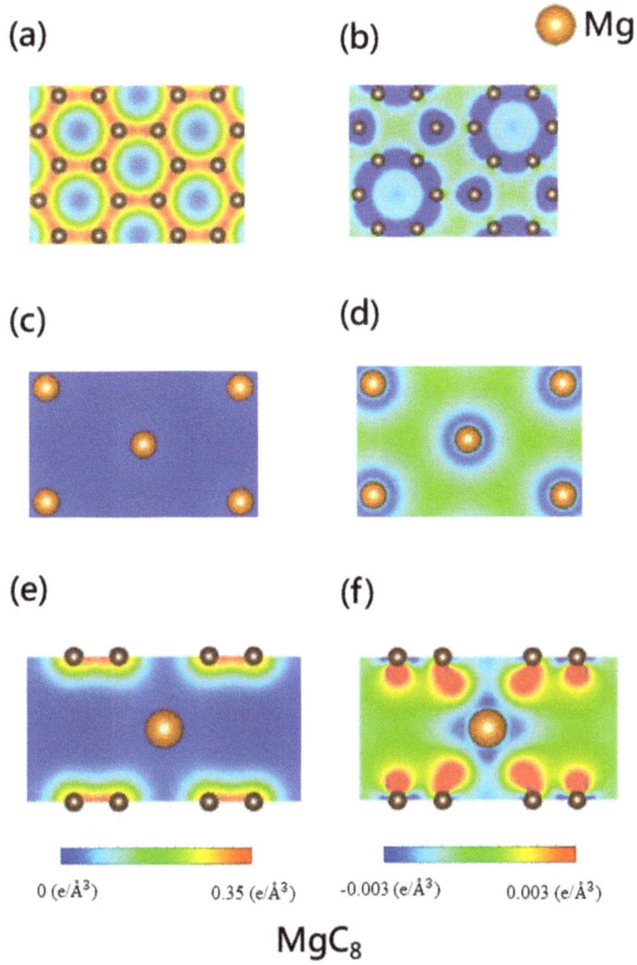

Figure 11.7. Similar plots to figure 11.5(a)–11.5(l), but associated with (a)/(c)/(e) and (b)/(d)/(f) and top view of intercalant layer MgC_8, and (g)/(i)/(k) and (h)/(j)/(l) MgC_{18}.

one effective orbital/two important orbitals for sodium and potassium/magnesium. This leads to the dominating/minor contributions from carbon-honeycomb lattices/ the intercalant layers under any intercalation cases (the green and red curves in figure 11.8(a), (d), (g), and (k), figure 11.9(a), (d), (g), and (k), and figure 11.10(a), (d), (g), and (k)). The main features of van Hove singularities mainly arise from the modified π and σ-electronic energy subbands [119, 238]. A local minimum density of states, a dip structure, corresponds to the corrected Dirac-cone structure, and its energy difference with the Fermi level of $E_F = 0$ is an observable blue shift associated with the electron transfer of metal adatoms. This important value is estimated to be, respectively, −1.50 eV/−1.49 eV/−1.48 eV for NaC_8/KC_8/MgC_8, −1.20 eV/−1.19 eV/−1.18 eV for NaC_{18}/KC_{18}/MgC_{18}, −1.0 eV/−0.99 eV/−0.98 eV for NaC_{24}/KC_{24}/MgC_{24}, as well

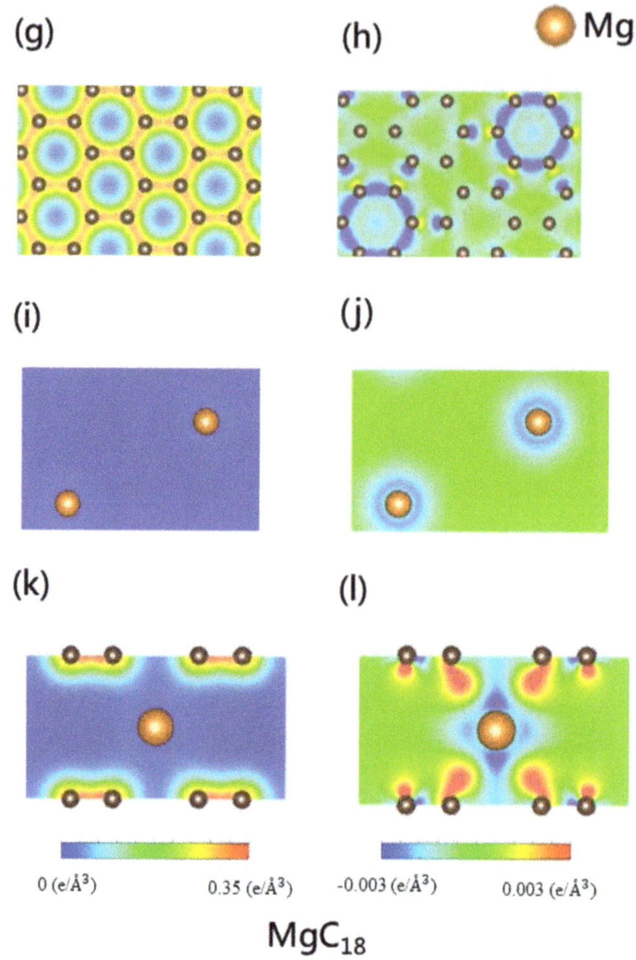

Figure 11.7. (Continued.)

as −0.90 eV/−0.89 eV/−0.88 eV for NaC_{32}/KC_{32}/MgC_{32}. Their magnitudes, which correspond to the 3D free electron densities, will have strong effects on the carrier excitations [214, 235] and transports [247], e.g., the optical plasmon modes of conduction electrons in the n-type graphite intercalation compounds [214, 235].

According to four-orbital dominances of carbon atoms, the main features in density of states could be classified into four specific energy ranges: e.g., for NaC_8 (figure 11.8(a)–(c)): (I) the σ-electronic energy spectrum of C-2s orbitals initiated from $E\sim-5.0$ eV and dominating at $E<-11.0$ eV (the purple curve), (II) the σ-valence subbands due to C-[$2p_x$, $2p_y$] orbitals driven from $E\sim-3.5$ eV and dominant within -11.0 eV $\leqslant E \leqslant -6.0$ eV (the red and blue curves), (III) the π-electronic valence spectrum of C-$2p_z$ orbitals in -11.0 eV$\leqslant$ E$\leqslant$$-6.0$ eV with dominance above −6.0 eV (the purple curve), and (IV) the π^* conduction spectrum higher than −1.50 eV. There exist three obvious van Hove singularities, corresponding to the dominant carbon-$2p_z$-orbital contributions, respectively, at −3.5 eV, −1.5 eV and 0.2 eV. Furthermore, the sodium 2s-orbital contributions are merged with them, clearly illustrating the $2p_z$–2s hybridization in each C–Na band. Similar chemical bondings are revealed in different intercalation cases and intercalants (other figures). The intercalant-decomposed contributions show the reduced amplitudes of van Hove singularities as their concentration decreases. This indicates 3s-3s/4s-4s/3s-3s single-orbital hybridizations for Na–Na/K–K/Mg–Mg bonds. The featured van Hove singularities have successfully identified the active orbital hybridizations of chemical

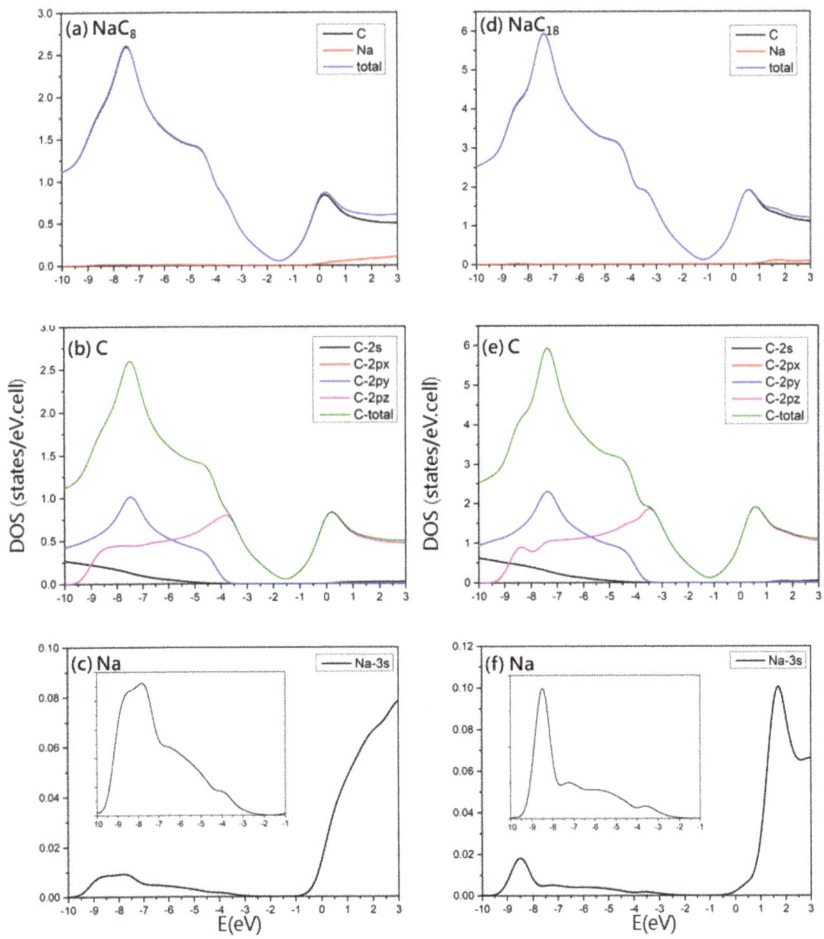

Figure 11.8. The energy-dependent density of states with atom-/carbon-orbital-/intercalant-orbital-decomposed van Hove singularities in Na-, K- and Mg-related graphite intercalation compounds: (a)/(b)/(c) NaC_8, (d)/(e)/(f) NaC_{18}, (g)/(h)/(i) NaC_{24}, as well as (j)/(k)/(l) NaC_{32}. The merged special structures represent the orbital hybridizations, being useful in determining the quasiparticle properties.

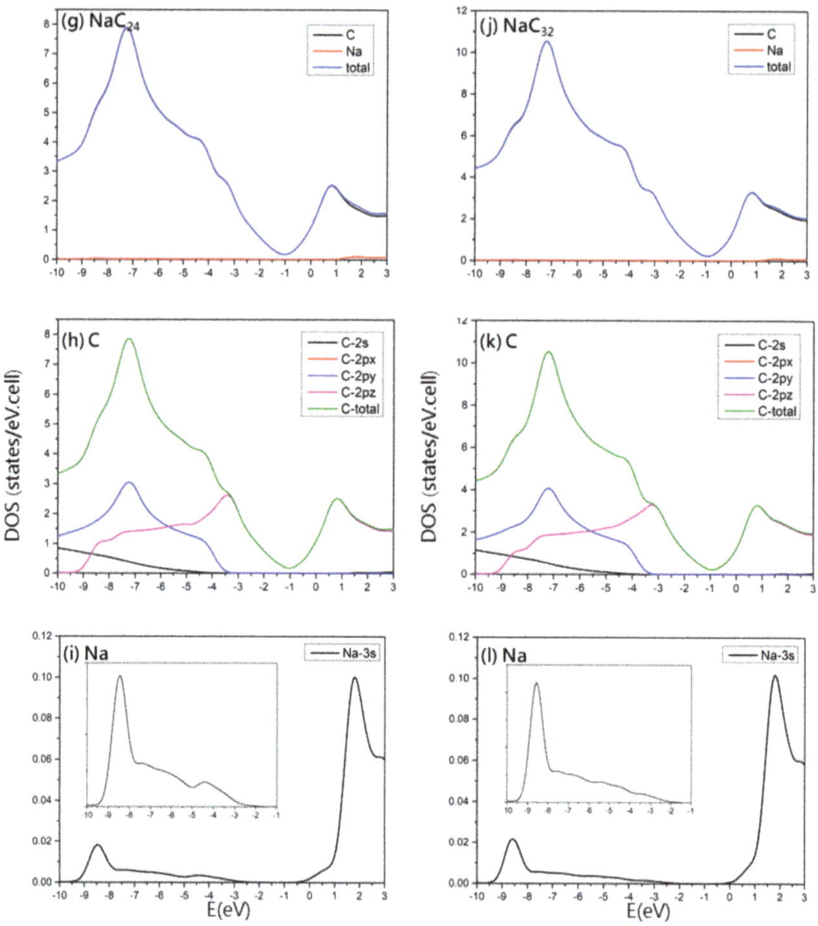

Figure 11.8. (Continued.)

bonds, being consistent with the previous development of quasiparticle framework [119, 238].

The main features of density of states are directly reflected in the essential properties, e.g., optical transitions [240, 248], Coulomb-field excitations [214, 233, 249], magneto-electronic properties [235, 250], quantum transports [251, 252], and heat capacities [253]. However, the experimental measurements about such physical quantities cannot fully be explained by the featured van Hove singularities, since each scattering event also relies on the initial and final states (the scattering amplitude of wave functions; [246, 254]). It is well known that STM and STS are very powerful in observing surface morphologies and density of states for low-dimension systems, respectively (details in chapter 3 section 3.2). For example, their high-precision measurements have very successfully been done on graphene-related carbon-sp^2 bonding systems, covering semi-metallic few-layer graphenes with distinct stacking configurations and layer numbers [244, 255–257], semiconducting or metallic carbon nanotubes [258, 259], and semiconducting graphene nanoribbons [235]. Interestingly, nanoscale crystal symmetries and electronic properties across the Fermi level are simultaneously determined by the delicate observations and analyses. Both measurements are also suitable for the chemically modified graphene-based materials, such as alkali adatom absorptions/intercalation on/inside graphene surfaces/carbon nanotubes [260]/[261]. STS would become useless for the observation of van Hove singularities in 3D graphite intercalation compounds because of the negligible tunneling current for a thick sample.

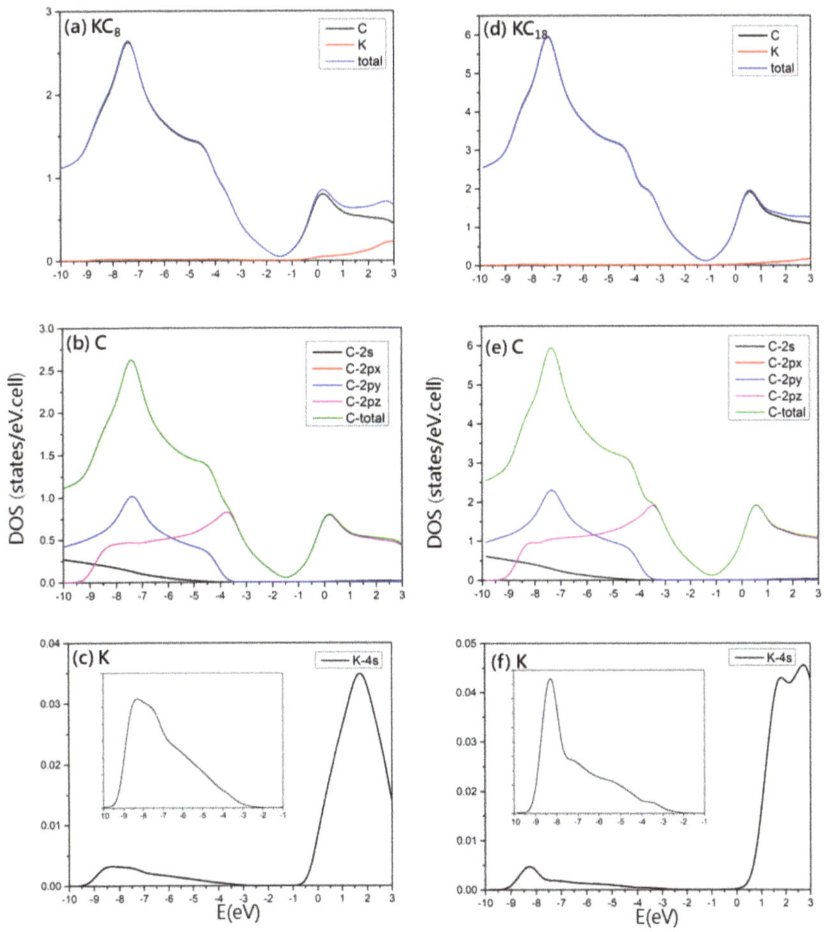

Figure 11.9. Similar plots to figure 11.8(a)–(l), but related to (a)/(b)/(c) KC_8, (d)/(e)/(f) KC_{18}, (g)/(h)/(i) KC_{24}, and (j)/(k)/(l) KC_{32}.

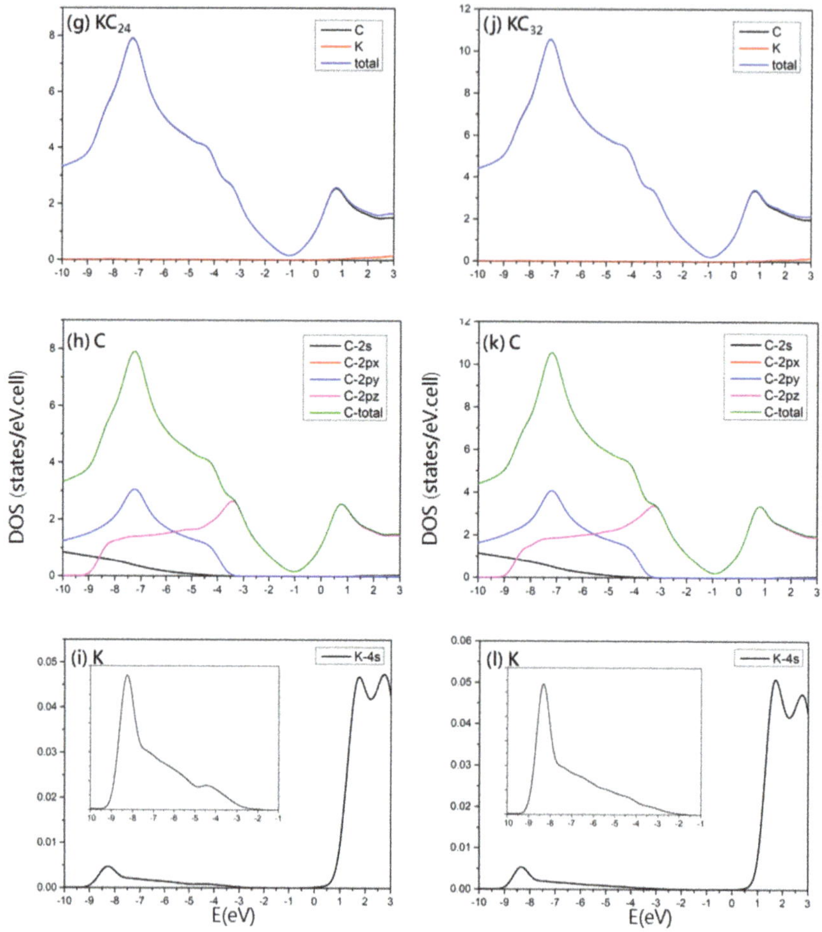

Figure 11.9. (Continued.)

11.4 Diversified essential properties by the distinct intercalations/de-intercalations

The lithium-, sodium-, potassium- and magnesium-graphite intercalation compounds (details in chapters 10 and 11) [119, 123, 170, 176, 238, 262] present significant similarities and differences in crystal symmetries and electronic properties and thus other fundamental properties [119, 123, 176, 238] and realized/potential applications [170, 263]. The critical multi-/single-orbital hybridizations, being thoroughly identified from the relevantly physical quantities (details in chapters 11.1–11.3; [119, 238]), cover 2pz-2pz and [2s, 2px, 2py]-[2s, 2px, 2py] mixings for C–C bonds, 2pz-2s/2pz-3s/2pz-4s/2pz-3s hybridizations of C–Li/C–Na/C–K/C–Mg bonds; 2s-2s/3s-3s/4s-4s/3s-3s ones through Li–Li/Na–Na/K–K/Mg–Mg bonds. The component, concentration and distribution way of intercalants [119, 238] can greatly diversify the geometric and electronic properties, such as the crystal symmetries, the periodical distance, the height and hollow site, the intralayer bond lengths, the observable modifications about the low-energy π-electronic energy spectra, the slight corrections on the σ valence bands (only due to the zone folding effects due to the specific intercalant distribution configuration; [119, 238]), the free conduction electron densities, and the atom- and orbital-dominated van Hove singularities. The rich quasiparticle behaviors are clearly associated with high-performance ion-based batteries, being under the current investigations.

Up-to-date chemical operations have been successfully realized in lithium- [264–266], sodium- [267, 268], potassium- [269], magnesium- [270–272], aluminum-ion-related batteries [125, 273, 274], as well as lithium phosphorus sulfur ones [275, 276]. When each battery is combined with the optimal cathode [272, 273, 277, 278], electrolyte [279–282] and anode materials [270, 283–287], their high-performance will be taken into

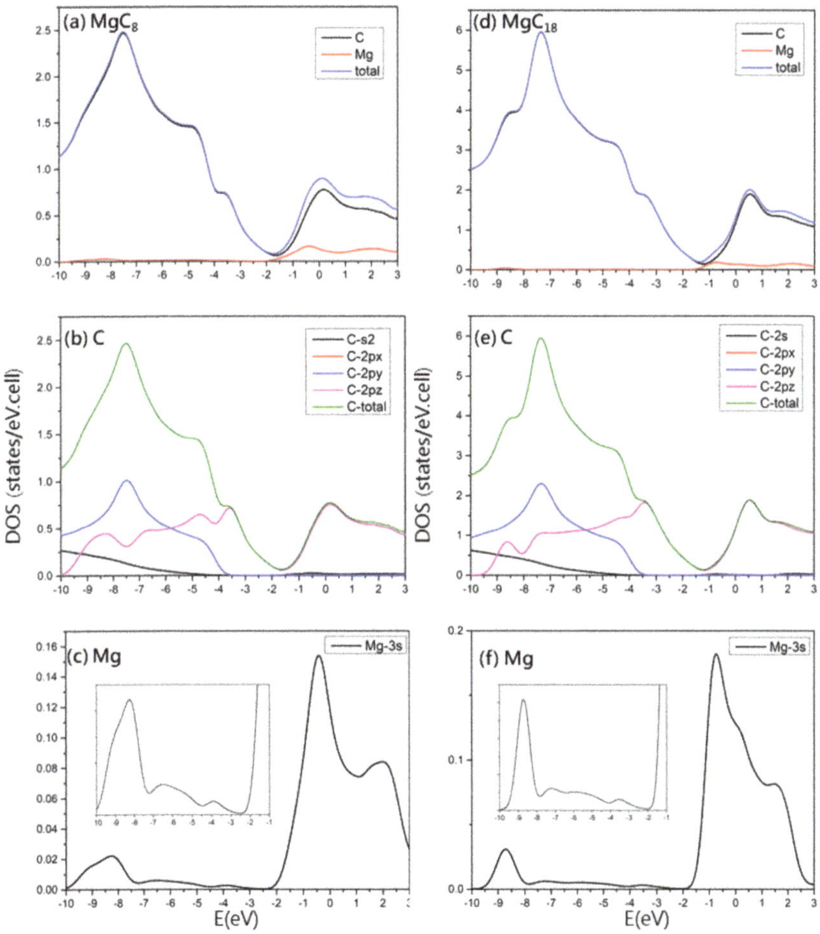

Figure 11.10. Similar plots to figure 11.8(a)–(l), but arising from (a)/(b)/(c) MgC_8, (d)/(e)/(f) MgC_{18}, (g)/(h)/(i) MgC_{24}, and (j)/(k)/(l) KC_{32}.

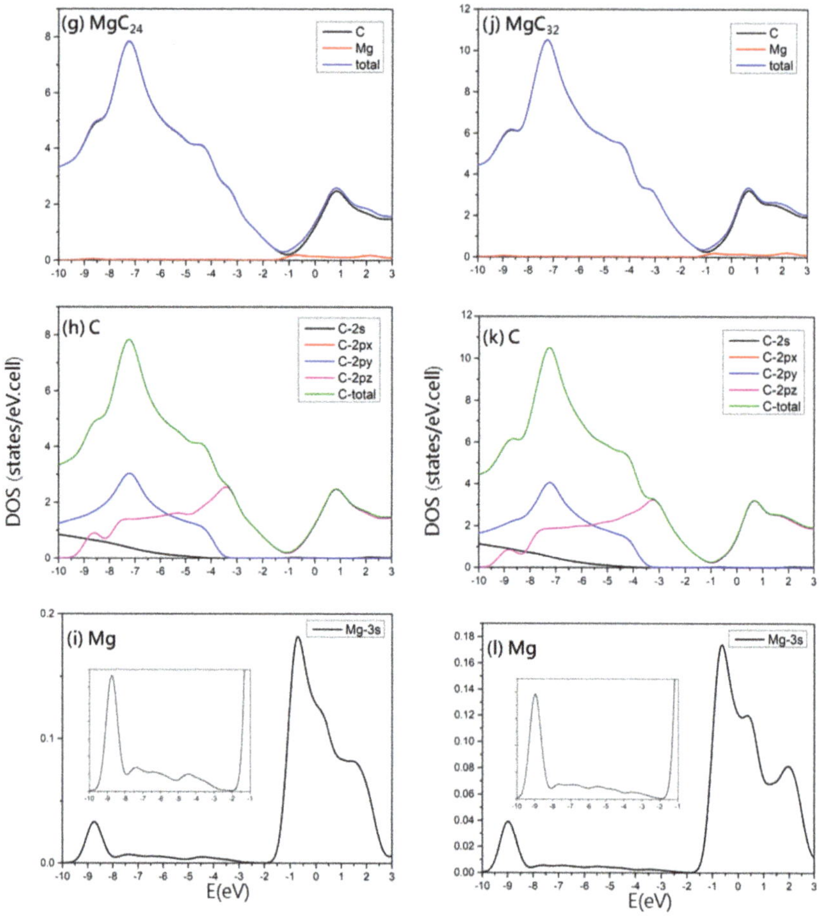

Figure 11.10. (Continued.)

commercial applications [275, 288]. Generally speaking, the concise merits and drawbacks are well characterized by the best transport property of the stationary ion current density [264, 273, 289], maximum ion capacity [264, 266, 275], largest voltage difference between two electrolytes [265, 266, 275], great reduction of charging time [264, 275] and normal chemical reactions under high temperatures/pressures [265, 273]. The strong competition or cooperation among the different batteries requires full support in the rich essential properties. This will be gradually finished in near-future book proposals, e.g., the unusual quasi-particle phenomena and commercial applications of Li-, P-, S- and O-related 3D/3D/1D compounds. Systematic investigations, which are conducted on open issues (chapter 18; [290]), are expected to create much progress in basic [290] and applied sciences [290, 291]. For example, 3D silicone compounds [292] (details in open issues of chapter 18) were easily synthesized many years ago [292], and recently they have been found to be an outstanding anode material in replacing graphite for greatly reduced charging time [292]. This discovery might lead to dramatic revolutions of electric vehicles.

However, the important differences between 3D silicone and layered graphite should be clearly clarified in terms of the unusual quasiparticle behaviors. The development of a quasiparticle framework [198, 199] will be thoroughly assisted by the explorations of ion-based batteries [293].

There also important differences between the meta-atom- (details in chapters 10 and 11; [119, 123, 238, 262]) and large-molecule-graphite intercalation compounds (chapter 5; [120]) in terms of the fundamental quasiparticle properties. Both of them present well-defined carbon honeycomb lattices and intercalant layers without any buckling, while their carbon–intercalant and intra–intercalant atomic interactions are in great contrast with each other. As a result, the former and the latter, respectively, exhibit free electron and hole transfers from the intercalant to carbon atoms; that is, blue and red shifts are observed in the experimental [294] and theoretical studies [119, 238]. Free carrier densities, as well as low-energy electronic properties, are very sensitive to the changes of intercalant concentrations and distribution configurations [119, 238]. It is relatively easy/difficult to establish the phenomenological models for the n-/p-type graphite intercalation compounds because of the concise/complicated orbital hybridizations in different chemical bonds (discussed earlier in section 11.3; [119, 238]). However, the opposite is true for p-type graphite intercalation systems [119, 238]. Of course, the metallic behaviors of conduction electrons are very suitable for the development of a quasiparticle framework [119, 238]. This will be helpful in a complete understanding of the ion transports of batteries. Up-to-date commercial battery products [264] clearly show that the donor- and acceptor-type graphite intercalation compounds frequently serve as the commercial anode [270, 284, 295] and cathode components [272, 276, 296], respectively. The intercalation and de-intercalation processes, being full of dynamic physics, chemistry, and materials [119, 123, 170, 176, 238, 262], need to be thoroughly clarified under current and near-future investigations. These might account for the battery performances of the various cathode, electrolyte and anode materials, e.g., greatly reduced charging time in the recently found emergent system of silicone (details in the open issues of chapter 18; [292, 293]).

11.5 Summary

First-principles simulations are utilized to enhance the development of the quasiparticle framework [198, 199, 215]. Sodium-, potassium- and magnesium-graphite intercalation compounds are verified to be outstanding samples in diversifying the quasiparticle behaviors [119, 238, 262], e.g., their quasi-stable and stable crystal structures under different concentrations and distribution configurations. The close relations between the crystal symmetries and the featured electronic properties are established through delicate analyses, being very helpful in building the exact phenomenological models. The concise intralayer and interlayer orbital hybridizations/atomic interactions [119, 238] are capable of presenting reliable one-site Coulomb potentials and nearest-neighbor hopping integrals in the tight-binding model [231]. And then, the generalized tight-binding model [231] is suitable for exploring the magnetic quantization phenomena in graphite intercalation compounds [235]. The 3D Landau subbands with the significant k_z dependences, are expected to induce rich magneto-Coulomb excitations [214, 235], magneto-optical selection rules [214] and unusual quantum transports [214, 235], with graphitic intercalations of large molecule ions, e.g., $FeCl_3^-$ (details in [297, 298]) and $AlCl_4^-$/$Al_2Cl_7^-$ (details in chapter 5; [120, 299]). The strengths of orbital mixings, which exist in intra-molecules and molecule–graphene interactions, might be very difficult to be accurately settled during numerical fittings [120, 299]. In general, n-type doping is relatively easy for comprehending their features through concise physical and chemical pictures [120, 299].

Delicate analyses of VASP simulations are conducted on the active organic environments, the atom-dominated electron energy spectra, the spatial charge densities and their changes from intercalations, as well as the merged van Hove singularities of the atom- and orbital-projected density of states within the specific energy ranges. These can achieve the critical multi-/single-orbital atomic interactions: 2pz-2pz and [2s, 2px, 2py]-[2s, 2px, 2py] mixings in C–C bonds, 2pz-3s/2pz-4s/2pz-3s hybridizations in C–Na/C–K/C–Mg bonds; 3s-3s/4s-4s/3s-3s ones in Na–Na/K–K/Mg–Mg bonds. Both intralayer and interlayer atomic interactions, which are fully driven by the intercalant concentrations and distributions, can account for the diversities of low-energy quasiparticle behaviors, such as the great enhancement of asymmetric hole and electron energy spectra relative to the Fermi level, an observable blue shift of E_F (n-type doping), free conduction electron density, modified Dirac-cone structure near the K or Γ valley, well-characterized, but enriched π and σ subbands, and few valence subbands associated with the intercalants. The featured band structures of sodium-, potassium- and magnesium-graphite intercalation compounds are expected to be well characterized by the tight-binding model. The reliable on-site Coulomb potentials and the orbital-related hopping integrals will be obtained from well-fitting with the VASP band structures. This is helpful in exploring the Coulomb- [235] and optical-field excitation phenomena [235]. Moreover, the generalized tight-binding model under a perpendicular magnetic field could further create diversities of quantization phenomena, e.g., the well-defined, perturbed and undefined Landau levels [235], magneto-plasmon modes [235], magneto-optical selection rules [235], and quantum transport properties [235]. As to the experimental examinations of Na-, K- and Mg-graphite intercalation compounds, crystal structures are observed under the specific stage-n configurations by the x-ray patterns [123, 170]. Such stackings could modulate the adatom/molecule concentrations by the number of graphitic layers n between two periodical intercalant layers, being different from those of the current through the dilute intercalant distribution ways. Whether the current geometric models could be successfully synthesized in experimental laboratories requires delicately thorough investigation. Very interestingly, the method of molecular dynamics [300] would be helpful in fully clarifying the optimal distributions of intercalant layers during the dynamic intercalations/de-intercalations. Up-to-date experiments, being conducted on 3D wave-vector-dependent band structures, spatial charge density distributions, as well as, energy-related van Hove singularities, are absent.

For the n-type metal-graphite intercalation compounds, the first and second physical properties would become meaningful by using low-energy band features and concise chemical bondings. There seems no way of solving the main features of van Hove singularities. Although such characteristics strongly affect the prominent structures in the essential properties, they cannot be thoroughly confirmed from the latter.

References

[1] Bachelet G B, Hamann D R and Schlïter M 1982 Pseudopotentials that work: from H to Pu *Phys. Rev.* B **26** 4199–228

[2] Dal C A 2014 Pseudopotentials periodic table: from H to Pu *Comput. Mater. Sci.* **95** 337–50

[3] Bekker H, Borschevsky A, Harman Z, Keitel C H, Pfeifer T, Schmidt P O, Crespo López-Urrutia J R and Berengut J C 2019 Detection of the $5p$–$4f$ orbital crossing and its optical clock transition in Pr^{9+} *Nat. Commun.* **10** 5651

[4] Rath M C 2018 Periodic table of elements revisited for accommodating elements of future years *Curr. Sci.* **115** 1644

[5] Villars P and Daams J L C 1993 Atomic-environment classification of the chemical elements *J. Alloys Compd* **197** 177–96

[6] Gu Z and Balbuena P B 2007 Chemical environment effects on the atomic oxygen absorption into Pt(111) subsurfaces *J. Phys. Chem.* C **111** 17388–96

[7] Guan H, Huang S, Ding J, Tian F, Xu Q and Zhao J 2020 Chemical environment and magnetic moment effects on point defect formations in CoCrNi-based concentrated solid-solution alloys *Acta Mater.* **187** 122–34

[8] Lin S Y, Chang S L, Chen H H, Su S H, Huang J C and Lin M F 2016 Substrate-induced structures of bismuth adsorption on graphene: a first principles study *Phys. Chem. Chem. Phys.* **18** 18978–84

[9] Solimannejad M, Kamalinahad S, Noormohammadbeigi M and Jouypazadeh H 2018 Chemisorption of pyrimidine nucleotide onto exterior surface of pristine $B_{12}N_{12}$ nano-cluster: a theoretical study *Phys. Chem. Res.* **6** 1–14

[10] Kwak K, Choi W, Tang Q, Kim M, Lee Y, Jiang D and Lee D 2017 A molecule-like $PtAu_{24}(SC_6H_{13})_{18}$ nanocluster as an electrocatalyst for hydrogen production *Nat. Commun.* **8** 14723

[11] Hojo H, Tsuda K, Kubouchi M and Kim D S 1998 Corrosion of plastics and composites in chemical environments *Met. Mater. Int.* **4** 1191–7

[12] Rostami R, Zarrebini M, Mandegari M, Mostofinejad D and Abtahi S M 2020 A review on performance of polyester fibers in alkaline and cementitious composites environments *Constr. Build. Mater.* **241** 117998

[13] Lee S-J, Eom A-H, Ryu S and Won J-P 2019 Resistance of an eco-friendly nano-polymer fireproof cementitious composite to physical and chemical environments *Compos. Struct.* **222** 110901

[14] De Yoreo J J *et al* 2015 Crystallization by particle attachment in synthetic, biogenic, and geologic environments *Science (80-.).* **349** aaa6760

[15] Siqueira K P F, Carvalho G B and Dias A 2011 Influence of the processing conditions and chemical environment on the crystal structures and phonon modes of lanthanide ortho-tantalates *Dalt. Trans.* **40** 9454–60

[16] Pukazhselvan D, Otero-Irurueta G, Pérez J, Singh B, Bdikin I, Singh M K and Fagg D P 2016 Crystal structure, phase stoichiometry and chemical environment of $Mg_xNb_yO_{x+y}$ nanoparticles and their impact on hydrogen storage in MgH_2 *Int. J. Hydrogen Energy* **41** 11709–15

[17] De Carvalho Oliveira R and Santelli R E 2010 Occurrence and chemical speciation analysis of organotin compounds in the environment: a review *Talanta* **82** 9–24

[18] Awater R H P and Dorenbos P 2017 The Bi^{3+} 6s and 6p electron binding energies in relation to the chemical environment of inorganic compounds *J. Lumin.* **184** 221–31

[19] Selvy P E, Lavieri R R, Lindsley C W and Brown H A 2011 Phospholipase D: enzymology, functionality, and chemical modulation *Chem. Rev.* **111** 6064–119

[20] Schlenoff J B 2014 Zwitteration: coating surfaces with zwitterionic functionality to reduce nonspecific adsorption *Langmuir* **30** 9625–36

[21] Webb P a 2003 Introduction to chemical adsorption analytical techniques and their applications to catalysis *MIC Tech. Publ.* **13** 1–4

[22] Bello A, Makgopa K, Fabiane M, Dodoo-Ahrin D, Ozoemena K I and Manyala N 2013 Chemical adsorption of NiO nanostructures on nickel foam-graphene for supercapacitor applications *J. Mater. Sci.* **48** 6707–12

[23] Iravani S, Korbekandi H, Mirmohammadi S V and Zolfaghari B 2014 Synthesis of silver nanoparticles: chemical, physical and biological methods *Res. Pharm. Sci.* **9** 385–406

[24] Akhtar M F, Hanif M and Ranjha N M 2016 Methods of synthesis of hydrogels … a review *Saudi Pharm. J.* **24** 554–9

[25] Huynh-Dinh T 1996 The logic of chemical synthesis *Biochimie* **78** 215

[26] Wei D, Liu Y, Zhang H, Huang L, Wu B, Chen J and Yu G 2009 Scalable synthesis of few-layer graphene ribbons with controlled morphologies by a template method and their applications in nanoelectromechanical switches *J. Am. Chem. Soc.* **131** 11147–54

[27] Nguyen D D, Tai N H, Chen S Y and Chueh Y L 2012 Controlled growth of carbon nanotube-graphene hybrid materials for flexible and transparent conductors and electron field emitters *Nanoscale* **4** 632–8

[28] Matsumoto R, Okabe Y and Akuzawa N 2015 Thermoelectric properties and performance of n-Type and p-Type graphite intercalation compounds *J. Electron. Mater.* **44** 399–406

[29] Meng X, Tongay S, Kang J, Chen Z, Wu F, Li S S, Xia J B, Li J and Wu J 2013 Stable p- and n-type doping of few-layer graphene/graphite *Carbon N. Y.* **57** 507–14

[30] Matsumoto R, Arakawa M, Yoshida H and Akuzawa N 2012 Alkali-metal-graphite intercalation compounds prepared from flexible graphite sheets exhibiting high air stability and electrical conductivity *Synth. Met.* **162** 2149–54

[31] Hashimoto A, Suenaga K, Gloter A, Urita K and Iijima S 2004 Direct evidence for atomic defects in graphene layers *Nature* **430** 870–3

[32] Warner J H, Rümmeli M H, Gemming T, Büchner B and Briggs G A D 2009 Direct imaging of rotational stacking faults in few layer graphene *Nano Lett.* **9** 102–6

[33] Geim A K and Novoselov K S 2009 The rise of graphene *Nanoscience and Technology: a Collection of Reviews from Nature Journals* (Singapore: World Scientific) pp 11–9

[34] Berdiyorov G R, Neek-Amal M, Peeters F M and Van Duin A C T 2014 Stabilized silicene within bilayer graphene: a proposal based on molecular dynamics and density-functional tight-binding calculations *Phys. Rev.* B **89** 24107

[35] Arafune R, Lin C L, Kawahara K, Tsukahara N, Minamitani E, Kim Y, Takagi N and Kawai M 2013 Structural transition of silicene on Ag(111) *Surf. Sci.* **608** 297–300

[36] Vogt P, De Padova P, Quaresima C, Avila J, Frantzeskakis E, Asensio M C, Resta A, Ealet B and Le Lay G 2012 Silicene: compelling experimental evidence for graphene-like two-dimensional silicon *Phys. Rev. Lett.* **108** 155501

[37] Bhimanapati G R *et al* 2015 Recent advances in two-dimensional materials beyond graphene *ACS Nano* **9** 11509–39

[38] Dávila M E, Xian L, Cahangirov S, Rubio A and Le Lay G 2014 Germanene: a novel two-dimensional germanium allotrope akin to graphene and silicene *New J. Phys.* **16** 95002

[39] Stephan R, Hanf M C, Derivaz M, Dentel D, Asensio M C, Avila J, Mehdaoui A, Sonnet P and Pirri C 2016 Germanene on al(111): interface electronic states and charge transfer *J. Phys. Chem.* C **120** 1580–5

[40] Derivaz M, Dentel D, Stephan R, Hanf M C, Mehdaoui A, Sonnet P and Pirri C 2015 Continuous Germanene layer on Al(111) *Nano Lett.* **15** 2510–6

[41] Zhu F, Chen W, Xu Y, Gao C, Guan D, Liu C, Qian D, Zhang S-C and Jia J 2015 Epitaxial growth of two-dimensional stanene *Nat. Mater.* **14** 1020–5

[42] Zhu F F, Chen W J, Xu Y, Gao C L, Guan D D, Liu C H, Qian D, Zhang S C and Jia J F 2015 Epitaxial growth of two-dimensional stanene *Nat. Mater.* **14** 1020–5

[43] Yuhara J, He B, Matsunami N, Nakatake M and Le Lay G 2019 Graphene's latest cousin: plumbene epitaxial growth on a 'Nano WaterCube' *Adv. Mater.* **31** 1901017

[44] Lee K W and Lee C E 2020 Spin–orbit coupling-induced band inversion and spin Chern insulator phase in plumbene and stanene *Curr. Appl Phys.* **20** 413–8

[45] Bihlmayer G, Sassmannshausen J, Kubetzka A, Blügel S, von Bergmann K and Wiesendanger R 2020 Plumbene on a magnetic substrate: a combined scanning tunneling microscopy and density functional theory study *Phys. Rev. Lett.* **124** 126401

[46] Yuhara J and Le Lay G 2020 Beyond silicene: synthesis of germanene, stanene and plumbene *Jpn. J. Appl. Phys.* **59** SN0801

[47] Bergeron H, Lebedev D and Hersam M C 2021 Polymorphism in post-dichalcogenide two-dimensional materials *Chem. Rev.* **121** 2713–75

[48] Cipriano L A, Di Liberto G, Tosoni S and Pacchioni G 2020 Quantum confinement in group III–V semiconductor 2D nanostructures *Nanoscale* **12** 17494–501

[49] Cornet C *et al* 2020 Strong electron-phonon interaction in 2d vertical homovalent III–V singularities *ACS Nano* **14** 13127–36

[50] Zhao J, Zeng H and Yao G 2021 Computational design of a polymorph for 2D III-V orthorhombic monolayers by first principles calculations: excellent anisotropic, electronic and optical properties *Phys. Chem. Chem. Phys.* **23** 3771–8

[51] Yang Z and Hao J 2019 Recent progress in 2D layered III–VI semiconductors and their heterostructures for optoelectronic device applications *Adv. Mater. Technol.* **4** 1900108

[52] Cai H, Gu Y, Lin Y-C, Yu Y, Geohegan D B and Xiao K 2019 Synthesis and emerging properties of 2D layered III–VI metal chalcogenides *Appl. Phys. Rev.* **6** 041312

[53] Chen J, Tan X, Lin P, Sa B, Zhou J, Zhang Y, Wen C and Sun Z 2019 Comprehensive understanding of intrinsic mobility in the monolayers of III-VI group 2D materials *Phys. Chem. Chem. Phys.* **21** 21898–907

[54] Du H *et al* 2020 Reversible transition between bipolar resistive switching and threshold switching in 2D layered III–VI semiconductor GaSe *Appl. Phys. Lett.* **116** 253102

[55] Guo F, Lyu Y, Jedrzejczyk M B, Zhao Y, Io W F, Bai G, Wu W and Hao J 2020 Piezoelectric biaxial strain effects on the optical and photoluminescence spectra of 2D III–VI compound α-In_2Se_3 nanosheets *Appl. Phys. Lett.* **116** 113101

[56] Joshi P S, Molyneaux B J, Feng L, Xie X, Macklis J D and Gan L 2008 Bhlhb5 regulates the postmitotic acquisition of area identities in layers II–V of the developing neocortex *Neuron* **60** 258–72

[57] Alyörük M M 2016 Piezoelectric properties of monolayer II–VI group oxides by first-principles calculations *Phys. Status Solidi Basic Res.* **253** 2534–9

[58] Zhou Y, Wang F and Buhro W E 2015 Large exciton energy shifts by reversible surface exchange in 2D II–VI nanocrystals *J. Am. Chem. Soc.* **137** 15198–208

[59] Zhou J 2019 Recent progress on 2D group II-VI binary chalcogenides ZnX and CdX (X = S, Se, Te): from a theoretical perspective *Adv. Theory Simul.* **2** 1900193

[60] Zheng H, Li X-B, Chen N-K, Xie S-Y, Tian W Q, Chen Y, Xia H, Zhang S B and Sun H-B 2015 Monolayer II–VI semiconductors: a first-principles prediction *Phys. Rev.* B **92** 115307

[61] Zhou J 2019 Recent progress on 2D group II-VI binary chalcogenides ZnX and CdX (X = S, Se, Te): from a theoretical perspective *Adv. Theory Simul.* **2** 1900061

[62] Rangappa D, Sone K, Kudo T and Honma I 2010 Directed growth of nanoarchitectured $LiFePO_4$ electrode by solvothermal synthesis and their cathode properties *J. Power Sources* **195** 6167–71

[63] Shi Y *et al* 2020 An overview and future perspectives of rechargeable zinc batteries *Small* **16** 7564–79

[64] Xia Y, Zhang W, Huang H, Gan Y, Tian J and Tao X 2011 Self-assembled mesoporous $LiFePO_4$ with hierarchical spindle-like architectures for high-performance lithium-ion batteries *J. Power Sources* **196** 5651–8

[65] Yang L, You W, Zhao X, Guo H, Li X, Zhang J, Wang Y and Che R 2019 Dynamic visualization of the phase transformation path in $LiFePO_4$ during delithiation *Nanoscale* **11** 17557–62

[66] Castro L, Dedryvère R, El Khalifi M, Lippens P E, Bréger J, Tessier C and Gonbeau D 2010 The spin-polarized electronic structure of $LiFePO_4$ and $FePO_4$ evidenced by in-lab XPS *J. Phys. Chem.* C **114** 17995–8000

[67] Liu H, Strobridge F C, Borkiewicz O J, Wiaderek K M, Chapman K W, Chupas P J and Grey C P 2014 Capturing metastable structures during high-rate cycling of $LiFePO_4$ nanoparticle electrodes *Science (80–.)* **344** 1252817

[68] Nagpure S C, Babu S S, Bhushan B, Kumar A, Mishra R, Windl W, Kovarik L and Mills M 2011 Local electronic structure of $LiFePO_4$ nanoparticles in aged Li-ion batteries *Acta Mater.* **59** 6917–26

[69] Kang B and Ceder G 2009 Battery materials for ultrafast charging and discharging *Nature* **458** 190–3

[70] Borah R, Hughson F R, Johnston J and Nann T 2020 On battery materials and methods *Mater. Today Adv.* **6** 100046

[71] Jiang Y, Alhassan S I, Wei D and Wang H 2020 A review of battery materials as cdi electrodes for desalination *Water (Switzerland)* **12** 1–32

[72] Fujiwara H, Fujimoto S, Tamakoshi M, Kato M, Kadowaki H, Miyadera T, Tampo H, Chikamatsu M and Shibata H 2017 Determination and interpretation of the optical constants for solar cell materials *Appl. Surf. Sci.* **421** 276–82

[73] Pana F and Put F 2013 A survey on the evolution of RSVP *IEEE Commun. Surv. Tutorials* **15** 1859–87

[74] Zhang T, Piao L, Zhao S, Xu Z, Yang L, Liu X and Ju S 2011 New progress in study of organic solar cell materials *Chinese J. Org. Chem.* **31** 260–72

[75] Zhou D, Zhou T, Tian Y, Zhu X and Tu Y 2018 Perovskite-based solar cells: materials, methods, and future perspectives *J. Nanomater.* **2018** 1–15
[76] Xue R, Zhang J, Li Y and Li Y 2018 Organic solar cell materials toward commercialization *Small* **14** 1801793
[77] Kanemitsu Y, Okano M, Phuong L Q and Yamada Y 2018 Review – light emission from thin film solar cell materials: an emerging infrared and visible light emitter *ECS J. Solid State Sci. Technol.* **7** R3102–10
[78] Kim M, Ham S, Cheng D, Wynn T A, Jung H S and Meng Y S 2020 Advanced characterization techniques for overcoming challenges of perovskite solar cell materials *Adv. Energy Mater.* 2001753
[79] Kanno S, Imamura Y and Hada M 2019 Alternative materials for perovskite solar cells from materials informatics *Phys. Rev. Mater.* **3** 075403
[80] De Wild J, Meijerink A, Rath J K, Van Sark W G J H M and Schropp R E I 2011 Upconverter solar cells: materials and applications *Energy Environ. Sci.* **4** 4835–48
[81] Filippov S P and Yaroslavtsev A B 2021 Hydrogen energy: development prospects and materials *Russ. Chem. Rev.* **90** 627
[82] Alekseeva O K, Pushkareva I V, Pushkarev A S and Fateev V N 2020 Graphene and graphene-like materials for hydrogen energy *Nanotechnol. Russ.* **15** 273–300
[83] Graetz J, Reilly J J, Yartys V A, Maehlen J P, Bulychev B M, Antonov V E, Tarasov B P and Gabis I E 2011 Aluminum hydride as a hydrogen and energy storage material: Past, present and future *J. Alloys Compd.* **509** S517–28
[84] Titus E, Campos Gil J and Chaudhari A 2019 State of the art materials for hydrogen energy *Int. J. Hydrogen Energy* **44** 1
[85] Kojima Y 2019 Hydrogen storage materials for hydrogen and energy carriers *Int. J. Hydrogen Energy* **44** 18179–92
[86] Ao Z M, Jiang Q, Zhang R Q, Tan T T and Li S 2009 Al doped graphene: a promising material for hydrogen storage at room temperature *J. Appl. Phys.* **105** 74307
[87] Yera R, Rossi N, Méndez C G and Huespe A E 2020 Topology design of 2D and 3D elastic material microarchitectures with crystal symmetries displaying isotropic properties close to their theoretical limits *Appl. Mater. Today* **18** 100456
[88] Li M, Hong M, Tang X, Sun Q, Lyu W-Y, Xu S-D, Kou L-Z, Dargusch M, Zou J and Chen Z-G 2020 Crystal symmetry induced structure and bonding manipulation boosting thermoelectric performance of GeTe *Nano Energy* **73** 104740
[89] Kaufmann K, Zhu C, Rosengarten A S, Maryanovsky D, Harrington T J, Marin E and Vecchio K S 2020 Crystal symmetry determination in electron diffraction using machine learning *Science (80-.).* **367** 564–8
[90] Xu Z-L, Lin S, Onofrio N, Zhou L, Shi F, Lu W, Kang K, Zhang Q and Lau S P 2018 Exceptional catalytic effects of black phosphorus quantum dots in shuttling-free lithium sulfur batteries *Nat. Commun.* **9** 4164
[91] Li C, He L, Luo J, Tang J, Cheng N and Chen Z 2020 Theoretical anchoring effect of new phosphorus allotropes for lithium-sulfur batteries *Nanoscale* **12** 11095–111
[92] Esfahanian V, Dalakeh M T and Aghamirzaie N 2019 Mathematical modeling of oxygen crossover in a lithium-oxygen battery *Appl. Energy* **250** 1356–65
[93] Guo H, Luo W, Chen J, Chou S, Liu H and Wang J 2018 Review of electrolytes in nonaqueous lithium-oxygen batteries *Adv. Sustain. Syst.* **2** 1700183

[94] Yang F, Song X, Dong G and Tsui K L 2019 A coulombic efficiency-based model for prognostics and health estimation of lithium-ion batteries *Energy* **171** 1173–82
[95] Zackrisson M, Avellán L and Orlenius J 2010 Life cycle assessment of lithium-ion batteries for plug-in hybrid electric vehicles-critical issues *J. Clean. Prod.* **18** 1519–29
[96] Li L, Jacobs R, Gao P, Gan L, Wang F, Morgan D and Jin S 2016 Origins of large voltage hysteresis in high-energy-density metal fluoride lithium-ion battery conversion electrodes *J. Am. Chem. Soc.* **138** 2838–48
[97] Xu J, Chen Y and Dai L 2015 Efficiently photo-charging lithium-ion battery by perovskite solar cell *Nat. Commun.* **6** 8103
[98] Cheng X-B, Huang J-Q and Zhang Q 2018 Review – Li metal anode in working lithium-sulfur batteries *J. Electrochem. Soc.* **165** A6058–72
[99] Liu Y, Weiss N O, Duan X, Cheng H-C, Huang Y and Duan X 2016 Van der Waals heterostructures and devices *Nat. Rev. Mater.* **1** 16042
[100] Bondi A 1964 Van der Waals volumes and radii *J. Phys. Chem.* **68** 441–51
[101] Engel E, Höck A and Dreizler R M 2000 Van der Waals bonds in density-functional theory *Phys. Rev.* A **61** 5
[102] Liu Y, Huang Y and Duan X 2019 Van der Waals integration before and beyond two-dimensional materials *Nature* **567** 323–33
[103] Thonhauser T, Cooper V R, Li S, Puzder A, Hyldgaard P and Langreth D C 2007 Van der Waals density functional: self-consistent potential and the nature of the van der Waals bond *Phys. Rev.* B **76** 125112
[104] Drozdov A D and deClaville Christiansen J 2020 Thermo-mechanical behavior of elastomers with dynamic covalent bonds *Int. J. Eng. Sci.* **147** 103200
[105] Huang S, Kong X, Xiong Y, Zhang X, Chen H, Jiang W, Niu Y, Xu W and Ren C 2020 An overview of dynamic covalent bonds in polymer material and their applications *Eur. Polym. J.* **141** 110094
[106] Prodjosantoso A K, Hertina A M and Irwanto 2019 The misconception diagnosis on ionic and covalent bonds concepts with three tier diagnostic test *Int. J. Instr.* **12** 1477–88
[107] Chakma P and Konkolewicz D 2019 Dynamic covalent bonds in polymeric materials *Angew. Chemie – Int. Ed.* **58** 9682–95
[108] Stephan R, Hanf M C and Sonnet P 2015 Molecular functionalization of silicene/Ag(111) by covalent bonds: a DFT study *Phys. Chem. Chem. Phys.* **17** 14495–501
[109] Eberhart M E 1996 The metallic bond: elastic properties *Acta Mater.* **44** 2495–504
[110] Cao K, Skowron S T, Biskupek J, Stoppiello C T, Leist C, Besley E, Khlobystov A N and Kaiser U 2020 Imaging an unsupported metal–metal bond in dirhenium molecules at the atomic scale *Sci. Adv.* **6** eaay5849
[111] Silvi B and Gatti C 2000 Direct space representation of the metallic bond *J. Phys. Chem.* A **104** 947–53
[112] Cui Y, Liang F, Yang Z, Xu S, Zhao X, Ding Y, Lin Z and Liu J 2018 Metallic bond-enabled wetting behavior at the liquid Ga/$CuGa_2$ interfaces *ACS Appl. Mater. Interfaces* **10** 9203–10
[113] Cui Y, Liang F, Xu S, Ding Y, Lin Z and Liu J 2019 Interfacial wetting behaviors of liquid Ga alloys/$FeGa_3$ based on metallic bond interaction *Colloids Surf.* A **569** 102–9
[114] Hu J-M and Eom C-B 2019 Magnetic-field control of ionic bonds on ferroelectric surfaces *Appl. Phys. Lett.* **114** 091601

[115] Jia S, Li C, Pan H, Wang M, Wang X and Lin Q 2021 Preparation and pore-forming mechanism of hydrogen bond and ionic bond double-driven chitosan-based mesoporous carbon *Int. J. Biol. Macromol.* **179** 519–31

[116] Lee H-H, Kim Y-W, Woo J, Park H, Hur K, Suo Z and Sun J-Y 2019 Fast healing of ionic bonds in tough hydrogels under an acoustic excitation *Extrem. Mech. Lett.* **33** 100572

[117] Wang X, Liang D and Cheng B 2020 Preparation and research of intrinsic self-healing elastomers based on hydrogen and ionic bond *Compos. Sci. Technol.* **193** 108127

[118] Sun C, Feng Y, Li Y, Qin C, Zhang Q and Feng W 2014 Solvothermally exfoliated fluorographene for high-performance lithium primary batteries *Nanoscale* **6** 2634–41

[119] Li W B, Lin S Y, Tran N T T, Lin M F and Lin K I 2020 Essential geometric and electronic properties in stage-: n graphite alkali-metal-intercalation compounds *RSC Adv.* **10** 23573–81

[120] Bhauriyal P, Mahata A and Pathak B 2017 The staging mechanism of $AlCl_4$ intercalation in a graphite electrode for an aluminium-ion battery *Phys. Chem. Chem. Phys.* **19** 7980–9

[121] Banhart F and Ajayan P M 1996 Carbon anions as nanoscopic pressure cells for diamond formation *Nature* **382** 433–5

[122] Caragiu M and Finberg S 2005 Alkali metal adsorption on graphite: a review *J. Phys. Condens. Matter* **17** R995

[123] Li Y, Lu Y, Adelhelm P, Titirici M M and Hu Y S 2019 Intercalation chemistry of graphite: alkali metal ions and beyond *Chem. Soc. Rev.* **48** 4655–87

[124] Yoon G, Kim H, Park I and Kang K 2017 Conditions for Reversible Na intercalation in graphite: theoretical studies on the interplay among guest ions, solvent, and graphite host *Adv. Energy Mater.* **7** 1601519

[125] Lin M C *et al* 2015 An ultrafast rechargeable aluminium-ion battery *Nature* **520** 325–8

[126] Manka D and Ivers-Tiffée E 2015 Electro-optical measurements of lithium intercalation/de-intercalation at graphite anode surfaces *Electrochim. Acta* **186** 642–53

[127] Noel M and Santhanam R 1998 Electrochemistry of graphite intercalation compounds *J. Power Sources* **72** 53–65

[128] Park J, Xu Z-L and Kang K 2020 Solvated ion intercalation in graphite: sodium and beyond *Front. Chem.* **8** 432

[129] Wang Y 2018 Na, K, Mg or Ca ion storage: a first-principle study *RSC Adv.* **8** 10848–54

[130] Yen M Y, Hsiao M C, Liao S H, Liu P I, Tsai H M, Ma C C M, Pu N W and Der G M 2011 Preparation of graphene/multi-walled carbon nanotube hybrid and its use as photo-anodes of dye-sensitized solar cells *Carbon N. Y.* **49** 3597–606

[131] Rahmani R and Antonov M 2019 Axial and torsional buckling analysis of single- and multi-walled carbon nanotubes: finite element comparison between armchair and zigzag types *SN Appl. Sci.* **1** 1134

[132] Han S W *et al* 2011 Band-gap transition induced by interlayer van der Waals interaction in MoS_2 *Phys. Rev.* B **84** 045409

[133] Chiu C-W, Shyu F-L, Lin M-F, Gumbs G and Roslyak O 2012 Anisotropy of π-plasmon dispersion relation of AA-stacked graphite *J. Phys. Soc. Japan* **81** 104703

[134] Lobato I and Partoens B 2011 Multiple dirac particles in AA-stacked graphite and multilayers of graphene *Phys. Rev.* B **83** 165429

[135] Chiu C W, Lee S H, Chen S C, Shyu F L and Lin M F 2010 Absorption spectra of AA-stacked graphite *New J. Phys.* **12** 083060

[136] Lee J-K *et al* 2016 The nature of metastable AA′ graphite: low dimensional nano- and single-crystalline forms *Sci. Rep.* **6** 39624
[137] Lee J K, Lee S C, Ahn J P, Kim S C, Wilson J I B and John P 2008 The growth of AA graphite on (111) diamond *J. Chem. Phys.* **129** 234709
[138] Loc Duong D, Kim G, Jeong H K and Hee Lee Y 2010 Breaking AB stacking order in graphite oxide: *Ab initio* approach *Phys. Chem. Chem. Phys.* **12** 1595–9
[139] Habib K M M, Sylvia S S, Ge S, Neupane M and Lake R K 2013 The coherent interlayer resistance of a single, rotated interface between two stacks of AB graphite *Appl. Phys. Lett.* **103** 243114
[140] Calvo H L, Vargas J E B and Torres L E F F 2020 Floquet boundary states in AB-stacked graphite *Phys. Rev.* B **101** 075424
[141] Rong Z Y and Kuiper P 1993 Electronic effects in scanning tunneling microscopy: moiré pattern on a graphite surface *Phys. Rev.* B **48** 17427–31
[142] Taut M, Koepernik K and Richter M 2014 Electronic structure of stacking faults in rhombohedral graphite *Phys. Rev.* B **90** 085312
[143] Taut M and Koepernik K 2016 Electronic structure of interfaces between hexagonal and rhombohedral graphite *Phys. Rev.* B **94** 035446
[144] Lu C L, Chang C P, Huang Y C, Ho J H, Hwang C C and Lin M F 2007 Electronic properties of AA- and ABC-stacked few-layer graphites *J. Phys. Soc. Japan* **76** 024701
[145] Charlier J C, Gonze X and Michenaud J P 1994 First-principles study of the stacking effect on the electronic properties of graphite(s) *Carbon N. Y.* **32** 289–99
[146] Yang Y *et al* 2019 Stacking order in graphite films controlled by van der Waals technology *Nano Lett.* **19** 8526–32
[147] Kurban Z, Lovell A, Jenkins D, Bennington S, Loader I, Schober A and Skipper N 2010 Turbostratic graphite nanofibres from electrospun solutions of PAN in dimethylsulphoxide *Eur. Polym. J.* **46** 1194–202
[148] Fayazi M, Liu B, Lei L, Shuai G, Odunmbaku O, Wang S, He Y and Boi F S 2019 Ferromagnetic hysteresis and structural recrystallization in turbostratic graphite *Mater. Res. Express* **6** 105612
[149] Wang J, Salim N, Fox B and Stanford N 2017 Anisotropic compressive behaviour of turbostratic graphite in carbon fibre *Appl. Mater. Today* **9** 196–203
[150] Kumar N, Dash S, Tyagi A K and Raj B 2011 Super low to high friction of turbostratic graphite under various atmospheric test conditions *Tribol. Int.* **44** 1969–78
[151] Binder C, Bendo T, Hammes G, Neves G O, Binder R, de Mello J D B and Klein A N 2017 Structure and properties of *in situ*-generated two-dimensional turbostratic graphite nodules *Carbon N. Y.* **124** 685–92
[152] Sedky A, Ali A M and Somaily H H 2020 Effect melting time on the excess conductivity and critical parameters of BSCCO copper oxide system *J. Supercond. Nov. Magn.* **33** 2963–9
[153] Kusakabe K, Wake A, Nagakubo A, Murashima K, Murakami M, Adachi K and Ogi H 2020 Interplanar stiffness in defect-free monocrystalline graphite *Phys. Rev. Mater.* **4** 043603
[154] Spanu L, Sorella S and Galli G 2009 Nature and strength of interlayer binding in graphite *Phys. Rev. Lett.* **103** 196401
[155] Lin M, Huang C and Chuu D 1997 Plasmons in graphite and stage-1 graphite intercalation compounds *Phys. Rev.* B **55** 13961–71

[156] Mak K F, Sfeir M Y, Misewich J A and Heinza T F 2010 The evolution of electronic structure in few-layer graphene revealed by optical spectroscopy *Proc. Natl. Acad. Sci. USA* **107** 14999–5004
[157] Oka T and Aoki H 2011 All optical measurement proposed for the photovoltaic hall effect *J. Phys. Conf. Ser.* 334 012060
[158] López-Lozano X, Noguez C and Meza-Montes L 2005 Electronic and optical properties of InAs(110) *Rev. Mex. Fis.* **51** 168–75
[159] Chiu C W 2018 Coulomb excitations in rhombohedral graphite *arXiv:* 1812.11050
[160] García-Vidal F J and Pitarke J M 2001 Optical absorption and energy-loss spectra of aligned carbon nanotubes *Eur. Phys. J.* B **22** 257–65
[161] Uda T 1970 Optical properties of graphite in the far-infrared region *J. Phys. Soc. Japan* **28** 946–54
[162] Egerton R F and Whelan M J 1974 Electron energy loss spectra of diamond, graphite and amorphous carbon *J. Electron Spectros. Relat. Phenomena* **3** 232–6
[163] Werner W S M, Bellissimo A, Leber R, Ashraf A and Segui S 2015 Reflection electron energy loss spectrum of single layer graphene measured on a graphite substrate *Surf. Sci.* **635** L1–3
[164] Azzolini M, Morresi T, Garberoglio G, Calliari L, Pugno N M, Taioli S and Dapor M 2017 Monte Carlo simulations of measured electron energy-loss spectra of diamond and graphite: role of dielectric-response models *Carbon N. Y.* **118** 299–309
[165] Ayria P, Tanaka S, Nugraha A R T, Dresselhaus M S and Saito R 2016 Phonon-assisted indirect transitions in angle-resolved photoemission spectra of graphite and graphene *Phys. Rev.* B **94** 075429
[166] Cheng C M, Hsu C J, Peng J L, Chen C H, Yuh J Y and Tsuei K D 2015 Tight-binding parameters of graphite determined with angle-resolved photoemission spectra *Appl. Surf. Sci.* **354** 229–34
[167] Guo Y, Smith R B, Yu Z, Efetov D K, Wang J, Kim P, Bazant M Z and Brus L E 2016 Li Intercalation into graphite: direct optical imaging and Cahn-Hilliard reaction dynamics *J. Phys. Chem. Lett.* **7** 2151–6
[168] Pan C J, Yuan C, Zhu G, Zhang Q, Huang C J, Lin M C, Angell M, Hwang B J, Kaghazchi P and Dai H 2018 An operando x-ray diffraction study of chloroaluminate anion-graphite intercalation in aluminum batteries *Proc. Natl. Acad. Sci. USA* **115** 5670–5
[169] Luo W *et al* 2015 Potassium ion batteries with graphitic materials *Nano Lett.* **15** 7671–7
[170] Xu J, Dou Y, Wei Z, Ma J, Deng Y, Li Y, Liu H and Dou S 2017 Recent progress in graphite intercalation compounds for rechargeable metal (Li, Na, K, Al)-ion batteries *Adv. Sci.* **4** 1700146
[171] Fiori S, Murata Y, Veronesi S, Rossi A, Coletti C and Heun S 2017 Li-intercalated graphene on SiC(0001): an STM study *Phys. Rev.* B **96** 125429
[172] Dunn B, Kamath H and Tarascon J M 2011 Electrical energy storage for the grid: a battery of choices *Science (80-.).* **334** 928–35
[173] Pramudita J C, Peterson V K, Kimpton J A and Sharma N 2017 Potassium-ion intercalation in graphite within a potassium-ion battery examined using *in situ* x-ray diffraction *Powder Diffr.* **32** S43–8
[174] Ziambaras E, Kleis J, Schröder E and Hyldgaard P 2007 Potassium intercalation in graphite: a van der Waals density-functional study *Phys. Rev.* B **76** 155425

[175] Zhao J, Zou X, Zhu Y, Xu Y and Wang C 2016 Electrochemical intercalation of potassium into graphite *Adv. Funct. Mater.* **26** 8103–10

[176] God C, Bitschnau B, Kapper K, Lenardt C, Schmuck M, Mautner F and Koller S 2017 Intercalation behaviour of magnesium into natural graphite using organic electrolyte systems *RSC Adv.* **7** 14168–75

[177] Kawaguchi M and Kurasaki A 2012 Intercalation of magnesium into a graphite-like layered material of composition BC_2N *Chem. Commun.* **48** 6897–9

[178] Prabhakar S, Raju V, Latha M and Rani J V 2020 Synthesis and application of magnesium diethylphosphate for rechargeable magnesium battery *ChemistrySelect* **5** 401–6

[179] Mele E J and Ritsko J J 1981 Low frequency plasmon spectra in stage I and stage II $FeCl_3$ intercalated graphite *Synth. Met.* **3** 89–98

[180] Yang H, Li H, Li J, Sun Z, He K, Cheng H M and Li F 2019 The rechargeable aluminum battery: opportunities and challenges *Angew. Chemie – Int. Ed.* **58** 11978–96

[181] Yensen N and Allen P B 2019 Open source all-iron battery for renewable energy storage *HardwareX* **6** e00072

[182] Hembram K P S S, Jung H, Yeo B C, Pai S J, Lee H J, Lee K R and Han S S 2016 A comparative first-principles study of the lithiation, sodiation, and magnesiation of black phosphorus for Li-, Na-, and Mg-ion batteries *Phys. Chem. Chem. Phys.* **18** 21391–7

[183] Kaneko T and Saito R 2017 First-principles study on interlayer state in alkali and alkaline earth metal atoms intercalated bilayer graphene *Surf. Sci.* **665** 1–9

[184] Sun J, Xue J, Liu X, Wang Z and Li L 2018 Simulation on the initial stage of sodium–graphite intercalation using first-principles calculation *Light Metals 2018Minerals, Metals and Materials Series* (Cham: Springer) pp 1337–43

[185] Huang T *et al* 2020 Boosting capacitive sodium-ion storage in electrochemically exfoliated graphite for sodium-ion capacitors *ACS Appl. Mater. Interfaces* **12** 52635–42

[186] Rosli A N, Wahab I F, Zabidi N A and Kassim H A 2015 The study of band structure of graphite intercalation compound containing sodium calculated using density functional theory *J. Phys. Conf. Ser.* 622 012049

[187] Nobuhara K, Nakayama H, Nose M, Nakanishi S and Iba H 2013 First-principles study of alkali metal-graphite intercalation compounds *J. Power Sources* **243** 585–7

[188] Wang Z, Ratvik A P, Grande T and Selbach S M 2015 Diffusion of alkali metals in the first stage graphite intercalation compounds by vdW-DFT calculations *RSC Adv.* **5** 15985–92

[189] Komaba S, Hasegawa T, Dahbi M and Kubota K 2015 Potassium intercalation into graphite to realize high-voltage/high-power potassium-ion batteries and potassium-ion capacitors *Electrochem. Commun.* **60** 172–5

[190] Pan Z-H, Camacho J, Upton M H, Fedorov A V, Howard C A, Ellerby M and Valla T 2011 Electronic structure of superconducting KC8 and nonsuperconducting LiC_6 graphite intercalation compounds: evidence for a graphene-sheet-driven superconducting state *Phys. Rev. Lett.* **106** 187002

[191] Starodub E, Bostwick A, Moreschini L, Nie S, El G F, McCarty K F and Rotenberg E 2011 In-plane orientation effects on the electronic structure, stability, and Raman scattering of monolayer graphene on Ir(111) *Phys. Rev.* **B 83** 125428

[192] Fujita D, Kumakura T, Onishi K, Sagisaka K, Ohgi T and Harada M 2004 Sprout-like growth of carbon nanowires on a carbon-doped Ni(1 1 1) surface *Surf. Sci.* **566–568** 361–6

[193] Fujita D, Ohgi T, Onishi K, Kumakura T and Harada M 2003 Discovery of carbon nanowires formed on a carbon-doped Ni(111) substrate by a bulk-to-surface precipitation process *Jpn. J. Appl. Physics* **42** 1391–4
[194] Kondo T *et al* 2012 Atomic-scale characterization of nitrogen-doped graphite: Effects of dopant nitrogen on the local electronic structure of the surrounding carbon atoms *Phys. Rev.* B **86** 35436
[195] Niimi Y, Matsui T, Kambara H, Tagami K, Tsukada M and Fukuyama H 2006 Scanning tunneling microscopy and spectroscopy of the electronic local density of states of graphite surfaces near monoatomic step edges *Phys. Rev.* B **73** 085421
[196] Chen R B, Lee C H, Chang C P and Lin M F 2007 Electronic and optical properties of finite carbon nanotubes in an electric field *Nanotechnology* **18** 508–15
[197] Lin C-Y, Do T-N, Huang Y-K and Lin M-F 2017 *Optical Properties of Graphene in Magnetic and Electric Fields* (Bristol: IOP Publishing)
[198] Lin S-Y, Liu H-Y, Nguyen D K, Tran N T T, Pham H D, Chang S-L, Lin C-Y and Lin M-F 2020 *Silicene-Based Layered Materials* (Bristol: IOP Publishing)
[199] Ngoc Thanh Thuy T, Lin S-Y, Lin C-Y and Lin M-F 2017 *Geometric and Electronic Properties of Graphene-Related Systems* (Boca Raton, FL: CRC Press)
[200] Howard C A, Dean M P M and Withers F 2011 Phonons in potassium-doped graphene: the effects of electron-phonon interactions, dimensionality, and adatom ordering *Phys. Rev.* B **84** 241404
[201] Wang Q, Zheng D, He L and Ren X 2019 Cooperative effect in a graphite intercalation compound: enhanced mobility of $AlCl_4$ in the graphite cathode of aluminum-ion batteries *Phys. Rev. Appl.* **12** 044060
[202] Kresse G and Furthmüller J 1996 Efficient iterative schemes for *ab initio* total-energy calculations using a plane-wave basis set *Phys. Rev.* B **54** 11169–86
[203] Liang H J, Hou B H, Li W H, Ning Q L, Yang X, Gu Z Y, Nie X J, Wang G and Wu X L 2019 Staging Na/K-ion de-/intercalation of graphite retrieved from spent Li-ion batteries: in operando x-ray diffraction studies and an advanced anode material for Na/K-ion batteries *Energy Environ. Sci.* **12** 3575–84
[204] Sutto T E, Duncan T T and Wong T C 2009 X-ray diffraction studies of electrochemical graphite intercalation compounds of ionic liquids *Electrochim. Acta* **54** 5648–55
[205] Horio Y, Yamazaki R, Yuhara J, Takakuwa Y and Yoshimura M 2018 Behavior of wave field on graphite surface observed using reflection high-energy electron diffraction technique *e-Journal Surf. Sci. Nanotechnol.* **16** 88–92
[206] Elsayed-Ali H E and Mourou G A 1988 Picosecond reflection high-energy electron diffraction *Appl. Phys. Lett.* **52** 103–4
[207] Ko W, Hus S M, Li X, Berlijn T, Nguyen G D, Xiao K and Li A-P 2018 Tip-induced local strain on MoS_2/graphite detected by inelastic electron tunneling spectroscopy *Phys. Rev.* B **97** 125401
[208] Rong Z Y 1994 Extended modifications of electronic structures caused by defects: scanning tunneling microscopy of graphite *Phys. Rev.* B **50** 1839–43
[209] Rong Z Y and Kuiper P 1993 Electronic effects in scanning tunneling microscopy: Moiré pattern on a graphite surface *Phys. Rev.* B **48** 17427–31
[210] Mao Y, Hao W, Wei X, Yuan J and Zhong J 2013 Edge-adsorption of potassium adatoms on graphene nanoribbon: a first principle study *Appl. Surf. Sci.* **280** 698–704

[211] Tran N T T, Lin S Y, Lin C Y and Lin M F 2017 *Geometric and Electronic Properties of Graphene-Related Systems: Chemical Bondings* (Boca Raton, FL: CRC Press)

[212] Ho J H, Chang C P and Lin M F 2006 Electronic excitations of the multilayered graphite *Phys. Lett. Sect. A Gen. At. Solid State Phys.* **352** 446–50

[213] Ho Y H, Wang J, Chiu Y H, Lin M F and Su W P 2011 Characterization of Landau subbands in graphite: a tight-binding study *Phys. Rev.* B **83** 121201

[214] Lin C Y, Do T N, Wu J Y, Shih P H, Lin S Y, Ho C H and Lin M F 2019 *Diverse Quantization Phenomena in Layered Materials* (Boca Raton, FL: CRC Press)

[215] Lin S Y, Tran N T T and Fa-Lin M 2018 *Structure- and Adatom-Enriched Essential Properties of Graphene Nanoribbons* (Boca Raton, FL,: CRC Press)

[216] Coletti C *et al* 2014 Revealing the electronic band structure of quasi-free trilayer graphene on SiC(0001) *Mater. Res. Soc. Symp. Proc.* 1693 155439

[217] Cattelan M and Fox N A 2018 A perspective on the application of spatially resolved ARPES for 2D materials *Nanomaterials* **8** 284

[218] Wartak M S and Fong C-Y 2019 Brillouin zones *Field Guide to Solid State Physics* (Bellingham, WA: SPIE)

[219] Li H *et al* 2020 Imaging moiré flat bands in 3D reconstructed WSe_2/WS_2 superlattices *arXiv:* 2007.06113

[220] Reidy K, Varnavides G, Thomsen J D, Kumar A, Pham T, Blackburn A M, Anikeeva P, Narang P, LeBeau J M and Ross F M 2021 Direct imaging and electronic structure modulation of moiré superlattices at the 2D/3D interface *Nat. Commun.* **12** 1290

[221] Lin M-F and Hsu W-D 2019 *Green Energy Materials Handbook* ed M-F Lin and W-D Hsu (Boca Raton, FL: CRC Press)

[222] Brachman M K 1954 Fermi level, chemical potential, and gibbs free energy *J. Chem. Phys.* **22** 1152

[223] Lin M and Chuu D 1998 Π plasmons in carbon nanotube bundles *Phys. Rev.* B **57** 10183–7

[224] Xu Z, Bando Y, Liu L, Wang W, Bai X and Golberg D 2011 Electrical conductivity, chemistry, and bonding alternations under graphene oxide to graphene transition as revealed by *in situ* TEM *ACS Nano* **5** 4401–6

[225] Hightower A, Ahn C C, Fultz B and Rez P 2000 Electron energy-loss spectrometry on lithiated graphite *Appl. Phys. Lett.* **77** 238–40

[226] Lin C Y, Huang B L, Gumbs G and Lin M F 2018 Geometry-diversified Coulomb excitations in trilayer AAB stacking graphene *Phys. Rev.* B **98** 195442

[227] Alfonsi J and Meneghetti M 2012 Excitonic properties of armchair graphene nanoribbons from exact diagonalization of the Hubbard model *New J. Phys.* **14** 53047

[228] Speakman S A 2013 Introduction to x-ray powder diffraction data analysis *Center for Materials Science and Engineering at MIT*

[229] Li J and Sun J 2017 Application of x-ray diffraction and electron crystallography for solving complex structure problems *Acc. Chem. Res.* **50** 2737–45

[230] Kerber S J, Barr T L, Mann G P, Brantley W A, Papazoglou E and Mitchell J C 1998 The complementary nature of x-ray photoelectron spectroscopy and angle-resolved x-ray diffraction Part I: background and theory *J. Mater. Eng. Perform.* **7** 329–33

[231] Foulkes W M C and Haydock R 1989 Tight-binding models and density-functional theory *Phys. Rev.* B **39** 12520–36

[232] Konschuh S, Gmitra M and Fabian J 2010 Tight-binding theory of the spin–orbit coupling in graphene *Phys. Rev.* B **82** 245412

[233] Lin C-Y, Wu J-Y, Chiu C-W and Lin M-F 2019 *Coulomb Excitations and Decays in Graphene-Related Systems* (Boca Raton, FL: CRC Press)

[234] Zhang X-G, Krstić P S and Butler W H 2003 Generalized tight-binding approach for molecular electronics modeling *Int. J. Quantum Chem.* **95** 394–403

[235] Chen S C, Wu J Y, Lin C Y and Lin M F 2017 *Theory of Magnetoelectric Properties of 2D Systems* (Bristol: IOP Publishing)

[236] Baltes H, Yacoby Y, Pindak R, Clarke R, Pfeiffer L and Berman L 1997 Measurement of the X-ray diffraction phase in a 2D crystal *Phys. Rev. Lett.* **79** 1285–8

[237] Tikhomirova K A *et al* 2020 Exotic two-dimensional structure: the first case of hexagonal NaCl *J. Phys. Chem. Lett.* **11** 3821–7

[238] bang L W, Tran N T T and yang L S 2019 Diverse fundamental properties in stage-n graphite alkali-intercalation compounds: anode materials of Li^+-based batteries *arXiv:* 2001.02042

[239] Kittel C 2004 *Introduction to Solid State Physics* 8th edn (New York: Wiley)

[240] Fox M and Bertsch G F 2002 Optical properties of solids *Am. J. Phys.* **70** 1269–70

[241] Sugawara E and Nikaido H 2014 Properties of AdeABC and AdeIJK efflux systems of Acinetobacter baumannii compared with those of the AcrAB-TolC system of *Escherichia coli Antimicrob. Agents Chemother.* **58** 7250–7

[242] Jiles D 2015 *Introduction to Magnetism and Magnetic Materials* (Boca Raton, FL: CRC Press)

[243] Kumar A, Poumirol J M, Escoffier W, Goiran M, Raquet B and Broto J M 2011 Electronic properties of graphene, few-layer graphene, and bulk graphite under very high magnetic field *Int. J. Nanosci.* **10** 43–7

[244] Vij D R 2006 *Handbook of Applied Solid State Spectroscopy* (Boston, MA: Springer)

[245] Kano S, Tada T and Majima Y 2015 Nanoparticle characterization based on STM and STS *Chem. Soc. Rev.* **44** 970–87

[246] Chi D, Wee A T S and Goh K E J 2020 *2*D Semiconductor Materials and Devices (Amsterdam: Elsevier)

[247] Mitchell B S 2004 An introduction to materials engineering and science for chemical and materials engineers *Choice Rev. Online* **41** 41-5327

[248] Horng J *et al* 2011 Intraband optical transitions in graphene *Optics InfoBase Conf. Papers* (Washington DC: Optical Society of America) p QFE1

[249] Ho J H, Lu C L, Hwang C C, Chang C P and Lin M F 2006 Coulomb excitations in AA- and AB-stacked bilayer graphites *Phys. Rev.* B **74** 085406

[250] Chen S C, Wang T S, Lee C H and Lin M F 2008 Magneto-electronic properties of graphene nanoribbons in the spatially modulated electric field *Phys. Lett.* A **372** 5999–6002

[251] Dubois S M M, Zanolli Z, Declerck X and Charlier J-C 2009 Electronic properties and quantum transport in graphene-based nanostructures *Eur. Phys. J.* B **72** 1–24

[252] Foa Torres L E F, Roche S and Charlier J-C 2013 *Introduction to Graphene-Based Nanomaterials* (Cambridge: Cambridge University Press)

[253] Kittel C, Kroemer H and Scott H L 1998 Thermal physics *Am. J. Phys.* **66** 164–7

[254] Damascelli A 2004 Probing the electronic structure of complex systems by ARPES *Phys. Scr. T.* **T109** 61–74

[255] Luican A, Li G and Andrei E Y 2009 Scanning tunneling microscopy and spectroscopy of graphene layers on graphite *Solid State Commun.* **149** 1151–6

[256] Le Quang T, Cherkez V, Nogajewski K, Potemski M, Dau M T, Jamet M, Mallet P and Veuillen J-Y 2017 Scanning tunneling spectroscopy of van der Waals graphene/semiconductor interfaces: absence of Fermi level pinning *2D Mater.* **4** 035019
[257] Flegler S L and Flegler S L 1997 *Scanning and Transmission Electron Microscopy* (Oxford: Oxford University Press)
[258] Lin M F, Shyu F L, Chang C P, Chen R B and Chiu C W 2003 Magnetoelectronic and optical properties of carbon nanotubes *Phys. Rev. B – Condens. Matter Mater. Phys.* **67** 45405
[259] Dresselhaus M S, Dresselhaus G and Saito R 1995 Physics of carbon nanotubes *Carbon N. Y.* **33** 883–91
[260] Fasel R and Osterwalder J 1995 Alkali-metal adsorption geometries on metal surfaces from photoelectron-diffraction experiments *Surf. Rev. Lett.* **02** 359–86
[261] Marulanda J M 2021 *Carbon Nanotubes – Redefining the World of Electronics* (Rijeka: IntechOpen)
[262] Okamoto Y 2014 Density functional theory calculations of alkali metal (Li, Na, and K) graphite intercalation compounds *J. Phys. Chem.* C **118** 16–9
[263] Laipan M, Xiang L, Yu J, Martin B R, Zhu R, Zhu J, He H, Clearfield A and Sun L 2020 Layered intercalation compounds: mechanisms, new methodologies, and advanced applications *Prog. Mater Sci.* **109** 100631
[264] Scrosati W A, van S and Scrosati B 2002 *Advances in Lithium-Ion Batteries* ed W A van Schalkwijk and B Scrosati (Boston, MA: Springer)
[265] Goodenough J B and Park K-S 2013 The li-ion rechargeable battery: a perspective *J. Am. Chem. Soc.* **135** 1167–76
[266] Nitta N, Wu F, Lee J T and Yushin G 2015 Li-ion battery materials: present and future *Mater. Today* **18** 252–64
[267] Yabuuchi N, Kubota K, Dahbi M and Komaba S 2014 Research development on sodium-ion batteries *Chem. Rev.* **114** 11636–82
[268] Slater M D, Kim D, Lee E and Johnson C S 2013 Correction: sodium-ion batteries *Adv. Funct. Mater.* **23** 3255
[269] Hosaka T, Kubota K, Hameed A S and Komaba S 2020 Research development on K-ion batteries *Chem. Rev.* **120** 6358–466
[270] Guo Q, Zeng W, Liu S-L, Li Y-Q, Xu J-Y, Wang J-X and Wang Y 2021 Recent developments on anode materials for magnesium-ion batteries: a review *Rare Met.* **40** 290–308
[271] Song J, Sahadeo E, Noked M and Lee S B 2016 Mapping the challenges of magnesium battery *J. Phys. Chem. Lett.* **7** 1736–49
[272] Saha P, Datta M K, Velikokhatnyi O I, Manivannan A, Alman D and Kumta P N 2014 Rechargeable magnesium battery: current status and key challenges for the future *Prog. Mater Sci.* **66** 1–86
[273] Das S K, Mahapatra S and Lahan H 2017 Aluminium-ion batteries: developments and challenges *J. Mater. Chem.* A **5** 6347–67
[274] Sun H, Wang W, Yu Z, Yuan Y, Wang S and Jiao S 2015 A new aluminium-ion battery with high voltage, high safety and low cost *Chem. Commun.* **51** 11892–5
[275] Li G, Chen Z and Lu J 2018 Lithium-sulfur batteries for commercial applications *Chem* **4** 3–7

[276] Yang L, Li Q, Wang Y, Chen Y, Guo X, Wu Z, Chen G, Zhong B, Xiang W and Zhong Y 2020 A review of cathode materials in lithium-sulfur batteries *Ionics (Kiel).* **26** 5299–318
[277] Li C, Zhang H P, Fu L J, Liu H, Wu Y P, Rahm E, Holze R and Wu H Q 2006 Cathode materials modified by surface coating for lithium ion batteries *Electrochim. Acta* **51** 3872–83
[278] Sha M, Liu L, Zhao H and Lei Y 2020 Review on recent advances of cathode materials for potassium-ion batteries *Energy Environ. Mater.* **3** 56–66
[279] Umeshbabu E, Zheng B and Yang Y 2019 Recent progress in all-solid-state lithium−sulfur batteries using high Li-ion conductive solid electrolytes *Electrochem. Energy Rev.* **2** 199–230
[280] Zhang S S 2006 A review on electrolyte additives for lithium-ion batteries *J. Power Sources* **162** 1379–94
[281] Wang Y, Song S, Xu C, Hu N, Molenda J and Lu L 2019 Development of solid-state electrolytes for sodium-ion battery–a short review *Nano Mater. Sci.* **1** 91–100
[282] Barghamadi M, Best A S, Bhatt A I, Hollenkamp A F, Musameh M, Rees R J and Rüther T 2014 Lithium–sulfur batteries – the solution is in the electrolyte, but is the electrolyte a solution? *Energy Environ. Sci.* **7** 3902–20
[283] Zuo X, Zhu J, Müller-Buschbaum P and Cheng Y-J 2017 Silicon based lithium-ion battery anodes: a chronicle perspective review *Nano Energy* **31** 113–43
[284] Perveen T, Siddiq M, Shahzad N, Ihsan R, Ahmad A and Shahzad M I 2020 Prospects in anode materials for sodium ion batteries – a review *Renew. Sustain. Energy Rev.* **119** 109549
[285] Qi S-H, Deng J-W, Zhang W-C, Feng Y-Z and Ma J-M 2020 Recent advances in alloy-based anode materials for potassium ion batteries *Rare Met.* **39** 970–88
[286] Zhang C, Zhao H and Lei Y 2020 Recent research progress of anode materials for potassium-ion batteries *Energy Environ. Mater.* **3** 105–20
[287] Zhao H, Deng N, Yan J, Kang W, Ju J, Ruan Y, Wang X, Zhuang X, Li Q and Cheng B 2018 A review on anode for lithium-sulfur batteries: progress and prospects *Chem. Eng. J.* **347** 343–65
[288] Pistoia G 2014 *Lithium-Ion Batteries* (Amsterdam: Elsevier)
[289] Ould Ely T, Kamzabek D, Chakraborty D and Doherty M F 2018 Lithium–sulfur batteries: state of the art and future directions *ACS Appl. Energy Mater.* **1** 1783–814
[290] Kudu Ö U, Famprikis T, Fleutot B, Braida M-D, Le Mercier T, Islam M S and Masquelier C 2018 A review of structural properties and synthesis methods of solid electrolyte materials in the Li_2S – P_2S_5 binary system *J. Power Sources* **407** 31–43
[291] Rosenberg S and Hintennach A 2014 Laser-printed lithium-sulphur micro-electrodes for Li/S batteries *Russ. J. Electrochem.* **50** 327–35
[292] Sardar V B, Rajhans N R, Pathak A and Prabhu T 2016 Development in silicone material for biomedical applications – a review *14th Int. Conf. on Humanizing Work and Work Environment HWWE-2016*
[293] Pilania G, Slenes K and Ramprasad R 2013 First principles study of the interface between silicone and undoped/doped $BaTiO_3$ *J. Appl. Phys.* **113** 064316
[294] Gunasekara N, Takahashi T, Maeda F, Sagawa T and Suematsu H 1988 Angle-resolved ultraviolet photoemission study of first stage alkali-metal graphite intercalation compounds *Z. Phys. B Condens. Matter* **70** 349–55
[295] Goriparti S, Miele E, De Angelis F, Di Fabrizio E, Proietti Zaccaria R and Capiglia C 2014 Review on recent progress of nanostructured anode materials for Li-ion batteries *J. Power Sources* **257** 421–43

[296] Wang D-Y *et al* 2017 Advanced rechargeable aluminium ion battery with a high-quality natural graphite cathode *Nat. Commun.* **8** 14283
[297] Qi X, Qu J, Zhang H-B, Yang D, Yu Y, Chi C and Yu Z-Z 2015 $FeCl_3$ intercalated few-layer graphene for high lithium-ion storage performance *J. Mater. Chem.* A **3** 15498–504
[298] Yan Z, Zhuxia Z, Tianbao L, Xuguang L and Bingshe X 2008 XPS and XRD study of $FeCl_3$–graphite intercalation compounds prepared by arc discharge in aqueous solution *Spectrochim. Acta Part A Mol. Biomol. Spectrosc.* **70** 1060–4
[299] Wu M S, Xu B, Chen L Q and Ouyang C Y 2016 Geometry and fast diffusion of $AlCl_4$ cluster intercalated in graphite *Electrochim. Acta* **195** 158–65
[300] Hansson T, Oostenbrink C and van Gunsteren W 2002 Molecular dynamics simulations *Curr. Opin. Struct. Biol.* **12** 190–6

Chapter 12

Essential electronic properties of armchair carbon and silicon nanotubes

Hsin-Yi Liu, Thi Dieu Hien Nguyen, Shih-Yang Lin, Hsien-Ching Chung, Wei-Bang Li, Ngoc Thanh Thuy Tran, Nguyen Thi Han, Hai Duong Pham and Ming-Fa Lin

12.1 Introduction

Tubular structures are one of the most fascinating materials since carbon nanotubes (CNTs) were created experimentally [1–7]. They have aroused a lot of interest due to their specific electronic properties. Normally, CNTs often refer to single-wall carbon nanotubes (sCNTs) of various diameters in the nanometer range. A CNT can be regarded as a quasi 1-D tubular structure which is set up in honeycomb lattices. An sCNT can present either semiconducting or metallic behavior due to its diameter and chiral vectors. The electronic properties of CNTs have been a topic of intense investigation since they were first discovered in 1991 [1]. Plenty of theoretical calculations on the electronic properties of CNTs have been produced via the tight-binding model [8, 9] and the first-principles method since the first accurate predictions in 1992 [10, 11]. Using these detailed calculations and analyses, all CNTs can be classified into three types: (I) metals with a finite density of states (DOS) at the Fermi level; (II) moderate-gap semiconductors with energy gaps inversely proportional to the radius; and (III) narrow-gap semiconductors with E_g α $1/r^2$ [12].

Graphene with its very high surface area and electrical conductivity can be applied in various areas. 1D CNTs can be thought of as rolled up graphene with a unique cylindrical symmetry; they are very suitable for studying the basic and applied sciences. Recent studies on CNTs-based anode material have been published [13–25]. The composites of graphene and CNTs exhibit synergistic effects between their graphite components [22–27]. The chirality- and radius-dependent semiconductors and metals are available in functionalized nano devices, such as electronic

doi:10.1088/978-0-7503-4685-6ch12

and optoelectronic devices. The intrinsic 1D band structures, with decoupled angular momenta, dominate the quantized electrical and optical properties, e.g., geometry-dependent resistance, capacitance, inductance, and optical conductance [28–30]. To date, many field-effect transistors (FETs) based on semiconducting CNTs have developed, mainly owing to their advantage of high carrier mobility under low scattering.

Silicon atoms classified in group IV of the period table have an electronic configuration similar to carbon atoms. Silicon nanotubes (SiNTs) present a hollow tube composed of silicon atoms. The geometry of SiNTs, due to the buckling angle, results in a deformation compared to CNTs. They present a gear-like structure from a cross-section view, quite different from CNTs. SiNTs were initially successfully synthesized in 2001 [31]. Soon after the initial synthesis, other research groups reported different growth processes [32–37]. Sha *et al* used nanochannel Al_2O_3 [32] and fabricated SiNTs by a method of chemical vapor deposition (CVD) [32, 38, 39]. After that, a variety of techniques were developed to create SiNTs, such as a method of molecular beam epitaxy (MBE) [33], a hydrothermal method [35] and a method of gas phase condensation [36, 40]. Ishai and Patolsky demonstrated that the diameter and wall thickness can be well-controlled, ranging from 1.5 nm to 500 nm for the former, which would be a great advantage in potential nanoscale electronics [41]. Yoo *et al* used a surface sol–gel reaction to obtain an optimal SiNT for energy storage devices [42]. Other studies which reported their own fabrication techniques of SiNTs, achieved better SiNTs properties for application in many biological, chemical and electrical fields, such as sensors, nanoscale electronics and optoelectronics [43–45]. Especially for an application in lithium-ion batteries, silicon is one of the most promising candidates to replace the currently available commercial materials for alloying anodes, i.e. graphite. Nanoscale SiNTs have overcome the disadvantages of volume expansion that lead to pulverization and a rapid loss of reversible capacity [43]. They show great performance compared to the capacity of graphite (372 math g^{-1}) [46, 47].

Lots of theoretical studies on the electronic properties of SiNTs have been carried out. Most of them are analyzed by the tight-binding model [48–54]. However, studies on the essential electronic properties of SiNTs that use the first-principle methods are relatively few. Before SiNTs had been created, Fagan and Bayle *et al* proposed a hypothetical SiNT to study its electronic and structural properties [55]. For the geometry of SiNTs a Si hexagonal nanotube (Si h-NT) and Si gear-like nanotubes (Si g-NTs) are taken into consideration. The Si h-NTs and Si g-NTs differ in their hybridization of the Si atoms and are produced by rolling up sp^2 and sp^3 Si atoms into a silicone sheet to form a nanotube. Besides, Si g-NTs are energetically more stable than Si h-NTs [48]. Like CNTs, SiNTs can exhibit metallic and semiconducting characteristics as well. The tubular structure composed of silicon atoms can be thought of as monolayer silicene rolled up into a cylindrical shape by itself. Monolayer silicene contains sp^2 and sp^3 hybridization. However, the orbital contributions of SiNTs are mostly dominated by sp^3 hybridization, because the curvature effect leads to a misorientation of the $3p_z$ orbitals and the sp^3 orbital hybridization on the cylindrical surface. Lin and Chuu report detailed calculations that show that the hopping integral on the dimer line is different from the other two, depending on the arrangement of the three kinds of chemical bonds [56].

In this chapter we focus on the diversified electronic properties of single-wall armchair CNTs and SiNTs, investigated by first-principles calculations. A relativistic interaction of a carbon atom's spin with its motion inside a potential is normally weak, as in the case of graphene. However, it is significant in a CNT due to the curvature surface [57]. The curved surface causes an asymmetric structure and leads to an intrinsic spin–orbit interaction in the CNT. The estimated splitting based on theoretical calculations is around 100 μeV; and the spin–orbit splitting measured by experimental methods is reasonably consistent with these theoretical predictions [58–61]. Armchair SiNTs and CNTs are taken into consideration to reveal that there are two basic kinds of geometric structures, which can be further modulated by varying the diameter and chiral vectors. This work presents the bond lengths b_1 and b_2, ground state energy as well as total energy E_0, buckling height Δh, energy bands, energy gap (E_g), charge density (ρ) and projected density of states (PDOSs) for both structures dependently on the chiral vectors and diameter that are also included in the calculations.

12.2 Methods

First-principles calculations are based on density functional theory (DFT) carried out by the Vienna *ab initio* simulation package (VASP) [62, 63]. The exchange-correlation energy derived from many-particle Coulomb interactions, is evaluated by use of the Perdew–Burke–Ernzerhof functional (PBE) [64] under generalized gradient approximation. Furthermore, the van der Waals (vdW) corrections are included via the semiemprical DFTD2 correction of Grimme [65]. A one-dimensional periodic boundary condition set along the z-direction and the vacuum distance at the x- and y-space is set to 15 Å to make sure that no resistance from neighboring SiNTs is included. The maximum cutoff energy of the wave function expanded by the plane wave is 500 eV. The Brillouin zone and geometric optimization is implemented with $1 \times 1 \times 15$ meshes and $1 \times 1 \times 500$ for the DOS calculations via the Gamma scheme. The convergence of the Helmann–Feymann force is less than 0.01 eV Å^{-1} during ionic relaxations.

12.2.1 Silicon nanotubes

Schematic single-wall armchair SiNTs are shown in figure 12.1(a)–(j), which exhibit the cross-section and side view for armchair Si (2,2) to Si (11,11) nanotubes, respectively. Compared with CNTs (figure 12.9), the gear-like shapes can be observed in the cross-section schema as result of sp^3 hybridization leading each neighboring Si atom to create a height difference. For the armchair structure a periodical bond length in a unit cell is $3b$ (b is a Si–Si bond length). After moderate adjustments of the structure, the optimal lattice constant and relatively lowest ground state energy E_0 is obtained, which are summarized in table 12.1. The ground state energy E_0 decreases with an increasing diameter of the armchair silicon nanotubes (aSiNTs), as shown in figure 12.3(a). Because of the buckling angle

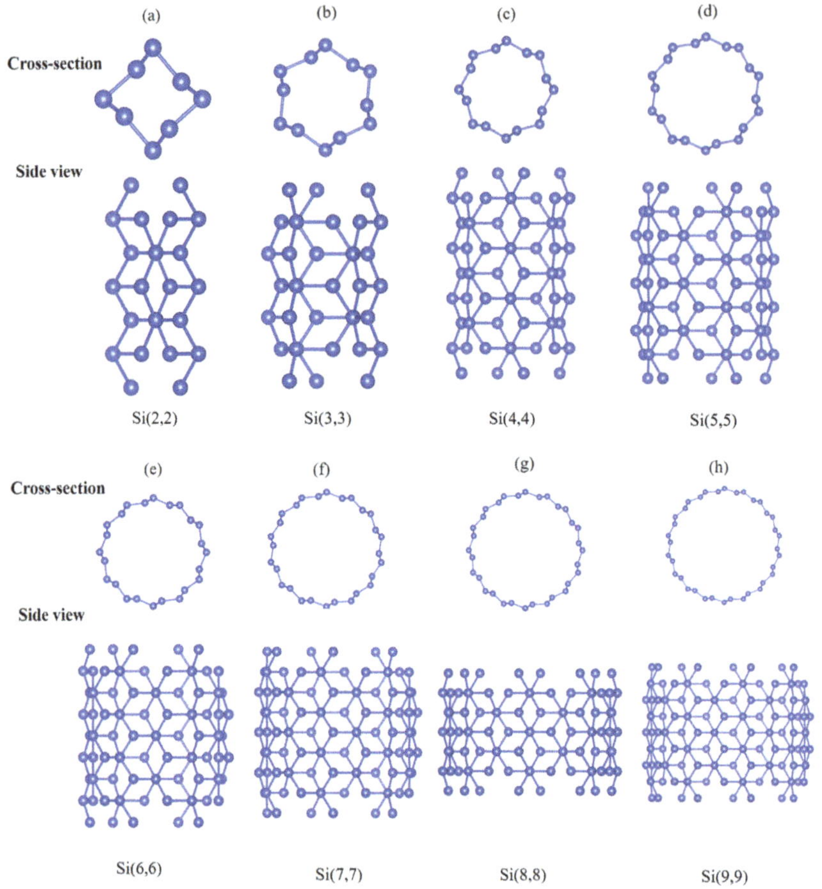

Figure 12.1 (a)–(d) (e)–(h) (i)–(j). Geometric structure of aSiNTs at side view and cross-section Si (2,2), Si (3,3), Si (4,4) and Si (5,5), Si (6,6), Si (7,7), Si (8,8) and Si (9,9), Si (10,10) and Si (11,11).

enhanced by the sp^3 hybridization, the SiNTs have a gear-like structure unlike CNTs. The gear-like structures cause a distance difference in two adjacent atoms to the center axis. This results in the two different diameters which are defined as D_1 and D_2 in figure 12.2. Meanwhile, the different types of Si–Si bonds are labeled b_1 and b_2. b_1 is a horizontal Si–Si bond and b_2 is a vertical one. Both of the bond lengths are altered depending on the tube size, especially for a small size of nanotubes. Table 12.1 shows the geometric parameters of the aSiNTs. The chiralities (m,m) are in the range from 2 to 11. The smallest sCNTs have a diameter of 4 Å, as determined by HR-TEM method; the possible structures can be armchair (3,3), zigzag (5,0) and chiral (4,2) [66]. Besides, in a multiwalled CNT a stable 3 Å of CNTs can be grown [67]. The results of the aSiNTs diameters for D_1 and D_2 are 3.452 to 22.728 Å and 5.397 to 23.515 Å, respectively. The buckling height Δh is determined by the distance difference of D_1 and D_2 and is dependent on the diameter and chirality of

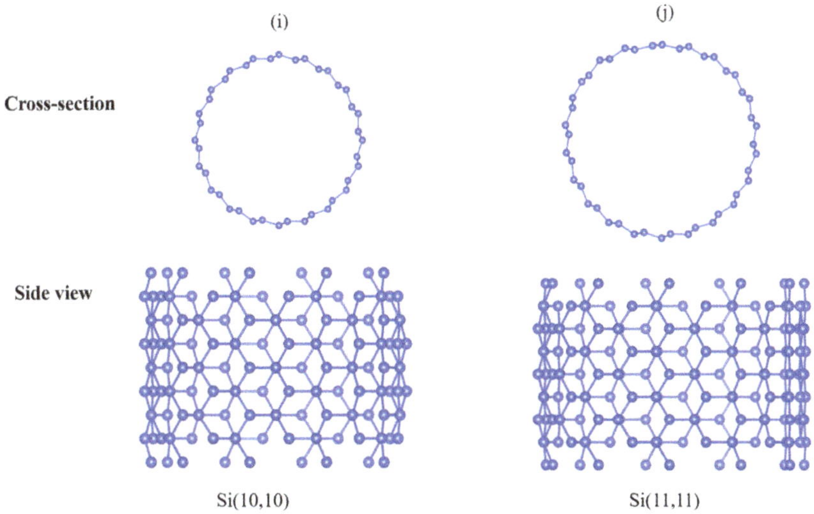

Figure 12.1 (a)–(d) (e)–(h) (i)–(j). (Continued.)

the SiNTs. The small silicon tubes lead to the large buckling distance. When the diameter and chirality of the SiNTs are increased, the buckling distance is reduced, as shown in figure 12.3(c). The Si–Si bond lengths of the SiNTs are different at the horizontal and vertical position and are named b_1 and b_2 in this chapter. b_1 and b_2 are highly dependent on the diameter and chirality, especially b_1. When the symbol m of chirality is small, the Si–Si bond length b_1 is extended. When the symbol m turns large, b_2 returns to a value close to the standard Si–Si bond length of b_0 (2.28 Å). A single-wall SiNT, as well as a CNT, can be thought of as unrolled silicene and graphene, as shown in figure 12.8(a). The chirality R_x is expressed by $R_x = ma_1 + na_2 \equiv (m,n)$, where a_1 and a_2 are the real space unit vectors. In the armchair case $m = n$, i.e. (m,m) and for the case of the zigzag tube $n = 0$, i.e. $(m,0)$. The value m is proportional to the atom numbers of the nanotubes. Small size SiNTs contain only a small number of Si atoms. When they roll up to form a tube, the distance between each Si atom may be lengthened. This phenomenon can be observed by b_2 of the Si–Si bonds as well. The values of the normalized bond length b_1/b_0 and b_2/b_0 are summarized in table 12.1 and illustrated in figure 12.3(b). The results are close to 1, when the diameter and the chirality m increase. The ground state energy E_0 obviously decreases as the diameters and the chirality m increase. The decreasing ground state energy E_0 reflects the stable geometric structure of SiNTs and could be due to the reduced curvature energy [68].

The band structures of aSiNTs with chiral vectors of (2,2), (3,3), (4,4), (5,5), (6,6), (7,7), (8,8), (9,9), (10,10) and (11,11) are shown in figure 12.4(a)–(j). The one-dimensional first Brillouin zone (1D BZ) of the nanotubes exhibits two symmetry points. One is in the zone boundary at the Γ point with $k = 0$, and the other is at the X point with $k = \pm\pi/a$, where a is the lattice constant. In addition to this, there is another important point K between Γ and X with $k = \pm 2\pi/3a$. SiNTs have one-dimensional band structures due to

Table 12.1. Parameters of aSiNTs ground state energy E_0, Si–Si bond length b_1 and b_2, diameter D_1 and D_2, buckling height Δh; $b_0 = 2.28$ Å.

aSiNTs	Lattice con. (Å)	E_o/unit cell (eV)	E_o/atom (eV)	b_1(Å)	b_2(Å)	b_1/b_0	b_2/b_0	D_1(Å)	D_2(Å)	Δh(Å)
Si (2,2)	3.540	−37.212	−4.652	2.291	2.385	1.004 825	1.046 053	3.452	5.397	0.9725
Si (3,3)	3.722	−56.289	−4.691	2.290	2.318	1.004 386	1.016 667	5.000	6.362	0.681
Si (4,4)	3.751	−75.523	−4.720	2.288	2.303	1.003 509	1.010 088	7.922	9.182	0.63
Si (5,5)	3.764	−94.759	−4.738	2.287	2.296	1.003 070	1.007 018	9.675	10.716	0.5205
Si (6,6)	3.766	−113.967	−4.749	2.285	2.291	1.002 193	1.004 825	12.323	13.371	0.524
Si (7,7)	3.772	−133.152	−4.755	2.283	2.288	1.001 316	1.003 509	14.099	15.086	0.4935
Si (8,8)	3.780	−152.319	−4.760	2.282	2.287	1.000 877	1.003 070	16.591	17.577	0.493
Si (9,9)	3.775	−171.474	−4.763	2.282	2.285	1.000 877	1.002 193	18.463	19.419	0.478
Si (10,10)	3.779	−190.619	−4.765	2.282	2.285	1.000 877	1.002 193	20.897	21.849	0.476
Si (11,11)	3.785	−209.612	−4.764	2.281	2.284	1.000 439	1.001 754	22.728	23.515	0.3935

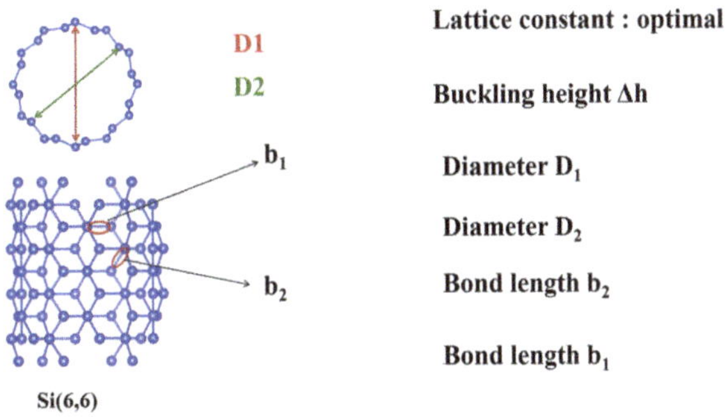

Figure 12.2. Illustration of SiNT parameters.

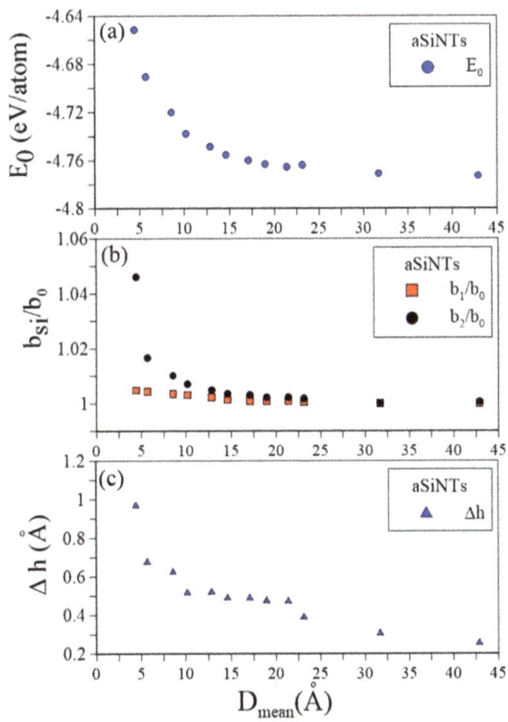

Figure 12.3. Ground state energy E_0, b_{Si}/b_0 bond length and buckling height Δh are illustrated as function of D_{mean} (average diameter of D_1 and D_2).

the buckling angle, honeycomb symmetry, edge structure and quantum confinement. Energy bands of each SiNT increase with an increasing chiral vector of the nanotubes. Most of the energy bands for aSiNTs are parabolically dispersed and some of the sub-bands are partially flat. The band-edge states located between $k = 0$ to $k = \pm\pi/a$ (from the Γ to the X point) present van Hove singularities in the DOS. The sub-band crossings are dominantly related to the sub-band anti-crossings. Each of the sub-bands corresponds to a different angular momentum J. The aSiNTs shown in figure 12.4(a)–(j) correspond to

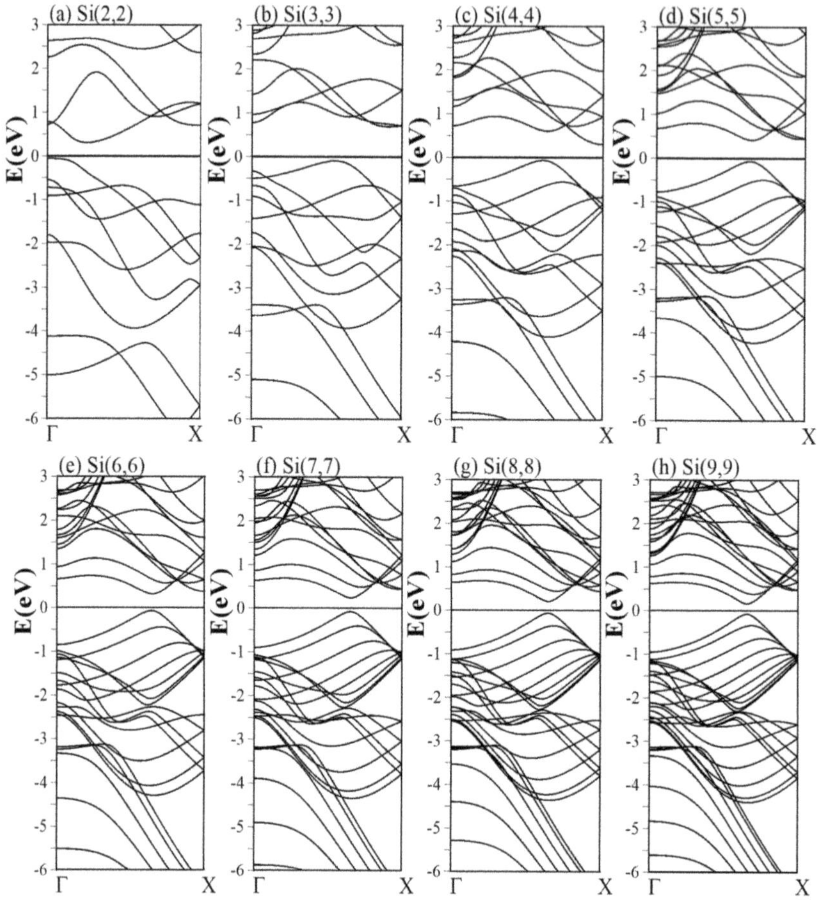

Figure 12.4(a)–(d) (e)–(h) (i)–(j). Band structure of armchair silicon nanotubes Si (2,2), Si (3,3), Si (4,4) and Si (5,5), Si (6,6), Si (7,7), Si (8,8) and Si (9,9), (10,10) and Si (11,11).

the chiral vectors of Si (2,2) to Si (11,11); they present semiconducting behaviors with energy gaps and Dirac cones located at the K points. According to the zone-folding approximation, the Dirac cones of the 1D BZ are folded at the K point ($k = \pm 2\pi/3a$), in agreement with previous results [8, 11, 69]. All of the electronic states measured from the Fermi level (E_F =0) and the energy bands of the occupied valence orbitals are asymmetric to the energy bands of unoccupied conduction ones. The reasons could be the sp^3 hybridization and periodic boundary conditions (PBCs) [10, 11]. The energy gap E_g of Si (2,2)-Si (11,11) are summarized in table 12.2 and illustrated in figure 12.5(a) and (b). The Si (2,2), Si (3,3) and Si (4,4) nanotubes shown in figure 12.4(a)–(c) possess indirect energy gaps of E_g = 0.350, 0.785 and 0.360 eV, respectively. In other words, the difference between the highest valence band (HVB) and the lowest conduction band (LCB) at the located 1D BZ point is not the same. The energy (eV) of the HVB point is at the Γ point, and the LCB point is around $k = \pm\pi/3a$. The other two indirect E_g energy bands of Si (4,4)

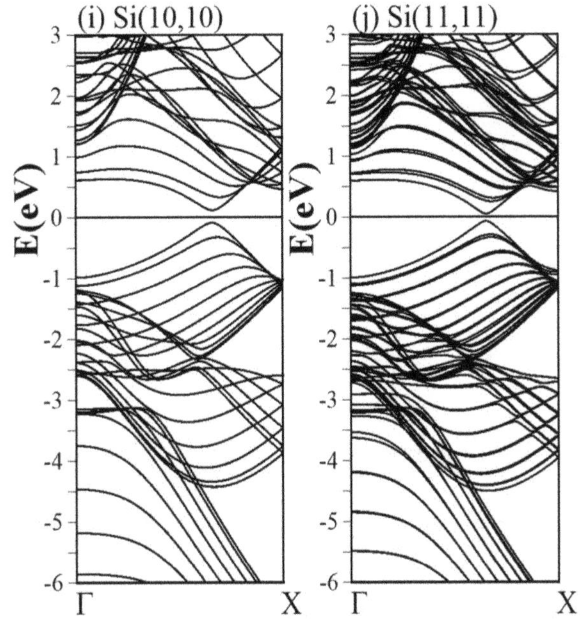

Figure 12.4(a)–(d) (e)–(h) (i)–(j). (Continued.)

Table 12.2. Energy gap E_g of aSiNTs.

aSiNTs	E_g(eV)
Si (2,2)	0.350
Si (3,3)	0.785
Si (4,4)	0.360
Si (5,5)	0.490
Si (6,6)	0.383
Si (7,7)	0.310
Si (8,8)	0.263
Si (9,9)	0.235
Si (10,10)	0.208
Si (11,11)	0.115

and Si (5,5) show that the HVB point is around the position of $k = \pm 2\pi/3a$ and the LCB point is at the X point ($k = \pm\pi/a$). However, the values for E_g show an irregularity for small values and radii of the SiNTs. These features could be due to the high buckling distances related to their neighboring atom and result in the high curvature energy of the geometric structure [68]. A high buckling distance appears in the gear-like structures and results in the energy band gap E_g [69]. There is a direct band gap with increasing chiral vector of Si (5,5) to Si (11,11). The anti-crossing dispersions take place between the

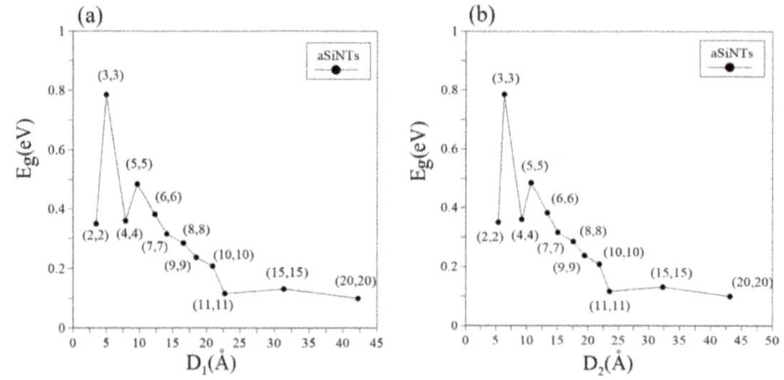

Figure 12.5(a)–(b). Energy gap E_g are illustrated as a function of diameter D_1 and D_2, respectively.

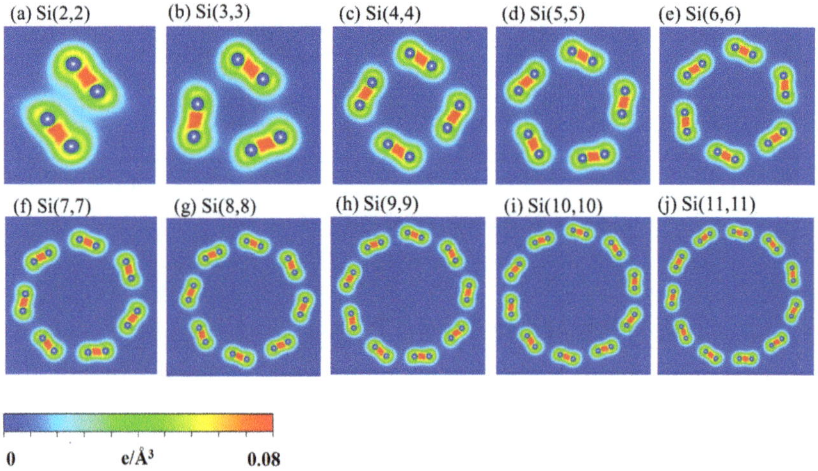

Figure 12.6(a)–(j). Charge density of armchair SiNTs from Si (2,2) to Si (11,11).

valence band and the conduction band located around the K point in the zone boundary of $k = \pm\pi/3a$. The E_g of Si (5,5) to Si (11,11) decreases, as the $1/r$ scaling law is applied with increasing chiral vectors and radii r, i.e., the large SiNTs exhibit less buckling and the geometry returns to one similar to the structure of CNTs. Some publications discuss silicon h-nanotubes (Si h-NTs) whose lattice is hexagonal and rolled up into nanotubes [48]. Si h-NTs can present metallic and semiconducting characteristics dependent on their size and chirality [48].

The charge density ρ is available for investigating the multi-/single-orbital hybridizations because of the various chemical bonds [70]. The results for the charge density ρ are illustrated in figure 12.6 with the $[x,y]$-plane projection calculated from first-principle calculations. The charge density ρ of aSiNTs represents clearly the chemical bonding as well as the charge transfer. In the pictures of figure 12.6(a)–(j), there are some red shapes similar to rectangles appearing between the silicon atoms. These are σ bonds. However, it is difficult to define whether they are the (3s, $3p_x$, $3p_y$) or (3s, $3p_x$, $3p_y$, $3p_z$) hybridization. In contrast to monolayer silicene, the results for the charge density ρ from aSiNTs have the similar intensity of the chemical bonding [71]. The armchair tubular geometric structure seems not to affect the distribution of the charge density. The results of bond length b_1 from Si (2,2) to Si (11,11) are consistent with this phenomenon. The bond length b_1 of Si (2,2) is 2.291 Å (see table 12.1), which is only 0.48 percent larger than the standard Si–Si bond length (2.28 Å). They then decrease to 2.281 Å when the bond length b_1 is Si (11,11).

The PDOSs as well as the 1D van Hove singularities of energy dependence of aSiNTs are clearly shown in figure 12.7(a)–(j), directly reflecting the primary features

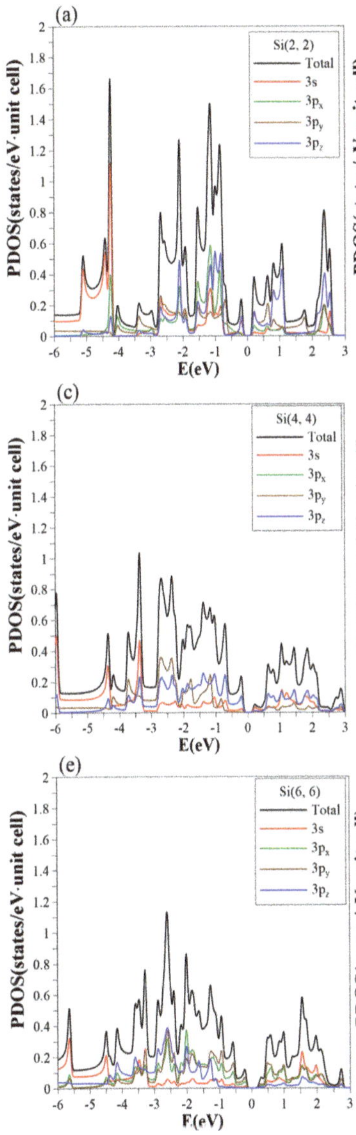

Figure 12.7(a) and (b) (c) and (d) (e) and (f) (g) and (h) (i) and (j). PDOSs of armchair silicon nanotubes Si (2,2) and Si (3,3), Si (4,4) and Si (5,5), Si (6,6) and Si (7,7), Si (8,8) and Si (9,9), Si (10,10) and Si (11,11).

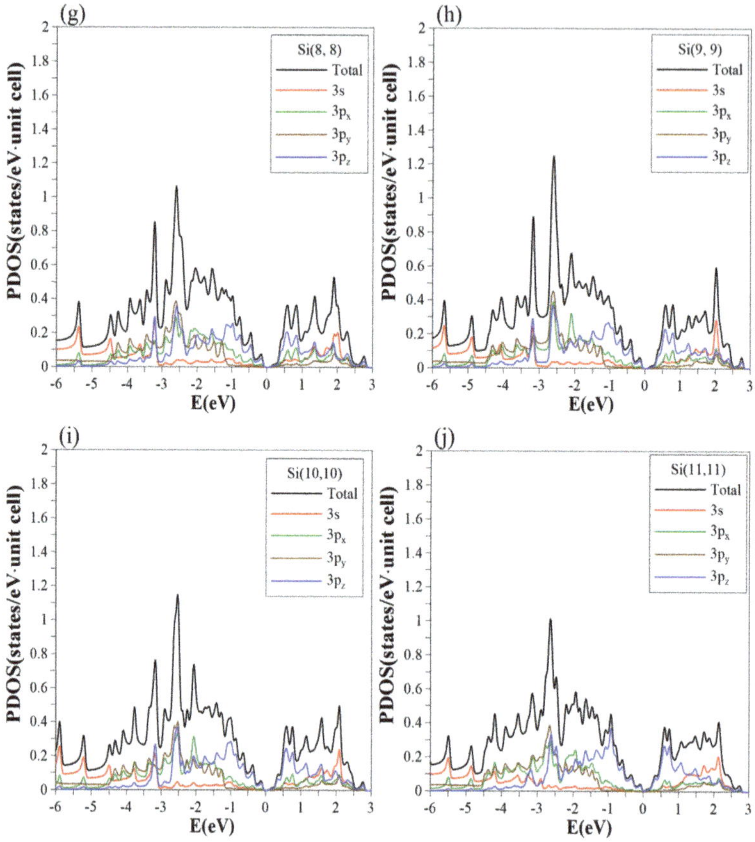

Figure 12.7(a) and (b) (c) and (d) (e) and (f) (g) and (h) (i) and (j). (Continued.)

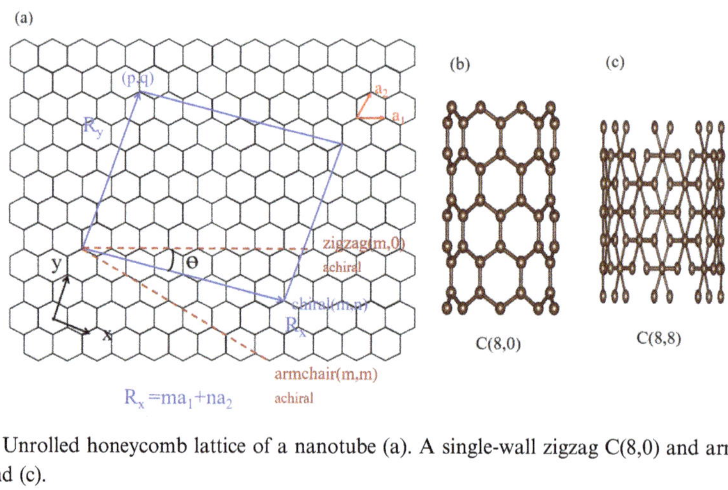

Figure 12.8. Unrolled honeycomb lattice of a nanotube (a). A single-wall zigzag C(8,0) and armchair C(8,8) CNTe (b) and (c).

of the electronic energy band structure. The partial orbital-PDOSs can be investigated fully to realize the orbital-decomposition contributions. For example, the results of 2D materials of monolayer silicene appear to be zero at $E_F = 0$. The contributions of the $3p_z$-orbital present a linear energy dependence at low energy, originating from the almost disappearing energy gap 0 eV and two symmetric logarithmic peaks at 1 and −1 eV, respectively, isotropic linear bands and the saddle points of π bands [71, 72]. Other features like a shoulder structure and a prominent peak correspond to the saddle point of the σ sub-bands at the M point. They appear in the middle energy range of the DOSs and are mainly contributed by $3p_x$ and $3p_y$ orbitals. In addition, $3p_z$-orbitals contribute partial (3s, $3p_x$, $3p_y$)-orbitals and yield a distinct peak contributed by the hybridization of the (3s, $3p_x$, $3p_y$, $3p_z$)-orbitals. The features of aSiNTs Si (2,2)-Si (11,11) in the DOSs are illustrated in figure 12.7(a)–(j). There are plenty of symmetric and asymmetric peaks that come from the parabolic and partially flat sub-bands. Interestingly, the DOSs of aSiNTs near the Fermi level E_F are located at $E = 0$. They have a zero value for semiconducting nanotubes. DOSs of armchair CNTs calculated by the tight binding model show mirror image peak between the valence and conduction regions [73]. However, the peaks of aSiNTs that occur between the valence and conduction regions are not symmetric, which could be a result of the buckling angles. Comparison of 1D DOSs for a single-wall SiNT and 2D DOSs for a silicene layer shows no independent orbitals contributed by the $3p_z$ orbitals of aSiNTs [71, 72]. The size of nanotubes reflects the strength of the curvature effect and further induces $3p_z$-orbitals hybridization. In the case of Si (2,2), $3p_z$ signals initially appear at ca. −0.2 eV near the Fermi level in the valence region. When chiral vectors increase to Si (3,3) and Si (4,4), the $3p_z$ peak in the valence region distributes at low energy increasingly. The reduced curvature effect leads the single $3p_z$-orbitals to rebuild a π bonding (from −1 eV to eV). Curved geometric structures influence $3p_z$-orbitals by an interaction of (3s, $3p_x$, $3p_y$)-orbitals and causes the (3s, $3p_x$, $3p_y$) and (3s, $3p_x$, $3p_y$, $3p_z$) hybridization, as well as sp^2 and sp^3 bonding. It is overall very difficult to distinguish the initial, middle (saddle-point structure) and final π/σ bands; thus, both armchair silicon nanotubes (figure 12.7(a)–(j)) are significantly different from each other. There are a lot of symmetric peaks, antisymmetric ones and shoulders over the entire energy range, mainly owing to the multi-orbital hybridizations, the oscillatory energy dispersions, and the frequent anticrossings.

12.2.2 Carbon nanotubes

The type of CNTs shown in figure 12.8(a) is characterized by two integers (m,n) and is formed by rolling a graphene sheet from the lattice vector $R_x = ma_1 + na_2$, where a_1 and a_2 are primitive lattice vectors of the 2D sheet. The radius r of a nanotube (m,n) behaves as $b\sqrt{3(m^2 + mn + n^2)}/2\pi$. The chiral angle θ, which is the angle between the vectors R_x and a_1, can characterize the non-chiral/chiral arrangements of carbon hexagons on a cylindrical surface. It is confined in the range $0^o \leqslant |\theta| \leqslant 30^o$ for all carbon nanotubes. Hence, the chiral angle θ yields an expression for $\tan\theta$, which is $\tan\theta = \sqrt{3}/(2m + n)$. Other types of carbon nanotubes are armchair and zigzag ones. They are both classified as achiral nanotubes. (m,m) and $(m,0)$ are shown in figure 12.8(a). The two dashed red lines are symbolized, respectively, the achiral armchair and zigzag systems with $\theta = 30^o$ and $\theta = 0^o$. In figure 12.8(b) and (c) are the rolled up honeycomb lattices of the zigzag and armchair CNTs, respectively. The number of carbon atoms N_u in a primitive unit cell is given by the equation $N_u = 4\sqrt{(m^2 + mn + n^2)(p^2 + pq + q^2)/3}$, where (p,q) is determined by the primitive lattice vector R_y along the nanotube axis. The lattice vector R_y can be described as $R_y = pa_1 + qa_2$, shown as a rectangle in figure 12.8(a). The axial wave vectors in the 1D first Brillouin zone are $| k_y | \leqslant \pi/| R_y |$.

Schematic single-wall armchair CNTs are shown in figure 12.9(a)–(j), which shows two types of the cross-section and side view for armchair C (3,3) to C (12,12) nanotubes, respectively. Compared with SiNTs (figure 12.1), no gear-like shapes are observed in the cross-section schema. The geometric parameters of the armchair carbon nanotubes (aCNTs) are summarized in table 12.3. The chiralities (m,m) are in the range of 3 to 12. After moderate adjustments of the structure, the optimal lattice constant and relatively lowest ground state energy E_0 are obtained, which are summarized in table 12.3. The ground state energy E_0 decreases with an increasing diameter of the aCNTs as shown in figure 12.10(a). The smallest single-wall CNTs have been found experimentally by the method of HR-TEM to be 4 Å in diameter. Possible geometric structures can be armchair (3,3), zigzag (5,0) and chiral (4,2) [66]. Besides, in a multiwalled CNT stable 3 Å carbon nanotubes can be grown [67]. In this chapter the diameter of the C (3,3) nanotube is calculated as 4.141 Å. This shows high precision compared with the experimental results. The other diameter results for C (4,4) to C (12,12) nanotubes are 5.532 Å to 16.352 Å, respectively. The total energy E_0 decreases when the CNT diameter increases. The higher total energy E_0 is observed at smaller radius and can be related to the curvature effects. The essential low-energy properties are dominated by the π bonding of the parallel $2p_z$ orbitals. The misorientation of the $2p_z$ orbitals and the sp^3 orbital hybridization on the cylindrical surface are not considered due to the curvature effects. The periodical boundary conditions play an important role in a determining metallic or semi-conducting behavior of the CNTs. When $2m + n = 3I$, CNTs are gapless metals. Linear bands of CNTs correspond to the Dirac-cone structure, or they can sample the Dirac point from graphene [8, 9]. Furthermore, for $2m + n = 3/I$, they are moderate-gap semiconductors. On the other hand, the full orbital hybridizations cause the non-armchair $2m + n = 3I$ CNTs to become a narrow-gap semiconductor except for $r < 2.5$ Å, and they create the metallic nanotubes with very small radii. Briefly, the metallic, moderate- and narrow-gap behavior of CNTs are characterized by the geometric structures: (I) $m = n$ or $r < 2.5$ Å, (e.g., (5,0) and (6,0) nanotubes) [68, 74] (II) $2m + n = 3/I$; ((7,0) and (8,0) nanotubes); (III) $2m + n = 3I$ and $m \neq n$ ((9,0) and (12,0) nanotubes). The three different energy gap types of CNTs have been verified by accurate STS measurements [75–77]. The (m,m) armchair and $(m,0)$ zigzag CNTs will be chosen for a model study to see the three types of electronic structures.

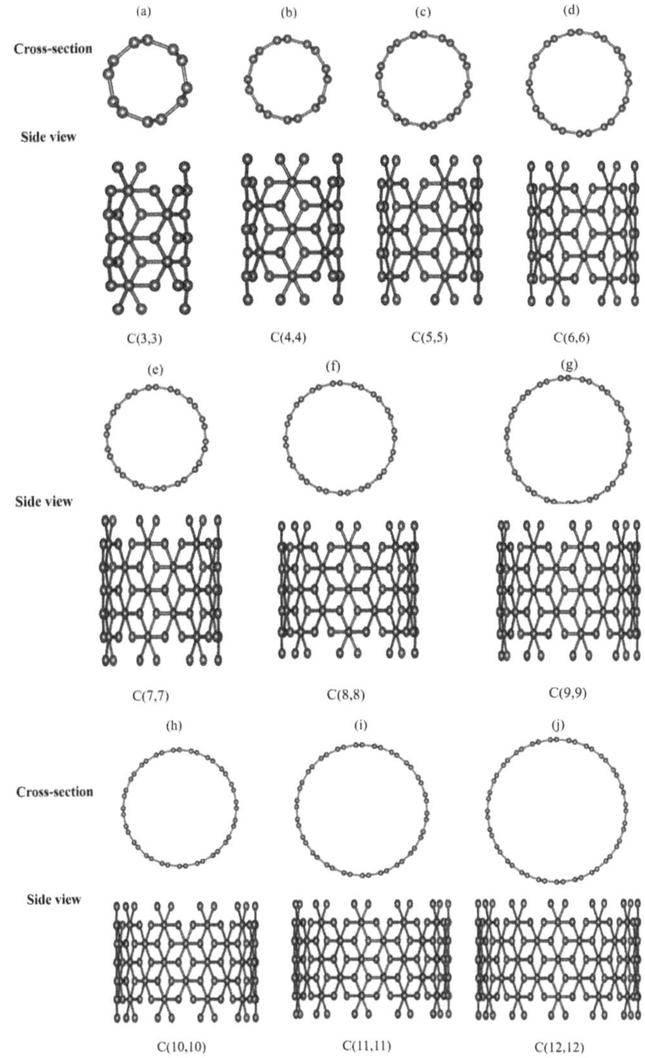

Figure 12.9(a)–(d) (e)–(g) (h)–(j). Geometric structure of aCNTs at side view and cross-section C (3,3), C (4,4), C (5,5) and C (6,6), C (7,7), C (8,8) and C (9,9), (10,10), C (11,11) and C (12,12).

The band structures of aCNTs with chiral vectors of (3,3), (4,4), (5,5), (6,6), (7,7), (8,8), (9,9), (10,10), (11,11) and (12,12) are shown in figure 12.11(a)–(j). They belong to 1D metals with a zero gap even in the presence of a significant curvature effect, and have k_x-dependent energy bands with strong dispersion relations, as shown in figure 12.11 for various armchair systems. Due to the cylindrical boundary condition, most of the energy bands are doubly degenerate, which is in agreement

Table 12.3. Parameters of armchair carbon nanotubes, ground state energy E_0, C–C bond length b_1 and b_2, diameter D; b_0 = 1.42 Å.

aCNTs	Lattice con. (Å)	E_o/unit cell (eV)	E_o/atom (eV)	b_1(Å)	b_2(Å)	b_1/b_0	b_2/b_0	D (Å)
C (3,3)	2.469	−105.832	−8.819	1.439	1.434	1.013 38	1.009 86	4.141
C (4,4)	2.466	−144.418	−9.026	1.431	1.429	1.007 75	1.006 34	5.532
C (5,5)	2.470	−182.419	−9.121	1.428	1.429	1.005 63	1.006 34	6.835
C (6,6)	2.468	−220.128	−9.172	1.428	1.427	1.005 63	1.004 93	8.222
C (7,7)	2.467	−257.677	−9.203	1.427	1.426	1.004 93	1.004 23	9.656
C (8,8)	2.468	−295.130	−9.223	1.426	1.426	1.004 23	1.004 23	11.206
C (9,9)	2.466	−332.501	−9.236	1.425	1.425	1.003 52	1.003 52	12.264
C (10,10)	2.468	−369.835	−9.246	1.425	1.426	1.003 52	1.004 23	13.863
C (11,11)	2.468	−407.138	−9.253	1.425	1.426	1.003 52	1.004 23	15.012
C (12,12)	2.469	−444.408	−9.259	1.425	1.426	1.003 52	1.004 23	16.352

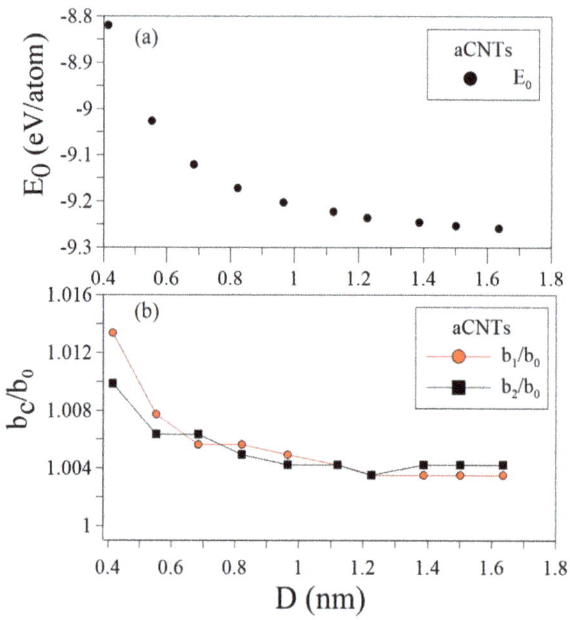

Figure 12.10. Ground state energy E_0 (a) and b_c/b_0 bond lengths (b) are illustrated as function of diameter D.

with the tight-binding model calculations. The band-edge states are located at $k_x = 0$ (Γ point) and/or near $k_x = 2/3$ as the unit of $\pi/|R_x|$. The energy dispersions relative to such critical points are in parabolic or linear form. For the larger armchair nanotubes with chiral vectors of (7,7) to (12,12) shown in figure 12.11(e)–(j), there are more low-lying energy bands initiating from the k_x = 2/3. Especially for the

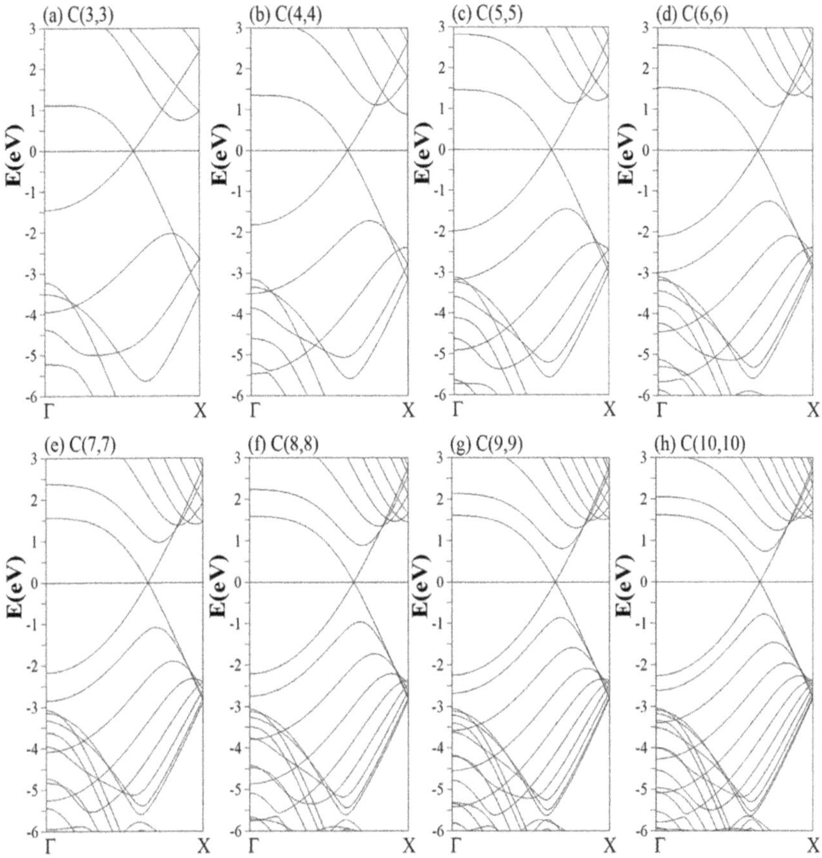

Figure 12.11 (a)–(d) (e)–(h) (i) and (j). Band structure of aCNTs C (3,3), C (4,4), C (5,5) and C (6,6), (7,7), C (8,8), C (9,9) and C (10,10), (11,11) and C (12,12).

first pair of valence and conduction bands, they are without the double degeneracy and linearly intersect at the Fermi level. These larger aCNTs indicate the 1D metallic behavior. The small radius of aCNT (3,3) situates the Fermi momentum around the middle. With increasing radius of the aCNTs, the Fermi momentum shifts to $k_x = 2/3$. Furthermore, large armchair carbon nanotubes, their Fermi velocity and slope of the linear band are close to that of monolayer graphene $3\gamma_0 b/2$, where γ_0 is the nearest-neighbor hopping integral of the $2p_z$ orbitals [78]. The effects of curvature on the Fermi-momentum states in the aCNTs can also be understood from the tight-binding model. The low-lying energy bands are dominated by the π bonding of the parallel $2p_z$ orbitals for a planar monolayer graphene. It presents an isotropic Dirac-cone structure near the K point, which is the corner of the hexagonal first Brillouin zone [71]. Since the honeycomb lattice has three uniform nearest-neighbor π bonding with the same hopping integral γ_0, the $2p_z$ orbitals perpendicular

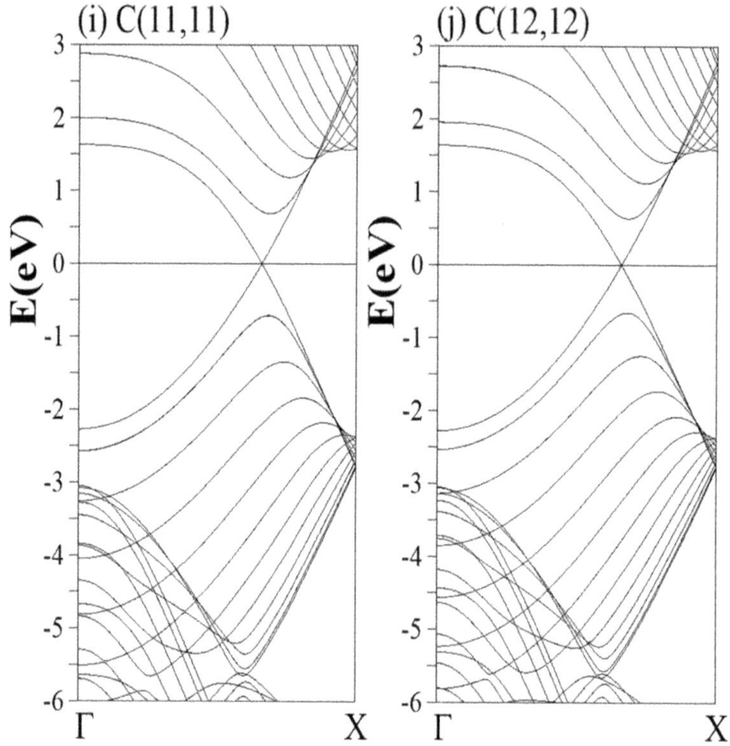

Figure 12.11 (a)–(d) (e)–(h) (i) and (j). (Continued.)

to a cylindrical surface are not parallel to one another. The angle between two neighboring $2p_z$ orbitals grows with the increasing curvature or with decreasing radius. The misorientation of the $2p_z$ orbitals can reduce the π bonding and create the σ bonding. The results of detailed calculations show that the hopping integral on the dimer line is different from the other two, depending on the arrangement of the three chemical bonds [56, 79]. The electronic states of the aCNTs could sample the K point of monolayer graphene under the periodical boundary condition, with modifications in the atomic interactions being consistent with the first-principles calculations [10, 11]. In small aCNTs, the sp^3 orbital hybridizations lead to significant contributions of the electronic states near E_F, while they do not completely alter the metallic behavior due to the linear bands.

Charge density ρ is available for investigating the multi-/single-orbital hybridizations because of the various chemical bonds [70]. Carbon atoms contribute four valence electrons to create two specific chemical bondings, namely the strong σ bonding of the (2s; 2px; 2py) orbitals and the π bonding of the $2p_z$ orbitals. However, the cylindrical carbon molecules exhibit another scenario which can be

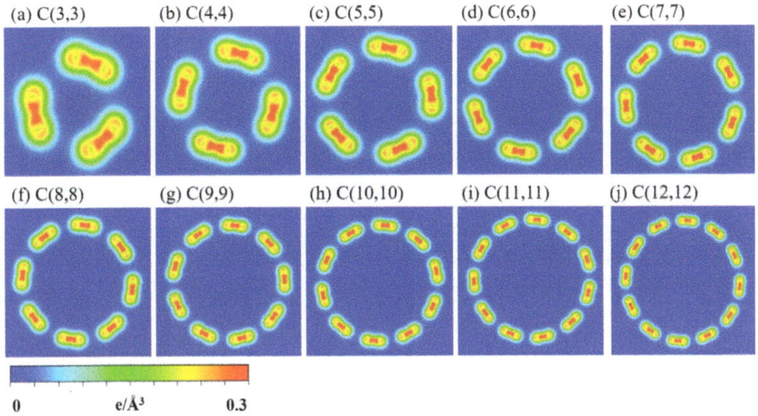

Figure 12.12. Charge density of armchair carbon nanotubes from C (2,2) to C (12,12).

described by the charge density method. The results of the charge density ρ for aCNTs are illustrated in figure 12.12 with the $[x,y]$-plane projection calculated from first-principle calculations. The charge density ρ of aCNTs represents clearly the chemical bonding, as well as the charge transfer. In the pictures of figure 12.12(a)–(j), there appear some red shapes similar to rectangles between the carbon atoms. These are σ bonds. However, it is difficult to define whether it is the (2s, $2p_x$, $2p_y$) or (2s, $2p_x$, $2p_y$, $2p_z$) hybridization. In contrast to monolayer graphene, the results for the charge density ρ from aCNTs have a similar intensity of chemical bonding [71]. The armchair tubular geometric structure seems not to effect the distribution of the charge density. The results of bond length b_1 from C (3,3) to C (12,12) are consistent with this phenomenon. The bond length b_1 of C (3,3) is 1.439 Å (see table 12.3), which is 1.33 percent larger than the standard C–C bond length (1.42 Å); it then decreases to 1.425 Å when the bond length b_1 is C (12,12).

The PDOSs as well as the 1D van Hove singularities of the orbital dependence of aCNTs are clearly shown in figure 12.13(a)–(j), directly reflecting the primary features of the electronic energy band structure. The partial orbital-PDOSs can be investigated fully to realize the orbital-decomposition contributions. Besides, they can provide useful information about the cooperative/competitive relations among the chiral angle, the periodical boundary condition, and the multi-orbital hybridizations. All aCNTs possess a finite DOS in the asymmetric plateau formed about the Fermi level. This directly reflects the metallic behavior of the 1D linear valence and conduction bands. At low-lying energy between −1 eV and 1 eV, the quasi mirror image spikes of optical transitions, i.e., $\nu_1 \rightarrow c_1$ and $\nu_2 \rightarrow c_2$, indicated for the case of C (11,11) are shown in figure 12.13(i). When the tube size is getting bigger, these mirror image spikes become more symmetric and the gap between $\nu_1 \rightarrow c_1$

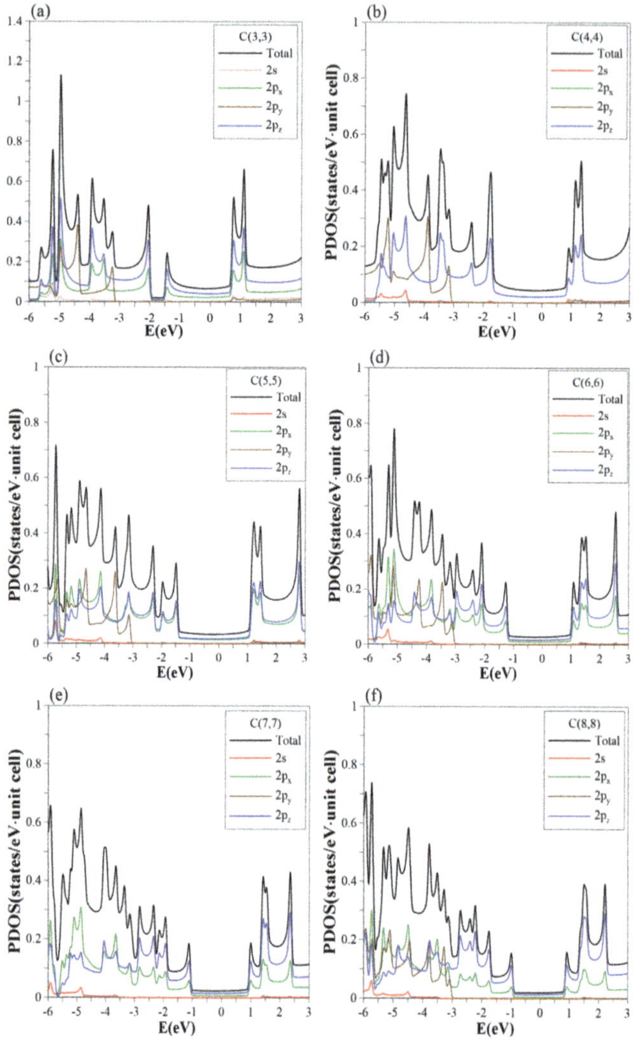

Figure 12.13(a) and (b) (c) and (d) (e) and (f) (g) and (h) (i) and (j). PDOSs of aCNTs C (3,3) and C (4,4), C (5,5) and C (6,6), C (7,7) and C (8,8), C (9,9) and C (10,10), C (11,11) and C (12,12).

tends to be narrow. These large aCNTs keep the plateau structure across E_F. For small size tubes like C (3,3) and C (4,4), the symmetric spikes are difficult to observe. A reason for this could be the curvature effect. Unlike graphene [71], the PDOSs of carbon nanotubes at low-lying energy range from −3 eV to 3 eV are contributed mostly by $2p_x$ and $2p_z$ orbitals. Contributions of 2s and $2p_y$ orbitals are observed below −3 eV.

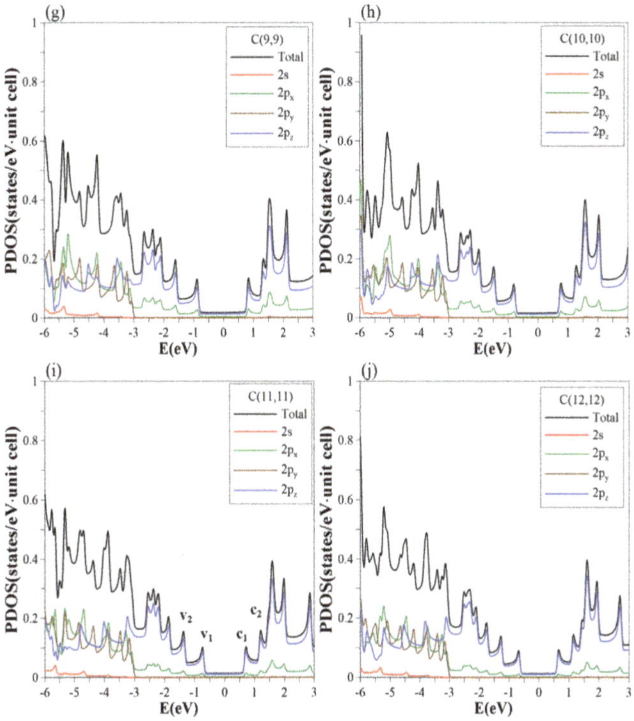

Figure 12.13(a) and (b) (c) and (d) (e) and (f) (g) and (h) (i) and (j). (Continued.)

References

[1] Iijima S 1991 Helical microtubules of graphitic carbon *Nature* **354** 56

[2] Journet C, Maser W K, Bernier P, Loiseau A, delaChapelle M L, Lefrant S, Deniard P, Lee R and Fischer J E 1997 Large-scale production of single-walled carbon nanotubes by the electric-arc technique *Nature* **388** 756

[3] Thess A *et al* 1996 Crystalline ropes of metallic carbon nanotubes *Science* **273** 483

[4] Yudasaka M, Komatsu T, Ichihashi T and Iijima S 1997 Single-wall carbon nanotube formation by laser ablation using double-targets of carbon and metal *Chem. Phys. Lett.* **278** 102

[5] Ebbesen T W and Ajayan P M 1992 Large-scale synthesis of carbon nanotubes *Nature* **358** 220

[6] Ebbesen T W, Hiura H, Fujita J, Ochiai Y, Matsui S and Tanigaki K 1993 Patterns in the bulk growth of carbon nanotubes *Chem. Phys. Lett.* **209** 83

[7] Seraphin S, Zhou D, Jiao J, Withers J C and Loutfy R 1993 Effect of processing conditions on the morphology and yield of carbon nanotubes *Carbon* **31** 685

[8] Saito R, Fujita M, Dresselhaus G and Dresselhaus M S 1992 Electronic structure of chiral graphene tubules *Appl. Phys. Lett.* **60** 2204

[9] Saito R, Fujita M, Dresselhaus G and Dresselhaus M S 1992 Electronic structure of graphene tubules based on C_{60} *Phys. Rev.* B **46** 1804

[10] Mintmire J, Dunlap B and White C 1992 Are fullerene tubules metallic? *Phys. Rev. Lett.* **68** 631

[11] Hamada N, Sawada S and Oshiyama A 1992 New one-dimensional conductors: graphitic microtubules *Phys. Rev. Lett.* **68** 1579

[12] Kane C L and Mele E 1997 Size, shape, and low energy electronic structure of carbon nanotubes *Phys. Rev. Lett.* **78** 1932

[13] Novoselov K S, Geim A K and Morozov S V 2004 Electric field effect in atomically thin carbon films *Science* **306** 666

[14] He L, Lu J Q and Jiang H Q 2009 Controlled carbon-nanotube junctions self-assembled from graphene nanoribbons *Small* **5** 2802

[15] Wang G X, Shen X P, Yao J and Park J S 2009 Graphene nanosheets for enhanced lithium storage in lithium ion batteries *Carbon* **47** 2049

[16] Lian P C, Zhu X F, Liang S Z, Li Z, Yang W S and Wang H H 2010 Large reversible capacity of high quality graphene sheets as an anode material for lithium-ion batteries *Electrochim. Acta* **55** 3909

[17] Wang G X, Wang B, Wang X L, Park J S and Dou S X 2009 Sn/graphene nanocomposite with 3D architecture for enhanced reversible lithium storage in lithium ion batteries *J. Mater. Chem.* **19** 8378

[18] Chen S Q, Chen P, Wu M, Pan D Y and Wang Y 2010 Graphene supported Sn-Sb@carbon core–shell particles as a superior anode for lithium ion batteries *Electrochem. Commun.* **12** 1302

[19] Chen S Q and Wang Y 2010 Microwave-assisted synthesis of a Co_3O_4-graphene sheeton-sheet nanocomposite as a superior anode material for Li-ion batteries *J. Mater. Chem.* **20** 9735

[20] Murugan A V, Muraliganth T and Manthiram A 2009 Rapid, facile microwavesolvothermal synthesis of graphene nanosheets and their polyaniline nanocomposites for energy strorage *Chem. Mater.* **21** 5004

[21] Gwon H, Kim H S, Lee K U, Seo D H, Park Y C, Lee Y S, Ahn B T and Kang K 2011 Flexible energy storage devices based on graphene paper *Energy Environ. Sci.* **4** 1277

[22] Tung V C, Chen L M, Allen M J, Wassei J K, Nelson K, Kaner R B and Yang Y 2009 Low-temperature solution processing of graphene-carbon nanotube hybrid materials for high-performance transparent conductors *Nano Lett.* **9** 1949

[23] Yu D S and Dai L M 2009 Self-assembled graphene/carbon nanotube hybrid films for supercapacitors *J. Phys. Chem. Lett.* **1** 467

[24] Fan Z, Yan J, Zhi L, Zhang Q, Wei T, Feng J, Zhang M, Qian W and Wei F 2010 A three-dimensional carbon nanotube/graphene sandwich and its application as electrode in super-capacitors *Adv. Mater.* **22** 3723

[25] Yoo E J, Kim J, Hosono E, Zhou H S, Kudo T and Honma I 2008 Large reversible Li storage of graphene nanosheet families for use in rechargeable lithium ion batteries *Nano Lett.* **8** 2277

[26] Su Q, Liang Y, Feng X and Mullen K 2010 Towards free-standing graphene/carbon nanotube composite films via acetylene-assisted thermolysis of organocobalt functionalized graphene sheets *Chem. Commun.* **46** 8279

[27] Chen S, Chen P and Wang Y 2011 Carbon nanotubes grown *in situ* on graphene nanosheets as superior anodes for Li-ion batteries *Nanoscale* **3** 4323

[28] Lin M F and Shung K W-K 1994 Plasmons and optical properties of carbon nanotubes *Phys. Rev.* B **50** 17744
[29] Pennington G and Goldsman N 2003 Semiclassical transport and phonon scattering of electrons in semiconducting carbon nanotubes *Phys. Rev.* B **68** 045426
[30] Duerkop T, Getty S, Cobas E and Fuhrer M 2004 Extraordinary mobility in semiconducting carbon nanotubes *Nano Lett.* **4** 35
[31] Schmidt O G and Eberl K 2001 Nanotechnology – thin solid films roll up into nanotubes *Nature* **410** 168
[32] Sha J, Niu J, Ma X, Xu J, Zhang X, Yang Q and Yang D 2002 Silicon nanotubes *Adv. Mater.* **14** 1219
[33] Jeong S Y, Kim J Y, Yang H D, Yoon B N, Choi S H, Kang H K, Yang C W and Lee Y H 2003 Synthesis of silicon nanotubes on porous alumina using molecular beam epitaxy *Adv. Mater.* **15** 1172
[34] Chen Y W, Tang Y H, Pei L Z and Guo C 2005 Self-assembled silicon nanotubes grown from silicon monoxide *Adv. Mater.* **17** 564
[35] Tang Y H, Pei L Z, Chen Y M and Guo C 2005 Self-assembled silicon nanotubes under supercritically hydrothermal conditions *Phys. Rev. Lett.* **95** 116102
[36] De Crescenzi M, Castrucci P, Scarcelli M, Diociauti M, Chaudhari P S, Balasubramanian C, Bhave T M and Bhoraskar S V 2005 Experimental imaging of silicon nanotubes *Appl. Phys. Lett.* **86** 231901
[37] Xie M, Wang J S, Fan Z Y, Lu J G and Yap Y K 2008 Growth of p-type Si nanotubes by catalytic plasma treatments *Nanotechnology* **19** 1
[38] Taghinejad M, Taghinejad H, Abdolahad M and Mohajerzadeh S 2013 A nickel-gold bilayer catalyst engineering technique for self-assembled growth of highly ordered silicon nanotubes (SiNT) *Nano Lett.* **13** 889
[39] Morata A, Pacios M, Gadea G, Flox C, Cadavid D, Cabot A and Tarancon A 2018 Large-area and adaptable electrospun silicon-based thermoelectric nanomaterials with high energy conversion efficiencies *Nat. Commun.* **9** 1
[40] Castrucci P, Scarselli M, De Crescenzi M, Diociaiuti M, Chaudhari P S, Balasubramanian C, Bhave T M and Bhoraskar S V 2006 Silicon nanotubes: synthesis and characterization *Thin Solid Films* **508** 226
[41] Ishai M B and Patolsk F 2009 Shape- and dimension-controlled single-crystalline silicon and SiGe nanotubes: toward nanofluidic FET devices *J. Am. Chem. Soc.* **131** 3679
[42] Yoo H *et al* 2018 Si nanocrystal-embedded SiO_x nanofoils: two-dimensional nanotechnology-enabled high performance Li storage *Materials Sci. Rep.* **8** 6904
[43] Kasavajjula U, Wang C S and Appleby A J 2007 Nano- and bulk-silicon-based insertion anodes for lithium-ion secondary cells *J. Power Sources* **163** 1003
[44] Chan C K, Peng H, Liu G, McIlwrath K, Zhang X F, Huggins R A and Cui Y 2008 High-performance lithium battery anodes using silicon nanowires *Nat. Nanotechnol.* **3** 31
[45] Zhu J, Yu Z, Burkhard G F, Hsu C M, Connor S T, Xu Y, Wang Q, McGehee M, Fan S and Cui Y 2009 Optical absorption enhancement in amorphous silicon nanowire and nanocone arrays *Nano Lett.* **9** 279
[46] Wakihara M and Yamamoto O 1998 *Lithium Ion Batteries: Fundamentals and Performance* (Weinheim: Wiley-VCH)
[47] Boukamp B A, Lesh G C and Huggins R A 1981 All-solid lithium electrodes with mixed-conductor matrix *Nano Lett.* **128** 725

[48] Chegel R and Behzad S 2013 Bandstructure modulation for Si-h and Si-g nanotubes in a transverse electric field: tight binding approach *Superlattices Microstruct.* **63** 79
[49] Ahmadi N, Shokri A A and Elahi S M 2016 Optical transition of zigzag silicon nanotubes under intrinsic curvature effect *Silicon* **8** 217
[50] Behzad S and Chegel R 2016 Magnetic field-induced splitting of optical spectra in silicon nanotubes: tight binding calculations *Silicon* **8** 43 2016
[51] Chegel R and Behzad S 2014 Electronic properties of SiNTs under external electric and magnetic fields using the tight-binding method *J. Electron. Mater.* **43** 329
[52] Zhang R Q, Lee H L, Li W K and Teo B K 2005 Investigation of possible structures of silicon nanotubes via density-functional tight-binding molecular dynamics simulations and *ab Initio* calculations *J. Phys. Chem.* B **109** 8605
[53] Seifert G, Koehler T, Urbassek H M, Hernandez E and Frauenheim T 2001 Tubular structures of silicon *Phys. Rev.* B **63** 193409
[54] Yang X and Ni J 2005 Electronic properties of single-walled silicon nanotubes compared to carbon nanotubes *Phys. Rev.* B **72** 195426
[55] Fagan S B, Baierle R J, Mota R, da Silva A J R and Fazzio A 2000 *Ab initio* calculations for a hypothetical material: silicon nanotubes *Phys. Rev.* B **61** 9994
[56] Lin M F and Chuu D S 1998 Persistent currents in toroidal carbon nanotubes *Phys. Rev.* B **57** 6731
[57] Steele G A, Pei F, Laird E A, Jol J M, Meerwaldt H B and Kouwenhoven L P 2013 Large spin–orbit coupling in carbon nanotubes *Nat. Commun.* **4** 1573
[58] Kuemmeth F, Ilani S, Ralph D and McEuen P 2008 Coupling of spin and orbital motion of electrons in carbon nanotubes *Nature* **452** 448
[59] Churchill H O H, Kuemmeth F, Harlow J W, Bestwick A J, Rashba E I, Flensberg K, Stwertka C H, Taychatanapat T, Watson S K and Marcus C M 2009 Relaxation and dephasing in a two-electron C-13 nanotube double quantum dot *Phys. Rev. Lett.* **102** 166802
[60] Jhang S H, Marganska M, Skourski Y, Preusche D, Witkamp B, Grifoni M, van der Zant H, Wosnitza J and Strunk 2010 Spin–orbit interaction in chiral carbon nanotubes probed in pulsed magnetic fields *Phys. Rev.* B **82** 041404
[61] Jespersen T S, Grove-Rasmussen K, Paaske J, Muraki K, Fujisawa T, Nygard J and Flensberg K 2011 Gate-dependent spin–orbit coupling in multielectron carbon nanotubes *Nat. Phys.* **7** 348
[62] Kresse G and Furthmueller J 1996 Efficient iterative schemes for *ab initio* total-energy calculations using a plane-wave basis set *Phys. Rev.* B **54** 11169
[63] Kresse G and Joubert D 1999 From ultrasoft pseudopotentials to the projector augmented-wave method *Phys. Rev.* B **59** 1758
[64] Perdew J P, Burke K and Ernzerhof M 1996 Generalized gradient approximation made simple *Phys. Rev. Lett.* **77** 3865
[65] Grimme S 2006 Semiempirical GGA-type density functional constructed with a longrange dispersion correction *J. Comput. Chem.* **27** 1787
[66] Guan L, Suenaga K and Iijima S 2008 Smallest carbon nanotube assigned with atomic resolution accuracy *Nano Lett.* **8** 459
[67] Zhao X, Liu Y, Inoue S, Suzuki T, Jones R O and Ando Y 2004 Smallest carbon nanotube Is 3 Å in diameter *Phys. Rev. Lett.* **92** 125502
[68] Guelseren O, Yildirim T and Ciraci S 2002 Systematic *ab initio* study of curvature effects in carbon nanotubes *Phys. Rev.* B **65** 153405

[69] Mintmire J W and White C T 1998 Universal density of states for carbon nanotubes *Phys. Rev. Lett.* **81** 2506
[70] Jacobsen H 2009 Chemical bonding in view of electron charge density and kinetic energy density descriptors *J. Comput. Chem.* **30** 1903
[71] Liu H Y, Lin S Y and Wu J Y 2020 Stacking-configuration-enriched essential properties of bilayer graphenes and silicenes *J. Chem. Phys.* **153** 154707
[72] Lin S Y, Chang S L, Thuy Tran N T, Yang P H and Lin M F 2015 H-Si bonding-induced unusual electronic properties of silicene: a method to identify hydrogen concentration *Phys. Chem. Chem. Phys.* **17** 26443
[73] Rao A M, Richter E, Bandow S, Chase B, Eklund P C and Dresselhaus M S *et al* 1997 Diameter-selective Raman scattering from vibrational modes in carbon nanotubes *Science* **275** 187
[74] Blase X, Benedict L X, Shirley E L and Louie S G 1994 Hybridization effects and metallicity in small radius carbon nanotubes *Phys. Rev. Lett.* **72** 1878
[75] Ouyang M, Huang J L, Cheung C L and Lieber C M 2001 Energy gaps in 'metallic' single-walled carbon nanotubes *Science* **292** 702
[76] Wilder J W, Venema L C, Rinzler A G, Smalley R E and Dekker C 1998 Electronic structure of atomically resolved carbon nanotubes *Nature* **391** 59
[77] Odom T W, Huang J-L, Kim P and Lieber C M 1998 Atomic structure and electronic properties of single-walled carbon nanotubes *Nature* **391** 62
[78] Wallace P R 1947 The band theory of graphite *Phys. Rev.* **71** 622
[79] Que Y, Xiao W, Chen H, Wang D, Du S and Gao H-J 2015 Stacking-dependent electronic property of trilayer graphene epitaxially grown on Ru (0001) *Appl. Phys. Lett.* **107** 263101

Chapter 13

Electronic and optical properties of lithium ion-based battery anode: lithium titanium oxide ternary $Li_4Ti_5O_{12}$

Thi Dieu Hien Nguyen, Shih-Yang Lin, Hsien-Ching Chung, Wei-Bang Li, Ngoc Thanh Thuy Tran, Hsin-Yi Liu, Hai Duong Pham and Ming-Fa Lin

This chapter has been adapted from [1] with permission of Elsevier.

Lithium-ion batteries (LIBs) attract much research interest because of their broad applications such as electric vehicles (EVs) [2, 3], hybrid EVs [2, 3], mobile devices [4], and also in the military and aerospace [5]. LIBs exhibit high performance energy resources, e.g., high power, energy densities, and high reliability, accompanied with affordability, long cycle life, and friendliness to the environment. Commonly, LIBs include three components, which are a cathode (positive electrode), an anode (negative electrode) and an electrolyte. In particular, the electrolyte plays an important role in the ionic transport of positive lithium ions (Li^+) between the electrodes [2, 3, 6]. Recent experimental progress shows that the anode can be formed by graphite and lithium titanium oxide. Graphite-based anode material is popular due to its high capacity and cheap price. However, this material faces a huge reduction of the volume expansion during many rapid charging and discharging processes [7]. Lithium titanium oxide serves as the negative electrode, and is known as a zero-strain material, thus it can overcome the disadvantages of graphitic materials, e.g., the drastic volume change. Besides, $Li_4Ti_5O_{12}$ has many merits, for example, a long life, rapid charging, high input/output power performance, excellent low-temperature operation, and a wide effective state of charge range. Recently, the charge and discharge process has presented a discharge capacity of 320 mAh g^{-1} and the reversible one of about 230 mAh g^{-1} in the range of 0.01 V to 2 V [8]. Moreover, lithium titanium oxide has a high reaction potential plateau at 1.55 V versus Li/Li^+,

doi:10.1088/978-0-7503-4685-6ch13

which will prevent the occurrence of lithium dendrites and whiskers. Therefore, this enhances the safety issue of lithium-ion batteries [9–11].

For theoretical study, first-principles analysis, which is based on Vienna *ab initio* simulation package (VASP) calculations, has proposed crucial mechanisms to thoroughly comprehend the geometric, electronic, and optical properties [12, 13]. The significant multi- or single-orbital hybridizations in diverse chemical bonds are thoroughly examined by optimal lattice symmetry, atom-dominated band structures, spatial charge densities, and atom- and orbital-decomposed density of states (DOS). Other emergent materials can use this developed framework in understanding their fundamental characteristics, which can be verified further by high-resolution experimental examination. Therefore, this systematic theoretical study can be reliable and suitable for studying the diversified geometric and electronic properties of cathode [7, 14, 15], anode [16, 17] and electrolyte materials [18, 19] of Li^+-ion based batteries. Also, linking between the geometric, electronic, and optical properties with essential properties is still absent. In this chapter, the useful connection of the above features under orbital-multi hybridization is studied in detail. In addition, the direct combination of numerical analysis (VASP calculations) with phenomenological models (generalized tight-binding models) [20] would be very useful to thoroughly comprehend the unusual and diverse properties.

13.1 An optimal crystal

The geometric structures, electronic and optical properties of the ternary compound $Li_4Ti_5O_{12}$ are analysed in the framework of density functional theory (DFT) [21], using the VASP [21]. The electron–electron Coulomb interactions are calculated by the many-particle exchange and correlation energies, constructed from the Perdew–Burke–Ernzerhof (PBE) functional under the generalized gradient approximation (GGA) [22]. Moreover, the electron–ion interactions are characterized by the projector-augmented wave (PAW) [23, 24] pseudopotentials. Plane waves with a kinetic energy cutoff of 520 eV are chosen as a complete set [25], being reliable and efficient for evaluating Bloch wave functions and electronic energy spectra. The first Brillouin zone is sampled $12 \times 12 \times 12$ k-point meshes within the Monkhorst–Pack scheme for electronic structures and optical properties. The convergence of the ground-state energy is 10^{-6} eV between two consecutive simulation steps, and the maximum Hellmann–Feynman force acting on each atom is less than 0.01 eV $Å^{-1}$ during the ionic relaxations.

$Li_4Ti_5O_{12}$ ternary compound presents an unusual geometric structure, which classifies into the triclinic symmetry, shown in figure 13.1(a) (where (Li, Ti, O) atoms are represented by the green, pink, and orange balls, respectively). A primitive unit cell has 42 atoms, including 8-Li, 10-Ti and 24-O atoms. The lattice constants are $a = 5.288$ Å, $b = 9.532$ Å and $c = 9.932$ Å, and the tilted angles of $\alpha = 73.122$, $\beta = 78.042$, and $\gamma = 78.913$ about x, y and z, respectively. This system belongs to Hermann Mauguin (P1), Hall (-P1), and point group (1) space group. The optimal geometry presents the LiO_6, TiO_4 and TiO_6 chemical bonds, indicating a highly non-uniform chemical environment. This 3D material exhibits as highly anisotropic. In our calculation, the

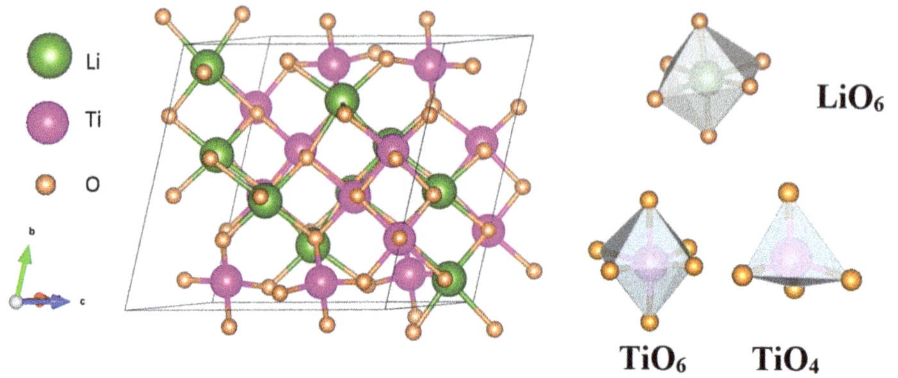

Figure 13.1. The geometry structure of $Li_4Ti_5O_{12}$ with triclinic symmetric, with the primitive unit cell consists of 42 atoms.

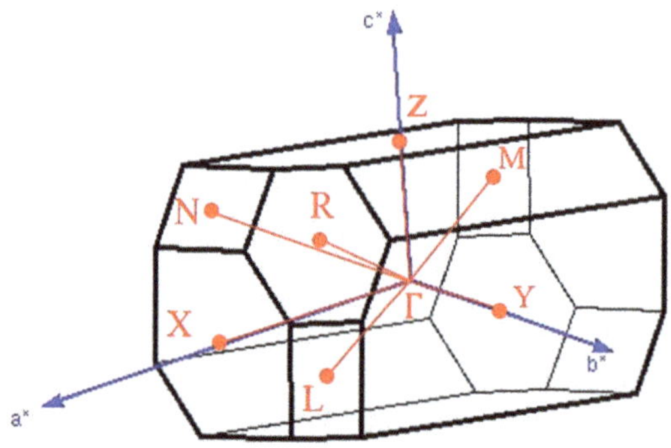

Figure 13.2. The first Brillouin zone with high-symmetric points.

band structure is achieved along the paths of high-symmetry points in the reciprocal lattice: X-Γ-Y| L-Γ-Z| N-Γ-M| R-Γ (presented in figure 13.2).

The fundamental properties can be determined by the unique chemical bonds, including Li–O and Ti–O bonds of the range 1.988–2.479 Å and 1.757–2.210 Å, respectively. Each Li, Ti atom is surrounded by six O atoms; Ti also has an extra case with connection with four O. This rich and unique phenomenon will be directly reflected in the distribution width of the spatial charge density (detailed in figure 13.4). Apparently, both Li–O and Ti–O bonds possess sufficiently large modulations (greater than 10%), clearly indicating non-uniform bonding lengths. This property plays a substantial role in easy intercalation of the Li^+ ions, which can enhance the efficiency of the Li-ion batteries. Furthermore, all the chemical bonds are originated from the multi-orbital hybridizations.

13.2 Orbital-hybridization-enriched electronic properties

$Li_4Ti_5O_{12}$ presents an unusual band structure, which is varied with the wave vector (figure 13.3(a)). The whole structure is highly anisotropic, where the occupied electron energy state and unoccupied hole state are asymmetric about the Fermi

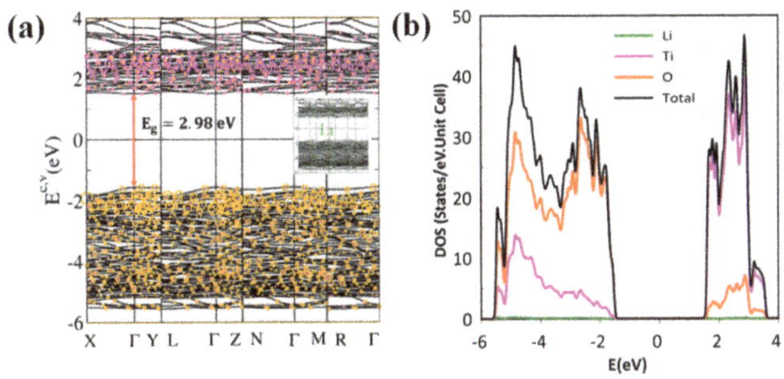

Figure 13.3. (a) The band structure with atom dominances (Li, Ti, O atoms corresponding to green, pink, and orange open circles, respectively) with (b) atom-project DOS.

level ($E_F = 0$). This is due to the different multi-orbital hybridizations in the non-uniform chemical bonds. At the same wave vector, an energy gap of $E_g \sim 2.98$ eV identified from the highest occupied valence-band state and the lowest unoccupied conduction-band state (red perpendicular arrow in figure 13.3(a)); leading to a direct-gap semiconductor. Comparing with some literature studies, this band gap is close to the theoretical calculations, which are 3.015 eV [26] and 2.52 eV [27], and to the experimental value of 2.95 eV [28]. From an experiment point of view, we can use the optical spectroscopy to test the theoretical prediction since this band gap is equivalent to the optical threshold absorption frequency. In this system, the spin-dependent interactions will be negligible because E_g is too wide to generate the spin-split electronic states across the Fermi level. The band structure consists of many valence and conduction subbands in the energy range -6.0 eV $< E^{v,c} <$ 4.0 eV with weak dispersions, due to many atoms and outer orbitals in a large unit cell. These diversified energy subbands exhibit non-crossing, crossing and anti-crossing behaviors, where the latter phenomenon will appear when two neighboring subbands possess comparable independent components within the anti-crossing region.

Rich electronic structures and chemical bonds can be identified through the atom dominances, which present in the valence- and conduction-band ranges of -5.50 eV $\leqslant E^v \leqslant -1.49$ eV and 1.49 eV $\leqslant E^c \leqslant$ 4.0 eV. Li atom distributions (green circles) are small but significant, revealed in the whole range of valence and conduction subbands (figure 13.3(a)). This phenomenon is consistent with the atom-decomposed density of state (DOS) in figure 13.3(b). Therefore, the essential properties should be associated with the Li–O bonds, where a single-orbital contribution of 2s is given by Li. The whole band structure can show the different contributions such as (Ti–O) co-dominance, Ti dominance, O dominance. Apparently, the unoccupied electronic states are dominated by titanium atoms, while oxygen contributions are prominent in occupied states. This phenomenon comes into the simultaneous existence of many O-related chemical bonds (100), with deeper (2s, $2p_x$, $2p_y$, $2p_z$)-orbital energies. Therefore, the partial critical multi-orbital hybridizations can be verified through atom dominance with the energy band revealing further absorption structures.

By the accuracy atom- and orbital-decomposed van Hove singularities, the orbital hybridizations can be examined further through the spatial charge distributions on different chemical bonds (figure 13.4(a)–(l)), e.g. Li–O and Ti–O bonds. This is directly related to the electron orbitals of each atom. From the shortest Li–O bonds (1.988 Å in figure 13.4(d)), the effective distribution range caused by only the 2s-orbital provides a dilute charge density around the lithium atom. The measured value is 0.52 Å from the deep blue region of the Li^+-ion core to the light region of the extended 2s-orbital. In this case, critical orbital hybridizations in the Li–O bonds are independent from the 1s orbitals. It is easy to observe the overlap of distinct orbital charges exists in the range of 1.20–1.60 Å from the O-core. Other significant O-orbitals corresponding to ($2p_x$, $2p_y$, $2p_z$) orbitals are shown by the light green and yellow colors. The O-2s orbitals can be negligible in the chemical bond contributions because they are far away from that of the lithium atom. When Li–O bond length increases, e.g. 2.479 Å and 2.249 Å in figure 13.4(e) and (f), respectively, the charge overlap behavior between Li and O atoms is noticeably reduced. Furthermore, the multi-orbital hybridizations of 2s–($2p_x$, $2p_y$, $2p_z$) in Li–O bonds exhibit various hopping integrals. Similarly to the Li–O bonds, Ti–O can be more effective by Ti atom than O atom due to the large atomic number. The charge densities could be classified into dark red, light red, and yellow-green regions, which, respectively, correspond to (3s, $3p_x$, $3p_y$, $3p_z$), 4s, and ($3d_{x^2-y^2}$, $3d_{xy}$, $3d_{yz}$, $3d_{xz}$, $3d_{z^2}$). The significant multi-orbital hybridizations in the Ti–O bonds come into existence from the two latter orbitals. The deformed carrier distributions between the Ti and O atoms, exhibited in the distinct bond length cases, clearly indicate the various charge transfers from the former to the latter. These analyses could be corresponding to many strong absorption peaks in further investigations.

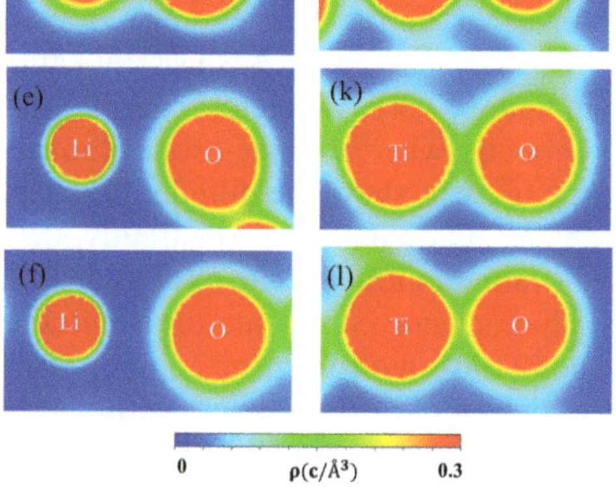

Figure 13.4. Spatial charge density distributions of (a)/(b)/(c) isolated atoms under the longest/middle/shortest Li–O for (d)/(e)/(f) and Ti–O for (i)/(k)/(l), respectively.

The atom- and orbital-projected DOS (in figure 13.3(b) and 13.6(b)–(d)) that is essentially calculated by the number of electronic states in a small range *dE*, can completely provide a full understanding about the multi-orbital hybridizations on the Li–O and Ti–O chemical bonds. In the experimental measurement, STS is one powerful method to investigate the significant characteristic of DOS (consider further). In the values from −1.49 to 1.49 eV, DOS is uncounted because this range corresponds to the large bandgap of the ternary compounds, e.g., $E^i_g = 2.98$ eV (figure 13.4(b)–(d)). DOS presents as remarkably asymmetric just above and below the Fermi level. The conduction and valence spectra exist in many specific forms such as shoulders and asymmetric/symmetric peaks in about 10 eV. These come from a lot of oscillatory energy subbands, with frequent crossing or anti-crossing in the compound. Also, the existence of the dense subbands might form the merged band-edge states which can create special structure. The unusual structures, van Hove singularities, originate from the zero-group velocity band-edge states of energy sub-bands. These critical points can classify into extrema (including local maximum and minimum), saddle points, and dispersionless points in the energy wave-vector subbands. They locate in the high symmetric points or somewhere along the wave vector energies. Very quirkily, the various van Hove singularities might depend on the different dimensions. For example, in 0D discrete energy levels, the delta-function-like peaks are prominent, the square-root asymmetric peaks/the plateau structures occur in the 1D parabolic/linear energy

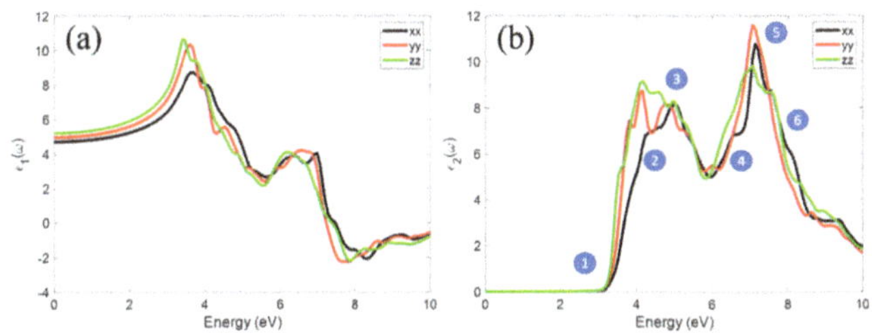

Figure 13.5. The complex dielectric functions, in which (a) the real part and (b) the imaginary part are along *xx*, *yy*, *zz* directions represented by black, red and green curves, respectively.

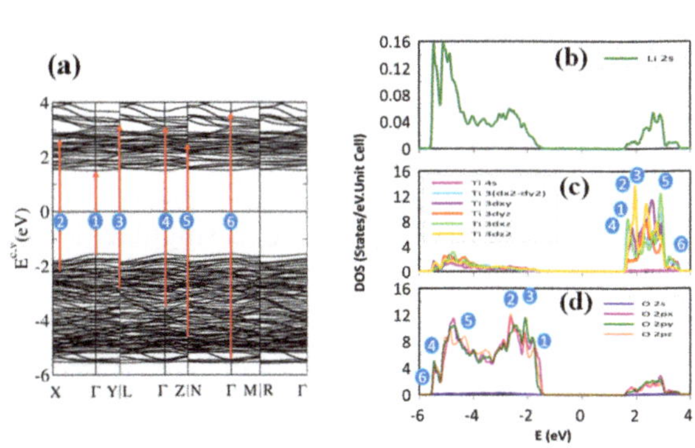

Figure 13.6. (a) Inter-transition with band-edge states and the orbital-project DOS of (b) Li (c) Ti, and (d) O associated with specific orbital interactions.

subbands, and the shoulder structures/V-shapes/logarithmic peaks/square-root asymmetric peaks belong to the 2D parabolic/Dirac-cone/saddle-point/constant-energy-loop band structures. In figure 13.3(b), the contribution lithium (green curve) is lowest, but very important in the whole occupied and unoccupied spectrum range. 1s core orbitals do not take part in DOS contribution. Consequently, the titanium and oxygen atoms' DOS play a dominating role in the energy range of -5.50 eV $\leqslant E^v \leqslant 1.49$ eV and 1.49 eV $\leqslant E^c \leqslant 4.0$ eV (pink and orange curves), respectively. As we discussed before in atom dominances, DOSs in figure 13.6(b) are totally in agreement with each other. These ranges are corresponding to the orbital bondings of 2s-(3s, $3p_x$, $3p_y$, $3p_z$) and 4s-($3d_{x^2-y^2}$, $3d_{xy}$, $3d_{yz}$, $3d_{xz}$, $3d_{z^2}$). These significant orbital identifications can be useful further in the strong optical responses (discussed later).

The orbital-projected van Hove singularities can be revealed through figure 13.6(b)–(d), and determine the important multi-orbital hybridizations in the Li–O and Ti–O. Apparently, the main orbital contributions create various van Hove singularities due to their number, energies, intensities. There are three groups, which correspond to lithium, titanium, and oxygen atoms, including (I) Li-2*s* orbital (green curve in figure 13.6(b)), (II) Ti-(4*s*, $3d_{x^2-y^2}$, $3d_{xy}$, $3d_{yz}$, $3d_{zx}$, $3d_{z^2}$) orbitals (pink, light blue, purple, red, light green, and yellow curves in figure 13.6(c)), and (III) O-(2*s*, $2p_x$, $2p_y$, $2p_z$) orbitals (violet, pink, green, and orange curves in figure 13.6(d)). Only the single Li-2s and three O-($2p_x$, $2p_y$, $2p_z$) contribute to the Li–O bonds, which is displayed by the DOS of these above atoms. Among the current orbitals, O-2s exhibits low DOS intensity, thus it can be negligible in the energy range (-6.0 eV $\leqslant E \leqslant 4.0$ eV). The essential reason comes from being fully filled in the occupied and unoccupied states associated with spin-up and spin-down configurations. In general, the rich chemical bonds Li–O is attributed to the complicated multi-orbitals 2*s*-($2p_x$, $2p_y$, $2p_z$), which imply multi-orbital hopping integrals of on-site Coulomb potentials.

Very interestingly, the oxygen atoms ($2p_x$, $2p_y$, $2p_z$) also make a significant hybridization with the transition-metal titanium atoms (4*s*, $3d_{x^2-y^2}$, $3d_{xy}$, $3d_{yz}$, $3d_{zx}$, $3d_{z^2}$) (figure 13.6(c)) because their van Hove singularities come into existence with each other. Oxygen atoms contribute higher intensity in the valence energy, while the opposite is true for titanium atoms. The orbital-dependent contributions are consistent in the energy spectrum except for the small 4*s* orbital density of state with the conduction-band range smaller than the valence one (pink curve). In summary, the 3D ternary compound consisting of the Li–O and Ti–O chemical bonds is created from multi-orbital hybridizations of 2*s*-($2p_x$, $2p_y$, $2p_z$), (4*s*, $3d_{x^2-y^2}$, $3d_{xy}$, $3d_{yz}$, $3d_{zx}$, $3d_{z^2}$)-($2p_x$, $2p_y$, $2p_z$), respectively.

As mentioned above, the high-resolution STS technique [29–31] provides an important measurement in identifying van Hove singularities of DOS due to the band-edge states from the tunneling current–voltage characteristic. Principally, the experimental measurement records *I*–*V* curves under a weak tunneling quantum current. The differential conductance $(dI/dV)/(I/V)$ is useful to achieve spatial distribution of DOS under current imaging tunneling spectroscopy (CITS). This technique is successful in comprehending the electronic properties of graphene-related systems, including a nearly symmetric V-shape structure vanishing at the

Fermi level for a monolayer system (a zero-gap semiconductor) [32], a gate-voltage-induced band gap in bilayer AB and tri-layer ABC stacking [33, 34], a prominent peak centered around the Fermi level due to partially flat bands (surface states) in tri-layer and penta-layer ABC stacking [35], a sharp dip structure near E_F combined with a pair of asymmetric peaks in tri-layer AAB stacking (a narrow-gap semiconductor with low-lying constant-energy loops) [36]. The experimental examinations, which directly apply to the 3D ternary $Li_4Ti_5O_{12}$ could confirm the theoretical calculations with van Hove singularities. This can lead to full understanding of the complicated multi-orbital hybridizations of Li–O and Ti–O.

Also, further combination with the phenomenological model by the reliable hopping integrals is worth examining. However, this method might face a difficulty due to the complicated structures of 3D $Li_4Ti_5O_{12}$ compound. For example, the primitive cell consists of a large moiré supperlattice (42 atoms) and a big number of chemical bonds (48 Li–O and 52 Ti–O bonds in figure 13.1). Besides, the band structures possess complicated subbands in various forms, e.g., weak dispersions, parabolic, associated with the band-edge states. The large bandgap $E_g \sim 2.98$ eV (figure 13.4(a)), nonhomogeneous spatial charge densities in diverse chemical bonds (figure 13.4(a)–(l)), and complicated multi-orbital hybridizations in van Hove singularities (figure 13.6(b)–(d)). As a result, due to the nonuniform hopping integral with diverse orbital hybridizations, the parameterized tight binding model might be difficult to involve in first-principles calculations. However, in the other Li^+-based anode [7, 16], cathode [3, 16], and electrolyte materials [2, 3], this combination can become a further investigation to understand the significant characteristic of the materials.

13.3 Diverse absorption phenomena

Significant couplings of electrons and photons are caused by the electromagnetic fields propagating in the condensed matter system. As a result, electrons are vertically excited from the occupied states to the unoccupied ones with the same wave vector because photons can only provide sufficient energies under the long wavelength limit. The valence electrons due to a specific orbital will effectively screen this perturbation, create the induced current density, and thus build the electric polarization. The transverse dielectric functions, which are determined by the specific ratio of the electric field before and after screening under the momentum and frequency space, are efficiently described by the vertical excitations pictures. Therefore, their real and imaginary parts, $Re(\epsilon(\omega))$ and $Im(\epsilon(\omega))$ can define all the optical physical quantities through the long wavelength limit (for further discussions; [37]). The optical absorption spectrum can be calculated by the number and intensity of the existing optical channels, meaning the joint DOSs of the valence and conduction subbands and the square of electric dipole moments. Therefore, $Im(\epsilon(\omega))$ can be given by

$$Im[\epsilon(\omega)] = \frac{4\pi e^2}{m^2\omega^2}\int_{BZ}\frac{2}{(2\pi)^3}\,|\langle i\rangle|^2\left(f(E^V) - f(E^C)\right)\delta\left(E_f - E_i - \omega\right)d^3k$$

Table 13.1. Prominent absorption structures: frequencies and identified orbital hybridizations.

Peak no.	Position (eV)	Transitions
1	3.103	O $2p_x/2p_y/2p_z$-Ti $3d_{xz}/\ 3d_{xz}$
2	4.302	O $2p_x/2p_z$-Ti $3d_{xy}$
3	5.008	O$2p_x/2p_y/2p_z$-Ti $3d_{xy}/3d_{yz}/3\ d_{z^2}$
4	6.559	O $2p_y$-Ti $3d_{yz}/3d_{xz}$
5	7.123	O $2p_z$-Ti 3 $d_{x^2-y^2}/3d_{xy}/3d_{yz}/3d_{xz}/d_{z^2}$
6	9.380	O $2p_x/2p_y/2p_z$-Ti $3d_{xy}/3d_{xz}$

Figure 13.5 provides the complex dielectric function, in which the real and imaginary part can be identified through the Kramers–Kronig relations [38]. Six pronounced peak/shoulder absorptions can be achieved in various values, namely 3.103 eV, 4.302 eV, 5.008 eV, 6.559 eV, 7.123 eV, and 9.380 eV (shown in table 13.1). The frequency-dependent vertical transitions and orbital hybridizations are identified by delicate analyses (figure 13.6). This can directly connect with the orbital-related response, mainly linking the band structure, charge density, and DOS of the ternary compounds [39]. For the first value of 3.103 eV, this closes to the direct band gap from the band structure (2.98 eV). The identified peaks from the band structure are specified by the various band-edge states from the unoccupied state to the occupied state about the Fermi level (indicated by the red arrows in figure 13.6.(a)). It is noticed that the first consideration is based on the x-direction, in which each absorption can be identified in the orbital responses. For example, 5.008 eV peak excited along Y|L in the first Brillouin zone is associated mainly from O $2p_x$/O $2p_y$/ O $2p_z$ to Ti 3d $d_{x^2-y^2}$/Ti $3d_{xy}$/Ti $3d_{yz}$/Ti $3d_{xz}$/Ti 3 d_{z^2}. Apparently, the Li-orbital also contributed to the current peak. However, due to their small states (about 1% compared with Ti and O orbitals), this contribution can be negligible in the whole transitions. Similar to the other polarization directions, e.g., y and z ones, the peak positions could appear at the certain energies with the variation of intensity. These phenomena come into existence because of the creation of different dipole matrices in the atomic arrangement. The crucial mechanisms for the various absorption structures can be achieved, being never revealed in the previous studies. This point of view has been successfully generated to anode and cathode materials of Li^+-based batteries.

The real part might appear as the normal and anomalous dispersions, in which the former is caused by an increase in $Re[\epsilon(\omega)]$. In general, the normal dispersion occurs everywhere, except the region close to the resonant frequency. Moreover, the real part of dielectric function may consist of three regions, including positive, zero and negative one. The black curve in figure 13.5 presents the real part ($\epsilon_1(\omega)$), in which the the static dielectric constant $Re(\epsilon(0))$ is 4.72, corresponding to zero frequency. It is noted that both $\epsilon_1(\omega)$ and $\epsilon_2(\omega)$ appear in many van Hove singularities in the energy range. These values be similar or different, being strongly co-related by the principal-value integration on the complex plane. Below the threshold frequency, $\epsilon_1(\omega)$ weakly depends on frequency ω, in which its value roughly lies within the range of 4.72–5.22 for *x*-, *y*-, *z*- polarizations. This will make a subsequent role in verifying the low-frequency reflectance spectrum (figure 13.7(c)) and the vanishing absorption coefficient (figure 13.7(a)).

The energy loss function (ELF) (Im[1/ϵ]), is the screened response ability of the significant valence charges of Li, Ti and O atoms under the Coulomb field perturbation. The prominent peaks in ELF exhibit collective excitation, meaning plasmon modes. Consequently, the coherent carrier oscillations in the optical transitions at long wave-length limit are caused by the charge screenings. In this ternary compound, the strongest peaks appear at 21.0/20.9/21.0 eVs for the *x*-/*y*-/*z*-direction electric polarizations, as shown in figure 13.7(b). These are caused by the contributions of the significant orbitals of Li-2s, O-[$2p_x$, $2p_y$, $2p_z$] and Ti-[4s, $3d_{x^2-y^2}$, $3d_{xy}$, $3d_{yz}$, $3d_{xz}$, $3d_{z^2}$]. The O-2s contribution for the most prominent plasma wave is ignored since its dominating valence states are small (the purple curve in figure 13.6(d)). Moreover, some minor plasmon peaks, for example, 10.5 and 19.5 eV positions are accompanied by the serious Landau dampings. The plasmon peaks mainly appear corresponding to the real part $\epsilon_1(\omega)$ close to 0.

Others optical properties are provided in figure 13.7 along the *x*-/*y*-/*z*-polarized direction, such as absorption, reflection, refraction, and extinction. These properties are related to the imaginary and real part of dielectric function. The absorption occurs during the propagation through the medium, in which the frequency of light is resonant with the transition frequencies of the atom. As a result, the absorption

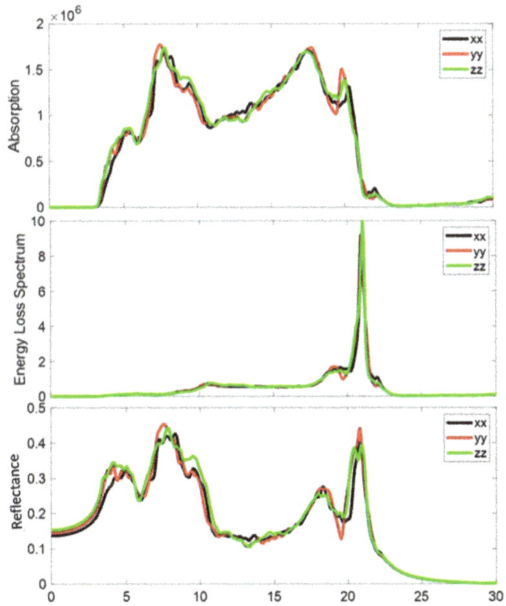

Figure 13.7. The various optical properties: (a) absorption coefficients, (b) energy loss functions, and (c) reflectance spectra.

spectrum mainly reveals different absorption peaks (figure 13.7(a)) (the same as with $\epsilon_2(\omega)$ analysis). Figure 13.7(c) presents the reflectance spectrum, which is caused by an electromagnetic wave propagating at a smaller velocity than in the free space. This quantity is characterized by $\left|\frac{\sqrt{\epsilon(\omega)}-1}{\sqrt{\epsilon(\omega)}+1}\right|^2$ [40]. Both reflectance and absorption coefficient can provide the main properties of a single particle, the inverse of propagation decay length in [41]. Reflectance is shown in figure 13.7(c) along x-, y-, and z-directions. When reflection is considered in a small frequency, the electron–hole excitation is vanishing, thus creating strong transmission due to the full EM wave propagation. For the higher value of frequency, the reflectance is compatible with the energy loss function at ω_p, which is shown as a large fluctuation under the various inter-band excitation transitions. In addition, the absorption creates a peak that is directly associated with the plasmon mode.

The theoretical predictions on the optimal geometric, electronic, and optical properties can be verified by the high-resolution measurements. Both x-ray diffractions [42, 43] and low-energy electron diffractions (LEED; [44]) are suitable in measuring the 3D lattice symmetries. Tunneling electron microscopy (TEM; [45]) and scanning electron microscopy (SEM, [45]) are used to identify the morphology for the nanoscaled top- and side-view structures. The above experimental technologies are very useful to comprehend the diversified lengths in Li–O and Ti–O bonds of 3D $Li_4Ti_5O_{12}$ material. As a result, the complicated multi-orbital hybridizations in all the chemical bonds can be revealed. In general, the wave-vector dependences of occupied valence states could be directly tested by angle-resolved photoemission spectroscopy (ARPES) [46]. There are a number of unusual energy subbands (figure 13.3) because a moiré superlattice would create high barriers in the ARPES measurements. Therefore, it is noted that this method will face difficulty in classifying the various subbands. To evaluate the band gap of the material, several common experimental measurements can be used such as ultraviolet-visible (UV–vis) absorption spectroscopy or photoluminescence spectra (PL) [47]. Measurement of dielectric properties relates to measurements of the complex relative permittivity (ε_r) and complex relative permeability (μ_r) of the materials. In general, the real part of the complex permittivity can identify the dielectric constant, which is a measure of the amount of energy from an external electrical field screening in the material. There are many methods developed for measuring the complex permittivity, which is based on the constraints in specific frequencies, materials and applications. Several methods can be counted such as transmission/reflection line method [48], open ended coaxial probe method [49], free space method [50] and resonant method [51]. Others optical properties, such as reflectance, absorptance, refractivity, excitation can be found through spectroscopy, for example, analysis of the specular reflection from the polished surface of a solid can give the complex index of refraction using the Kramers–Kronig method [37]. Apparently, the diverse optical properties in cathode/electrolyte/anode materials of Li^+-based batteries are worthy of systematic investigations both experimentally and theoretically.

The geometric, electronic, and optical properties are investigated in $Li_4Ti_5O_{12}$ ternary compound, based on first-principles calculation. The critical orbital hybridizations in chemical bonds are thoroughly studied to understand the essential properties of this material. As a potential anode material, $Li_4Ti_5O_{12}$ exhibits rich and unusual properties. An anisotropic and highly non uniform environment might play a critical role in the intercalation and deintercalation in the charging and discharging process of ion-batteries. Other electronic and optical properties are also included in this chapter, such as the lithium-, titanium-, or oxygen-dominated energy bands, the orbital-induced spatial charge densities, and the atom- and orbital-decomposed van Hove singularities, absorption, energy loss function, reflectance. Furthermore, the creation of a prominent optical peak can be verified by the different band-edge states and the specific orbital interactions. The direct gap is the same as the optical gap $E_g^d = 2.98$ eV. The coherent charge oscillation phenomenon creates and an optical plasmon peak at greater than 20 eV, which depends on the electric field direction. The further investigation based on this theoretical framework could apply into LiTiO-related system, e.g., $LiTiO_2$, $Li_2Ti_2O_4$, $Li_4Ti_7O_{12}$, and also anode/cathode/electrolyte materials of ion-based batteries.

References

[1] Nguyen T D H, Vo K D, Pham H D, Huynh T M D and Lin M-F 2021 Electronic and optical excitation properties of vanadium pentoxide V_2O_5 *Comput. Mater. Sci.* **198** 110675

[2] Goodenough J B and Park K-S 2013 The Li-ion rechargeable battery: a perspective *J. Am. Chem. Soc.* **135** 1167–76

[3] Nitta N, Wu F, Lee J T and Yushin G 2015 Li-ion battery materials: present and future *Mater. Today* **18** 252–64

[4] Voelker P 2014 Trace degradation analysis of lithium-ion battery components *R&D Mag.* (April 2014)

[5] Oman H 2002 Aerospace and military battery applications *IEEE Aerosp. Electron. Syst. Mag.* **17** 29–35

[6] Tarascon J-M and Armand M 2011 *Materials for Sustainable Energy: A Collection of Peer-Reviewed Research and Review Articles from Nature Publishing Group* (Singapore: World Scientific) pp 171–9

[7] Lu J, Chen Z, Pan F, Cui Y and Amine K 2018 High-performance anode materials for rechargeable lithium-ion batteries *Electrochem. Energy Rev.* **1** 35–53

[8] Liu H, Zhu Z, Huang J, He X, Chen Y, Zhang R, Lin R, Li Y, Yu S and Xing X 2019 Elucidating the limit of Li insertion into the spinel $Li_4Ti_5O_{12}$ *ACS Mater. Lett.* **1** 96–102

[9] Armand M and Tarascon J-M 2008 Building better batteries *Nature* **451** 652–7

[10] Ouyang D, Chen M, Liu J, Wei R, Weng J and Wang J 2018 Investigation of a commercial lithium-ion battery under overcharge/over-discharge failure conditions *RSC Adv.* **8** 33414–24

[11] Wang X, Sone Y, Segami G, Naito H, Yamada C and Kibe K 2006 Understanding volume change in lithium-ion cells during charging and discharging using *in situ* measurements *J. Electrochem. Soc.* **154** A14

[12] Tran N T T, Lin S-Y, Lin C-Y and Lin M-F 2017 *Geometric and Electronic Properties of Graphene-Related Systems: Chemical Bonding Schemes* (Boca Raton, FL: CRC Press)

[13] Shih-Yang Lin H-Y L, Duy Khanh Nguyen, Tran N T T, Pham H D, Chang S-L, Lin C-Y and Lin M-F 2020 *Silicene-Based Layered Materials: Essential Properties* (Bristol: IOP Publishing)
[14] Ouyang X, Lei M, Shi S, Luo C, Liu D, Jiang D, Ye Z and Lei M 2009 First-principles studies on surface electronic structure and stability of $LiFePO_4$ *J. Alloys Compd.* **476** 462–5
[15] Singh D P, Mulder F M, Abdelkader A M and Wagemaker M 2013 Facile micro templating $LiFePO_4$ electrodes for high performance Li-ion batteries *Adv. Energy Mater.* **3** 572–8
[16] Mekonnen Y, Sundararajan A and Sarwat A I 2016 A review of cathode and anode materials for lithium-ion batteries *SoutheastCon 2016* (Piscataway, NJ: IEEE) 1–6
[17] Li H, Huang X, Chen L, Wu Z and Liang Y 1999 A high capacity nano Si composite anode material for lithium rechargeable batteries *Electrochem. Solid State Lett.* **2** 547
[18] Croce F, Appetecchi G, Persi L and Scrosati B 1998 Nanocomposite polymer electrolytes for lithium batteries *Nature* **394** 456–8
[19] Balbuena P B and Wang Y 2004 *Lithium-Ion Batteries: Solid-Electrolyte Interphase* (London: Imperial College Press)
[20] Konschuh S, Gmitra M and Fabian J 2010 Tight-binding theory of the spin–orbit coupling in graphene *Phys. Rev.* B **82** 245412
[21] Hafner J 2008 Ab-initio simulations of materials using VASP: density-functional theory and beyond *J. Comput. Chem.* **29** 2044–78
[22] Perdew J P, Burke K and Ernzerhof M 1996 Generalized gradient approximation made simple *Phys. Rev. Lett.* **77** 3865
[23] Blöchl P E 1994 Projector augmented-wave method *Phys. Rev.* **B 50** 17953
[24] Rangel T, Caliste D, Genovese L and Torrent M 2016 A wavelet-based projector augmented-wave (PAW) method: reaching frozen-core all-electron precision with a systematic, adaptive and localized wavelet basis set *Comput. Phys. Commun.* **208** 1–8
[25] Poulet G, Sautet P and Artacho E 2003 Comparison between plane-wave and linear-scaling localized basis sets for structural calculations of microporous molecular sieves *Phys. Rev.* B **68** 075118
[26] Jain A, Ong S P, Hautier G, Chen W, Richards W D, Dacek S, Cholia S, Gunter D, Skinner D and Ceder G 2013 Commentary: The Materials Project: A materials genome approach to accelerating materials innovation *APL Mater.* **1** 011002
[27] Liu Y, Lian J, Sun Z, Zhao M, Shi Y and Song H 2017 The first-principles study for the novel optical properties of $LiTi_2O_4$, $Li_4Ti_5O_{12}$, $Li_2Ti_2O_4$ and $Li_7Ti_5O_{12}$ *Chem. Phys. Lett.* **677** 114–9
[28] Özen S, Şenay V, Pat S and Korkmaz Ş 2016 Optical, morphological properties and surface energy of the transparent $Li_4Ti_5O_{12}$ (LTO) thin film as anode material for secondary type batteries *J. Phys.* D **49** 105303
[29] Bussolotti F, Chi D, Goh K J, Huang Y L and Wee A T 2020 *2D Semiconductor Materials and Devices* (Amsterdam: Elsevier) pp 199–220
[30] Hipps K 2006 *Handbook of Applied Solid State Spectroscopy* (Berlin: Springer) pp 305–50
[31] Kano S, Tada T and Majima Y 2015 Nanoparticle characterization based on STM and STS *Chem. Soc. Rev.* **44** 970–87
[32] Deshpande A, Bao W, Miao F, Lau C N and LeRoy B J 2009 Spatially resolved spectroscopy of monolayer graphene on SiO_2 *Phys. Rev.* **B 79** 205411
[33] Gao L 2014 Probing electronic properties of graphene on the atomic scale by scanning tunneling microscopy and spectroscopy *Graphene 2D Mater.* **1** 23–46

[34] Lu C, Chang C-P, Huang Y-C, Lu J, Hwang C-C and Lin M-F 2006 Low-energy electronic properties of the AB-stacked few-layer graphites *J. Phys. Condens. Matter* **18** 5849
[35] Pierucci D, Sediri H, Hajlaoui M, Girard J-C, Brumme T, Calandra M, Velez-Fort E, Patriarche G, Silly M G and Ferro G 2015 Evidence for flat bands near the Fermi level in epitaxial rhombohedral multilayer graphene *ACS Nano* **9** 5432–9
[36] Mak K F, Shan J and Heinz T F 2010 Electronic structure of few-layer graphene: experimental demonstration of strong dependence on stacking sequence *Phys. Rev. Lett.* **104** 176404
[37] Kittel C 1996 *Introduction to Solid State Physics* (New York: Wiley)
[38] Grundmann M 2010 *The Physics of Semiconductors* (Berlin: Springer) pp 775–6
[39] Nguyen T D H, Pham H D, Lin S Y and Lin M F 2020 Featured properties of Li^+-based battery anode: $Li_4Ti_5O_{12}$ *RSC Adv.* **10** 14071–9
[40] Jackson J D 2007 *Classical Electrodynamics* (New York: Wiley)
[41] Fox M 2002 *Optical Properties of Solids* (College Park, MD: American Association of Physics Teachers)
[42] Li J and Sun J L 2017 Application of X-ray diffraction and electron crystallography for solving complex structure problems *Acc. Chem. Res.* **50** 2737–45
[43] Iwashita N 2016 *Materials Science and Engineering of Carbon* (Amsterdam: Elsevier) pp 7–25
[44] van Hove M A, Weinberg W H and Chan C-M 2012 *Low-Energy Electron Diffraction: Experiment, Theory and Surface Structure Determination* vol 6 (Cham: Springer)
[45] Flegler S L and Flegler S L 1997 *Scanning and Transmission Electron Microscopy* (Oxford: Oxford University Press)
[46] Lv B, Qian T and Ding H 2019 Angle-resolved photoemission spectroscopy and its application to topological materials *Nat. Rev. Phys.* **1** 609–26
[47] Kumar C S 2013 *UV-VIS and Photoluminescence Spectroscopy for Nanomaterials Characterization* (Berlin: Springer)
[48] Shenhui J, Ding D and Quanxing J 2003 Measurement of electromagnetic properties of materials using transmission/reflection method in coaxial line *Asia-Pacific Conf. on Environmental Electromagnetics, 2003. CEEM 2003. Proc.* (Piscataway, NJ: IEEE) 590–5
[49] K-m L, Wu Y-q, Qian C and Du G-p 2011 *Electrical Power Systems and Computers* (Berlin: Springer) pp 779–84
[50] Ghodgaonkar D K, Varadan V V and Varadan V K 1990 Free-space measurement of complex permittivity and complex permeability of magnetic materials at microwave frequencies *IEEE Trans. Instrum. Meas.* **39** 387–94
[51] Dressel M, Klein O, Donovan S and Grüner G 1996 High frequency resonant techniques for studying the complex electrodynamic response in solids *Ferroelectrics* **176** 285–308

Chapter 14

Bulk and monolayer transition metal dichalcogenides group IV MX_2 (M = Zr, Hf; X = S, Se, or Te)

Thi My Duyen Huynh, Hai Duong Pham, Thi Dieu Hien Nguyen, Vo Khuong Dien, Shih-Yang Lin, Hsien-Ching Chung, Wei-Bang Li, Ngoc Thanh Thuy Tran, Nguyen Thi Han, Hsin-Yi Liu and Ming-Fa Lin

This chapter has been adapted with permission from [1].

14.1 Group IV transition metal dichalcogenides

With materials, dimensionality is one of the most fundamental parameters, and it defines the atomic structures and properties [2]. After the discovery of graphene when it was isolated from graphite in 2004, 2D materials, particularly graphene, emerged as an attractive field in research and applications [2–5]. The 2D materials family as listed in table 14.1 ranges from insulators, semiconductors to semi-metallic ones including carbon materials, transition metal dichalcogenides (TMDs) and layered metal oxides [2]. Among them, TMDs exhibit a diversity of structures and orientation of crystallography from hexagonal sulfides to orthorhombic tellurides [6]. They can be constructed as monolayers due to their unique features which offer the forefront of solid states research. Particularly, they also have some electronic applications as an anode [7–12] or electrode [13] in lithium-ion batteries (LIBs) or sodium-ion batteries (SIBs) [14–19].

As an emergence in the various TMDs available, MoS_2 has been studied presenting most detail for electrochemical energy storage applications [7, 14, 19]. Its layers structurally degrade after the first lithiation cycle itself [20] and act as the electrochemically active material. MoS_2 has a high capacity as a negative electrode and low interlayer spacing while $MoSe_2$ has a narrower bandgap and greater

doi:10.1088/978-0-7503-4685-6ch14

Table 14.1. 2D materials family based on kind of compounds [2].

Carbon materials	Graphene	hBN	BCN	Fluorographene Graphene oxide
TMDs	MoS_2, $MoSe_2$, WS_2, WSe_2	Semiconducting dichalcogenides: $MoTe_2$, WTe_2, …		Metallic dichalcogenides: NbS_2, $NbSe_2$, $ZrTe_2$, $HfTe_2$, … Layered semiconductors: GaSe, GaTe, …
Layered metal oxides	Micas, Bismuth strontium calcium copper oxides (BSCCO)	MoO_3, WO_3	Perovskites: $LaNb_2O_7$, $Bi_4Ti_3O_{12}$, …	Hydroxides: $Ni(OH)_2$, $Eu(OH)_2$, …
	Layered Cu oxides	TiO_2, MnO_2, V_2O_5, …		Others

interlayer spacing, indicating its performance as an anode in SIBs [9]. Likewise, $MoTe_2$ shows large interlayer spacing to be able to accommodate larger ions [21], performing as an anode [22] in SIBs. As a whole, in MoX_2 (X = S, Se, or Te) systems, Te illustrates greater electronic conductivity than both S and Se [21] predicting improved electronic conductivity. In addition, both WS_2 [11, 12] and VS_2 [23] have demonstrated interest as anodes in LIBs. WTe_2 has promised to be a good substitute delivering a specific capacity [10, 13]. With regard to electrochemical energy storage systems, they all provide evidence that beyond-MoS_2 TMDs hold considerable promise as anode or electrode materials.

Among beyond-MoS_2 TMDs, group IV TMDs have become well-known and distinguished for their chemical and physical properties and applications, including ultrahigh mobility [24–26], superconducting transition [27], electrochemical energy storage [17, 28]. In this group, MX_2 (M = Zr, Hf; X = S, Se, or Te) emerge as promising materials for applications in optical and electronic devices [29–33] with abundant properties that shown in previous analyses. ZrX_2 and HfX_2 (X = S, Se, or Te) are two classes in the group IV chalcogenides of the 2D-TMDs family whose general properties were reported in previous studies [34–43]. The thermoelectric properties of single-layer ZrX_2 (X = S, Se, or Te) are investigated [25, 44] indicating the parameters are varied due to tensile strain. Theoretical study on electronic and optical properties of ZrX_2 [37] reveals good agreement with experimental investigation related to band gap direction, reflection and absorption coefficient. By using photoelectrons and X-ray absorption spectroscopy, $ZrSe_2$ and $ZrTe_2$ [45] show that the change in the splitting in the crystal field bases on the ratio change of covalent and ionic contribution to chemical bond. Especially, $ZrTe_2$ is a candidate in the successful growth of thin film material [46, 47]. The abrupt linear dispersions are exhibited in electronic band energy of $ZrTe_2$ grown in InAs substrate [47] by angle resolved photoelectron spectroscopy. $ZrTe_2$ shows a large magnetoresistance and high electron mobility [46] of their magneto-transport properties.

Moreover, experimental investigations [24, 48–54] on Hf-based materials have been conducted in bulk and monolayer. HfS_2 is anticipated to have high electron mobility and sheet current density, proposing a candidate for electronic devices [55, 56]. Moreover, HfS_2 was experimentally investigated in a flake form as promising for using in transistors [26, 50, 51, 57–59]. Monolayer and bilayer $HfSe_2$, exhibit low lattice thermal conductivity, suggesting a great potential for thermoelectric devices [24, 52, 54, 59, 60]. Epitaxial thin films of $HfTe_2$ are grown in AlN substrates using angle-resolved photoemission spectroscopy (ARPES) [49] provided that $HfTe_2$/AlN is an epitaxial topological semimetal. Simultaneously, theoretical studies via first-principles calculations on HfS_2 show a slow increase in the bandgap with increasing tensile strain and thickness [40, 61, 62]. Similar electronic characteristics are also observed in $HfSe_2$ [63]. In-depth investigations of the electronic band structures of HfS_2 and $HfSe_2$ [38] reveal that sulfur and selenium atoms have a primary effect on the valence band, while the hafnium atoms play a dominant role in the conduction band. In general, the 1T phase structures are found in among the possible phases ZrX_2 [25, 34, 37, 38, 40, 45] and HfX_2 [34, 38–42, 61] could exist in, and predict that T phase is the most favorable phase. Previous studies show that a monolayer of 1T-MTe_2 is semi-metallic [46–49], while those of 1T-MS_2 and 1T-MSe_2 [25, 35, 37, 39–41, 50] are indirect-gap semiconductors. They all belong to brittle materials which are indicated in elastic modulus [38]. In an attempt to tune the catalytic properties of group IV, many perturbations are disclosed by electrochemical oxidation or reduction [16, 19, 64, 65].

From a theoretical point of view, the first-principles calculations for layered materials pose challenges in investigating structural properties because of interactions between layers that are dominated by van der Waals force. Furthermore, the intra-layer chemical bonding (M–X bonding), which is mainly covalent in nature, predicts that the interactions play important roles in the structural features [66]. Even though many 3D TMDs have been reported and numerous research papers published on 2D-TMDs, especially MoS_2 [59, 67–71], the available information on geometric and electronic properties of layered group IV chalcogenides is severely limited. While most numbers of the group IV MX_2 family that have been constructed to date show promise for applications, the detailed characteristics might be lacking. In this chapter, we focus on bulk and monolayer MX_2 (M = Zr, Hf; X = S, Se, or Te) using VASP simulation. This group material is beyond-MoS_2 which is predicted to apply in LIBs as electrodes. The geometric and electronic properties of these materials will be investigated to verify their related features for explaining electrochemical energy.

14.2 Bulk structures MX_2

Transition metal dichalcogenides (TMDs) are characterized by layered structures labeled MX_2 where M represents transition metal and X is chalcogen atoms. They form a hexagonal packing in which each layer has a sandwich-like structure of M and X atoms. In a molecular unit, transition metal atoms are six-fold coordinated (figure 14.1). Depending on the relative position of chalcogen atoms, various

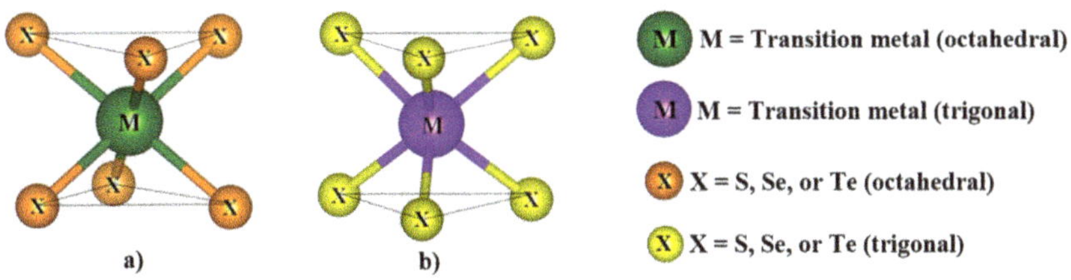

Figure 14.1. The local coordination of the two types in (a) octahedral and (b) trigonal prismatic symmetries.

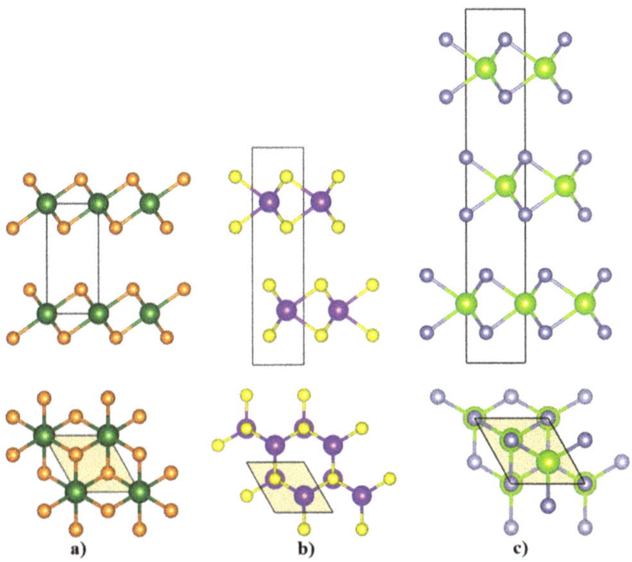

Figure 14.2. The structural polymorphs from top to bottom side view, top view of bulk TMDs (a) 1T, (b) 2H and (c) 3R phases. The yellow-filled lozenges highlight the spatial position of the transition metal and chalcogen atoms in a unit cell from the top view.

polymorphs can be exhibited which follow the octahedral or trigonal prismatic coordination [6]. In these symmetries, the two chalcogen planes form a stagger in the octahedral arrangement (figure 14.1(a)), while a slab is directly stacked one above the other in the trigonal prismatic arrangement (figure 14.1(b)). As a result, the lopsided coordination around chalcogen atoms is perpendicular to the symmetry axis that causes the cleavage properties. Furthermore, the surface is very stable and nonreactive because of the absence of dangling bonds.

Bulk TMDs are usually found in three phases of 1T, 2H, and 3R depending on their coordination, where the number stands for number of layers per unit cell and the letter represents the symmetry (T – trigonal, H – hexagonal, and R – rhombohedral) [6, 72]. These polymorphs are described in figure 14.2. A unit cell is defined with the z-axis (c parameter) perpendicular to the layers and the x and y-axes (a, b parameters) along the minimal X–X distance. Sometimes, the same TMDs can be found in the multiple polymorphs like MoS_2 which has been found in the T and H as well as the distorted phase [73, 74]—T′ phase. However, group IV MX_2 (M = Zr, Hf; X = S, Se, or Te) have all most been found in the T phase [25, 37, 41, 42] which is discussed in the next section.

With regard to theoretical point of views, the formation energy is one of the factors to verify the stability of a material which is calculated using equation (14.1).

$$E_f = \frac{E_{tot} - n_M\mu_M - n_X\mu_X}{N} \tag{14.1}$$

where E_f is the formation energy; E_{tot} is the total energy of MX_2 in bulk; and n_M, n_X are, respectively, the number of atoms of zirconium, hafnium and chalcogen atoms per unit cell; and μ_M, μ_X are, respectively, the chemical potential of zirconium, hafnium and chalcogen atoms in their bulk phases; N is the number of layers per unit cell in bulk phase (T, H and R have respectively one, two and three layers per unit cell). Three phases of 1T, 2H and 3R combine with the ground state energy to examine the stability of MX_2 as listed in table 14.1. These energies in the 1T phase are slightly smaller than two others indicating that the 1T phase is the favorable one in constructing bulk structures. This finding is in good agreement with experimental studies [25, 49, 57, 58] in these materials.

As mentioned, the 1T phase is the favorable phase, it is therefore focused to explain the geometric and electronic properties. For the crystal optimization of group IV MX_2 family in the T phase as shown in figure 14.2(a), M atom (Zr and Hf) and two chalcogen atoms (X) form the X–M–X layered unit per unit cell, where the M atom is sandwiched between two X atoms. In other words, each layer consists of three atomic planes that include one M plane and two chalcogen planes through ionic-covalent interaction (see side view). In some cases, the sandwiched layer is called the triple layer or single layer [6]. From the top view, the spatial position of the chalcogen atoms is additionally filled with the light-yellow triangles that show the different stacking in the two forms. The structural arrangements are constructed in the T phase with the period layer and trigonal symmetry being formulated by the octahedrally-coordinated stacking.

From the point of view of physical chemistry, van der Waals interaction (vdW interaction) is typically used to describe the sum of the attractive or repulsive forces between molecular entities [75]. With TMDs, in most cases the vdW interaction is bonding between the triple layers being not covalent and significantly weaker than the covalent intralayer bonds. However, group IV MX_2 exists with strong covalence in triple layers and weak vdW interaction between the layers. It is well-known that the vdW interaction is necessary for describing the geometric properties of layered systems [76]. Therefore, the structural properties of bulk structures are investigated using DFT-D3 approximation, as shown table 14.2. These calculated values are in good agreement with the experimental one [46, 47, 77–81].

To clarify the chemical bonds in these materials and the contribution of these bonds to electronic properties, the chemistry of each kind of atom is individually discussed. In the highest valence shell, the electron configuration of chalcogens is ns^2p^4 or $ns^2px^1py^1pz^2$ which means there are two electrons located on the atomic s-orbital and two of three p-orbitals. It indicates that unpaired electrons possess while the third orbital is occupied by a pair of electrons called lone-pair (LP)

Table 14.2. Total energy (E_{tot}, eV) and formation energy (E_f, eV) of three common phases in bulk structures.

	1T		2H		3R	
MX_2	E_{tot}(eV)	E_f(eV)	E_{tot}(eV)	E_f(eV)	E_{tot}(eV)	E_f(eV)
ZrS_2	−22.927	−4.77	−44.547	−4.12	−66.801	−4.11
$ZrSe_2$	−20.979	−4.17	−40.909	−3.65	−61.252	−3.62
$ZrTe_2$	−18.677	−2.48	−36.407	−2.01	−54.882	−2.1
HfS_2	−21.439	−4.72	−41.749	−4.16	−62.587	−4.15
$HfSe_2$	−19.564	−4.2	−38.317	−3.79	−57.369	−3.76
$HfTe_2$	−17.471	−2.71	−34.149	−2.32	−51.255	−2.33

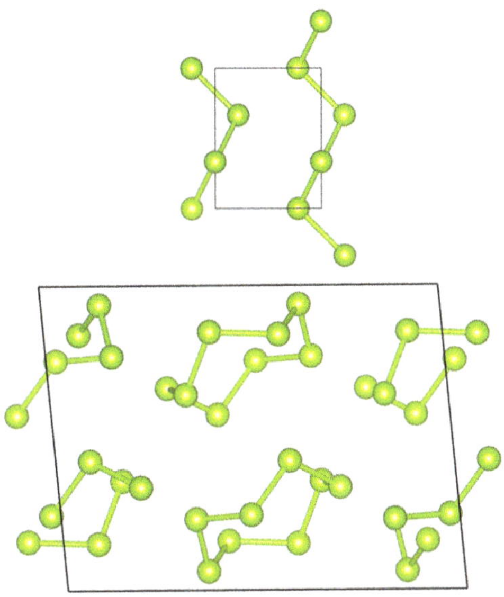

Figure 14.3. Chains (top) and rings (bottom) form selenium in the trigonal and monoclinic phase [6].

electrons. In most cases, the s-electrons are also chemically inactive. In fact, only the two unpaired p-electrons take part in the formation of covalent bonds, the pair of s-electrons and the p-electrons LP remain inert. Although sulfur, selenium and tellurium atoms have the same electronic configuration in the outer shell, the lighter sulfur atoms tend to form rings while the heavier tellurium atoms favor chains and the selenium can be rings or chains (figure 14.3). It should be noted that the ratio of the nearest distances between the chalcogen atoms within the same chain and between chains decrease as one goes from S and Se to Te, indicating a stronger covalent component in the latter case. Meanwhile, in transition metal chemistry Zr and Hf, the atom has a partially filled d subshell or an incomplete d subshell. These metals have the electronic configuration of $4d^2 5s^2$ and $5d^2 6s^2$, respectively. At the same time, an electron is transferred from the s-shell into d-shell because the (n-1)d and ns shells are rather close in energy and half-filled or filled shells are characterized by an increased stability [6]. The coordination number of these metals is determined by the size of the central metal ion and the number of d electrons. When six ligands coordinate to a central metal, octahedral coordination is the most stable geometry. Properties of coordination complexes are determined not only by the chemical nature of the ligands but also by their spatial arrangement around the transition metal.

The chemical bonds d_{M-X} are listed in table 14.2. When the chalcogen atoms go from S to Se and Te, the bonding is increased in both ZrX_2 and HfX_2. Furthermore, HfX_2 has the bonding which is larger than that of ZrX_2. The d-orbital of transition metal atoms and the LP electrons of chalcogen atoms determine the rich chemistry of these materials [6]. In other words, when chalcogen atoms are involved in chemical bonding, the atomic electronic configuration changes depending on the transition metal present in the compound. In the ideal case, each chalcogen atom forms two covalent bonds with its neighbors. In IV crystals with heavier metal elements, all metal–chalcogen distances are equal, all atoms are properly octahedrally coordinated. While TMDs bonding is usually considered between empty orbitals of the transition metals and LPs of ligands; the metal atoms provide four electrons to fill the bonding states. The transition metal and chalcogens can be ascribed a formal charge of +4 and −2, respectively. Furthermore, the chalcogen atoms are sp^3 hybridized with LP electrons and atomic orbitals of a transition metal characterize the geometry and properties of TMDs.

An example of such hybridization is sp^3 hybridization where each atom is fourfold covalently coordinated. In fact, the electrons of chalcogen atoms are located on sp^3-hybridized orbitals which lead to the coordination around chalcogen atoms being lopsided. In transition metal, six orbitals of s, p, and d can be hybridized being similar to sp^3 in chalcogens that generates six equivalent orbitals directed to the octahedral vertices. When an octahedral bonding geometry is considered, the d_{xy} lobes lie in between the x and the y axes, d_{xz} lobes lie in between the x and the z axes, d_{yz} lobes line in between the y and the z axes, $d_{x^2-y^2}$ lobes lie in between the x and the y axes and d_{z^2} is represented by two lobes along the z axis and a donut-shape ring that lies on the xy plane around the other two lobes (figure 14.4).

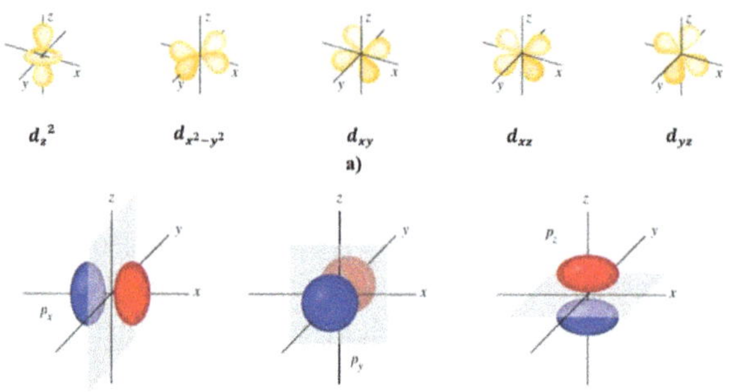

Figure 14.4. The representations of (a) the five d- and (b) three p-orbitals [6, 82].

Obviously, the electron configuration of these compounds affects the geometry that leads to the essential properties in electronic spectra. In the Brillouin zone, the electronic band structures have been calculated along with the high symmetry point Γ-M-K-Γ-A-L-H-A belonging to the hexagonal lattice. The system band gaps of monolayer MX_2 are shown in table 14.2. Both bulk MS_2 and MSe_2 are semi-conducting while bulk MTe_2 shows semimetal properties. In detail, ZrS_2 and HfS_2 show the middle-gap of 1.11 and 1.24 eV, respectively. $ZrSe_2$ and $HfSe_2$ remain semi-conducting and belong to a narrow gap of 0.33 and 0.56 eV, respectively. In contrast, both bands of $ZrTe_2$ and $HfTe_2$ reveal the overlap between the conduction and valence band. The system band energy spectra of bulk MX_2 are shown in figure 14.5, which presents the atom dominances of zirconium, hafnium and chalcogens. Two groups of these MX_2 share a common feature that the unoccupied conduction bands are asymmetric to the occupied valence bands about the Fermi level. The valence band maximum (VBM) is located at Γ point (k = 0, 0, 0) and the conduction band minimum (CBM) is at L point (k = ½, 0, ½) of the high symmetry k-points.

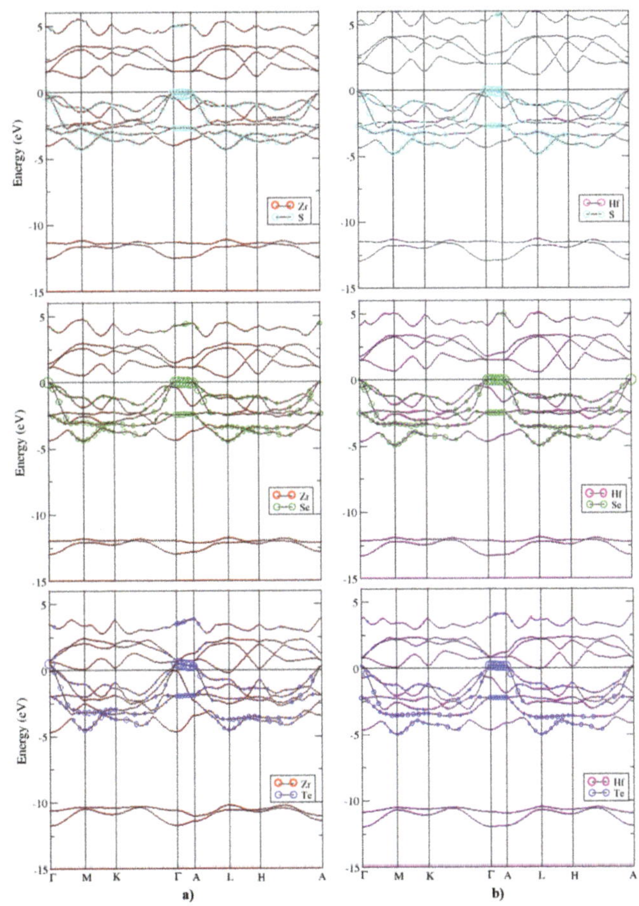

Figure 14.5. Band energy structure of bulk MX_2 (M = Zr, Hf; X = S, Se, or Te) with atoms dominance.

Most energy bands have a strong dispersion relation, with the parabolic dispersion being exhibited in both the conduction and the valence bands. However, each material is predicted to have an unusual energy band.

The energy bands of bulk ZrX_2 and HfX_2 are shown in figure 14.5(a) (left panel) and 14.5(b) (right panel). The red and magenta circles represent the zirconium and hafnium atoms, respectively, while the cyan, green and blue circles, respectively, describe sulfur, selenium and tellurium atoms. In figure 14.5(a), the asymmetry between occupied holes bands and unoccupied electron bands is clearly illustrated with various dispersions. Both ZrS_2 and $ZrSe_2$ show the indirect gap semiconductors where the direction of gap shifts from Gamma to L point. The σ and $\sigma*$ bands (in the valence and conduction bands) of ZrS_2 initiate from Γ point at −2.6, 0 and 1.6 eV, while the σ and $\sigma*$ of $ZrSe_2$ are at −2.5, 0 and 1.25 eV. There exist the special valleys located in the low-lying valence band and the flat bands are between Gamma and A points near the Fermi level in both conduction and valence bands. The energy band of $ZrTe_2$ reveals contrasting features compared to ZrS_2 and $ZrSe_2$ that there is an overlap between conduction and valence bands indicating a semimetal one. The two σ and two $\sigma*$, respectively, initiate from Γ point at −2, 0 eV and 0.5, 0.7 eV. These features show the difference between two subgroups in ZrX_2 that are semiconductors and semimetal. With regard to atom dominance in bands, the chalcogen atoms in all structures dominate in the valence band ranging from 0 to −5 eV, while the contribution of zirconium atoms is smaller than in this band and it is dominated in a low-lying band ranging from −10.5 to −12 eV. Zirconium contributes to the conduction band more than the valence band. Both zirconium and chalcogens contribute to the whole band with ranges from −15 to 6 eV, but chalcogens show a sharper dominance in valence band while zirconium dominates in conduction band.

Similarly to ZrX_2, bulk HfX_2 reveals two separated subgroups that are semiconductors (HfS_2 and $HfSe_2$) and semimetal ($HfTe_2$) as shown in figure 14.5(b). Both HfS_2 and $HfSe_2$ are semiconducting while $HfTe_2$ is a gapless semimetal. The unoccupied valence bands are also asymmetric to occupied conduction bands. There are two σ and a $\sigma*$ band initiating from Γ point at −2.6, −0.1 and 2 eV for HfS_2; −2.5, 0 and 1.5 eV for $HfSe_2$ while $HfTe_2$ has a σ initiate at −2.4 eV and two $\sigma*$ are at 0.1 and 1 eV. Apart from the special valleys located in the low-lying valence bands, the flat bands are between Gamma and A points near the Fermi level in both conduction and valence bands. The energy band of $HfTe_2$ reveals contrasting features compared to HfS_2 and $HfSe_2$ with more parabolic dispersion in whole bands. Like ZrX_2, the chalcogen atoms in all structures dominate in valence band ranging from 0 to −5 eV, while the contribution of hafnium atoms is smaller than in this band and it is dominated in the low-lying band ranging from −10.5 to −12 eV. Hafnium also contributes to the conduction band. With ranges from −15 to 6 eV, both hafnium and chalcogens contribute to the whole band.

When combining with chalcogen atoms, zirconium and hafnium compounds show the common features as previously mentioned. However, there are some differences in general. With semiconducting properties, the band gap of HfX_2 is larger than that of ZrS_2. Besides, ZrX_2 has crossing bands in the valence band that are obviously prominent compared to HfX_2. With regard to the semimetal one,

$ZrTe_2$ presents clearly the feature of a semimetal with the overlap between the conduction and valence bands while the band gap of $HfTe_2$ is approximately zero.

To further verify the various chemical bondings and examine the curvature effect, the orbital-projected density of states (DOS) could be used as described in figures 14.6 and 14.7. The figures show the local DOS with multi-orbital hybridization monolayer ZrX_2 and HfX_2. They show that the main contributions are due to two s, d orbitals of the zirconium, hafnium (M atoms) and two s, p orbitals of the chalcogen atoms. With ZrX_2 structures, the local DOS with multi-hybridization of zirconium in the compounds (figure 14.6(a)) and chalcogen atoms (figure 14.6(b)) are shown in the DOS range of (0, 3) and (0, 5), respectively. In the conduction band, zirconium atoms are dominated by 4 d_{z^2} and 4 $d_{x^2-y^2}$ orbitals, as clearly indicated in figure 14.6(a) while chalcogen atoms unremarkably contribute to this band. For valence band, both s, d orbitals of zirconium and s, p orbitals chalcogen atoms contribute but the latter shows obvious dominance. The DOS in this band depends on the bonding of ($3p_x$,$3p_y$), ($4p_x$,$4p_y$) and ($5p_x$,$5p_y$) of ZrS_2, $ZrSe_2$ and $ZrTe_2$, respectively (figure 14.6(b)). Especially, at low-lying energy ranging from −12 to −11 eV, −13 to −12 eV and −12 to −10 eV for ZrS_2, $ZrSe_2$ and $ZrTe_2$, respectively, 3s, 4s and 5s uniquely dominate without other orbitals. Moreover, figure 14.7 describes the local DOS with multi-hybridization of hafnium in the compounds and chalcogen atoms. In the conduction band, the chalcogen atoms unremarkably contribute while hafnium atoms are dominated by 5 d_{z^2} and 5 $d_{x^2-y^2}$ orbitals, as

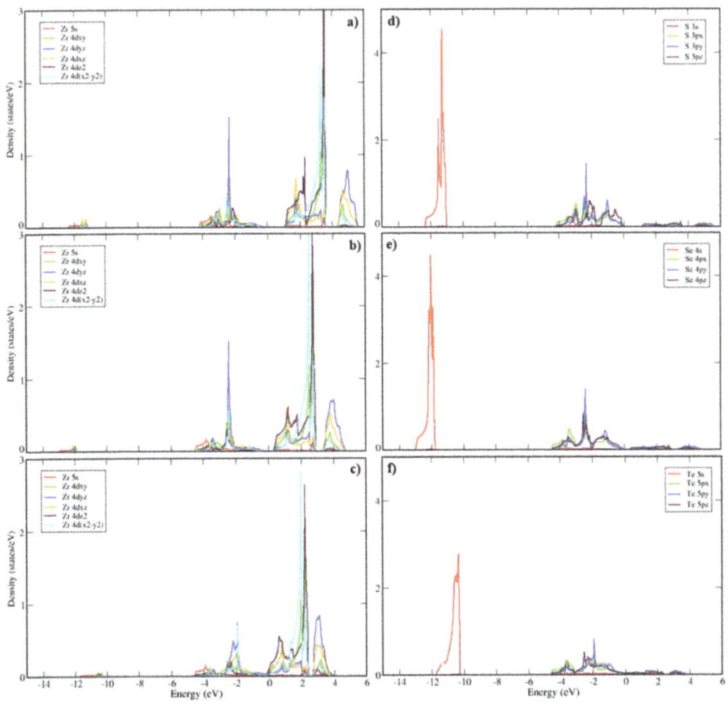

Figure 14.6. The orbital projected DOS of Zr, S, Se, and Te atoms in bulk ZrX_2.

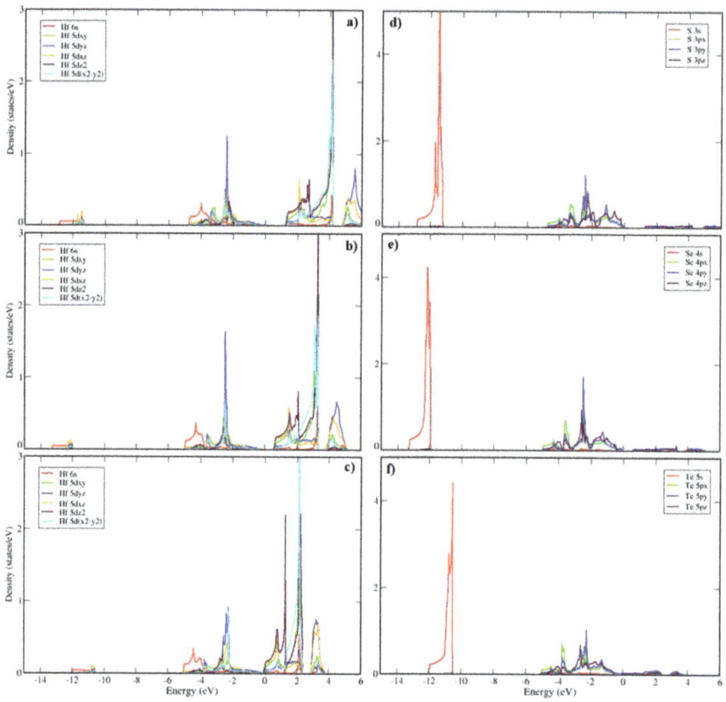

Figure 14.7. The orbital projected DOS of Hf, S, Se, and Te atoms in bulk HfX_2.

clearly demonstrated in figure 14.7(a). Similar to ZrX_2, both s, d orbitals of hafnium and s, p orbitals chalcogen atoms contribute to the valence band but the latter shows obvious dominance. The DOS in this band depends on the bonding of ($3p_x$,$3p_y$), ($4p_x$,$4p_y$) and ($5p_x$,$5p_y$) of HfS_2, $HfSe_2$ and $HfTe_2$, respectively (figure 14.7(b)). Especially, the 3s, 4s and 5s of HfS_2, $HfSe_2$ and $HfTe_2$, respectively, uniquely dominates without other orbitals at low-lying energy being smaller than -10 eV with the DOS slightly larger than that of ZrX_2.

In conclusion, bulk MX_2 (M = Zr, Hf; X = S, Se, or Te) reveal interesting and unique properties. They share common features and are differentiated by distinct properties simultaneously. The electronic properties strongly depend on the geometry of such electron configuration, chemical bonding, van der Waals interaction. These materials suggest themselves as potential candidates and show promise for applications. However, with the development of 2D materials beyond graphene, it is possible to investigate the properties of monolayers of these materials as discussed in the next section.

14.3 Monolayer MX_2

Monolayer MX_2 has only two polymorphs namely the T and H phases depending on their hexagonal or trigonal prismatic coordination. According to stability which is confirmed in the bulk structures, the T phase is the stable one that is also favorable. With layered structures in this work, the crystal structures are only conducted in the T phase, as shown in figure 14.7. The crystal optimization for three compounds in

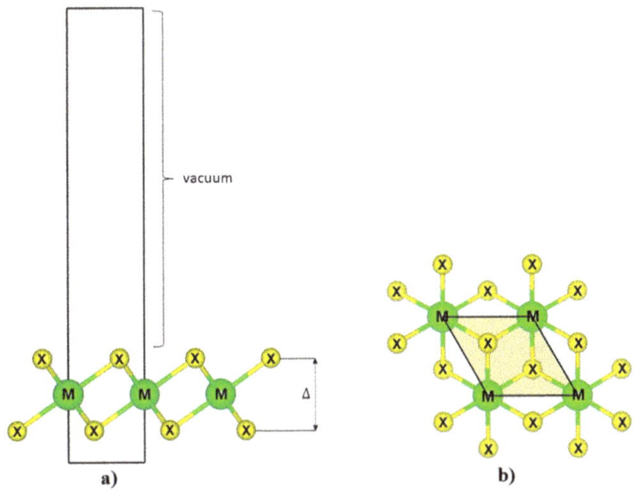

Figure 14.8. Optimized structures of (a) side and (b) top view of monolayer MX_2 (M = Zr, Hf; X = S, Se, or Te) in the T phase.

monolayer MX_2 is performed with the top (*xy* plane) and side (*yz* plane) views. A vacuum of 15 Å is added to avoid the interaction between layers.

Moreover, monolayer HfX_2 exists in a buckled form (delta) (figure 14.8) of approximate 2.92, 3.13, and 3.58 Å, respectively, for monolayer ZrS_2, $ZrSe_2$, and $ZrTe_2$ while those of HfS_2, $HfSe_2$, and $HfTe_2$ are, respectively, 2.89, 3.12, and 3.53 Å. The Hf-X interaction suppresses the height fluctuation of the hafnium atom in these monolayers. As a result, the buckling in monolayer MX_2 is sharper than in silicene [83], which can be attributed to the sandwiched structure of these materials. Notice that the height fluctuations (buckling) are presented at equilibrium that provides the broad distribution of height displacement. This phenomenon predicts the phenomenological theories of thermal fluctuations in flexible membranes.

In the Brillouin zone, the electronic band structures have been calculated along with the high symmetry point Γ-M-K-Γ belonging to the hexagonal lattice. The system band gaps of monolayer MX_2 are shown in table 14.3. Both MS_2 and MSe_2 are semi-conducting with each material exhibiting distinct properties. ZrS_2 and HfS_2 show the middle-gap of 1.12 and 1.2 eV, respectively. $ZrSe_2$ and $HfSe_2$ remain semi-conducting and belong to a narrow gap of 0.79 and 0.86 eV, respectively. In contrast, both $ZrTe_2$ and $HfTe_2$ are semi-metallic. The values of approximately zero reveal them to be a gapless semimetal similar to graphene [84]. The experimental gaps of their monolayer are also listed in table 14.3. However, these values are smaller than the calculated band gaps, e.g, there respectively exists a difference of approximate 0.6 and 0.3 eV for HfS_2 and $HfSe_2$. In this work, the band gaps are obtained from a Perdew–Burke–Ernzerhof (PBE) functional, and notes that the approaches in calculating might cause the differences that are significant when comparing them to experimental values.

The system band energy spectra of monolayer MX_2 are shown in figure 14.9, which presents the atom dominances of zirconium, hafnium and chalcogens. The left

Table 14.3. Lattice parameter (a, c (Å)), the chemical bonding between transition metal and chalcogen atoms (d_{M-X}(Å)) and band gap in bulk MX_2.

	a (Å)		c (Å)			E_g(eV)	
MX_2	Cal.	Exp.	Cal.	Exp.	d_{M-X}(Å)	Cal.	Exp.
ZrS_2	3.68	3.66[a]	6.6	5.84[a]	2.57	1.11	1.68[b]
$ZrSe_2$	3.77	3.77[b]	6.8	6.16[ab]	2.71	0.33	1.1[c]
$ZrTe_2$	4.0	3.95[d]	7.06	6.63[d]	2.92	−0.71	–
HfS_2	3.64	3.63[b]	6.59	5.88[b]	2.55	1.24	1.8[e]
$HfSe_2$	3.77	3.74[b]	6.93	6.14[b]	2.681	0.56	1.13[b]
$HfTe_2$	3.99	3.95[f]	7.66	6.69[f]	2.887	−0.24	–

[a]: Ref. [77];
[b]: Ref. [78];
[c]: Ref. [79];
[d]: Ref. [46, 47];
[e]: Ref. [80];
[f]: Ref. [81];
– No experimental report

Table 14.4. Lattice constant (a, (Å)), chemical bonding (d_{M-X} (Å)), buckling (Δ, (Å)) and band gap (E_g(eV)) of monolayer MX_2.

MX_2	a (Å)	d_{M-X}(Å)	Δ(Å)	E_g(eV)
ZrS_2	3.671	2.573	2.917	1.12
$ZrSe_2$	3.817	2.7	3.135	0.79
$ZrTe_2$	3.996	2.92	3.581	−0.53
HfS_2	3.643	2.552	2.892	1.2
$HfSe_2$	3.781	2.684	3.122	0.86
$HfTe_2$	3.976	2.896	3.531	0.02

panel (figure 14.9(a)) describes the system band of ZrX_2, and the right panel (figure 14.9(b)) shows the system band of HfX_2. All three materials share as a common feature that the VBM is located at Γ point (k = 0, 0, 0) and the CBM is at M point (k = ½, 0, 0) of the high symmetry k-points. The unoccupied conduction bands are asymmetric to the occupied valence bands about the Fermi level. Most energy bands have a strong dispersion relation, with the parabolic dispersion being exhibited in both the conduction and the valence bands. However, each material is predicted to have an unusual energy band. There are many distinguishing features between ZrX_2 and HfX_2 in the electronic band structures that are discussed below.

As previously mentioned, monolayer MX_2 (M = Hf, Zr; X = S, Se, or Te) are constructed with the T phase following the bulk structures that reveal distinct properties. The electronic bands of ZrX_2 and HfX_2 with atom dominance are, respectively, shown in figure 14.9(a) (left panel) and 14.9(b) (right panel). They all show an indirect bandgap with the direction of the gap shifted from Γ point to M point.

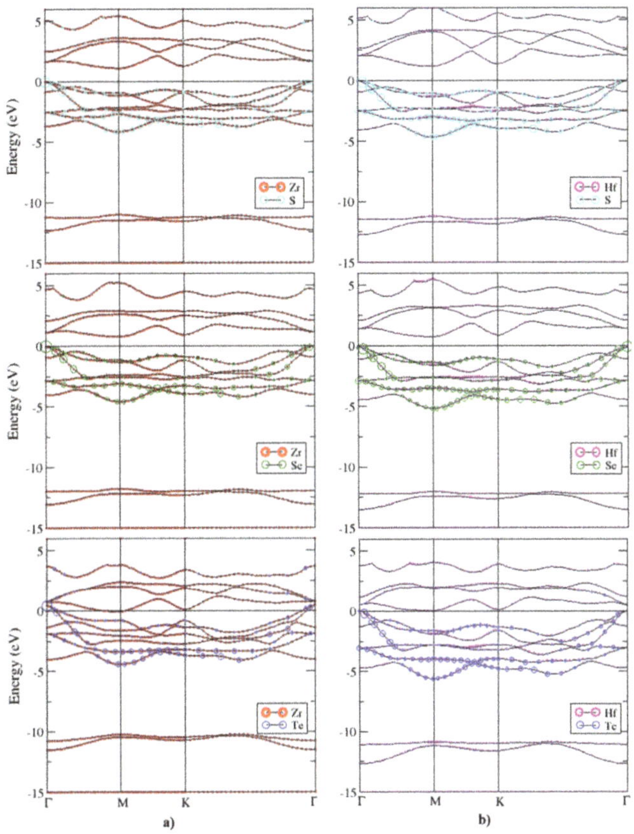

Figure 14.9. Band structure of monolayer MX_2 (M = Zr, Hf; X = S, Se, or Te) with atoms dominance.

The occupied hole band is strongly asymmetric to the unoccupied electron band. The low-lying valence bands exhibit valleys between Γ and M point, as well as K and Γ point. The low-lying valence bands also change into parabolic bands. The parabolic dispersions have band-edge states at M point belonging to saddle points, which results in van Hove singularities (vHs).

In figure 14.9(a), the σ and $\sigma*$ bands (in the valence and conduction bands) of ZrS_2 initiate from Γ point at −2.5, 0 and 2 eV, while the σ and $\sigma*$ of $ZrSe_2$ are at −3, 0 and 1.5 eV. The energy band of $ZrTe_2$ reveals contrasting features compared to ZrS_2 and $ZrSe_2$. Apart from the special valleys located in the low-lying valence band that is similar to ZrS_2 and $ZrSe_2$, there exist flat bands in the low-lying valence bands. There are σ and $\sigma*$ that initiate from Γ point at −2, 0.5, and 1 eV. There is an overlap between valence band and conduction band of $ZrTe_2$, which presents a gapless material. This feature illustrates the difference between two subgroups in ZrX_2 that are semiconductors and semimetal. With regard to atom dominance in bands, the chalcogen atoms in all structures dominate in the valence band ranging from 0 to −15 eV, while the contribution of zirconium atoms is smaller than in this band. In contrast, zirconium contributes to the conduction band more than the valence band. In valence band, this atom dominates in bands of a scale of 0 to −2 eV. Both hafnium and chalcogens contribute to the whole band with ranges from −15 to 6 eV, but chalcogens show a sharper dominance compared to hafnium.

Similarly, in figure 14.9(b), the σ and $\sigma*$ bands of HfS_2 initiate from Γ point at −2.5, 0 and 2 eV, while the σ and $\sigma*$ of $HfSe_2$ are at −3, 0 and 1.5 eV. Like $ZrTe_2$, a semi-metallic property of $HfTe_2$ has been shown, which is investigated by the pressure-dependence conductivity and Hall coefficient [49]. Apart from the special valleys located in the low-lying valence band that is similar to HfS_2 and $HfSe_2$, there exist flat bands in the low-lying valence bands. There are σ and $\sigma*$ that initiate from Γ point at −3, 0, and 0.5 eV. The difference between $HfTe_2$ and both HfS_2 and $HfSe_2$ is that all of the VBMs and CBMs are nearly at the Fermi level, which presents a gapless material. Besides, the chalcogen atoms in all structures dominate in valence band ranging from 0 to −15 eV, while the contribution of hafnium atoms is smaller than in this band. In contrast, hafnium contributes to the conduction band more than the valence band. Compared to ZrX_2, the chalcogen atoms in HfX_2 show the stronger contribution with the larger dominant ranges in valence band (figure 14.9).

In general, the band energy spectra exhibit a lot of asymmetric peaks and few symmetric ones due to the parabolic form of energy bands. A pair of asymmetric peaks exists near the Fermi level that defines the indirect energy gaps. The bandgap opening is associated with the quantum confinement effect. Although all three materials are members of the same chalcogen group, MS_2, MSe_2, and MTe_2 are different in their low-lying peaks and degenerate with varying the stacking configuration between the two phases. Moreover, most vHs come from the parabolic energy dispersions. The peaks near the Fermi level are found to arise from the band-edge states mainly. These peaks could be a signature feature unique to vHs.

The orbital-projected DOS is used as described in figures 14.10 and 14.11 for verifying the configuration effect. The figures show the local DOS with multi-orbital hybridization monolayer ZrX_2 and HfX_2. They show that the main contributions are due to two orbitals of the zirconium, hafnium (M atoms) and two orbitals of the chalcogens, corresponding to the s, d orbitals of M atoms, and the s, p orbitals of chalcogens. With monolayer ZrX_2, the local DOS with multi-hybridization is shown in figure 14.11, presenting the dominance of the multi-orbital zirconium and chalcogen atoms. As shown in figure 14.11(a), hafnium atoms are dominated by $4d_{xy}$, 4 d_{z^2} and 4 $d_{x^2-y^2}$ orbitals in conduction band while the DOS depends on the bonding of ($3p_x$,$3p_y$,$3p_z$), ($4p_x$,$4p_y$,$4p_z$) and ($5p_x$,$5p_y$,$5p_z$) of ZrS_2, $ZrSe_2$ and $ZrTe_2$, respectively (figure 14.11(b)) in valence band. These features are the same as those of bulk structures. However, the DOS covers in the larger range compared to their bulk phase. With regard to multi-hybridization of the orbital-projected DOS, monolayer HfX_2 shows similar properties to monolayer ZrX_2 where $5d_{xy}$, 5 d_{z^2} and 5 $d_{x^2-y^2}$ orbitals of hafnium and ($3p_x$,$3p_y$,$3p_z$), ($4p_x$,$4p_y$,$4p_z$) and ($5p_x$,$5p_y$,$5p_z$) of sulfur, selenium, tellurium, respectively, dominate in conduction band and valence band. In addition, there is the dominance of 3s, 4s and 5s of sulfur, selenium and tellurium, respectively, located at low-lying energy.

In summary, their contributions are comparable to each other in that the feature-rich energy bands are mainly determined by the d and p orbitals. There exist strong

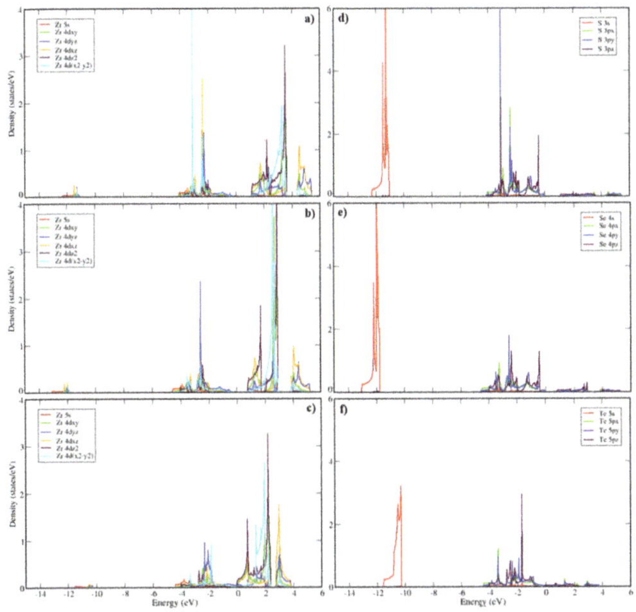

Figure 14.10. The orbital projected DOS of Zr, S, Se, and Te atoms in monolayer ZrX_2.

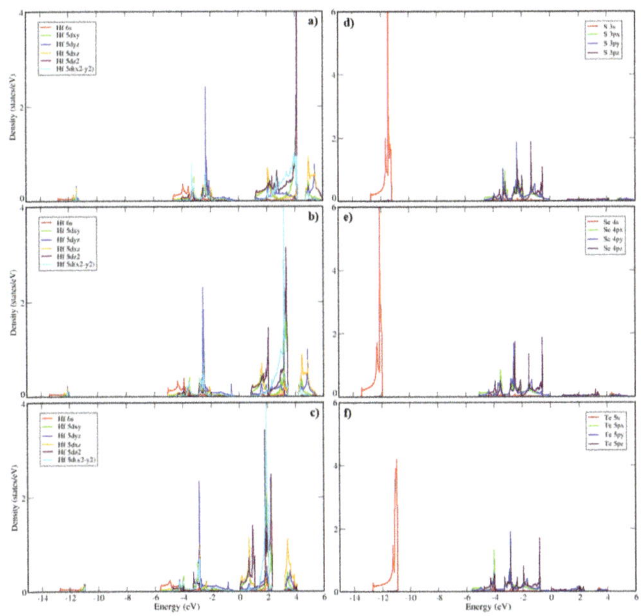

Figure 14.11. The orbital projected DOS of Hf, S, Se, and Te atoms in monolayer HfX_2.

hybridizations of ($4d_{xy}$,4 d_{z^2},4 $d_{x^2-y^2}$), ($5d_{xy}$,5 d_{z^2},5 $d_{x^2-y^2}$) Zr-, Hf-orbitals and ($3p_x$,$3p_y$,$3p_z$), ($4p_x$,$4p_y$,$4p_z$) and ($5p_x$,$5p_y$,$5p_z$) chalcogen-orbitals. In addition, the p-orbitals of the chalcogens make significant contributions to the bonding. These orbitals are evident for the multi-orbital hybridization in these materials.

Additionally, the DOS ranges from −4 to 4 eV and shows a pair of asymmetric peaks centered at the Fermi level, which characterizes the energy gap. The large energy difference between the valence and conduction peaks obviously demonstrates the asymmetry of the two bands. These structures originate from linear bands, parabolic bands near saddle points, and initial band-edge states of the parabolic bands. Also, DOS exhibits a logarithmic divergence at middle energy and shoulder structures at deeper and higher energy. Partially flat energy is exhibited in the valence bands of all structures and in the conduction bands of MSe_2 and MTe_2, while shoulder structures evidently appear in the conduction band of MS_2. Moreover, the linear straights in the DOS near the Fermi level, which are parallel lines, indicate the feature of vHs accompanied with the parabolic energy dispersions and saddle points. This vHs is peculiar in the corresponding region of Brillouin zone [6], as neither conduction band nor valence band have singularities in the DOS.

14.4 Remarks

In brief, ZrX_2 and HfX_2 reveal that 1T phase is the favorable phase being in good agreement with experimental studies. They all show the d-orbitals and p-orbitals of the M atoms (zirconium and hafnium) and X atoms (chalcogen atoms), respectively, take part in the chemical bonding of M–X. All monolayer structures exist with a sharp buckle that shows height fluctuations at equilibrium. The energy gap trend suggests that the atomic species play a dominant role. A lot of asymmetric peaks and few symmetric ones due to the parabolic form are shown in the energy bands and DOS in both bulk and monolayer. Especially, the vHs mainly arise from the band-edge states in the DOS. The logarithmic divergence at middle energy and shoulder structures at lower and higher energy are exhibited in the DOS corresponding to the $3p_z$, $4p_z$, $5p_z$ and ($3p_x$,$3p_y$), ($4p_x$,$4p_y$), ($5p_x$,$5p_y$) of chalcogen atoms in the compounds. In addition, these orbitals demonstrate multi-orbital hybridizations, as shown in the orbital-projected DOS. These materials reveal the essential properties that are similar to MoS_2 or WS_2 promising for applications in electrochemical energy storage. They present many potential properties for these applications such as large surface areas, mechanical stability, nontoxic nature [14]. However, there exist some challenges for application of these materials that are low electronic conductivity, large volume changes upon ion insertion and re-stacking of exfoliated 2D TMD nanosheets. Besides, this group of TMDs materials is still new in storage areas because of lack of detail analytic reports. There is also opportunity for further research related to these materials to enhance the range of application.

References

[1] Huynh T M D, Nguyen D K, Nguyen T D H, Dien V K, Pham H D and Lin M-F 2021 Geometric and electronic properties of monolayer HfX_2 (X = S, Se, or Te): a first-principles calculation *Front. Mater.* **7** 569756

[2] Gupta A, Sakthivel T and Seal S 2015 Recent development in 2D materials beyond graphene *Prog. Mater Sci.* **73** 44–126

[3] Rosenkranz A, Liu Y Q, Yang L and Chen L 2020 2D nano-materials beyond graphene: from synthesis to tribological studies *Appl. Nanosci.* **10** 3353–88

[4] Mas-Balleste R, Gomez-Navarro C, Gomez-Herrero J and Zamora F 2011 2D materials: to graphene and beyond *Nanoscale* **3** 20–30
[5] Li G, Zhang Y Y, Guo H, Huang L, Lu H L, Lin X, Wang Y L, Du S X and Gao H J 2018 Epitaxial growth and physical properties of 2D materials beyond graphene: from monatomic materials to binary compounds *Chem. Soc. Rev.* **47** 6073–100
[6] Kolobov A V and Tominaga J 2016 *Two-dimensional Transition-metal Dichalcogenides* (Berlin: Springer)
[7] Barik G and Pal S 2019 Defect induced performance enhancement of monolayer MoS_2 for Li- and Na-ion batteries *J. Phys. Chem.* C **123** 21852–65
[8] Srinivaas M, Wu C Y, Duh J G and Wu J M 2019 Highly rich 1T metallic phase of few-layered WS_2 nanoflowers for enhanced storage of lithium-ion batteries *ACS Sustain. Chem. Eng.* **7** 10363–70
[9] Zhang Y, Liu Z Q, Zhao H Y and Du Y P 2016 $MoSe_2$ nanosheets grown on carbon cloth with superior electrochemical performance as flexible electrode for sodium ion batteries *RSC Adv.* **6** 1440–4
[10] Hong M L, Li J, Zhang W F, Liu S T and Chang H X 2018 Semimetallic 1T′WTe_2 nanorods as anode material for the sodium ion battery *Energy Fuels* **32** 6371–7
[11] Kim I, Park S W and Kim D W 2019 Onion-like crystalline WS_2 nanoparticles anchored on graphene sheets as high-performance anode materials for lithium-ion batteries *Chem. Eng. J.* **375** 122033
[12] Bhandavat R, David L and Singh G 2012 Synthesis of surface-functionalized WS_2 nanosheets and performance as Li-ion battery anodes *J. Phys. Chem. Lett.* **3** 1523–30
[13] Soares D M and Singh G 2020 Superior electrochemical performance of layered WTe_2 as potassium-ion battery electrode *Nanotechnology* **31** 455406
[14] Soares D M, Mukherjee S and Singh G 2020 TMDs beyond MoS_2 for electrochemical energy storage *Chem. Eur. J.* **26** 6320–41
[15] Wang H, Feng H B and Li J H 2014 Graphene and graphene-like layered transition metal dichalcogenides in energy conversion and storage *Small* **10** 2165–81
[16] Seman R, Azam M A and Ani M H 2018 Graphene/transition metal dichalcogenides hybrid supercapacitor electrode: status, challenges, and perspectives *Nanotechnology* **29** 502001
[17] Yun Q B, Li L X, Hu Z N, Lu Q P, Chen B and Zhang H 2020 Layered transition metal dichalcogenide-based nanomaterials for electrochemical energy storage *Adv. Mater.* **32** 1903826
[18] Chen D Y, Ji G, Ding B, Ma Y, Qu B H, Chen W X and Lee J Y 2014 Double transition-metal chalcogenide as a high-performance lithium-ion battery anode material *Ind. Eng. Chem. Res.* **53** 17901–8
[19] Zhang Y X, Zhang L, Lv T A, Chu P K and Huo K F 2020 Two-dimensional transition metal chalcogenides for alkali metal ions storage *ChemSusChem* **13** 1114–54
[20] Stephenson T, Li Z, Olsen B and Mitlin D 2014 Lithium ion battery applications of molybdenum disulfide (MoS_2) nanocomposites *Energy Environ. Sci.* **7** 209–31
[21] Panda M R *et al* 2019 Blocks of molybdenum ditelluride: a high rate anode for sodium-ion battery and full cell prototype study *Nano Energy* **64** 103951
[22] Mukherjee S, Schuppert N, Bates A, Lee S C and Park S 2017 Novel mesoporous microspheres of Al and Ni doped LMO spinels and their performance as cathodes in secondary lithium ion batteries *Int. J. Green Energy* **14** 656–64
[23] Wang D S, Liu Y H, Meng X, Wei Y J, Zhao Y Y, Pang Q and Chen G 2017 Two-dimensional VS_2 monolayers as potential anode materials for lithium-ion batteries and beyond: first-principles calculations *J. Mater. Chem.* A **5** 21370–7

[24] Yin L, Xu K, Wen Y, Wang Z X, Huang Y, Wang F, Shifa T A, Cheng R, Ma H and He J 2016 Ultrafast and ultrasensitive phototransistors based on few-layered $HfSe_2$ *Appl. Phys. Lett.* **109** 213105

[25] Zhen Y X, Yang M, Zhang H, Fu G S, Wang J L, Wang S F and Wang R N 2017 Ultrahigh power factors in P-type 1T-ZrX_2 (X = S, Se) single layers *Sci. Bull.* **62** 1530–7

[26] Xu K, Wang Z X, Wang F, Huang Y, Wang F M, Yin L, Jiang C and He J 2015 Ultrasensitive phototransistors based on few-layered HfS_2 *Adv. Mater.* **27** 7881–7

[27] Zibouche N, Kuc A, Musfeldt J and Heine T 2014 Transition-metal dichalcogenides for spintronic applications *Ann. Phys.* **526** 395–401

[28] Pumera M, Sofer Z and Ambrosi A 2014 Layered transition metal dichalcogenides for electrochemical energy generation and storage *J. Mater. Chem.* A **2** 8981–7

[29] Novoselov K S, Mishchenko A, Carvalho A and Neto A H C 2016 2D materials and van der Waals heterostructures *Science* **353**

[30] Zhou J D *et al* 2018 A library of atomically thin metal chalcogenides *Nature* **556** 355

[31] Lebegue S, Bjorkman T, Klintenberg M, Nieminen R M and Eriksson O 2013 Two-dimensional materials from data filtering and *Ab Initio* calculations *Phys. Rev.* **3** 031002

[32] McDonnell S J and Wallace R M 2016 Atomically-thin layered films for device applications based upon 2D TMDC materials *Thin Solid Films* **616** 482–501

[33] Kaul A B 2014 Two-dimensional layered materials: structure, properties, and prospects for device applications *J. Mater. Res.* **29** 348–61

[34] Toh R J, Sofer Z and Pumera M 2016 Catalytic properties of group 4 transition metal dichalcogenides (MX_2; M = Ti, Zr, Hf; X = S, Se, Te) *J. Mater. Chem.* A **4** 18322–34

[35] Kar I, Chatterjee J, Harnagea L, Kushnirenko Y, Fedorov A V, Shrivastava D, Buchner B, Mahadevan P and Thirupathaiah S 2020 Metal-chalcogen bond-length induced electronic phase transition from semiconductor to topological semimetal in ZrX_2 (X = Se and Te) *Phys. Rev.* B **101** 165122

[36] Kumar A, He H Y, Pandey R, Ahluwalia P K and Tankeshwar K 2015 Pressure and electric field-induced metallization in the phase-engineered ZrX_2 (X = S, Se, Te) bilayers *Phys. Chem. Chem. Phys.* **17** 19215–21

[37] Reshak A H and Auluck S 2004 Theoretical investigation of the electronic and optical properties of ZrX_2 (X = S, Se and Te) *Phys. B-Condens. Matter* **353** 230–7

[38] Zhao Q Y, Guo Y H, Si K Y, Ren Z Y, Bai J T and Xu X L 2017 Elastic, electronic, and dielectric properties of bulk and monolayer ZrS_2, $ZrSe_2$, HfS_2, $HfSe_2$ from van der Waals density-functional theory *Phys. Status Solidi* B **254** 1700033

[39] Abdulsalam M and Joubert D P 2016 Optical spectrum and excitons in bulk and monolayer MX_2 (M = Zr, Hf; X = S, Se) *Phys. Status Solidi* B **253** 705–11

[40] Salavati M 2019 Electronic and mechanical responses of two-dimensional HfS_2, $HfSe_2$, ZrS_2, and $ZrSe_2$ from first-principles *Front. Struct. Civil Eng.* **13** 486–94

[41] Reshak A H and Auluck S 2005 *Ab initio* calculations of the electronic and optical properties of 1T-HfX_2 compounds *Phys. B-Condens. Matter* **363** 25–31

[42] Chen Q Y, Liu M Y, Cao C and He Y 2019 Engineering the electronic structure and optical properties of monolayer 1T-HfX_2 using strain and electric field: a first principles study *Phys. E-Low-Dimens. Syst. Nanostruct.* **112** 49–58

[43] Sun X L and Wang Z G 2017 *Ab initio* study of adsorption and diffusion of lithium on transition metal dichalcogenide monolayers *Beilstein J. Nanotechnol.* **8** 2711–8

[44] D'Souza R, Mukherjee S and Ahmad S 2019 Strain induced large enhancement of thermoelectric figure-of-merit (ZT similar to 2) in transition metal dichalcogenide monolayers ZrX_2 (X = S, Se, Te) *J. Appl. Phys.* **126** 214302
[45] Shkvarin A S, Merentsov A I, Shkvarina E G, Yarmoshenko Y M, Pis I, Nappini S and Titov A N 2018 Electronic structure of ZrX_2 (X = Se, Te) *J. Chem. Phys.* **148** 124707
[46] Wang H C, Chan C H, Suen C H, Lau S P and Dai J Y 2019 Magnetotransport properties of layered topological material $ZrTe_2$ thin film *ACS Nano* **13** 6008–16
[47] Tsipas P, Tsoutsou D, Fragkos S, Sant R, Alvarez C, Okuno H, Renaud G, Alcotte R, Baron T and Dimoulas A 2018 Massless Dirac fermions in $ZrTe_2$ semimetal grown on InAs (111) by van der Waals epitaxy *ACS Nano* **12** 1696
[48] Mangelsen S *et al* 2017 Large nonsaturating magnetoresistance and pressure-induced phase transition in the layered semimetal $HfTe_2$ *Phys. Rev.* B **96** 205148
[49] Aminalragia-Giamini S, Marquez-Velasco J, Tsipas P, Tsoutsou D, Renaud G and Dimoulas A 2017 Molecular beam epitaxy of thin $HfTe_2$ semimetal films *2D Mater.* **4** 015001
[50] Glebko N, Aleksandrova I, Tewari G C, Tripathi T S, Karppinen M and Karttunen A J 2018 Electronic and vibrational properties of TiS_2, ZrS_2, and HfS_2: periodic trends studied by dispersion-corrected hybrid density functional methods *J. Phys. Chem.* C **122** 26835–44
[51] Kanazawa T, Amemiya T, Ishikawa A, Upadhyaya V, Tsuruta K, Tanaka T and Miyamoto Y 2016 Few-layer HfS_2 transistors *Sci. Rep.* **6** 22277
[52] Yue R Y *et al* 2015 $HfSe_2$ thin films: 2D transition metal dichalcogenides grown by molecular beam epitaxy *Acs Nano* **9** 474–80
[53] Tsai H S *et al* 2018 Photoluminescence characteristics of multilayer $HfSe_2$ synthesized on sapphire using ion implantation *Adv. Mater. Interfaces* **5** 1701619
[54] Aretouli K E, Tsipas P, Tsoutsou D, Marquez-Velasco J, Xenogiannopoulou E, Giamini S A, Vassalou E, Kelaidis N and Dimoulas A 2015 Two-dimensional semiconductor $HfSe_2$ and $MoSe_2/HfSe_2$ van der Waals heterostructures by molecular beam epitaxy *Appl. Phys. Lett.* **106** 143105
[55] Fiori G, Bonaccorso F, Iannaccone G, Palacios T, Neumaier D, Seabaugh A, Banerjee S K and Colombo L 2014 Electronics based on two-dimensional materials *Nat. Nanotechnol.* **9** 768–79
[56] Zhang W X, Huang Z S, Zhang W L and Li Y R 2014 Two-dimensional semiconductors with possible high room temperature mobility *Nano Res.* **7** 1731–7
[57] Traving M, Seydel T, Kipp L, Skibowski M, Starrost F, Krasovskii E E, Perlov A and Schattke W 2001 Combined photoemission and inverse photoemission study of HfS_2 *Phys. Rev.* B **63** 035107
[58] Nie X R, Sun B Q, Zhu H, Zhang M, Zhao D H, Chen L, Sun Q Q and Zhang D W 2017 Impact of metal contacts on the performance of multilayer HfS_2 field-effect transistors *ACS Appl. Mater. Interfaces* **9** 26996–7003
[59] Mirabelli G *et al* 2016 Air sensitivity of MoS_2, $MoSe_2$, $MoTe_2$, HfS_2, and $HfSe_2$ *J. Appl. Phys.* **120** 125102
[60] Kang M, Rathi S, Lee I, Lim D, Wang J, Li L, Khan M A and Kim G H 2015 Electrical characterization of multilayer $HfSe_2$ field-effect transistors on SiO_2 substrate *Appl. Phys. Lett.* **106** 143108
[61] Wu N H, Zhao X, Ma X, Xin Q Q, Liu X M, Wang T X and Wei S Y 2017 Strain effect on the electronic properties of 1T-HfS_2 monolayer *Phys. E-Low-Dimension. Syst. Nanostruct.* **93** 1–5

[62] Kreis C, Werth S, Adelung R, Kipp L, Skibowski M, Krasovskii E E and Schattke W 2003 Valence and conduction band states of HfS_2: from bulk to a single layer *Phys. Rev.* B **68** 235331
[63] Setiyawati I, Chiang K R, Ho H M and Tang Y H 2019 Distinct electronic and transport properties between 1T-$HfSe_2$ and 1T-$PtSe_2$ *Chin. J. Phys.* **62** 151–60
[64] Chen B, Chao D L, Liu E Z, Jaroniec M, Zhao N Q and Qiao S Z 2020 Transition metal dichalcogenides for alkali metal ion batteries: engineering strategies at the atomic level *Energy Environ. Sci.* **13** 1096–131
[65] Zhao T F, Shu H B, Shen Z H, Hu H M, Wang J and Chen X S 2019 Electrochemical lithiation mechanism of two-dimensional transition-metal dichalcogenide anode materials: intercalation versus conversion reactions *J. Phys. Chem.* C **123** 2139–46
[66] Kumar A and Ahluwalia P K 2012 Electronic structure of transition metal dichalcogenides monolayers 1H-MX_2 (M = Mo, W; X = S, Se, Te) from ab-initio theory: new direct band gap semiconductors *Eur. Phys. J.* B **85** 186
[67] Zeng H L, Dai J F, Yao W, Xiao D and Cui X D 2012 Valley polarization in MoS_2 monolayers by optical pumping *Nat. Nanotechnol.* **7** 490–3
[68] Zeng H L and Cui X D 2015 An optical spectroscopic study on two-dimensional group-VI transition metal dichalcogenides *Chem. Soc. Rev.* **44** 2629–42
[69] Xiao D, Liu G B, Feng W X, Xu X D and Yao W 2012 Coupled spin and valley physics in monolayers of MoS_2 and other group-VI dichalcogenides *Phys. Rev. Lett.* **108** 196802
[70] Lin Y K, Chen R S, Chou T C, Lee Y H, Chen Y F, Chen K H and Chen L C 2016 Thickness-dependent binding energy shift in few-layer MoS_2 grown by chemical vapor deposition *ACS Appl. Mater. Interfaces* **8** 22637–46
[71] Hu J Q, Shi X H, Wu S Q, Ho K M and Zhu Z Z 2019 Dependence of electronic and optical properties of MoS_2 multilayers on the interlayer coupling and Van Hove singularity *Nanoscale Res. Lett.* **14** 288
[72] Katzke H, Toledano P and Depmeier W 2004 Phase transitions between polytypes and intralayer superstructures in transition metal dichalcogenides *Phys. Rev.* B **69** 134111
[73] Komsa H P, Kotakoski J, Kurasch S, Lehtinen O, Kaiser U and Krasheninnikov A V 2012 Two-dimensional transition metal dichalcogenides under electron irradiation: defect production and doping *Phys. Rev. Lett.* **109** 035503
[74] Castellanos-Gomez A, Poot M, Steele G A, van der Zant H S J, Agrait N and Rubio-Bollinger G 2012 Mechanical properties of freely suspended semiconducting graphene-like layers based on MoS_2 *Nanoscale Res. Lett.* **7** 1–4
[75] Zallen R, Slade M L and Ward A T 1971 Lattice vibrations and interlayer interactions in crystalline As_2S_3 and As_2Se_3 *Phys. Rev.* B **3** 4257
[76] Wilson J A and Yoffe A D 1969 Transition metal dichalcogenides discussion and interpretation of observed optical, electrical and structural properties *Adv. Phys.* **18** 193
[77] Hodul D T and Stacy A M 1984 Anomalies in the properties of $Hf(S_{2-x}Te_x)_{1-y}$ and $Hf(Se_{2-x}Te_x)_{1-y}$ near the metal-insulator-transition *J. Solid State Chem.* **54** 438–46
[78] Greenaway D L and Nitsche R 1965 Preparation and optical properties of Group 4–62 chalcogenides having CdI_2 structure *J. Phys. Chem. Solids* **26** 1445
[79] Starnberg H I, Brauer H E and Hughes H P 1996 Photoemission studies of the conduction band filling in $Ti_{1.05}S_2$ and Cs-intercalated TiS_2 and $ZrSe_2$ *J. Phys.-Condens. Matter* **8** 1229–34
[80] Terashima K and Imai I 1987 Indirect absorption-edge of ZrS_2 and HfS_2 *Solid State Commun.* **63** 315–8

[81] Bayliss S C and Liang W Y 1982 Symmetry dependence of optical-transitions in group-4b transition-metal dichalcogenides *J. Phys. C-Solid State Phys.* **15** 1283–96
[82] Petrucci R H *et al* 2010 *General Chemistry—Principles and Modern Applications* 10th edn (Toronto: Pearson)
[83] Meng L *et al* 2013 Buckled silicene formation on Ir(111) *Nano Lett.* **13** 685–90
[84] Castro Neto A H, Guinea F, Peres N M R, Novoselov K S and Geim A K 2009 The electronic properties of graphene *Rev. Mod. Phys.* **81** 109–62

Chapter 15

Electric characteristics evaluation of Li-ion repurposed batteries based on UL 1974

Hsien-Ching Chung, Thi Dieu Hien Nguyen, Shih-Yang Lin, Wei-Bang Li, Ngoc Thanh Thuy Tran, Nguyen Thi Han, Hsin-Yi Liu, Hai Duong Pham and Ming-Fa Lin

Based on the popularization of electric vehicles worldwide and the development of renewable energy, lithium-ion (Li-ion) batteries are widely applied in many fields, from small-scale personal mobile products to large-scale energy storage systems. Recently, enormous power batteries have reached their end-of-life (EOL) and should be retired. Such a situation causes environmental protection threats and waste of resources. Since most of the retired power battery modules still possess about 80% of their initial energy capacity, the second use of these batteries becomes a possible way to overcome the emergent problem. Safety and performance become important judgments when applying these second-use repurposed batteries. Underwriters Laboratories (UL) recently published the standard for evaluating the safety and performance of repurposed batteries, i.e., UL 1974. In this chapter, the test procedures designed according to UL 1974 are discussed with a case study.

15.1 Emergent problem causing by retired Li-ion power batteries

The rapid development of wind, solar photovoltaic, and electric vehicle industries promoted the application of Li-ion batteries in fields of various scales, from small-scale 3C (computing, communication, and consumer) products to large-scale battery energy storage systems (BESSs) [1]. The Li-ion battery demonstrates the capabilities of electrochemical energy storage, e.g., high power density, high energy density, very short response time, and appropriate for various size scales. For instance, Tesla and Neoen have installed a big Li-ion battery (100 MW/129 MWh) in South Australia in 2017 [2, 3]. Several weeks after installation, a power shortage happened as the coal-fired Loy Yang power plant in Victoria failed and the backup electricity provided by

doi:10.1088/978-0-7503-4685-6ch15

the big battery kicked in and delivered as large as 100 MW into the national grid in less than 140 ms [4, 5]. Such responding time is even faster than the coal-fired backups. These shock absorber- and emitter-type capacities can help us to avoid blackouts that might occur. The batteries can provide inertia services and fast responses (for example, FCAS, frequency control ancillary services) to the electricity grid, giving the route for potential regulatory changes and revenue streams to incentivize grid-scale batteries [6–8]. Large-scale batteries are starting to show that they have crucial roles in supply electric networks.

As the promotion of commercialization and the popularization of electric cars worldwide [9], Li-ion batteries are widely applied in the industry. Meanwhile, the great demand for Li-ion batteries causes fundamental researches, industry development, as well as standards and policymaking in the field of Li-ion power batteries. Recently, the number of retired power batteries has been largely grown, carrying out environmental protection threats and resource-wasting problems. Hence, re-utilization and recycling of such retired power batteries have been proposed [10, 11]. Some retired power batteries still possess about 80% of their initial energy capacity [12–14]. This means that they can be repurposed and applied once again, e.g., to serve as batteries in low-power stationary energy storage systems [15–17]. Governments in various areas have recognized this important issue and prepared to publish their policies to deal with the recovery and reuse of repurposed batteries, such as coding principles, manufacturing factory guidelines, dismantling process guidelines, traceability management systems, residual energy measurements, rebates, federal and state tax credits, as well as other financial support [18–20].

Safety and performance are important in using the repurposed battery cells. In 2018, Underwriters Laboratories (UL, a global safety certification company established in 1894) published the standard for evaluating the safety and performance of repurposed batteries, i.e., UL 1974 [21, 22]. In this chapter, the designed test procedures for Li-ion batteries based on UL 1974 are described and a case study on LFP batteries is demonstrated.

15.2 Methods suggested by UL 1974

The UL 1974 standard [21, 22] provides the sorting and grading processes of cells, modules, and packs of battery (electrochemical capacitors) that were originally applied for other aims, e.g., main power for electric vehicles [23], vehicle supply power [24], and light electric rail applications [25]. Meanwhile, the focused aims are intended for repurposed usages, such as for use in energy storage systems [26] and other applications for cells, modules, and packs of battery (electrochemical capacitors). UL 1974 also includes application-specific requirements for repurposed battery systems and battery systems using repurposed modules, cells, and other components.

According to UL 1974, the battery module can be disassembled into battery cells and other used components. The used components in the battery systems, e.g., the battery management system (BMS), thermal management system (TMS), and other auxiliary systems (ASs), and battery enclosure, should not be considered for repurposing. Because those components have already been used for a long time (over the calendar expiration date specified by the manufacturer). The disassembled battery cells preparing for repurposing will experience the performance test and sorting. UL 1974 recommends the repurposed manufacturer to apply the following test procedures as part of the routine analysis: (1) incoming open circuit voltage (OCV) measurements in section 19.2 of UL 1974; (2) incoming high voltage isolation check in section 19.3 of UL 1974; (3) capacity check in section 19.4 of UL 1974; (4) internal resistance check in section 19.5 of UL 1974; (5) check of BMS controls and protection components in section 19.6 of UL 1974; (6) discharge and charge cycle test in section 19.7 of UL 1974; and (7) self-discharge in section 19.8 of UL 1974.

The measurement of electric characteristics of batteries according to section 19 of UL 1974 will be discussed. The measurement can be divided into two procedures. The first procedure includes sections 19.2 and 19.4 of UL 1974. The second procedure includes sections 19.5, 19.7, and 19.8 of UL 1974. The major parameters in these procedures are described as follows.

In the OCV measurements (section 19.2 of UL 1974), the battery cells' OCVs are detected. These OCVs can be compared to the minimum acceptable voltage for the cell. In addition, the OCVs are detected for a short time for checking the OCV stability. The value of OCV should stay constant. If this condition fails, the cell cannot pass this step. The incoming high-voltage isolation check is often skipped for most of the low-voltage battery systems (voltage < 60 V), since the battery module is directly disassembled into cells (it becomes unnecessary to check the insulation breakdown in such a system).

The battery cell capacity check is designed following the instructions of section 19.4 of UL 1974. The battery cell is entirely charged by the standard constant–current constant voltage (CC–CV) charge process. Then, the battery cell is entirely discharged by the standard constant current (CC) discharge process. The discharge ampere hour capacity can be obtained after the completion of the entire discharge process. At last, the battery cell is entirely charged again for the next test step, and the charge ampere hour capacity is available.

The battery cells need capacity sorting while going to the next procedure. The internal resistance check is designed according to section 19.5 of UL 1974. The internal resistance is measured by two-tier direct current (DC) load method under CC-mode discharge at two different states of charge (SOC). In this step, internal resistances at 85% and 20% SOC are suggested. The former stands for internal resistance when the battery contains most of the electricity. The latter represents the internal resistance for low power conditions [27].

The cycle tests of discharge and charge under normal and maximum loadings are suggested according to section 19.7 of UL 1974. It gives users another criterium for performance evaluation. According to section 19.8 of UL 1974, the self-discharge test can be taken at the end as part of the determination of the state of health (SOH).

15.3 Case study: LFP batteries

The lithium iron phosphate battery ($LiFePO_4$ battery) or lithium ferrophosphate battery (LFP battery), is one type of Li-ion battery with $LiFePO_4$ as the cathode material and graphitic carbon as the anode [28]. LFP batteries possess the features of high power, high energy density, high thermal stability, high cycle life, low cost, low toxicity, and low self-discharge, making them play important roles in vehicle usages [29], large-scale stationary energy storage applications [30], and backup power supplies [31]. The test procedures designed according to UL 1974 can be applied to evaluate the performance and safety of the repurposed LFP batteries. Recently, Chung has published the charge and discharge profile dataset of repurposed LFP batteries [27], carrying out the procedure described in UL 1974 on an LFP repurposed battery [32].

The first-life applications of these repurposed LFP battery cells are power battery modules installed in golf carts. The golf course is a good choice for design verification of the power battery module. The flat roads are suitable for continuous power output tests and some gentle slopes are appropriate for the up and downhill tests. The power battery modules normally discharge in two conditions. The first one is instant high-power output for motor start and the second one is continuous medium power output for continuously pushing the golf cart. These LFP battery modules have been used for 1–2 years, closely reaching the EOL.

The batteries under repurposing are 15 Ah 40138-LFP battery cells originally manufactured by C-life Technologies, Inc. (40138 indicates that the dimensions of the cylinder battery cell, i.e., 40 mm diameter and 138 mm height). The detailed specification of the battery cells and a brief description of the module disassembly procedure are available in the data repository (https://doi.org/10.17605/OSF.IO/PFH3G) [33].

The repurposed battery with serial number, MAP150919190000004, in the data repository [33] is taken as an example to demonstrate the test procedures according to UL 1974. The voltage, current, capacity, and temperature variations under test procedure 1 are shown in figure 15.1. The initial OCV is measured, $OCV_{ini} = 3.327$ V. The battery cell is discharged under CC mode with C-rate = 0.5 h^{-1} and cutoff voltage = 2.5 V, obtaining a discharge ampere hour capacity, $Cap_D = 14.392$ Ah. Then the battery cell is entirely charged under CC–CV mode with C-rate = 0.5 h^{-1} and threshold voltage = 3.5 V.

The voltage, current, capacity, and temperature variations under test procedure 2 are shown in figure 15.2. When discharging the battery cell to 85% SOC, the internal resistance $R_{85} = 0.0095\ \Omega$ is obtained by applying the two-tier DC load method. As the battery keeps discharge to 20% SOC, the internal resistance $R_{20} = 0.0158\ \Omega$ is measured. Later, the discharge ampere hour capacity under normal conditions (Cap_{DN}) and maximum loading conditions (Cap_{DM}) are measured, i.e., $Cap_{DN} = 14.147$ Ah and $Cap_{DN} = 14.037$ Ah. Eventually, the OCVs after an entire charge for 5 minutes (OCV_{5m}), 1 hour (OCV_{1h}), and 24 hours (OCV_{24h}) are available, i.e., $OCV_{5m} = 3.3654$ V, $OCV_{1h} = 3.3393$ V, and $OCV_{24h} = 3.3351$ V.

It should be noted that the key values obtained according to UL 1974 give us some valuable judging criteria. $OCV_{ini} = 3.327$ V within the normal range (generally 2.5~3.4 V) indicates that the rest voltage of the battery cell is in normal conditions, and $Cap_D = 14.392$ Ah < the nominal value 15 Ah shows the capacity reduction

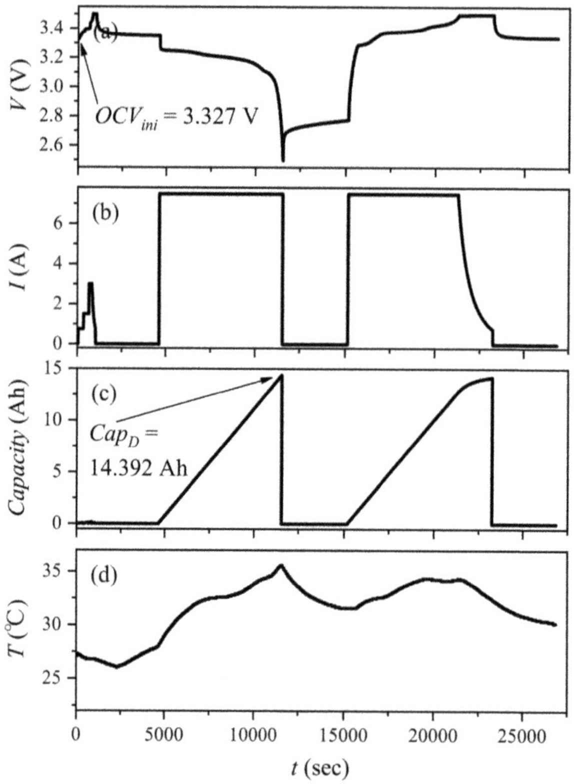

Figure 15.1. The voltage, current, capacity, and temperature variations under test procedure 1 according to UL 1974.

feature of repurposed battery. $R_{85} < R_{20}$ indicates that the internal resistance is larger for low SOC, reflecting the feature of internal resistance variation of Li-ion battery. Also, R_{85} and R_{20} give users a reference to judge the battery ability of continuous output, which are parameters for sorting battery cells. Cap_{DM} and Cap_{DN} carry out the capacity for maximum and normal conditions. In general, Cap_{DM} is always smaller than Cap_{DN} owing to heat caused by internal resistance. For Li-ion batteries in normal conditions, the difference between Cap_{DM} and Cap_{DN} is less than 10%. When the battery is at rest after the entire charge, the voltage will gradually decrease, and the reducing rate varies from high to low. Hence, $OCV_{5m} > OCV_{1h} > OCV_{24h}$ will be observed. On the other hand, in all procedures, the measured curves (such as voltage and temperature) are piecewise smooth (no sudden jump or drop), exhibiting a stable condition [34]. The temperature variation is under normal conditions, i.e., the temperature rises (decreases) in the normal range during charging and discharging (rest). In this case study, we have seen that UL 1974 provides an efficient route to obtain the key values for judging the performance and safety of battery cells.

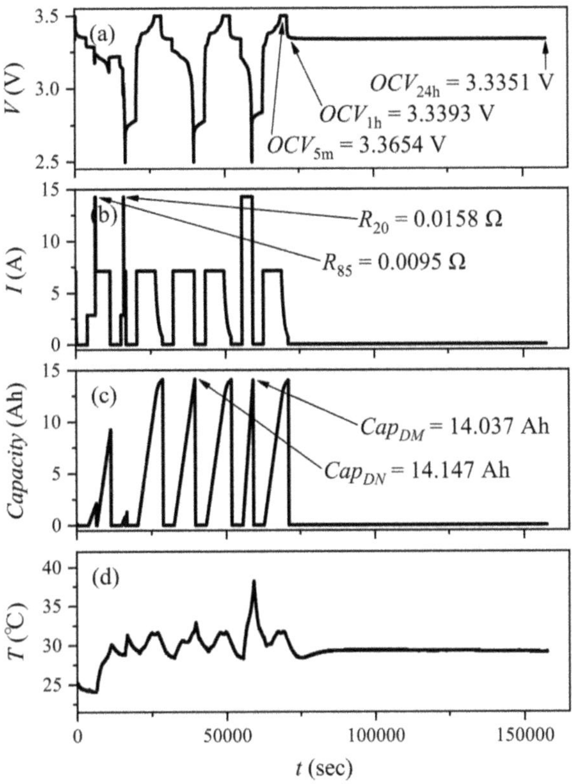

Figure 15.2. The voltage, current, capacity, and temperature variations under test procedure 2 according to UL 1974.

Acknowledgements

H C Chung would like to thank all the contributors to this article for their valuable discussions and recommendations, especially Alan Lo, Daniel Lo, Yaw-Chung Cheng, Hui-Wen Feng, Chaur-Yuh Chen, Bao-Sheng Huang, Kuan-Lin Shih, Ming-Yi You, Jung-Feng Lin, Hsiao-Wen Yang, Yen-Kai Lo, An-De Andrew Chung. The author (H C Chung) thanks Pei-Ju Chien for English discussions and corrections as well as Ming-Hui Chung, Su-Ming Chen, Lien-Kuei Chien, Mi-Lee Kao, and Fu-Long Chen for financial support. This work was supported in part by the Industrial Energy Technology Program of Bureau of Energy (BOE), Ministry of Economic Affairs (MOEA) of Taiwan under contract number 107-D0720.

References

[1] Chung H-C 2020 Engineering integrations, potential applications, and outlooks of Li-ion batteries *engrXiv* https://doi.org/10.31224/osf.io/swcyg

[2] Keck F, Lenzen M, Vassallo A and Li M Y 2019 The impact of battery energy storage for renewable energy power grids in Australia *Energy* **173** 647–57

[3] Radcliffe J C 2018 The water energy nexus in Australia—The outcome of two crises *Water-Energy Nexus* **1** 66–85

[4] Ye Y D, Qiao Y and Lu Z X 2019 Revolution of frequency regulation in the converter-dominated power system *Renew. Sustain. Energy Rev.* **111** 145–56

[5] Zeng H, Sun F and Ge W *et al* 2018 Introduction of Australian 100MW Storage Operation and its Enlightenment to China *2018 China Int. Conf. on Electricity Distribution (CICED)* **2895**

[6] Fernandez-Munoz D, Perez-Diaz J I, Guisandez I, Chazarra M and Fernandez-Espina A 2020 Fast frequency control ancillary services: An international review *Renew. Sustain. Energy Rev.* **120** 109662

[7] Pagliaro M 2019 Renewable energy systems: enhanced resilience, lower costs *Energy Technol.* **7** 1900791

[8] Aziz A, Oo A M T and Stojcevski A 2018 Issues and mitigations of wind energy penetrated network: Australian network case study *J. Mod. Power Syst. Clean Energy* **6** 1141–57

[9] IEA 2020 *Global EV Outlook 2020*

[10] Martinez-Laserna E, Gandiaga I, Sarasketa-Zabala E, Badeda J, Stroe D I, Swierczynski M and Goikoetxea A 2018 Battery second life: Hype, hope or reality? A critical review of the state of the art *Renew. Sustain. Energy Rev.* **93** 701–18

[11] Ahmadi L, Yip A, Fowler M, Young S B and Fraser R A 2014 Environmental feasibility of re-use of electric vehicle batteries *Sustain. Energy Technol. Assess.* **6** 64

[12] Casals L C, Garcia B A and Canal C 2019 Second life batteries lifespan: Rest of useful life and environmental analysis *J. Environ. Manage.* **232** 354–63

[13] Podias A, Pfrang A, Di Persio F, Kriston A, Bobba S, Mathieux F, Messagie M and Boon-Brett L 2018 Sustainability assessment of second use applications of automotive batteries: ageing of Li-ion battery cells in automotive and grid-scale applications *World Electr. Veh. J.* **9** 24

[14] Wood E, Alexander M and Bradley T H 2011 Investigation of battery end-of-life conditions for plug-in hybrid electric vehicles *J. Power Sources* **196** 5147–54

[15] Kamath D, Shukla S, Arsenault R, Kim H C and Anctil A 2020 Evaluating the cost and carbon footprint of second-life electric vehicle batteries in residential and utility-level applications *Waste Manage.* **113** 497–507

[16] Quinard H, Redondo-Iglesias E, Pelissier S and Venet P 2019 Fast electrical characterizations of high-energy second life lithium-ion batteries for embedded and stationary applications *Batteries* **5** 33

[17] Heymans C, Walker S B, Young S B and Fowler M 2014 Economic analysis of second use electric vehicle batteries for residential energy storage and load-levelling *Energy Policy* **71** 22–30

[18] Chung H-C and Cheng Y-C 2019 Action planning and situation analysis of repurposing battery recovery and application in China *J. Taiwan Energy* **6** 425

[19] Hossain E, Murtaugh D, Mody J, Faruque H M R, Sunny M S H and Mohammad N 2019 A comprehensive review on second-life batteries: current state, manufacturing considerations, applications, impacts, barriers and potential solutions, business strategies, and policies *IEEE Access* **7** 73215–52

[20] Gur K, Chatzikyriakou D, Baschet C and Salomon M 2018 The reuse of electrified vehicle batteries as a means of integrating renewable energy into the European electricity grid: A policy and market analysis *Energy Policy* **113** 535–45

[21] Laboratories U 2018 UL 1974—Standard for evaluation for repurposing batteries (Northbrook, IL: Underwriters Laboratories)
[22] Chung H-C and Cheng Y-C 2020 Summary of safety standards for repurposing batteries *Monthly J. Taipower's Eng.* **860** 35
[23] Iclodean C, Varga B, Burnete N, Cimerdean D and Jurchis B 2017 Comparison of different battery types for electric vehicles *IOP Conf. Ser. Mater. Sci. Eng.* **252** 012058
[24] Fathabadi H 2019 Combining a proton exchange membrane fuel cell (PEMFC) stack with a Li-ion battery to supply the power needs of a hybrid electric vehicle *Renew. Energy* **130** 714–24
[25] Ghaviha N, Campillo J, Bohlin M and Dahlquist E 2017 Review of Application of Energy storage devices in railway transportation *Energy Proc.* **105** 4561–8
[26] White C, Thompson B and Swan L G 2020 Repurposed electric vehicle battery performance in second-life electricity grid frequency regulation service *J. Energy Storage* **28** 101278
[27] Chung H C 2021 Charge and discharge profiles of repurposed $LiFePO_4$ batteries based on the UL 1974 standard *Scientific Data* **8** 165
[28] Padhi A K, Nanjundaswamy K S and Goodenough J B 1997 Phospho-olivines as positive-electrode materials for rechargeable lithium batteries *J. Electrochem. Soc.* **144** 1188
[29] Carrilero I, Gonzalez M, Ansean D, Viera J C, Chacon J and Pereirinha P G 2018 Redesigning European public transport: impact of new battery technologies in the design of electric bus fleets *Transp. Res. Proc.* **33** 195
[30] Lian B, Sims A, Yu D M, Wang C and Dunn R W 2017 Optimizing $LiFePO_4$ Battery energy storage systems for frequency response in the UK system *IEEE Trans. Sust. Energy* **8** 385–94
[31] Stan A, Swierczynski M, Stroe D, Teodorescu R, Andreasen S J and Moth K 2014 A comparative study of lithium ion to lead acid batteries for use in UPS applications *IEEE 36th Int. Telecommunications Energy Conf. (INTELEC)*
[32] Zhu J E, Mathews I and Ren D S *et al* 2021 End-of-life or second-life options for retired electric vehicle batteries *Cell Rep. Phys. Sci.* **2** 100537
[33] Chung H-C 2019 *Technology development and field verification of innovative home energy storage system* https://doi.org/10.17605/OSF.IO/PFH3G
[34] Chung H-C 2018 Failure mode and effects analysis of LFP battery module *engrXiv* https://doi.org/10.31224/osf.io/acxsp

IOP Publishing

Thi Dieu Hien Nguyen, Shih-Yang Lin, Hsien-Ching Chung, Ngoc Thanh Thuy Tran and Ming-Fa Lin

Chapter 16

Engineering integrations, potential applications, and outlooks of Li-ion battery industry

Hsien-Ching Chung, Thi Dieu Hien Nguyen, Shih-Yang Lin, Wei-Bang Li, Ngoc Thanh Thuy Tran, Nguyen Thi Han, Hsin-Yi Liu, Hai Duong Pham and Ming-Fa Lin

Lithium-ion (Li-ion) batteries are one of the most widely used secondary battery systems in the world. Compared to other rechargeable batteries, e.g., nickel-cadmium, nickel-metal hydride, and lead-acid batteries, Li-ion batteries are featured with higher operating voltages, higher energy density, low self-discharging, and lower maintenance requirements [1–3].

In this chapter, how to combine and control a set of Li-ion battery cells are described in the 'Engineering integrations' section. Some potential applications in various scales are demonstrated in the 'Potential applications' section. The outlook provides a wide view of the market, cost, safety, reuse and recycle, energy storage as a service, as well as next-generation materials.

16.1 Engineering integrations

As we study extensive fundamental knowledge emerges about a single Li-ion battery cell. It seems very simple to build a battery system in various scales, just combine the battery cells together and they will work together perfectly as we thought. In fact, the truth is not so beautiful, and the industry paid a high prices to study using them together properly. The most important architectures such as the battery management system (BMS) and the energy management system (EMS) will be described briefly, demonstrating how industry integrates Li-ion battery cells and uses them safely in an energy storage system (ESS). In general, the architectures can be applied in other types of energy sources, such as fuel cell [4–9] and vanadium flow battery [10–15].

doi:10.1088/978-0-7503-4685-6ch16

16.1.1 Battery management system

16.1.1.1 An extremely simple charge/discharge strategy for a single battery cell

It is simple to use a single battery cell as a power source. In the charge process, a charger is used with a charge voltage equal to or a little lower than the maximum charge voltage of the battery cell, $V^{cell}_{ch,\ cut}$. In the discharge process, a relay is used to cut the power off when the voltage of the battery cell is equal to or a little above the cutoff discharge voltage of the battery cell, $V^{cell}_{dis,\ cut}$. With these appropriate charge/discharge processes, we can easily use the battery cell, just as we use a cell phone every day. The above strategy on charge/discharge processes can be treated as a very simple BMS.

16.1.1.2 Problems of the extremely simple charge/discharge strategy for multi-battery cells

Many people might think that the above simple BMS can be used for all battery systems with multicells. For example, a battery system is built by two battery cells in a series connection. In the charge process, a charger is used with charge voltage equal to or a little lower than the maximum charge total voltage of the battery cells in series $V^{tot}_{ch,\ cut}$, i.e., $V^{tot}_{ch,\ cut} = 2V^{cell}_{ch,\ cut}$. In the discharge process, a relay is used to cut the power off when the voltage of the battery cell is equal to or a little above the cutoff discharge total voltage of the battery cells in series $V^{tot}_{dis,\ cut}$, i.e., $V^{tot}_{dis,\ cut} = 2V^{cell}_{dis,\ cut}$. These charge/discharge processes seem reasonable and easy to predict. We just double the upper and lower cutoff voltages, and everything will be OK. Indeed, the battery system actually can work under such brutal BMS, but only for a short time. The user will encounter serious problems in the battery system, i.e., over charge and over discharge. These problems result in the reduction of capacity and battery life, and even more dangerously, fire and explosion (thermal runaway).

16.1.1.3 Model battery system

The true story is very different when there is more than one battery cell. This is because it is very hard to obtain battery cells with perfect charge/discharge characteristics, and then use them to build a dream battery system. To show the real case, a model battery system with nine battery cells is given as an example as illustrated in figure 16.1(a). If we want to build a battery system with a triple voltage of the single cell and a triple capacity of the single cell, we should use a series connection of three cells (3S) to triple the voltage and a parallel connection of three cells (3P) to triple the capacity. (*n*S/*n*P indicates *n* battery cells in a series/parallel connection.) The general 3S3P configuration is shown in figure 16.1(b). The cells are connected in parallel first to enlarge the capacity, and then connected in series to enlarge the voltage[1]. The voltage of the cell is denoted by V_{mn}, and the voltages of

[1] Parallel connection first or series connection first is always a design problem and open question in the real system. In a small scale system (e.g., figure 16.1(a)), parallel connection first is an efficient and low-cost choice (i.e., figure 16.1(b)). On the other hand, series connection first can give the system durability because the battery modules in various series can provide power independently. Such a system is very expensive.

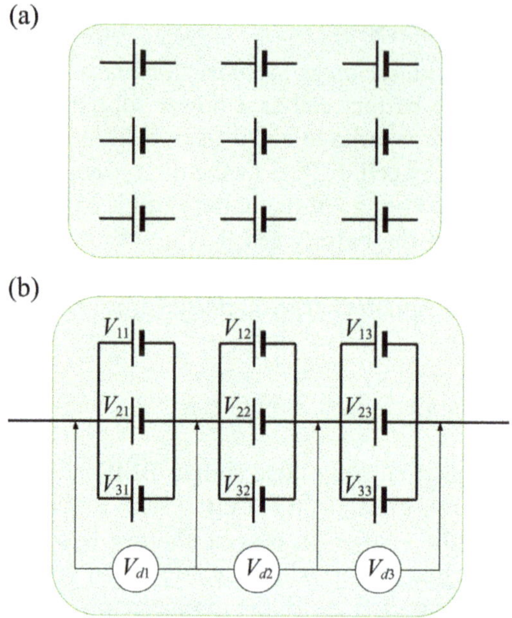

Figure 16.1. Battery cells in 3S3P configuration. (a) The battery system with 9 battery cells. (b) The battery system in 3S3P configuration. The cell voltages are denoted by V_{mn}, and the three voltmeters are denoted by V_{d1}, V_{d2}, and V_{d3}.

the cells in the *i*th series (parallel) connection are V_{in} (V_{mi}). The voltages of the cells in the first/second/third series connection are (V_{11}, V_{12}, V_{13})/(V_{21}, V_{22}, V_{23})/(V_{31}, V_{32}, V_{33}). The voltages of the cells in the first/second/third parallel connection are (V_{11}, V_{21}, V_{31})/(V_{12}, V_{22}, V_{32})/(V_{13}, V_{23}, V_{33}). The cell voltages in the first/second/third parallel connection are detected by voltmeters V_{d1}, V_{d2}, and V_{d3}. We will find that $V_{d1} = V_{11} = V_{21} = V_{31}$, $V_{d2} = V_{12} = V_{22} = V_{32}$, $V_{d3} = V_{13} = V_{23} = V_{33}$ due to rapidly dynamic balance among the parallel battery cells[2].

Since the real battery system is built by the real battery cells with slightly different charge/discharge characteristics, the over charge/discharge problems will happen if we use the system total voltage V_{tot} as the only reference. In the charge process, although the total voltage of the battery system is under the maximum total charge voltage ($V_{tot} < V^{tot}_{ch,\ cut}$), one or some battery cells might exceed the cell charge cutoff voltage ($V_{mn} > V^{cell}_{ch,\ cut}$). For example, in the battery system with 3S3P configuration (figure 16.1(b)), the total voltage of the battery system ($V_{tot} = 2.9V^{cell}_{ch,\ cut}$) is under the maximum total charge cutoff voltage ($V^{tot}_{ch,\ cut} = 3V^{cell}_{ch,\ cut}$), but $V_{d1} = 0.8V^{cell}_{ch,\ cut}$, $V_{d2} = 1.2V^{cell}_{ch,\ cut}$, and $V_{d3} = 0.9V^{cell}_{ch,\ cut}$, indicating that the cell voltages V_{12}, V_{22},

[2] In the parallel connection, the battery cells will automatically balance their voltages dynamically, and we will get a single voltage value eventually. In other words, the battery cells with higher voltages will discharge to charge the battery cells with lower voltages. Their voltages will approach the balance voltage. Usually, the dynamic balance will finish very quickly (less than 0.1 sec) when the charge/discharge rate is small.

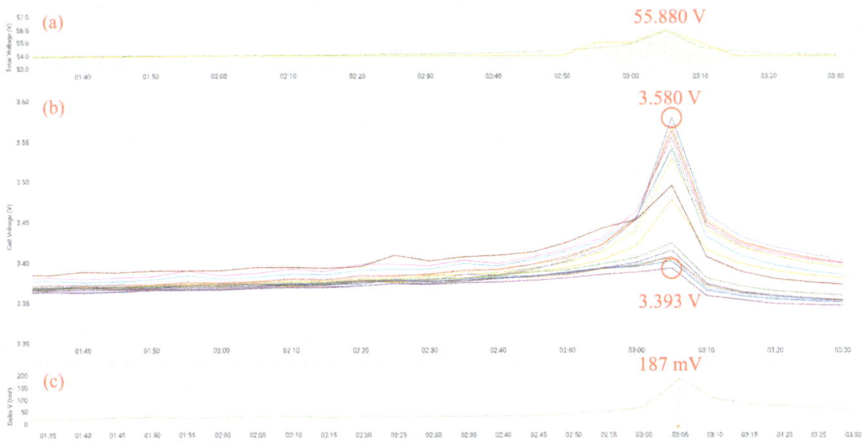

Figure 16.2. Demonstration of over voltage in a real battery system. (a) Variation of the total voltage. (b) Variation of the voltages in various parallel connections. (c) Variation of the ΔV between max and min voltages in various parallel connections.

and V_{32} are over the cell charge cutoff voltage $V^{cell}_{ch,\ cut}$. In the discharge process, although the total voltage of the battery system is above the minimum total discharge voltage ($V_{tot} < V^{tot}_{dis,\ cut}$), one or some battery cells might come under the cell discharge cutoff voltage ($V_{mn} < V^{cell}_{dis,\ cut}$). For example (figure 16.1(b)), the total voltage of the battery system ($V_{tot} = 3.1\,V^{cell}_{dis,\ cut}$) is above the minimum total discharge voltage ($V_{tot} = 3V^{cell}_{ch,\ cut}$), but $V_{d1} = 1.2V^{cell}_{ch,\ cut}$, $V_{d2} = 0.8V^{cell}_{ch,\ cut}$, and $V_{d3} = 1.1V^{cell}_{ch,\ cut}$, indicating that the cell voltages V_{12}, V_{22}, and V_{32} are below the cell discharge cutoff voltage $V^{cell}_{dis,\ cut}$. In these situations, we do not encounter the explicit 'over charge/discharge event of the battery system', but the implicit 'over charge/discharge event of the cell' happens. The cell over charge/discharge problems result in the reduction of capacity of the system and battery life, and even more dangerously, fire and explosion (thermal runaway). This example shows that the naive battery management strategy based on simply enlarging the total charge/discharge voltage limitations is not suitable.

The voltage variations of the real battery system in the Super Double Power Technology Co., Ltd, Taiwan (SDP) during charging is shown in figure 16.2. This data is obtained from a lithium iron phosphate battery (LFP battery) system with 16S5P configuration, using the GreenIce 50 lFP battery cells produced by Chang Hong Energy Technology Co., Ltd, Taiwan. Since the system has been used for more than one year, the difference among charge/discharge characteristics of battery cells becomes larger, and is suitable for demonstrating the over charge situation. The difference among charge/discharge characteristics of battery cells can be observed directly from the voltage difference between max and min cell voltages ΔV (figure 16.2(c)). When the system is near full charge state, ΔV gradually increases from 30 to 187 mV, indicating that the voltages in various parallel connections gradually deviate and reaching the maximum deviation at $\Delta V = 187$ mV.

The highest voltage is 3.580 V pointed by the upper circle in figure 16.2(b). The lowest voltage is 3.393 V indicated by the lower circle in figure 16.2(b). If $V_{ch,\ cut}^{cell}$ is set at 3.5 V, there are eight voltages above/below $V_{ch,\ cut}^{cell}$, indicating that 'over charge event of the cell' happens. However, the total voltage of the system $V_{tot} = 55.880$ V remains lower than the $V_{ch,\ cut}^{tot} = 16V_{ch,\ cut}^{cell} = 56$ V (figure 16.2(a)).

16.1.1.4 Functionalities of the battery management system

To use the battery safely, a well-designed BMS is required. The BMS is an integrated electronic circuit system with optional sensors and control components that manages a battery cell/pack/module/system, providing functionalities such as ensuring the battery operates within its safety region, monitoring the current state of the battery, recording primary data, calculating secondary data, reporting statistical data, controlling the environment, and balancing the cell voltages. Some functionalities of BMS are stated briefly below.

16.1.1.5 Physical and electric quantities detection functions

The basic physical and electric quantities are measured, such as voltage, current, temperature, and pressure.

1. **Total voltage of the system.** The detection points of the total voltage of the battery system can be set before/after the power relay. In the shutdown mode (i.e., the battery system is turned off), the detected voltage before the power relay equals the total voltage of the battery, while it is zero after the power relay.
2. **Voltage in parallel connection (cell voltage).** Voltages in every parallel connection are detected. In the industrial habit, this term is commonly called 'cell voltage.' In fact, it's very hard to know the exact voltage of a single battery cell in the battery system. Especially when the cells are in parallel connection, the exact voltages are affected by the rapidly dynamic balance effect.
3. **Current.** Charge current and discharge current of the battery system are detected. The current can be measured by shunt resistances, Hall current sensors, etc [16–19].
4. **Temperature.** The temperature sensors can be put on any hot spot, e. g., metal-oxide-semiconductor field-effect transistor (MOSFET), different positions in the battery configuration, and environment.
5. **Pressure.** The pressure in the battery system container or a closed box.

16.1.1.6 Calculation functions

With the measured basic physical and electric quantities, further quantities can be simply obtained by comparators, subtractors, multipliers, integrators, and even more complex model calculations.

1. **Max voltage in parallel connection (max cell voltage).** Obtained by the comparator.

2. **Min voltage in parallel connection (min cell voltage).** Obtained by the comparator.
3. **Voltage difference between the max and min voltages in parallel connection** (ΔV). Obtained by the subtractor on the max and min voltages in parallel connection.
4. **Power.** Obtained by the multiplier on the total voltage and current.
5. **Max temperature.** Obtained by the comparator.
6. **Min temperature.** Obtained by the comparator.
7. **Temperature difference between the max and min temperatures** (ΔT). Obtained by the subtractor on the max and min temperatures.
8. **Max pressure.** Obtained by the comparator.
9. **Min pressure.** Obtained by the comparator.
10. **Pressure difference between the max and min pressures** (ΔP). Obtained by the subtractor on the max and min pressures.
11. **State of charge (SOC).** Obtained by the integrator. SOC as an indicator for the remaining capacity t ratio of the battery is defined as, where SOC(t_0) is the previous SOC of the battery, *Cap* is the ampere hour capacity of the fully charged battery, and $I(\tau)$ is the current with positive (negative) value for discharge (charge) [20, 21]. The *Cap* could be chosen as the nominal ampere hour capacity (Cap_N), the latest capacity, or the capacity at a given time for a specific purpose.
12. **State of health (SOH).** Complex model calculation. A simple definition on SOH is given by the ratio of the maximum ampere hour capacity of an aged battery (Cap_{Aged}) to the nominal ampere hour capacity of the new battery (Cap_N), i.e., SOH = Cap_{Aged}/Cap_N [21].
13. **Impedance/Internal resistance.** Complex model calculation. Many techniques are used to measure the impedance/internal resistance of batteries, such as the tests conducted according to the IEC 61951–1 standard [22], IEC 61960–3 standard [23], and ISO 12405–4 standard [24]. In UL 1974, the two-tier direct current (DC) load method is adopted, offering an alternative method by applying two sequential discharge loads of different currents and time durations [25].
14. **Total charge ampere hour (Ah).** Obtained by the integrator.
15. **Total discharge Ah.** Obtained by the integrator.
16. **Total charge energy.** Obtained by the integrator.
17. **Total discharge energy.** Obtained by the integrator.

16.1.1.7 Protection functions

When one or more measured basic physical and electric quantities reach the critical values [26], the BMS should take action to maintain safety and protect the battery.

1. **Over total voltage protection (OTVP).** Possible situation: In charge process, the total voltage over the OTVP value. Possible protection action strategy: Stop charge.

2. **Under total voltage protection (UTVP).** Possible situation: In discharge process, the total voltage under the UTVP value. Possible protection action strategy: Stop discharge.
3. **Over cell voltage protection (OCVP).** Possible situation: In charge process, the cell voltage over the OCVP value. Possible protection action strategy: Stop charge.
4. **Under cell voltage protection (UCVP).** Possible situation: In discharge process, the cell voltage over the UCVP value. Possible protection action strategy: Stop discharge.
5. **Over charge current protection (OCCP).** Possible situation: In charge process, the current over the OCCP value. Possible protection action strategy: (PA 1) Reduce the charge current to the value under the OCCP value. If (PA 1) is not available, stop charge immediately.
6. **Over discharge current protection (ODCP).** Possible situation: In discharge process, the current over the ODCP value. Possible protection action strategy: (PA 1) Reduce the discharge current to the value under the ODCP value. If (PA 1) is not available, stop discharge immediately.
7. **Over charge power protection (OCPP).** Possible situation: In charge process, the power over the OCPP value. Possible protection action strategy: (PA 1) Reduce the charge current to adjust the power under the OCPP value. If (PA 1) is not available, stop charge immediately.
8. **Over discharge power protection (ODPP).** Possible situation: In discharge process, the power over the ODPP value. Possible protection action strategy: (PA 1) Reduce the discharge current to adjust the power under the ODPP value. If (PA 1) is not available, stop discharge immediately.
9. **Short circuit protection (SCP).** Possible situation: Mistouch the positive and negative terminals of the battery system. Possible protection action strategy: Stop discharge immediately.
10. **Over cell temperature protection (OCTP).** Possible situation: (PS 1) In charge/discharge pro- cess, the cell temperature over the OCTP value. (PS 2) In a too hot environment. Possible protection action strategy: (PA 1) Reduce the charge/discharge current to reduce the temperature under the OCTP value. (PA 2) Use a cooler to cool down the battery system. If (PA 1) and/or (PA 2) is not available, stop charge/discharge immediately.
11. **Under cell temperature protection (UCTP).** Possible situation: In a too cold environment. Possible protection action strategy: (PA 1) Use a small charge/discharge current to increase the system temperature over the UCTP value. (PA 2) Use a heater to warm up the battery system. If (PA 1) and/or (PA 2) is not available, stop charge/discharge immediately.
12. **Over environment temperature protection (OETP).** Possible situation: In a too hot environment. Possible protection action strategy: (PA 1) Reduce the charge/discharge current to reduce the temperature under the OCTP value. (PA 2) Use a cooler to cool down the battery system. If (PA 1) and/or (PA 2) is not available, stop charge/discharge immediately.

13. **Under environment temperature protection (UETP).** Possible situation: In a too cold environment. Possible protection action strategy: (PA 1) Use a small charge/discharge current to increase the system temperature over the UCTP value. (PA 2) Use a heater to warm up the battery system. If (PA 1) and/or (PA 2) is not available, stop charge/discharge immediately.
14. **Over voltage difference protection (OVDP).** Possible situation: (PS 1) Too old battery system. (PS 2) Sudden system failure. (PS 3) Wrong assembly of battery cells with different nominal capacity/remaining capacity. Possible protection action strategy: For (PS 1), reduce the charge/discharge current to decrease ΔV under the OVDP value. For (PS 2) and (PS 3), stop charge/discharge immediately. If the BMS cannot judge the situation among (PS 1), (PS 2), and (PS 3), stop charge/discharge immediately.
15. **Over temperature difference protection (OTDP).** Possible situation: (PS 1) Heat accumulation during charge/discharge processes. (PS 2) Electrolyte leakage and/or cell explosion. Possible protection action strategy: stop charge/discharge immediately.
16. **Over pressure difference protection (OPDP).** Possible situation: Possible situation: (PS 1) gas leakage from the cells during charge/discharge processes. (PS 2) Cell explosion. Possible protection action strategy: stop charge/discharge immediately.

16.1.1.8 Alert functions

When the system encounters any of the protection conditions, the alert functions should be triggered to inform the administrator/user. Several inform routes are listed.

1. **Light-emitting diode (LED)** Type: light. Range: short, for on-site personnel.
2. **On-screen message** Type: light. Range: short, for on-site personnel.
3. **Buzzer** Type: sound. Range: short, for on-site personnel.
4. **Bluetooth** Type: digital signal. Range: short, for on-site personnel using bluetooth devices.
5. **Alert signal to control center** Type: digital signal. Range: long, for remote personnel using computers in the control center.
6. **Social media message** Type: digital signal. Range: long, for remote personnel using comput- ers/smart phone. For example: LINE (https://line.me/), Twitter, (https://twitter.com/), Telegram (https://telegram.org/), Google hangouts chat (https://hangouts.google.com/), Microsoft teams (https://www.microsww/microsoft-365/microsoft-teams/group-chat-software), etc.
7. **E-mail** Type: digital signal. Range: long, for remote personnel using computers/smart phone.

16.1.1.9 Physical control functions

When the system encounters serious protection conditions, the charge/discharge processes can be terminated by relays.

1. **Charge relay.** Control the input current.

2. **Discharge relay.** Control the output current.
3. **Dry contact.** Dry contacts for other purposes, e.g., fan control, cooler control, heater control.

16.1.1.10 Communication functions

The BMS can provide the signal to other devices or the control center by various types of communication ports and/or protocols.

1. **RS-232.** Protocol: modubs (RTU, ASCII, TCP), user defined.
2. **RS-485.** Protocol: modubs (RTU, ASCII, TCP), user defined.
3. **RS-422.** Protocol: modubs (RTU, ASCII, TCP), user defined.
4. **CAN bus.** Protocol: CAN bus.
5. **BACnet.** Protocol: BACnet.
6. **USB.**
7. **Ethernet.** Long-range connectivity through internet.
8. **Bluetooth.** Wireless connectivity.
9. **Wi-Fi.** Wireless connectivity.
10. **Zigbee.** Wireless connectivity.

16.1.1.11 Memory functions

The counters in the BMS can be used to record the number of some key events for evaluating the system status as the database is not available.

1. **Cycles of charge/discharge.**
2. **Times of over voltage.**
3. **Times of under voltage**
4. **Times of over current.**
5. **Times of short circuit.**
6. **Times of over temperature.**
7. **Times of under temperature.**

16.1.1.12 Database functions

Database (DB) has the ability to record numerous data, which can give detail operation status.

1. **Data history.** All measured physical and electric quantities are recorded in a time-series database.
2. **Event log.** All events are recorded, such as charge/discharge events, warning events, and protection events.
3. **Basic statistical data.** The time-variation basic statistical data. There are many databases.

1. **Relational databases (RDB).**
   (a) **Oracle.** https://www.oracle.com/database/
   (b) **MySQL.** https://www.mysql.com/
   (c) **PostgreSQL.** https://www.postgresql.org/
   (d) **SQLite.** https://www.sqlite.org/
   (e) **Microsoft Access.** https://www.microsoft.com/en-us/microsoft-365/access

2. **Time-series databases (TSDB).**
   (a) **InfluxDB.** https://www.influxdata.com/
   (b) **Prometheus.** https://prometheus.io/
   (c) **TimescaleDB.** https://www.timescale.com/
   (d) **OpenTSDB.** http://opentsdb.net/
   (e) **Graphite.** https://graphiteapp.org/

16.1.1.13 Energy saving functions

The BMS itself is an energy-consuming device. How to control its energy consumption is a problem.

1. **Normal mode.** Normal energy-consuming situation. The BMS provides full functions when the battery system is under usage.
2. **Sleeping mode.** Low energy-consuming situation. The BMS provides partial functions. When the battery system is unused for a short time, the system can be woken up immediately by charge or discharge.
3. **Shutdown mode.** Zero energy-consuming situation. The BMS provides partial functions. When the battery system is unused for a long time, e.g., during transportation, the system cannot be woken up directly by charge or discharge. The input/output of electricity is locked for safety. The system can be woken up by power on the BMS.

16.1.1.14 Balance functions

Some BMS can balance the voltages in various parallel connections.

1. **Passive balance.** A common design is to use resistors, reducing the voltages of high-voltage cells. Such a design is energy-consuming.
2. **Active balance.** The ideal design is to move the energy from high-voltage cells to low-voltage ones. The dream design will cost zero energy consumption. Technically, it is very hard to achieve.

The aforementioned description of the functionality of BMS is a reference for designing a safety battery system. We should mention one thing: 'Not all functionalities are required in the BMS.' The design should choose the functions according to system safety, usage scenarios, design costs, etc. For example, the over discharge current protection (ODCP) for power application and stationary application is different. For stationary application, the discharge current rate C_R is often lower than 0.3 h^{-1}, the ODCP is set at $I = 0.3Cap_N$ h^{-1}[3].For power application, the discharge current rate is often higher than 3 h^{-1} for a short time. If the ODCP is set at $I = 0.3Cap_N$ h^{-1}, the power will be cut off without warning during battery usage. It is a failed design and may put users in danger. If the battery is installed in an electric car, sudden power loss will cause a car accident. In some usage scenarios, some protection functions should be turned off to ensure the power provided. For example, it is suggested to turn off the ODCP while designing a vehicle starter.

[3] C_R, also called C rate, is the normalized current per unit of nominal ampere hour capacity Cap_N, i.e., $I = C_R Cap_N$

16.1.1.15 Real case

In building an LFP battery system with 16S5P configuration, SDP chose to use a BMS providing measurements on 16 voltages in parallel connections, 4 temperatures, and 1 current (BMS architecture is shown in figure 16.3). The 16 voltages in parallel connection in the battery module are measured by 17 detection points[4].The temperature sensors contact the surfaces of battery cells. The charge/discharge current is detected by the bidirectional Hall current sensor [16–19], providing safer no-contact sensing[5].The battery system can be charged by the charger. The charge current passes through the charge relay (R1), the diode (D1), and the fuse. If something is wrong during charge, the charge relay will be turned off immediately by the BMS. The D1 diode guarantees that the current cannot feed back to the charger, avoiding damage by the feedback current. On the other hand, the battery system can discharge to the load. The discharge current passes through the discharge relay (R2),

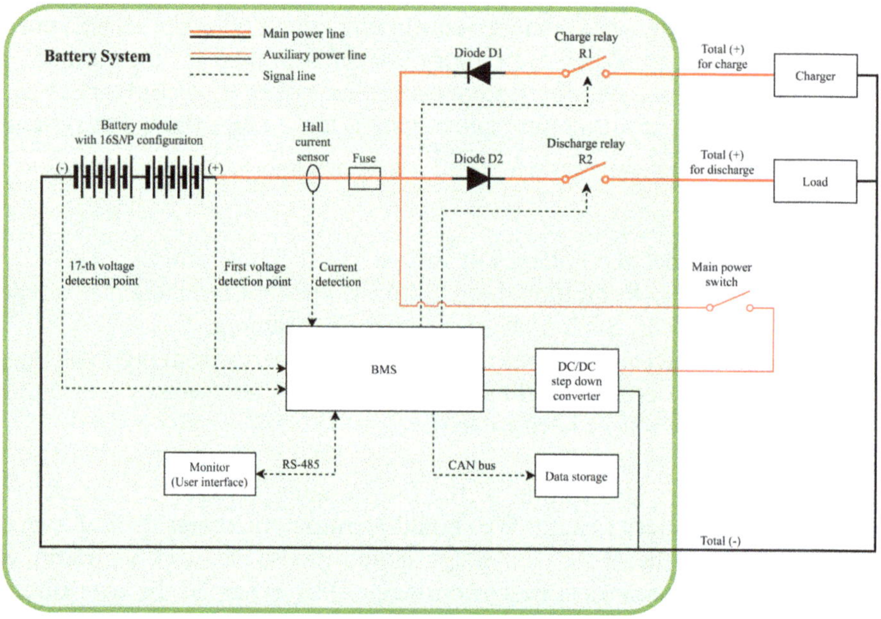

Figure 16.3. Architecture of the BMS for a battery system with 16S*N* P configuration. The positive and negative main power lines are denoted by red and black heavy lines, respectively. The positive and negative auxiliary power lines are denoted by red and black light lines, respectively. The signal lines are denoted by the dashed lines with arrows indicating signal directions.

[4] We do not need 32 (16 × 2 = 32) detection points to measure the 16 voltages, 17 (16 + 1 = 17) detection points are sufficient. Because the voltage potential of the negative terminal of the *m*th cells in parallel connection is almost equal to the voltage potential of the positive terminal of the $(m + 1)$-th cells in parallel connection as the charge/discharge current is small, the number of voltage detection points can be reduced from $2M$ to $(M + 1)$, where M is the number of total cells in series connection.

[5] The charge/discharge current detection can also be achieved by two mono-directional Hall current sensors. In this BMS, we choose the bidirectional one.

the diode (D2), and the fuse. If something is wrong during discharge, the discharge relay will be turned off immediately by the BMS. The D2 diode guarantees that the current cannot feed back to the load. The fuse is used on the main power line as a passive over current protection device. It is backup protection when charge and/or discharge relays do not work normally. The auxiliary power of the BMS is obtained from the DC/DC step down converter, which converts the high-voltage power to the low-voltage auxiliary power[6].The main power switch is used to control the power on/off of the whole battery. When the main power switch is off, there is no auxiliary power passing through the BMS (almost zero power consumption), and the charge and discharge relays are off (shutdown mode, no current in and out). When the main power switch is on, the auxiliary power passes through the BMS, and the charge/discharge relays are turned on and monitored by the BMS (normal mode, current can in and out). The BMS provides bidirectional serial communication RS-485 to the touch panel monitor. Some parameters can be set from the user interface of the touch panel monitor. The detecting and calculating data are sent to the local/remote database by CAN bus.

16.1.1.16 Software for monitoring the battery system

Software for monitoring the battery system plays an important role in displaying the crucial data from the BMS. The remote dashboard of the BMS developed by SDP is shown in figure 16.4. In the panel of basic information, some important physical and electric quantities as well as the alert messages are displayed.

(a) **Total voltage.** The total voltage is calculated from the sum of all voltages in parallel connection. It does not come from the direct measurement on the positive and negative terminals of the battery system.

(b) **Current.** The current measured by the Hall current sensor. The charge (discharge) current is chosen as a positive (negative) value.

(c) **Power.** The power is calculated by multiplying the total voltage and current. The charge (discharge) power is chosen as a positive (negative) value.

(d) **Max cell voltage.** Max voltage in parallel connection.

(e) **Min cell voltage.** Min voltage in parallel connection.

(f) ΔV. The difference between the max and min voltages in parallel connection (in unit mV).

(g) **Max temperature.** Max temperature of the 4 temperatures.

(h) **SOC.** Remaining capacity.

[6] The high-voltage power should be obtained from the positive and negative terminals of the battery system instead of obtaining from partial voltages in parallel connection. Otherwise, the electricity will run out partially. Chung has encountered a special case. An LFP battery system with 8S4P configuration ran out its electricity after one month of rest. It is strange since the self-discharge rate of the LFP battery is about 2% per year. After checking all voltages in parallel connection, the 1st ~ 5th voltages were still in the normal working range, while the 6th ~ 8th voltages were under the discharge cutoff voltage. Since there was no load connected, the energy consumption role was attributed to the BMS. After circuit checking, we knew that the auxiliary power of BMS is obtained from the 6th ~ 8th voltages. This is the wrong demonstration. Such a design of auxiliary power will cause voltage unbalancing in the battery system, reduced life span, and safety problems.

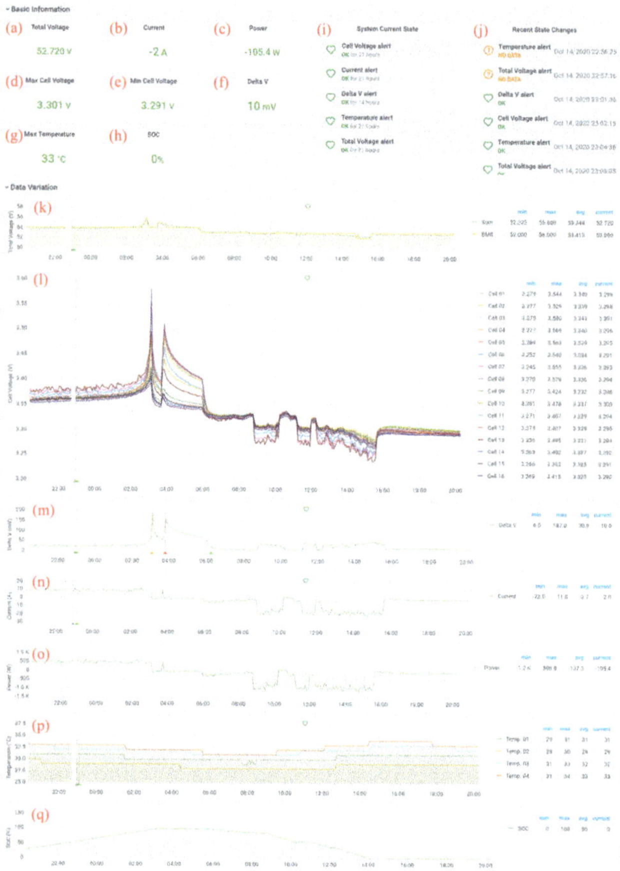

Figure 16.4. Dashboard of BMS. Basic information: (a) Total voltage. (b) Current. (c) Power. (d) Max cell voltage. (e) Min cell voltage. (f) Voltage difference between max and min cell voltages. (g) Max temperature. (h) SOC (i) System current state. (j) Recent state change. Data variation: (k) Total voltage. (l) Cell voltage. (m) Delta V. (n) Current. (o) Power. (p) Temperature. (q) SOC.

(i) **System current state.** ‘OK’ indicates that the data is in the normal range. When data is outside the normal range for a specific time, alert messages will appear.

(j) **Recent state change.** The historical log for shown recent state change.

In the panel of data variation, the variations of some important physical and electric quantities are shown.

(k) **Total voltage.** Two values of the total voltage. ‘Sum’ indicates that the value is calculated from the sum of all voltages in parallel connection. ‘BMS’ indicates that the value is given from the BMS directly.

(l) **Cell voltage.** The 16 voltages in parallel connection.

(m) ΔV. The difference between the max and min voltages in parallel connection (in unit mV).

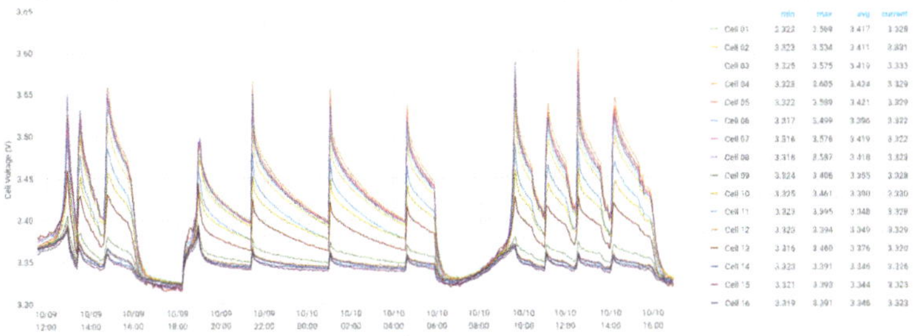

Figure 16.5. Demonstration of the OCVP. Owing to the OCVP of BMS, the highest cell voltage will not be higher than the OCVP value (3.600 V in this case).

(n) **Current.** The current measured by the Hall current sensor. The charge (discharge) current is chosen as a positive (negative) value.
(o) **Power.** The power is calculated by multiplying the total voltage and current. The charge (discharge) power is chosen as a positive (negative) value.
(p) **Temperature.** The temperatures obtained from the 4 sensor are displayed.
(q) **SOC.** Remaining capacity.

The maximum (max), minimum (min), average (avg), and current values in the monitor time range are given on the right side.

With the well-designed BMS, the over voltage condition during charge has been controlled and the battery system works in a safe region. When the highest cell voltage is larger than the over cell voltage protection (OCVP) value (3.600 V in this case), the charge relay is turned off by the BMS and the highest cell voltage will not go over the OCVP value (figure 16.5). The OCVP will continue until the highest cell voltage gradually down to the release point of OCVP (3.400 V in this case). After the release point of OCVP, the charge relay will turn on and the battery system is ready for charge. If the charger providing current after OCVP release immediately, the cell voltages will rapidly increase to the OCVP value, and then the OCVP start to function again[7].

16.1.2 Energy management system and software

16.1.2.1 Energy management system

An energy management system (EMS) provides hardware and software to monitor, control, and optimize the energy systems. Also, EMSs can be used in various scales of systems, such as home energy management [27–29], automated control in buildings [30–32], and microgrids [33–38].

[7] Readers might notice that the highest cell voltage of cell 04 is 3.605 V, which is over the OCVP value. Does it mean that the OCVP does not function? Actually, the OCVP does function. However, there is a time duration from voltage detection, voltage comparison, to protection action. The cell voltage might have the possibility of over the OCVP value. Hence, how to tune the OCVP requires trial and error to keep the cell voltage really in the safe working region.

In the early 1990s, EMSs were commonly developed based on proprietary hardware and operating system software. Some EMS companies developed their proprietary hardware, e.g., Control Data Corporation (CDC), General Electric (GE), Hitachi, Siemens, and Toshiba. For EMS companies that did not construct their hardware, they usually relied on hardware developed by other companies, e.g., Digital Equipment Corporation (DEC), Gould Electronics, and Modular Computer Systems, Inc. (MODCOMP). While developing EMSs, many companies found that it is expensive to use the proprietary systems and began to search for more economical solutions by applying standard industrial hardware, e.g., those provided by DEC, International Business Machines Corporation (IBM), and Sun Microsystems, Inc. (Sun).

The typical operating systems used in EMSs were UNIX and DEC OpenVMS. Since 2004, many EMS companies (e.g., ASEA Brown Boveri (ABB), Alstom, and Open Systems International, Inc. (OSI)) had begun to use Windows-based operating systems. Hence, customers had a chance to choose a UNIX-, Linux, or Windows-based EMS solution. Recently, owing to the development and popularity of Linux-based systems, open-source software, and cloud technology, EMS companies started to extend their services to the cloud, developing the cloud-native open-source EMS on the Linux-based system. In advancing energy efficiency, EMS can optimize the energy system by reporting detailed energy usage of individual equipment. Furthermore, current cloud-based EMSs possess the capability to control heating, ventilation, and air conditioning (HVAC)8[8] and other energy devices remotely from a place outside the field. The detailed real-time energy report for each piece of equipment is collected for users to realize the actual energy consumption of their energy system, guiding users to find the best energy savings opportunities [42]. The EMSs developed for control and monitoring of the electromechanical facilities in a building (e.g., HVAC and lighting devices) can also be applied as part of an automated control system in buildings. The system scale of EMS can range from a single building to a group of buildings. For example, from personal rooms/houses/offices to universities or factories. Not only can electricity be managed by EMSs, but also gas, oil, and water consumption. The reported data obtained from various sensors (such as voltage, current, gas flow, oil flow, and water flow sensors) can provide a generalized energy usage analysis and give improvement suggestions [43].

16.1.2.2 Energy management software

Energy management software (EMSW)9[9] plays an important role in the EMS, providing many functionalities, such as real-time measuring data display, utility bill record for tracking, control systems of building HVAC and lighting, energy

---

[8] Heating, ventilation, and air conditioning (HVAC) is the technology that controls indoor and vehicular environmental comfort. Its primary attempt is to give indoor thermal comfort and acceptable air quality. The design of an HVAC system requires multidisciplinary knowledge, e.g., mechanical engineering, thermodynamics, fluid mechanics, and heat transfer, automatic control, and system integration [39–41].

[9] The abbreviation of the 'energy management software' is chosen as EMSW to avoid using the same abbreviation for the abbreviation of the 'energy management system.'

consumption modeling and simulation, carbon footprint calculation, and sustainability reporting, remote communication, demand response, and energy consumption audits [44]. For buildings or communities, EMSWs often provide tools for energy consumption analysis, load analysis, figuring out unusual energy usage, and cost-down suggestions. EMSWs collect energy-related data and analyze the data for three primary goals, i.e., reporting, monitoring, and engagement. Reporting may include energy-related data collection and verification, energy system benchmarking, and figuring out objects for energy reduction. Monitoring may include energy consumption tracking and tendency analysis for searching and suggesting cost-saving opportunities. Engagement may include automated/manual real-time actions to control the energy system. Meanwhile, communication between system managers and users is initiated to improve energy conservation. One engagement method that has recently gained popularity is the real-time energy consumption display available in internet applications or an on-site energy dashboard/display. EMSWs collect real-time data for showing the current state of the energy system and preserve historic data for data analysis after a later time. The time interval of the real-time data ranges from second-by-second, minute-by-minute smart meter acquisitions to quarterly billing records. The data is collected from various sources, such as interval/external meters (volt, current, temperature, and flow meters), building automation systems (BAS), energy data from utilities/facilities. Past bills recorded in EMSW can be used as checkpoints to verify the benefit of EMS by comparison of energy consumption between before and after EMS installation. For example, a monthly (weekly) electricity bill can provide 12 (52) checkpoints for energy-saving efficiency checking. The energy data analysis of EMSWs helps users to deal with complex data flow and apply complex mathematical formulas and computer algorithms for analyzing, forecasting, and tracking energy systems. Energy data analysis assists energy managers to combine energy and non-energy data for making crucial performance indicators, such as carbon footprint, greenhouse gas (GHG), and renewable heat incentive (RHI). EMSWs give these crucial indicators for energy managers to meet governmental climate change policies and energy efficiency certifications. Recent energy analysis further introduces artificial intelligence and machine learning to analyze the energy consumption of buildings, establish a memory of energy usage patterns, learn/identify the good/bad energy usage behaviors, and figure out abnormal energy usage conditions. Reporting tools of EMSWs are developed for owners and executives to effectively automate energy consumption and carbon emissions auditing. Energy consumption and cost data from many buildings/facilities/energy systems can be collected and compared within a specific EMSW, avoiding time-consuming human data gathering and reporting. Especially, EMSWs offer more detailed energy information with shorter time intervals (often minute-by-minute) than regular utility billing with longer time intervals (often monthly or weekly). An additional advantage of EMSWs is that outside and indirect factors influencing energy consumption (e.g., environmental temperature/humidity, weather conditions, and office/building/facility occupancy) can be combined in the report for further analysis. Bill verification and analysis are part of the main features in EMSWs. The energy consumption measured by the internal meters of the energy system can be compared to the billed energy consumption. First, the energy consumption data from the bill can serve as checkpoints to adjust the accuracy of EMSs. Second, the billed data can be an indicator to show the benefit of the installation of EMSs. Third, an EMS with high resolution and accuracy data can be an assisted tool to confirm the bill, avoiding errors from the utilities

Monitoring tools of EMSWs give real-time measured data display and historical data tracking. Various tools for benchmarking are also developed, e.g., energy consumption per person, weather/temperature normalization, complex computer algorithms, and various energy models. A well-developed EMSW can assist facility or energy managers to identify anomalous energy consumption (or energy leaking) and figuring out savings opportunities.

EMSWs can also be used to discover malfunctioning or inefficient equipment, and then suggest users remove or replace the equipment. For instance, an unexpected energy peak at a specific time every day might be resulting from an improper device (energy leaking object) or malfunctioning timer. The energy leaking can be turned off by replacing the device or adjusting the timer.

EMSWs may offer alerts or warnings, through light, sound, text, and/or email, as values exceed pre-defined thresholds according to energy consumption or cost. Owing to the popularity of smartphones and mobile devices in recent years, EMSW begins to support offering alerts through social media messaging applications, such as LINE, Twitter, WeChat, Telegram, Skype, etc. These thresholds for alerting can be set at absolute levels, or be set at relative levels provided by energy models. Recently, portable devices (such as smartphones and tablets) have become mainstream platforms for EMSW use in small-scale home EMSs [45, 46].

Engagement of EMSWs indicates data-based responses. The data can be obtained from the collected energy consumption or indicators given by model analysis. The responses can be some actions taken by human or computer systems. For example, the HVAC can respond to environmental temperature fluctuation. When cooperating with the EMS, the balance between comfort air condition and energy consumption will be reached. On the other hand, engagement can link the daily choices of users with building energy consumption. By showing real-time energy consumption information, users can realize the immediate effect of their actions, avoiding unnecessary energy consumption actions next time.

16.1.2.3 Demonstration of the energy management system

The off-grid solar energy system developed by SDP includes four parts, including (1) the photovoltaic (PV) panels generating the energy, (2) the ESS storing the energy, (3) the load indicating the energy consumption, and (4) the diesel generators providing the backup energy. The basic working strategy is described below. The PV panels generate energy for the load and/or ESS. When the power of PV is more than the power consumed by the load, the rest energy will be stored in the ESS. When the power of PV is less than the power consumed by the load, the insufficient energy will be supplied by the ESS. If the ESS is out of energy, the diesel generators will take over the supply work.

The EMS developed by SDP can control the energy input of PV, the energy input/output of the ESS, the energy output to the load, and the backup power from the diesel generators. The EMS can work with default parameters to ensure the functionality of the basic working strategy. Also, the parameters and sensors in the EMS can be adjusted to fit some special cases. For example, we provided part of the EMS to Industrial Technology Research Institute (ITRI), Taiwan, who has an idea to build an off-grid solar energy system. The EMS parameters will dynamically be changed according to the weather database in Central Weather Bureau, Taiwan. Solar energy is expected to be largely used based on the predicted weather conditions.

16.1.2.4 Demonstration of the energy management software

SDP develops the EMSW for the EMS based on the Linux operating system, an open-source Unix-like operating system, as shown in figure 16.6. The EMSW monitors the energy input of PV panels, the operation status of the ESS, and the energy output to the load. In the energy generation part, the power, voltage, and current of PV panels are monitored. In the energy consumption part, the power, voltage, and current of AC load are monitored. In the operation status of the ESS, the total voltage, current, max/min cell voltages, and SOC are monitored. Their recent variations of important parameters are displayed for users to realize the energy generation time as well as for users to analyze/judge their energy usage habits/conditions. For example, the PV panels began to generate energy at 6:00, and the power gradually grew up to the highest power about 12:00 13:00, and then gradually decreased to zero at 18:30 (figure 16.6(b)). (The zigzag of the curve is due to clouds

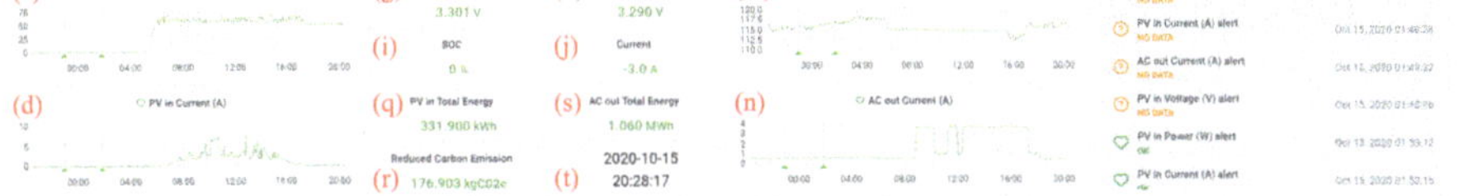

Figure 16.6. Dashboard of the EMSW of solar ESS. Basic information about PV in: (a) power of PV in displayed by gauge. (b) Power variation of PV in. (c) Voltage variation of PV in. (d) Current variation of PV in. Basic information of ESS: (e) Total voltage of ESS displayed by gauge. (f) Total voltage variation of ESS. (g) Max cell voltage of ESS. (h) Min cell voltage of ESS. (i) SOC of ESS. (j) Current of ESS. Basic information of AC out: (k) Power of AC out displayed by gauge. (l) Power variation of AC out. (m) Voltage variation of AC out. (n) Current variation of AC out. Alert message: (o) System current state. (p) Recent state change. Statistic information: (q) total energy of PV power generation. (r) Reduced carbon emission calculated from the total energy generated by PV. (s) Total energy of AC power consumption. Other information: (t) current date and time.

covering the PV panels.) The ESS was in normal working status, e.g., the total voltage and max/min cell voltages were in the normal range (figure 16.6(f)–(h))[10]. Our normal energy consumption is about 125 W, this energy is used for the working light (figure 16.6(l)). If the weather is too hot, the HVAC will be turned on, and the energy consumption will go up to 800 W.

Alert messages are shown on the right side of the monitor dashboard. If there is something wrong, the system's current status will change from 'OK' to 'Alert,' or other hint messages. The recent state changes are also given for users to track the system status. For example, all system current states show 'OK,' indicating that the system is working in good condition (figure 16.6(o)). However, there was a signal loss event at 1:48 ('No data' shows in figure 16.6(p)). This example also demonstrates the importance of event log. The state change history will tell us something special/abnormal even we are not on-site. (At 1:48, we slept at home, and we are not in the company.)

The input total energy of PV panels and the output total energy of loads are given as a brief report of the energy system (figures 16.6(q) and (s)). The reduced carbon emission displays the reduced carbon footprint, which is calculated from the total energy generated by PV panels (figure 16.6(r)).

The EMSW also provides a brief statistical report on the input/output energies. In the basic information, the energy generated by PV panels today, this week, this month, this year, and all-time, are shown in figures 16.7(a)–(e), respectively. The energy consumption by loads today, this week, this month, this year, and all-time, are shown in figures 16.7(f)–(j), respectively. This information gives users a brief understanding of the input/output energies. In the statistic report, the input/output energies today, this week, this month, and this year are given. From the hourly energy today, the report reflects the facts that the solar energy is generated during the day (no solar energy is generated at night) as shown in figure 16.7(k), and the energy consumption is larger (almost constant) during (outside) the working time as shown in figure 16.7(p). The weekly, monthly, and annual reports will reflect the seasonal fluctuation of solar energy (the feature of renewable energy), the seasonal usage of HVAC, as well as the sudden reduction in holidays and weekends.

16.2 Potential applications

Roughly speaking, the design of battery systems deviates in two directions, i.e., power and stationary applications. For power applications, the battery systems are designed to output high power for a short time, and the product life cycle is short. For stationary applications, the battery systems are designed to output low power for a long time, and the product life cycle is long. Some potential power and stationary applications in various scales will be demonstrated below.

[10] The too detailed data of the ESS will not be displayed in the EMSW. The detailed data is monitored by the BMS. In the EMSW, it is sufficient to show several important data as a brief status reference.

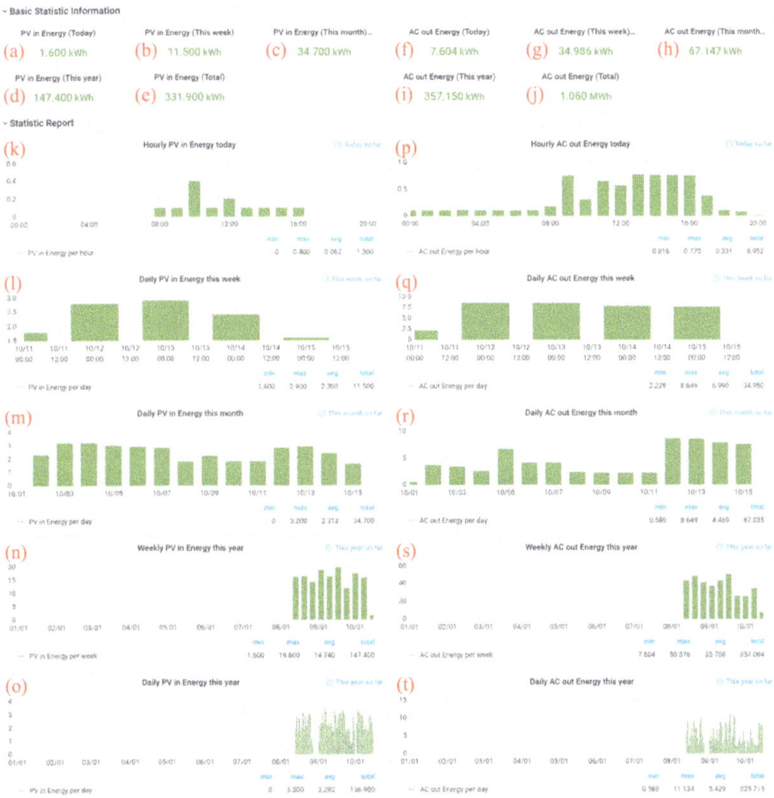

Figure 16.7. Energy usage analysis of the solar ESS. Basic statistic information about PV in and AC out: (a) PV energy generation today. (b) PV energy generation this week. (c) PV energy generation this month. (d) PV energy generation this year. (e) Total PV energy generation. (f) AC energy consumption today. (g) AC energy consumption this week. (h) AC energy consumption this month. (i) AC energy consumption this year. (j) Total AC energy consumption. Statistic report of PV in and AC out: (k) hourly energy of PV in today. (l) Daily energy of PV in this week. (m) Daily energy of PV in this month. (n) Weekly energy of PV in this year. (o) Daily energy of PV in this year. (p) Hourly energy of AC out today. (q) Daily energy of AC out this week. (r) Daily energy of AC out this month. (s) Weekly energy of AC out this year. (t) Daily energy of AC out this year.

16.2.1 Small-scale power applications

16.2.1.1 Two- and three-wheelers

Two- and three-wheelers are the fastest growing transport mode in many low and middle-income countries. Although Asia has the largest share of the global motorcycle fleet, many African countries have the highest motorcycle growth rates in the world. Electric micromobility options have expanded rapidly since their emergence in 2017. Shared electric scooters (e-scooters), electric-assist bicycles (e-bikes), and electric mopeds are now available in over 600 cities across more than 50 countries worldwide. In 2020, an estimated stock of 350 million electric two/three-wheelers worldwide, the majority of which are in China (up 25%), driven by bans in many Chinese cities on two-wheelers with internal combustion engines [47].

Experts agree that two/three-wheelers are the first priority in moving to electric mobility. Scenario calculations using the UN Environment eMob calculator exhibit that assuming a steep and global shift to 90% battery-electric motorcycles sales by 2030 could lead to reductions of about 11 billion tons CO_2 emissions between now and 2050. Meanwhile, overall monetary savings resulting from lower fuel, lower maintenance costs, and higher purchase price of electric motorcycles could amount to about USD 350 billion by 2050 [48].

According to statistics from IEK of ITRI, Taiwan, the sales volume of electric motorcycles reached 82 000 in 2018, an increase of 2.3 times compared with 2017, and reached 150 000 in 2019, with an annual growth rate of 82.9%. In 2019, Gogoro still ranked first in electric motorcycles, with a market share over 90%, followed by manufacturers such as Kymco, CMC e-moving, and Yamaha. Although the proportions of some motorcycle manufacturers have declined compared with 2018, the increase in overall sales has increased profits. The output value of electric vehicles in 2019 has a growth rate of about 55%.

Taiwan's electric motorcycle industry has independent development capabilities. From 2019 to 2020, manufacturers have successively introduced heavy-duty electric motorcycle models. Under the development trend of government policy, market, and technology, sales of the electric motorcycle are growing. In addition to business-to-customer (B2C) sales, the business-to-business (B2B) sales of electric motorcycles will rise with the increase in business cooperation, and the related output value will also be affected by the increase in domestic demand and international exports. The overall observation has a trend of increasing year by year.

Since 2018, the number of new listings for electric and diesel motorcycles purchased by Taiwanese consumers is getting closer every year. From the second half of 2018 to 2019, in addition to the release of new models by electric motorcycle manufacturers, fuel motorcycle manufacturers have also stepped into the development of electric motorcycles. The price of vehicles ranges from USD 1300 to 3300, of which light-duty models fall between USD 1300 and 2000 and heavy-duty models about USD 2300 to 3300. The entry of fuel motorcycle manufacturers has injected more options into the electric motorcycle market and has gradually shifted the consumer market. Looking back on 2016, the market share of electric motorcycles was only 2.7%, and the market share in 2017 was 4.9%; in 2018, the market share has increased to 10.9%; in 2019, it reached 15%. It is estimated that in 2020, the number of listed diesel motorcycles will be 646 000, and the number of electric motorcycles will be 176 000, accounting for 21.4% of the total motorcycle sales.

16.2.2 Medium-scale power applications

16.2.2.1 Electric car

Electric vehicles (EVs) are the technology of choice for eco-friendly transport mode because they have no emissions tied to their operation. They can be powered with clean, renewable energy, and even when charged with power from nonrenewable sources. The inherent fuel efficiency of electric motors versus internal combustion engines means they release less carbon per km.

Sales of electric cars topped 2.1 million globally in 2019, surpassing the record in 2018, boosting the stock to 7.2 million electric cars. Electric cars, accounting for 2.6% of global car sales and about 1% of global car stock in 2019, registered a 40% year-on-year increase. After entering commercial markets in the first half of the decade, electric car sales have risen very rapidly. In 2010, only about 17 000 electric cars were on the world's roads. By 2019, that number had grown to 7.2 million, with 47% in China; nine countries had more than 100 000 electric cars on the road, and at least 20 countries reached market shares above 1% [47].

The 2.1 million electric car sales in 2019 demonstrate a 6% growth from the previous year, down from year-on-year sales growth at least above 30% since 2016. Three underlying reasons explain this trend [47].

Car markets contracted. Total passenger car sales volumes were depressed in 2019 in many key countries. In the 2010s, fast-growing markets (e.g., China and India) for all types of vehicles had lower sales in 2019 than in 2018. Against this backdrop of sluggish sales in 2019, the 2.6% market share of electric cars in worldwide car sales constitutes a record. In particular, China (at 4.9%) and Europe (at 3.5%) achieved new records in EV market share in 2019.

The levels of EV purchase subsidies were decreased in many major markets. China reduced the EV purchase subsidies by about 50% in 2019 (as part of a gradual phase-out of direct incentives set out in 2016). The US federal tax credit program ran out for major EV manufacturers (e.g., General Motors and Tesla). These actions caused obvious drops in EV sales in China and the United States over 2019.

About a decade ago, the consumer profile in the EV market is adopters and technophile purchasers. Now, the consumer profile evolves to the general public. The expectations of EV from consumers improve the EV technology and new models. In 2018–2019, several common EV models show higher energy density of the power battery (about 20%–100%) than their counterparts in 2012. On the other hand, battery costs have largely lowered (~85% compares to 2010) owing to the advance of Li-ion technologies and manufacturing techniques. Some mass-adoption models (e.g., the Tesla Model 3) lead to a peak in sales in major markets (e.g., the United States) in 2018. Recently, EV Manufacturers have revealed rich choices of EVs, and many of them are predicted to enter the market in 2020 or 2021. For the next five years, EV manufacturers have prepared to release about 200 new EV models (many models are designed for the popular sport-utility EV market portion).

16.2.3 Large-scale power applications

16.2.3.1 Electric bus

The electric bus market has shrunk after a sales peak globally in 2016. In 2019, the registration number of new electric buses reached 75 000, which is reduced ~20% from 93 000 vehicles in 2018. In 2019, the total registration number is ~513 000 worldwide, which is gradually grew up ~17% from 2018 [47]. Currently, China is the largest market for electric buses. About 95% of registered electric buses were manufactured and sold in China. However, the annual registration number declined by ~20% in 2019, owing to the drop of governmental purchase subsidy (about 40% compared to 2016 [49]) and the market shrinkage.

In Europe and the United States, the electric bus registration number has largely grown. However, this number is lower than that in China by about two orders, leading to a slight distribution decrease of the Chinese market. Up to now, the majority of electric buses still work in China, i.e., about 500 000 buses, about 98% of the global registration number.

In India, after the implementation of the Phase II of the Faster Adoption and Manufacturing of Electric Vehicles (FAME II) programme [50], electric buses were quickly introduced in 2019. Now, we can see electric bus fleets in many cities, e.g., Bengaluru, Kolkata, Mumbai, and Pune. Over a year, the registration number of new buses has doubled, approaching 450 units and resulting in the number of total electric buses over 800 in 2019.

In Europe, 1900 electric buses were registered in 2019. This number is double that of 2018. Most of Europe's 4500 electric buses are used in France (600 units), Germany (450 units), Netherlands (800 units), and United Kingdom (800 units). It should be noted that the electric bus registration number in Europe is over that of natural gas buses (an alternative fuel bus for public transportation [51]. In 2019, the registration number of electric buses in Europe was bigger than that in North America (~2250 units) [52]. Recently, 63 new electric buses have been introduced into Mexico City [53–55].

The Taiwan government announced that urban buses will be fully electrified in 2030 [56, 57]. However, since 2011, Taipei Bus Co., Ltd took the lead in introducing two electric buses [58, 59]. In the past ten years, there have been no more than 500 electric buses in Taiwan. The main reason is that the passenger transport industry feels that the cost of electric buses is more expensive and the cruising mileage does not meet the operating mileage.

Faced with these two issues, Great City Life Bus took the lead in introducing fast-charging electric buses based on operational needs. Each charging time only takes 15 to 30 min. In the case of short charging times, it can reduce the mileage requirements, and thereby reduces the battery capacity as well as effectively cutting down the battery cost. Hence, the past slow-charging induced phenomenon 'one electric vehicle is equipped with one charger' is changed to 'multiple electric vehicles are equipped with one charger,' which saves the purchase cost of chargers and the land scale of charging stations. In addition to the reduction in cost, because the charging speed is fast, it can be charged multiple times a day, so the daily operating mileage is not limited by the cruising mileage, which can appropriately meet the operating needs [60].

Under the advantages of fast-charging electric buses, Great City Life Bus introduced two vehicles in 2017. After half a year of trial operation, it introduced electric buses in batches and put them into operation in January 2018. The proper rate during the operation period reached 99%, creating many national firsts in Taiwan: No. 1 in the country in terms of 'number,' currently, there are 68 electric buses; No. 1 in the country for 'operating mileage,' exceeding 300 000 kilometers per month; No. 1 in the country for 'carrying capacity' with more than 8 million passengers [61, 62]. Also, in order to promote the domestic electric bus industry, Great City Life Bus has combined companies such as Tong Ying Body Co., Ltd, Amita Technologies Inc., Solpower Machine Electronic Corp., and eEasy Technology Co., Ltd With the assistance of ITRI, Taiwan, it is currently conducting the Ministry of Economic 'A$^+$ Industrial innovative R&D program' to build the domestically produced electric bus with the highest self-made rate.

The operating mileage of passenger electric buses in the city has reached 5 million kilometers, reducing carbon dioxide emissions by 150 tons, which is equivalent to planting 160 000 trees. In addition to the effect of energy saving and carbon reduction, the operating cost of electric buses is much lower than that of diesel buses. 5 million kilometers of operating mileage reduced operating costs by more than USD 1.3 million [62, 63].

In order to promote electric buses, the Taiwan government issued the 'Key Points of Public Transport Subsidy for Electric Buses,' which will be implemented through an inter-ministerial meeting with the re- sources of the Ministry of Transportation and Communications (MOTC) and the Environmental Protection Administration (EPA). The MOTC subsidizes the car body (electric bus body without battery cost) and charging stations, and the EPA subsidizes the battery and additional subsidy to the car body [64].

16.2.4 Small-scale stationary applications

In recent years, most portable devices, such as smartphones (iPhone, Android cell phone), laptops, tablets, and portable power supplies, have used Li-ion batteries as their power sources. The advantages of using Li-ion batteries include durability, more energy storage, light weight, easy charging, and low maintenance cost.

The Li-ion batteries charge faster, last longer, and have a higher power density for more battery life in a lighter package. A Li-ion battery can store up to 150 Wh of electricity in 1 kg of battery [1, 3]. This is much more than the nickel-metal hydride battery, or other kinds of battery technology can store. While the nickel-metal battery stores about 60–70 Wh kg^{-1} [3], a lead-acid battery can only store only 25 Wh kg^{-1} [3]. Though not the lightest, Li-ion batteries are generally much lighter than most types of batteries of the same capability. Many manufacturers prefer them because they make smartphones very light. The electrodes of a Li-ion battery are made of lightweight lithium and carbon which makes them very light. Lithium is a highly reactive element, making it possible to store a lot of energy in its atomic bonds. The Li-ion battery makes charging a device easier. This indicates that you can charge your smartphone with Li-ion batteries whenever you want. There is no need to wait for it to discharge to 0% of SOC before recharging, your device could remain plugged into electricity for a long period without side memory effects. Li-ion batteries can handle hundreds of charge/discharge cycles and still maintain the same long-lasting quality. The maintenance cost of Li-ion batteries is low. The self-discharge for Li-ion batteries falls to around 2% per month, less than half compared to nickel–cadmium, making Li-ion batteries perfect for devices [1, 3].

Table 16.1. Numbers of teams and people climbing Mt. Jade. Lots of climbers around the world come every year. The number of teams and people has gradually increased in recent years. Ref: Yushan National Park Headquarters.

Year	Number of teams	Number of people
2019	14 744	58 124
2018	12 437	54 466
2017	11 461	52 245

16.2.5 Medium-scale stationary applications

16.2.5.1 Off-grid solar energy storage system in Paiyun Lodge in Mt. Jade

Mt. Jade (also known as Mount Jade, Jade Mountain, or Yushan, Yu-Shan in Taiwanese) is the highest mountain in Taiwan at 3952 m (12 966 ft) above sea level, giving Taiwan the fourth-highest maximum elevation of any island in the world. It is also the highest point in the western Pacific region [65].

Tens of thousands of climbers around the world come every year (over 58 thousand people in 2019 according to the official report, as shown in table 16.1). Conquering Mt. Jade has been one of the most fascinating and difficult challenges for mountain hikers at home and abroad.

Paiyun Lodge, located at 3402 m (11 161 ft) on the trail to the Main Peak of Mt. Jade, has been the important overnight boarding base for hikers planning to reach the summit of Mt. Jade since 1943. Here you can see the sea of clouds, sunrise, sunset, night view, Milky Way Galaxy, stars, and other special scenery.

It was redesigned and rebuilt with modern facilities and then reopened in 2013. The new lodge is much larger and now provides more comfortable bedding, hot meals, as well as search and rescue services. Now, the lodge serves over 30 thousand people every year (according to the official report).

In the past half-century, the electricity was provided by a diesel generator with lead-acid batteries, which cause high carbon emission and high pollution. In order to preserve the ecology of Mt. Jade, the old system must be replaced. The power supply improvement plan was awarded to Wu Sia-Syong Architect office. The city power plan will cost about 6.7 million USD to build an 8.5 km (5.3 miles) power line from Tataka Saddle along the ridge or trail to Paiyun Lodge [66]. The city power is expected to provide high stability and sufficient power, but with high construction cost, high maintenance cost, and impact on the ecological environment. This is the first time Paiyun Lodge has proposed a utility power plan since its establishment in 1943, but it has also aroused controversy in all circles [67–71][11].There were still other

[11] The city power plan has attracted much attention from hikers and civil groups who guard against bringing modern comforts to Yushan's formidable heights at the expense of the environment. To fully explore public opinion, the government set up an online page on the Public Policy Network Participation Platform (http://goo.gl/mT8s2a) on Sep. 23th, 2015. Concerned citizens could go to the discussion site and voice their opinion before Oct. 23th, 2015.

options under consideration, such as hydroelectric power, wind power, and fuel cell. Eventually, an off-grid solar energy storage system was chosen from many options and replaced the city power plan. A more detailed comparison is listed in table 16.2.

The solar energy storage system was provided by SDP in 2016.The electricity is purely generated by solar panels and can be saved in the energy storage system. Hence, the entire system guarantees green, zero- carbon-emission, eco-friendly, and sufficient energy. The rest of the electricity can be preserved in the energy storage system constructed by an LFP battery of 60 kWh capacity. If solar energy is not available, the system can provide electric power for about one week.

Now, the solar energy storage system not only provides electricity for daily usage but also gives a stable power supply for the automated external defibrillator (AED), medical oxygen generators for sudden mountain sickness, other basic medical equipment, as well as ambulance communications.

16.2.5.2 Off-grid solar energy storage system of solar tree in SSGESC

Shalun Smart Green Energy Science City (SSGESC), a demonstration site, is located in a specific high- speed rail (HSR) area, Tainan, Taiwan, covering an area of nearly 7.5 hectares. It is one of the very important topics in Taiwan national project '5 plus 2 industries.' It not only emphasizes the demand drives innovation and innovation drives industrial development, but will also become a research, experiment, and development base in the future. SSGESC includes the convention and exhibition zone, core zone, the peripheral living and residential zones near HSR Tainan Station, and Academia Sinica Biotechnology Center in Southern Taiwan. SSGESC is built to innovate green energy research and design clusters, link with locality and sustainable development, forming a cluster on the 30 parks within 45 km in radius from SSGESC. There are several goals of SSGESC: activation of existing green energy industries, innovation of green energy industries, as well as a linkage among

Table 16.2. Comparison between the city power plan and off-grid solar energy storage system for Paiyun Lodge.

	City power plan	Off-grid solar energy storage system
Funding	about 6.7 million USD	less than 0.3 million USD
Power supply	8.5 km (5.3 miles) power line from Tataka Saddle along the ridge or trail to Paiyun Lodge	Solar panels installed on the roof
Advantages	High stability and sufficient power	Low construction cost, low maintenance cost, eco-friendly renewable energy, carbon- footprint reduction
Disadvantages	High construction cost, high maintenance cost, impact on the ecological environment	Intermittent renewable energy

industry, academia, and research sources to build an innovative ecosystem of green energy industries [72–74].

In the SSGESC, there is a solar tree with a very special shape. It is constructed from steel sheet components, PV panel, and ESS. It will be regularly scheduled every night. The colorful show with LED lights, brilliant and dazzling, has become the most eye-catching new landmark recently. This solar tree is 20 meters high and 4 meters higher than the HSR track. When the light show is launched at night, it is quite eye-catching. As the north-south train passes by, it attracts the attention of passengers. They are curious about this wonderful landmark. The solar tree is composed of three flexible tree paths representing economy, environment, and education, as well as a large canopy with photovoltaic panels. It is hoped that this will highlight the characteristics of the green energy industry and the significance of the SSGESC having the advantages of the three. In particular, there are 156 high-efficiency monocrystalline silicon module boards of 300 W_p on the tree canopy, providing 46.8 kW_p peak power to generate electricity in the daytime. The 15 kWh ESS provided by SDP can supply the LED colorful show for more than one hour at night. The LED lighting system includes 23 sets of 150 W multi-color LED and 36 sets of 150 W white LED. The colorful show will be staged in accordance with the time of the HSR trains, so that more people can appreciate the beauty of solar trees [75]. The light show was officially opened on Dec. 16th, 2019.

16.2.6 Large-scale stationary applications

The installation of the world's largest Li-ion battery (100 MW, 129 MWh from Tesla and Neoen) was completed in Hornsdale, South Australia in 2017 [76, 77]. After several weeks, when the Loy Yang coal-fired power plant in Victoria failed, leading to a power shortage, the backup battery connected and delivered as much as 100 MW into the national electricity grid in 140 ms [78, 79], responding even more quickly than the coal-fired backups that were supposed to provide emergency power. That shock absorber-type/emitter- type capacity is what helps us to stop a blackout that would otherwise occur. The batteries are capable of providing inertia services and rapid frequency responses (e.g., frequency control ancillary services; FCAS) to the grid, paving the way for potential regulatory changes and revenue streams to incentivize further grid- scale batteries [80–82]. The facility saved nearly USD 40 million in its first year alone and helped to stabilize and balance the region's unreliable grid. Large-scale batteries are beginning to show that they have a role in supply electric networks.

ESS is transforming the global electric grid and is an increasingly important element of the world's transition to sustainable energy. Utility-scale ESS can provide many applications, such as smoothing out the intermittency of renewable energies by storing and dispatching when needed, postponing costly grid infrastructure upgrades by supplying power at a distributed location to defer the need to upgrade aging infrastructure, injecting and absorbing reactive power to maintain local grid voltage levels, discharging at times of peak capacity to reduce demands on distribution and transmission infrastructure, building a localized grid that can disconnect from the main power grid, providing service to the grid in response to signals sent by system operators, and maintaining grid stability by rapidly changing charge/discharge power in response to grid frequency changes.

16.3 Outlooks

16.3.1 Growing market and falling cost

The battery market is USD 25 billion in 2019 and is predicted to be worth USD 28 billion in 2020, showing a growing market tendency. By 2030 the battery market will be worth USD 116 billion annually according to BloomNEF's forecasts [83], and this does not include investment in the supply chain.

A decade ago, Li-ion batteries were a costly proposition. Li-ion battery packs cost 1183 kWh^{-1} in 2010. Nine years later, the price had dropped nearly tenfold to USD 156 kWh^{-1} in 2019, fallen 87% in real terms, driving a rapidly expanding market for EVs. Manufacturers are closing in on a point where EVs will reach cost parity with their fossil fuel-powered cousins at around USD 100 kWh^{-1}. That price is widely seen as a sweep point in the sector, where consumers will no longer regard EVs as pricey options.

Cost reductions in 2019 are attributed to increasing order size, growth in EV sales, and the continued penetration of high energy density cathodes. The advance in pack designs and falling manufacturing costs will drive prices down. BloombergNEF forecast that Li-ion battery costs will fall under USD 100 kWh^{-1} in 2024 and hitting around USD 60 kWh^{-1} by 2030 [83].

16.3.2 Safety

With the widespread use of Li-ion batteries in recent years, the safety issues of lithium batteries have gradually surfaced. The following provides a brief history: Dell laptop computer battery thermal runaway in 2006 and 2007 [84], Chevrolet Volt battery thermal runaway after a crash testing in 2011 [85], thermal runaway, smoke, and fire onboard the hybrid tugboat Campbell Foss in 2013 [86], Boeing Dreamliner battery thermal runaway in 2013 [87, 88], Samsung Galaxy Note 7 thermal runaway in 2017 [89], ESS Fire in Taiwan in 2020 [90, 91], and more than 29 ESS Fire events in South Korea in from 2017 [92–94].

16.3.2.1 Recent event study: McMicken ESS fire event

On Apr. 19th, 2019, the McMicken storage facility in Surprise, Arizona started smoking. After consulting with the contractor who built the facility and other experts, firefighters opened a door to the steel structure that housed Li-ion batteries to investigate the source of the smoke [95]. A few minutes later, a jet of flame shot out an estimated 23 meters (75 feet) from the door. Four firefighters were hospitalized, two of them with traumatic brain injuries.

After investigation [96, 97], the initial overheating was found to be caused by a 'thermal runaway' within one of the battery cells buried deep in an apparatus that held 27 separate racks containing a total of 10 584 li-ion battery cells[12]. Because the batteries were stacked close together, the thermal runaway spread to other batteries as a chain reaction, called 'cascading thermal runaway.' The National Fire Protection Association also warned that such chain reactions are possible in a 2016 report [98][13].

The explosion at the ESS facility made apparent another hazard: when Li-ion batteries heat up, they emit flammable electrolyte and gases. The explosion was triggered as firefighters agitated the air inside the structure housing the batteries, causing flammable gases to combust. Even though a fire-suppression system inside the structure went off as designed, the released aerosol could not stop a chemical chain reaction under thermal runaway. The system design did not allow any ventilation that could release the build-up of flammable gases.

The lessons the industry has learned from these incidents is that Li-ion batteries are inherently fragile. Any electrical, thermal, or mechanical abuse, along with internal defects, can potentially initiate cell failure and thermal runaway. The danger of Li-ion batteries is real but can be mitigated and will be as scientists learn more about the reasons for thermal runaways [95, 102, 103].

16.3.3 Reuse and recycle of repurposing batteries

Owing to the vigorous promotion of commercialization and the popularization of EV worldwide [47], the widely used Li-ion batteries result in fundamental researches, industry development, as well as standard and policymaking in the field of Li-ion power batteries. Recently, retirement of power batteries has largely increased, causing waste of resources and environmental protection threats. Hence, recycling and utilization of such retired batteries have been promoted [104, 105]. Some retired power batteries still possess about 80% initial capacity [106–109]. So they can be repurposed and utilized once again, e.g., serving as the battery modules in stationary energy storage systems [110–113]. Governments in various countries have acknowledged this emergent issue and prepared to launch their policies to deal with the recovery and reuse of repurposing batteries, such as coding principles, traceability management system, manufacturing factory guidelines, dismantling process guidelines, residual energy measurement, federal and state tax credits, rebates, and other financial support [114–116].

Safety and performance are important in using the retired power batteries, i.e., repurposing batteries. Underwriters Laboratories (UL), a global safety certification company established in 1894, published the standard for evaluating the safety and performance of repurposing batteries in 2018 (UL 1974) [117, 118]. The charge/discharge profiles of Li-ion repurposing batteries could be measured based on UL 1974. The test procedures designed according to UL 1974 could be used to evaluate

[12] A thermal runaway is a violent chemical reaction that produces heat, but no visible flames. Typically, the reaction is caused by defects of Li-ion batteries.

[13] LG Chem Ltd, the Korean company that manufactured the suspect Li-ion batteries, submitted its investigative report that concluded the explosion was possibly caused by a heat source such as external electrical arcing [99–101].

the safety and performance of the Li-ion repurposing batteries. The charge/discharge profile dataset in the study provided researchers/engineers the characteristic curves in estimating the repurposing batteries under UL 1974 [25].

16.3.4 Energy storage as a service

Energy storage as a service (ESaaS) allows a facility to benefit from the advantages of an ESS by entering into a service agreement without purchasing the system. Energy storage systems provide a range of services to generate revenue, create savings, and improve electricity resiliency. The operation of the ESaaS systems is a unique combination of advanced ESSs, EMSs, and service contracts that can deliver value to a business by providing reliable power more economically.

The most common ESSs used for ESaaS are Li-ion [119–122] or flow [123–125] batteries due to their compact size, non-invasive installation, high efficiencies, and fast reaction times. Other storage mediums may be used such as pumped hydroelectric storage [126, 127], capacitor/supercapacitor [128–130], compressed air energy storage [131–133], fuel cell [134–136], solar fuel [137–140], superconducting magnetic energy storage [141–144], flywheel [145], and thermal energy storage [146]. The batteries are sized based on the facility's needs and are paired with a power inverter, converting the DC power to AC power to connect directly to the facility's electricity supply.

Current ESaaS is used to provide a variety of services, such as power quality, coincident peak management, demand response, power factor correction, back-up power, peak saving, energy arbitrage, market ancillary services, and transmission support. There are several benefits of ESaaS, such as the system benefactor does not require installation capital, ESS is operated by a third-party system operator, and it is environmentally friendly.

16.3.5 Next-generation energy materials

In the 1970s, M Stanley Whittingham worked on developing methods that could lead to fossil-fuel-free energy technologies. He started to research superconductors and discovered an extremely energy-rich material, titanium disulfide (TiS_2), which he used to create an innovative cathode in a lithium battery (voltage $\sim$ 2.5 V) [147–152]. However, metallic lithium in the anode is reactive and the battery was too explosive to be commercially viable. John B Goodenough predicted that the cathode would have even greater potential if it was made by a metal oxide instead of a metal sulfide. After systematic research, in 1980 he exhibited that cobalt dioxide (CoO_2) with intercalated lithium ions can produce a voltage as much as 4 V. This was an important breakthrough and would lead to much more powerful and energetic batteries [153, 154]. In 1985, based on Goodenough's research, Akira Yoshino created the first commercially viable Li-ion battery. Instead of using reactive lithium in the anode, he used petroleum coke (a carbon material) to intercalate lithium ions [155, 156]. The advantages of Li-ion batteries have been revealed, i.e., they are not based on chemical reactions that break down the electrodes, but rely on Li-ions flowing back and forth between the anode and cathode. The result was a lightweight, high-energy, and hardwearing battery that could be charged several hundreds of times before its performance deteriorated. Li-ion batteries entered the market in 1991 [158] and have revolutionized our lives since then. They have laid the foundation of a wireless, fossil-fuel-free society, and are of the greatest benefit to humankind. The 2019 Nobel Prize for Chemistry [158–163] was awarded to John B Goodenough, M Stanley Whittingham, and Akira Yoshino 'for the development of lithium-ion batteries.'

The focus of continuous development of various batteries is still on the improvement of energy density, power density, and safety. When the energy density is higher, more electricity can be stored per unit volume, the endurance of electric vehicles can be continuously improved, and the space required for the energy storage system can be smaller. Currently, engineers and researchers are still trying to test various types of anode materials [164–167], cathode materials [168–173], separators [174–176], and electrolytes [177–183] to improve the energy density, power density, and safety of Li-ion battery. Some practices include increasing the silicon-based lithium-ion battery anodes [184–189], nickel content of lithium ternary batteries to more than 80% [190–193] (for example, NMC 811),[14] developing electrode materials that can chemically react with Li-ions (or Li-atoms) [194–199], and discarding flammable organic solvents in the electrolyte and using non-flammable solid electrolyte materials such as ceramics [200–203]. In recent years, the rapid progress of nano and low-dimensional materials, e.g., graphene [204–223], silicene [224], and other two-dimensional materials [225–228], has provided many choices for the next generation of Li-ion battery materials [229]. New materials and associated production technologies will help manufacturers to reduce costs, reduce energy use, reduce pollution, improve product quality, and increase competitiveness.

Acknowledgements

The author (H C Chung) thanks Professor Ming-Fa Lin for the book chapter invitation and for inspiring him to study this topic. H C Chung thanks Jung-Feng Jack Lin, Chairman of Super Double Power Technology Co., Ltd, Taiwan, for many valuable discussions, recommendations, information about ESSs in various scales and real case examples. Chung thanks Professor Yaw-Chung Cheng and Mr Hui-Wen Feng for their valuable discussions on the fuel cells; Dr Yao-Ming Wang, Metal Industries Research and Development Centre (MIRDC), Taiwan, for discussions on the vanadium flow batteries; Kevin Yang, Chairman of Ecolo- has Energy Technology Co., Ltd, Taiwan, for discussions on Paiyun Lodge; Donald Lee, eEasy Technology Co., Ltd, Taiwan, for discussions on Taiwan's electric buses operation strategy in recent years; Hsueh-Lung Lu, Acting Manager of Certified Senior Industrial Analyst, Industry, Science and Technology International Strategy Center, ITRI, Taiwan, for his valuable discussions on the Korea ESS fires; Wen-Hua Lin, CEO of Apogee Power Inc., for his valuable discussions on the McMicken ESS fire event; Professor Chin-Lung Kuo (Department of Materials Science and

[14] NMC 811 is a cathode composition with 80% nickel, 10% manganese, and 10% cobalt.

Engineering, National Taiwan University, Taiwan), Professor W Kavee- vivitchai (Department of Chemical Engineering, National Cheng Kung University, Taiwan), Professor N T Tien (College of Natural Science, Can Tho University, Vietnam), and Dr N T T Tran (Hierarchical Green-Energy Materials (Hi-GEM) Research Center, National Cheng Kung University, Taiwan) for their valuable suggestions on the next-generation energy materials (stated in the order of the chapter). H C Chung would like to thank the contributors to this chapter for their valuable discussions and recommendations, Hsiao-Wen Yang, Yen-Kai Lo, and An-De Andrew Chung. The author (H C Chung) thanks Pei-Ju Chien for English discussions and corrections as well as Ming-Hui Chung, Su-Ming Chen, Lien-Kuei Chien, and Mi-Lee Kao for financial support. This work was supported in part by Super Double Power Technology Co., Ltd, Taiwan, under the project 'Development of Cloud-native Energy Management Systems for Medium-scale Energy Storage Systems (https://osf.io/7fr9z/)' (Grant number: SDP-RD-PROJ-001–2020).

References

[1] Scrosati B and Garche J 2010 Lithium batteries: status, prospects and future *J. Power Sources* **195** 2419–30

[2] Armand M and Tarascon J M 2008 Building better batteries *Nature* **451** 652–7

[3] Tarascon J M and Armand M 2001 Issues and challenges facing rechargeable lithium batteries *Nature* **414** 359–67

[4] Debe M K 2012 Electrocatalyst approaches and challenges for automotive fuel cells *Nature* **486** 43–51

[5] Logan B E, Hamelers B, Rozendal R A, Schrorder U, Keller J, Freguia S, Aelterman P, Verstraete W and Rabaey K 2006 Microbial fuel cells: methodology and technology *Environ. Sci. Technol.* **40** 5181–92

[6] Lin M, Cheng Y and Yen S 2005 Evaluation of PEMFC power systems for UPS base station applications *J. Power Sources* **140** 346–9

[7] Winter M and Brodd R J 2004 What are batteries, fuel cells, and supercapacitors? *Chem. Rev.* **104** 4245–69

[8] Cheng Y C, Yuen W H, Lin A S and Tsay J T 1995 Automation and control for the pafc fuel cell components and the power plant *Proc. IEEE Conf. on Industrial Automation and Control Emerging Technology Applications*

[9] Minh N Q 1993 Ceramic fuel cells *J. Am. Ceram. Soc.* **76** 563–88

[10] Bayeh A W, Kabtamu D M, Chang Y C, Chen G C, Chen H Y, Liu T R, Wondimu T H, Wang K C and Wang C H 2019 Hydrogen-treated defect-rich $W_{18}O_{49}$ nanowire-modified graphite felt as high-performance electrode for vanadium redox flow battery *ACS Appl. Energy Mater.* **2** 2541–51

[11] Kabtamu D M, Bayeh A W, Chiang T C, Chang Y C, Lin G Y, Wondimu T H, Su S K and Wang C H 2018 $TiNb_2o_7$ nanoparticle-decorated graphite felt as a high-performance electrode for vanadium redox flow batteries *Appl. Surf. Sci.* **462** 73–80

[12] Bayeh A W, Kabtamu D M, Chang Y C, Chen G C, Chen H Y, Lin G Y, Liu T R, Wondimu T H, Wang K C and Wang C H 2018 Synergistic effects of a $tiNb_2o_7$-reduced graphene oxide nanocomposite electrocatalyst for highperformance all-vanadium redox flow batteries *J. Mater. Chem.* A **6** 13908–17

[13] Ulaganathan M, Aravindan V, Yan Q Y, Madhavi S, Skyllas-Kazacos M and Lim T M 2016 Recent advancements in all-vanadium redox flow batteries *Adv. Mater. Interfaces* **3** 1500309

[14] Kim K J, Park M S, Kim J H, Hwang U, Lee N J, Jeong G and Kim Y J 2012 Novel catalytic effects of Mn_3O_4 for all vanadium redox flow batteries *Chem. Commun.* **48** 5455–7

[15] Rychcik M and Skyllas-Kazacos M 1988 Characteristics of a new all-vanadium redox flow battery *J. Power Sources* **22** 59–67

[16] Chen K L and Chen N M 2011 A new method for power current measurement using a coreless Hall effect current transformer *IEEE Trans. Instrum. Meas.* **60** 158–69

[17] Ripka P 2010 Electric current sensors: a review *Meas. Sci. Technol.* **21** 1–23

[18] Ziegler S, Woodward R C, Iu H H C and Borle L J 2009 Current sensing techniques: a review *IEEE Sens. J.* **9** 354–76

[19] Xiao C, Zhao L, Asada T, Odendaal W G and van Wyk J D 2003 An overview of integratable current sensor technologies *38th IAS Annual Meeting on Conf. Record of the Industry Applications Conf., 2003*

[20] Piller S, Perrin M and Jossen A 2001 Methods for state-of-charge determination and their applications *J. Power Sources* **96** 113–20

[21] Coleman M, Hurley W G and Lee C K 2008 An improved battery characterization method using a two-pulse load test *IEEE Trans. Energy Convers.* **23** 708–13

[22] IEC 2017 IEC 61951-1 2017 Secondary cells and batteries containing alkaline or other non-acid elec- trolytes – secondary sealed cells and batteries for portable applications – part 1: Nickel-cadmium

[23] IEC 2017 IEC 61960-3 2017 Secondary cells and batteries containing alkaline or other non-acid elec- trolytes – secondary lithium cells and batteries for portable applications – part 3: Prismatic and cylindrical lithium secondary cells and batteries made from them

[24] ISO 2018 ISO 12405-4 2018 Electrically propelled road vehicles – test specification for lithium-ion traction battery packs and systems – part 4: Performance testing

[25] Chung H C 2021 Charge and discharge profiles of repurposed $LiFePO_4$ batteries based on the UL 1974 standard *Sci. Data* **8** 165

[26] Chung H C 2018 Failure mode and effects analysis of LFP battery module *engrXiv*

[27] Shafie-Khah M and Siano P 2018 A stochastic home energy management system considering satisfaction cost and response fatigue *IEEE Trans. Industr. Inform.* **14** 629–38

[28] Al-Ali A R, Zualkernan I A, Rashid M, Gupta R and AliKarar M 2017 A smart home energy management system using IoT and big data analytics approach *IEEE Trans. Consum. Electron.* **63** 426–34

[29] Zhao Z, Lee W C, Shin Y and Song K B 2013 An optimal power scheduling method for demand response in home energy management system *IEEE Trans. Smart Grid* **4** 1391–400

[30] Thomas D, Deblecker O and Ioakimidis C S 2018 Optimal operation of an energy management system for a grid-connected smart building considering photovoltaics' uncertainty and stochastic electric vehicles' driving schedule *Appl. Energy* **210** 1188–206

[31] Miller C, Nagy Z and Schlueter A 2015 Automated daily pattern filtering of measured building performance data *Autom. Constr.* **49** 1–17

[32] Jain S, Vinoth Kumar N, Paventhan A, Kumar Chinnaiyan V, Arnachalam V and Pradish M 2014 Survey on smart grid technologies-smart metering, IoT and EMS *2014 IEEE Students' Conf. on Electrical, Electronics and Computer Science*

[33] Saleh M, Esa Y and Mohamed A A 2019 Communication-based control for DC microgrids *IEEE Trans.Smart Grid* **10** 2180–95
[34] Saleh M, Esa Y, Mohamed A A, Grebel H and Rojas-Cessa R 2017 Energy management algorithm for resilient controlled delivery grids *2017 IEEE Industry Applications Society Annual Meeting*
[35] Shi W B, Li N, Chu C C and Gadh R 2017 Real-time energy management in microgrids *IEEE Trans.Smart Grid* **8** 228–38
[36] Su W and Wang J 2012 Energy management systems in microgrid operations *Electr. J.* **25** 45–60
[37] Jimeno J, Anduaga J, Oyarzabal J and de Muro A G 2011 Architecture of a microgrid energy management system *Euro. Trans. Electr. Power* **21** 1142–58
[38] Celli G, Pilo F, Pisano G and Soma G G 2005 Optimal participation of a microgrid to the energy market with an intelligent EMS *2005 Int. Power Engineering Conf.*
[39] Lu N 2012 An evaluation of the HVAC load potential for providing load balancing service *IEEE Trans.Smart Grid* **3** 1263–70
[40] Wang S W and Ma Z J 2008 Supervisory and optimal control of building HVAC systems: a review *HVAC R Res.* **14** 3–32
[41] Fong K F, Hanby V I and Chow T T 2006 HVAC system optimization for energy management by evolutionary programming *Energy Buildings* **38** 220–31
[42] Liasi S G and Bathaee S M T 2017 Optimizing microgrid using demand response and electric vehicles connection to microgrid *2017 Smart Grid Conf. (SGC)*
[43] Brambley M R, Haves P, McDonald S C, Torcellini P, Hansen D, Holmberg D R and Roth K W 2005 Advanced sensors and controls for building applications: market assessment and potential RD pathways *OSTI* 859997
[44] Mahmud K and Town G E 2016 A review of computer tools for modeling electric vehicle energy requirements and their impact on power distribution networks *Appl. Energy* **172** 337–59
[45] Han D M and Lim J H 2010 Smart home energy management system using IEEE 802.15.4 and ZIGBEE *IEEE Trans. Consum. Electron.* **56** 1403–10
[46] Son Y S, Pulkkinen T, Moon K D and Kim C 2010 Home energy management system based on power line communication *IEEE Trans. Consum. Electron.* **56** 1380–6
[47] IEA 2020 Global EV outlook 2020 report https://iea.org/reports/global-ev-outlook-2020
[48] Programme U E Electric two and three wheelers
[49] China Government 2015 Notice on financial support for new energy vehicles 2016–20
[50] India G O 2019 Publication of notification in gazette of india (extraordinary) regarding phase-II of fame India scheme
[51] European Alternative Fuels Observatory (EAFO) 2020 Electric vehicle charging infrastructure
[52] EV-Volumes 2020 EV data center (accessed March and April 2020)
[53] TaaS News – Transportation as a Service Magazine 2019 Mexico city to be the new home for 63 Yutong dual-powered trolleybuses
[54] Sustainable Bus 2019 Yutong begins the delivery of 63 new trolleybuses for Mexico city
[55] Yutong news 2019 63 Yutong dual-source trolleybuses enter Mexico across the ocean!
[56] The Liberty Times 2018 Improve air pollution – Premier Lai of the Executive Yuan: starting this year, 10,000 buses will be gradually replaced by electric vehicles

[57] Executive Yuan of Taiwan 2017 Premier Lai of the Executive Yuan: inter-ministerial will promote the 'air pollution prevention action plan' and local governments make every effort to improve air quality
[58] Taipei Bus Information 2011 The opening ceremony of the country's first electric bus
[59] Trans. Taipei – Bus Information Net 2011 Information on new buses on the road in Taipei (2011)
[60] DIGITIMES 2019 Electric buses on the roads in the urban area – Great City Life Bus: 72 kWh is enough
[61] The Liberty Times 2019 Electric buses of great city life bus, 5 national firsts in Taiwan
[62] CVNews 2019 The cumulative mileage of the electric buses of Great City Life Bus reached 5 million kilometers, setting the first record in 5 Taiwan
[63] Great City Life Bus 2019 Great City Life Bus has transformed into a model of electric buses
[64] Automotive Research and Testing Center (ARTC), Taiwan 2018 Introduction to the promotion of electric buses in our country
[65] Wikipedia 2020 List of islands by highest point
[66] Wu Sia-Syong Architect office 2015 Paiyun Lodge of Yushan National Park – Symposium on feasibility evaluation of power supply
[67] Taiwan Environmental Protection Union (TEPU) 2015 Paiyun Lodge does not emit carbon: green energy, energy saving plus energy storage, saving money, no shortage of electricity for conservation
[68] PeoPo 2015 Environmental groups demand no carbon emissions at Paiyun Lodge: green energy, energy saving and energy storage, saving money, and protecting no shortage of electricity
[69] Commercial Times 2015 Expert fax – Paiyun Lodge's 'zero energy consumption' green energy vision
[70] Taiwan People News 2015 Spend 200 million to pull electric wires to Paiyun Lodge – environmental groups criticize the destruction of Yushan ecology
[71] The Liberty Times 2015 Plan to use 200 million to improve the power supply of Yushan 'Paiyun Lodge' – Yushan National Park administration is blamed
[72] Industrial Technology Monthly 2020 Shalun Smart Green Energy Science City opened *Industrial Technology Monthly* **337** 66–9
[73] Media Center of ITRI 2019 Build a green gold settlement in Taiwan – Shalun Smart Green Energy Science City (SSGESC) opened
[74] Central News Agency 2019 Build an industrial settlement – Shalun Smart Green Energy Science City (SSGESC) opened
[75] The Liberty Times 2020 New landmark of Quynhon Tainan! LED lights combined with solar trees – colorful show every night
[76] Keck F, Lenzen M, Vassallo A and Li M Y 2019 The impact of battery energy storage for renewable energy power grids in Australia *Energy* **173** 647–57
[77] Radcliffe J C 2018 The water energy nexus in Australia – the outcome of two crises *Water-Energy Nexus* **1** 66–85
[78] Ye Y D, Qiao Y and Lu Z X 2019 Revolution of frequency regulation in the converter-dominated power system *Renew. Sustain. Energy Rev.* **111** 145–56
[79] Zeng H *et al* 2018 Introduction of Australian 100 MW storage operation and its enlightenment to China 2018 China Int. Conf. on Electricity Distribution (CICED) 2895

[80] Fernandez-Munoz D, Perez-Diaz J I, Guisandez I, Chazarra M and Fernandez-Espina A 2020 Fast frequency control ancillary services: an international review *Renew. Sustain. Energy Rev.* **120** 109662
[81] Pagliaro M 2019 Renewable energy systems: enhanced resilience, lower costs *Energy Technol.* **7** 1900791
[82] Aziz A, Oo A M T and Stojcevski A 2018 Issues and mitigations of wind energy penetrated network: Australian network case study *J. Mod. Power Syst. Clean Energy* **6** 1141–57
[83] BloombergNEF 2019 Battery pack prices fall as market ramps up with market average at $156/kWh in 2019
[84] Darlin D 2006 Dell recalls batteries because of fire threat *The New York Times*
[85] Smith B 2012 Chevrolet volt battery incident overview report *US Department of Transportation, National Highway Traffic Safety Administration (NHTSA)* DOT HS 811 573
[86] Tyler D A 2012 Battery-related fire damages famed hybrid tug, puts it out of service *Professional Mariner Magazine*
[87] National Transportation Safety Board (NTSB) 2014 Auxiliary power unit battery fire Japan airlines Boeing 787-8, JA829J, Boston, Massachusetts January 7, 2013 NTSB/AIR–14/01, PB2014–108867
[88] Mouawad J 2014 Report on Boeing 787 dreamliner battery flaws finds lapses at multiple points *The New York Times*
[89] News B 2017 Samsung confirms battery faults as cause of Note 7 fires
[90] The Liberty Times 2020 Containers in Taipei rushed to the fire, storing a large number of Li-ion batteries, the fire was fierce
[91] SETN 2020 Stock up the battery! the container under the Mincyuan Bridge suddenly burst into flames—emergency water jet rescue
[92] MOTIE News 2019 The cause of the ESS fire was revealed
[93] Ministry of Trade, Industry and Energy (MOTIE), South Korea 2019 ESS accident cause investigation results and safety enhancement measures announced
[94] Shin J H 2019 Ministry postpones ESS fire probe report, industry suffers setback *The Korea Herald*
[95] McKinnon M B, DeCrane S and Kerber S 2020 Four firefighters injured in lithium-ion battery energy storage system explosion – Arizona *UL Firefighter Safety Research Institute*
[96] Hill D 2020 Mcmicken battery energy storage system event technical analysis and recommendations
[97] Center A S 2020 Update – utility solar battery fire in Arizona
[98] Blum A F and Thomas Long R Jr 2016 Hazard assessment of lithium ion battery energy storage systems *National Fire Protection Association (NFPA)*
[99] Swart J, White K and Cundy M 2020 APS McMicken progress report
[100] Burdick G 2020 LG Chem refutes alleged cause of explosion at APS battery storage facility *Utility Dive*
[101] Spector J 2020 What sparked the Arizona battery fire? LG Chem has a different version *GTM*
[102] Herh M 2020 Investigation committee says defective batteries are to blame for ESS fires *BusinessKorea*
[103] Herh M 2020 Probe blames faulty batteries for ESS fires *BusinessKorea*

[104] Martinez-Laserna E, Gandiaga I, Sarasketa-Zabala E, Badeda J, Stroe D I, Swierczynski M and Goikoetxea A 2018 Battery second life: hype, hope or reality? a critical review of the state of the art *Renew. Sustain. Energy Rev.* **93** 701–18

[105] Ahmadi L, Yip A, Fowler M, Young S B and Fraser R A 2014 Environmental feasibility of re-use of electric vehicle batteries *Sustain. Energy Technol. Assess.* **6** 64

[106] Casals L C, Garcia B A and Canal C 2019 Second life batteries lifespan: rest of useful life and environmental analysis *J. Environ. Manage.* **232** 354–63

[107] Podias A, Pfrang A, Di Persio F, Kriston A, Bobba S, Mathieux F, Messagie M and Boon-Brett L 2018 Sustainability assessment of second use applications of automotive batteries: ageing of Li-ion battery cells in automotive and grid-scale applications *World Electr. Veh. J.* **9** 24

[108] Tong S, Fung T, Klein M P, Weisbach D A and Park J W 2017 Demonstration of reusing electric vehicle battery for solar energy storage and demand side management *J. Energy Storage* **11** 200–10

[109] Wood E, Alexander M and Bradley T H 2011 Investigation of battery end-of-life conditions for plug-in hybrid electric vehicles *J. Power Sources* **196** 5147–54

[110] Kamath D, Shukla S, Arsenault R, Kim H C and Anctil A 2020 Evaluating the cost and carbon footprint of second-life electric vehicle batteries in residential and utility-level applications *Waste Manage.* **113** 497–507

[111] Quinard H, Redondo-Iglesias E, Pelissier S and Venet P 2019 Fast electrical characterizations of high- energy second life lithium-ion batteries for embedded and stationary applications *Batteries* **5** 33

[112] Heymans C, Walker S B, Young S B and Fowler M 2014 Economic analysis of second use electric vehicle batteries for residential energy storage and load-levelling *Energy Policy* **71** 22–30

[113] Casals L C and Garcia B A 2017 Second-life batteries on a gas turbine power plant to provide area regulation services *Batteries* **3** 10

[114] Chung H C and Cheng Y C 2019 Action planning and situation analysis of repurposing battery recovery and application in China *J. Taiwan Energy* **6** 425

[115] Hossain E, Murtaugh D, Mody J, Faruque H M R, Sunny M S H and Mohammad N 2019 A com- prehensive review on second-life batteries: current state, manufacturing considerations, applications, impacts, barriers and potential solutions, business strategies, and policies *IEEE Access* **7** 73215–52

[116] Gur K, Chatzikyriakou D, Baschet C and Salomon M 2018 The reuse of electrified vehicle batteries as a means of integrating renewable energy into the european electricity grid: a policy and market analysis *Energy Policy* **113** 535–45

[117] Laboratories U 2018 Ul 1974 – standard for evaluation for repurposing batteries

[118] Chung H C and Cheng Y C 2020 Summary of safety standards for repurposing batteries *Monthly J. Taipower's Eng.* **860** 35

[119] Yang Y Q, Bremner S, Menictas C and Kay M 2018 Battery energy storage system size determination in renewable energy systems: a review *Renew. Sustain. Energy Rev.* **91** 109

[120] Hill C A, Such M C, Chen D M, Gonzalez J and Grady W M 2012 Battery energy storage for enabling integration of distributed solar power generation *IEEE Trans. Smart Grid* **3** 850–7

[121] Teleke S, Baran M E, Huang A Q, Bhattacharya S and Anderson L 2009 Control strategies for battery energy storage for wind farm dispatching *IEEE Trans. Energy Convers.* **24** 725–32

[122] Divya K C and Ostergaard J 2009 Battery energy storage technology for power systems-an overview *Electr. Power Syst. Res.* **79** 511
[123] Hu B, DeBruler C, Rhodes Z and Liu T L 2017 Long-cycling aqueous organic redox flow battery (AORFB) toward sustainable and safe energy storage *J. Am. Chem. Soc.* **139** 1207–14
[124] Kear G, Shah A A and Walsh F C 2012 Development of the all-vanadium redox flow battery for energy storage: a review of technological, financial and policy aspects *Int. J. Energy Res.* **36** 1105–20
[125] Li L Y *et al* 2011 A stable vanadium redox-flow battery with high energy density for large-scale energy storage *Adv. Energy Mater.* **1** 394–400
[126] Silva G D E and Hendrick P 2016 Pumped hydro energy storage in buildings *Appl. Energy* **179** 1242–50
[127] Rehman S, Al-Hadhrami L M and Alam M M 2015 Pumped hydro energy storage system: a techno- logical review *Renew. Sustain. Energy Rev.* **44** 586
[128] Banerjee P, Perez I, Henn-Lecordier L, Lee S B and Rubloff G W 2009 Nanotubular metal-insulator-metal capacitor arrays for energy storage *Nat. Nanotechnol.* **4** 292–6
[129] Kinjo T, Senjyu T, Urasaki N and Fujita H 2006 Output levelling of renewable energy by electric double-layer capacitor applied for energy storage system *IEEE Trans. Energy Convers.* **21** 221–7
[130] Conway B E 1991 Transition from supercapacitor to battery behavior in electrochemical energy storage *J. Electrochem. Soc.* **138** 1539–48
[131] Wang J D, Lu K P, Ma L, Wang J H, Dooner M, Miao S H, Li J and Wang D 2017 Overview of compressed air energy storage and technology development *Energies* **10** 991
[132] Budt M, Wolf D, Span R and Yan J Y 2016 A review on compressed air energy storage: basic principles, past milestones and recent developments *Appl. Energy* **170** 250
[133] Cavallo A 2007 Controllable and affordable utility-scale electricity from intermittent wind resources and compressed air energy storage (CAES) *Energy* **32** 120–7
[134] Jang M and Agelidis V G 2011 A minimum power-processing-stage fuel-cell energy system based on a boost-inverter with a bidirectional backup battery storage *IEEE Trans. Power Electron.* **26** 1568–77
[135] Smith W 2000 The role of fuel cells in energy storage *J. Power Sources* **86** 74–83
[136] Mitlitsky F, Myers B and Weisberg A H 1998 Regenerative fuel cell systems *Energy Fuels* **12** 56–71
[137] Sun K, Shen S H, Liang Y Q, Burrows P E, Mao S S and Wang D L 2014 Enabling silicon for solar-fuel production *Chem. Rev.* **114** 8662–719
[138] Borjesson K, Dzebo D, Albinsson B and Moth-Poulsen K 2013 Photon upconversion facilitated molec- ular solar energy storage *J. Mater. Chem.* A **1** 8521–4
[139] Moore G F and Brudvig G W 2011 Energy conversion in photosynthesis: a paradigm for solar fuel production *Annu. Rev. Condens. Matter Phys.* **2** 303–27
[140] Gust D, Moore T A and Moore A L 2009 Solar fuels via artificial photosynthesis *ACC. Chem. Res.* **42** 1890–8
[141] Vulusala G V S and Madichetty S 2018 Application of superconducting magnetic energy storage in electrical power and energy systems: a review *Int. J. Energy Res.* **42** 358
[142] Dondapati R S, Kumar A, Kumar G R, Usurumarti P R and Dondapati S 2017 Superconducting magnetic energy storage (SMES) devices integrated with resistive type superconducting fault current limiter (SFCL) for fast recovery time *J. Energy Storage* **13** 287–95

[143] Aly M M, Abdel-Akher M, Said S M and Senjyu T 2016 A developed control strategy for mitigating wind power generation transients using superconducting magnetic energy storage with reactive power support *Int. J. Electr. Power Energy Syst.* **83** 485–94
[144] Boenig H J and Hauer J F 1985 Commissioning tests of the bonneville power administration 30 MJ superconducting magnetic energy storage unit *IEEE Trans. Power Syst.* **104** 302–12
[145] Pullen K R 2019 The status and future of flywheel energy storage *Joule* **3** 1394–9
[146] Sarbu I and Sebarchievici C 2018 A comprehensive review of thermal energy storage *Sustainability* **10** 191
[147] Whittingham M S 2012 History, evolution, and future status of energy storage *Proc. IEEE* **100** 1518
[148] Whittingham M S 1978 Chemistry of intercalation compounds – metal guests in chalcogenide hosts *Prog. Solid State Chem.* **12** 41
[149] Whittingham M S 1976 Electrical energy-storage and intercalation chemistry *Science* **192** 1126
[150] Whittingham M S and Gamble F R 1975 The lithium intercalates of the transition metal dichalcogenides *Mater. Res. Bull.* **10** 363
[151] Whittingham M S 1975 Batterie a base de chalcogenures Belgian patent no. 819672
[152] Whittingham M S 1974 Electrointercalation in transition-metal disulfides *J. Chem. Soc. Chem. Com- mun.* **1974** 328–9
[153] Goodenough J B and Mizushima K 1982 Fast ion conductors US Patent No. 4357215
[154] Mizushima K, Jones P C, Wiseman P J and Goodenough J B 1980 Li_xCoO_2 $(0<x<-1)$: a new cathode material for batteries of high energy density *Mater. Res. Bull.* **15** 783
[155] Yoshino A, Sanechika K and Nakajima T 1987 Secondary battery US patent no. 4668595
[156] Yoshino A, Sanechika K and Nakajima T 1985 Secondary battery Japanese patent no. 1989293
[157] Nishi Y 2001 The development of lithium ion secondary batteries *Chem. Rec.* **1** 406
[158] Nobel Prize Organisation 2019 The Nobel Prize in chemistry 2019
[159] Guy J 2019 Nobel Prize in chemistry awarded for work on lithium ion batteries that 'revolutionized our lives'
[160] Rincon P 2019 Nobel chemistry Prize: lithium-ion battery scientists honoured
[161] Davis N and Devlin H 2019 Nobel Prize in chemistry awarded for work on lithium-ion batteries
[162] Staff S N 2019 Nobel Prize in chemistry goes to development of lithium batteries
[163] Castelvecchi D and Stoye E 2019 Chemistry Nobel honours world-changing batteries
[164] Tsai P C, Nasara R N, Shen Y C, Liang C C, Chang Y W, Hsu W D, Tran N T T and Lin S K 2019 Ab initio phase stability and electronic conductivity of the doped-$Li_4Ti_5O_{12}$ anode for li-ion batteries *Acta Mater.* **175** 196–205
[165] de las Casas C and Li W Z 2012 A review of application of carbon nanotubes for lithium ion battery anode material *J. Power Sources* **208** 74–85
[166] Ji L W, Lin Z, Alcoutlabi M and Zhang X W 2011 Recent developments in nanostructured anode materials for rechargeable lithium-ion batteries *Energy Environ. Sci.* **4** 2682–99
[167] Wang H L, Cui L F, Yang Y A, Casalongue H S, Robinson J T, Liang Y Y, Cui Y and Dai H J 2010 Mn_3O_4-graphene hybrid as a high-capacity anode material for lithium ion batteries *J. Am. Chem. Soc.* **132** 13978–80

[168] Chang C H, Li A C, Popovs I, Kaveevivitchai W, Chen J L, Chou K C, Kuo T S and Chen T H 2019 Elucidating metal and ligand redox activities of a copper-benzoquinoid coordination polymer as the cathode for lithium-ion batteries *J. Mater. Chem.* A **7** 23770–4

[169] Pan T Y, Tran N T T, Chang Y C and Hsu W D 2020 First-principles study on the initial reactions at $LiNi_{1/3}Co_{1/3}Mn_{1/3}O_2$ cathode/electrolyte interface in lithium-ion batteries *Appl. Surf. Sci.* **507** 144842

[170] Kaveevivitchai W and Jacobson A J 2015 Exploration of vanadium benzenedicarboxylate as a cathode for rechargeable lithium batteries *J. Power Sources* **278** 265–73

[171] Fergus J W 2010 Recent developments in cathode materials for lithium ion batteries *J. Power Sources* **195** 939–54

[172] Wang Y and Cao G Z 2008 Developments in nanostructured cathode materials for high-performance lithium-ion batteries *Adv. Mater.* **20** 2251–69

[173] Lu Z H, MacNeil D D and Dahn J R 2001 Layered cathode materials $Li[NixLi_{(1/3-2x/3)}Mn_{(2/3-x/3)}]O_2$ for lithium-ion batteries *Electrochem. Solid State Lett.* **4** A191–4

[174] Lee H, Yanilmaz M, Toprakci O, Fu K and Zhang X W 2014 A review of recent developments in membrane separators for rechargeable lithium-ion batteries *Energy Environ. Sci.* **7** 3857–86

[175] Huang X S 2011 Separator technologies for lithium-ion batteries *J. Solid State Electrochem.* **15** 649–62

[176] Venugopal G, Moore J, Howard J and Pendalwar S 1999 Characterization of microporous separators for lithium-ion batteries *J. Power Sources* **77** 34–41

[177] Dinh T T A, Huynh T T K, Le L T M, Truong T T T, Nguyen O H, Tran K T T, Tran M V, Tran P H, Kaveevivitchai W and Le P M L 2020 Deep eutectic solvent based on lithium bis (trifluo- romethyl)sulfonyl imide (LiTFSI) and 2,2,2-trifluoroacetamide (TFA) as a promising electrolyte for a high voltage lithium-ion battery with a $LiMn_2O_4$ cathode *ACS Omega* **5** 23843–53

[178] Lin F W, Tran N T T and Hsu W D 2020 Effect of 1,3-propane sultone on the formation of solid electrolyte interphase at Li-ion battery anode surface: a first-principles study *ACS Omega* **5** 13541–7

[179] Bachman J C *et al* 2016 Inorganic solid-state electrolytes for lithium batteries: mechanisms and properties governing ion conduction *Chem. Rev.* **116** 140–62

[180] Kaveevivitchai W and Jacobson A J 2013 High capacity microporous molybdenum-vanadium oxide electrodes for rechargeable lithium batteries *Chem. Mater.* **25** 2708–15

[181] Zhang S S 2006 A review on electrolyte additives for lithium-ion batteries *J. Power Sources* **162** 1379–94

[182] Song J Y, Wang Y Y and Wan C C 1999 Review of gel-type polymer electrolytes for lithium-ion batteries *J. Power Sources* **77** 183–97

[183] Meyer W H 1998 Polymer electrolytes for lithium-ion batteries *Adv. Mater.* **10** 439

[184] Feng K, Li M, Liu W W, Kashkooli A G, Xiao X C, Cai M and Chen Z W 2018 Silicon-based anodes for lithium-ion batteries: from fundamentals to practical applications *Small* **14** 1702737

[185] Zuo X X, Zhu J, Muller-Buschbaum P and Cheng Y J 2017 Silicon based lithium-ion battery anodes: a chronicle perspective review *Nano Energy* **31** 113–43

[186] Casimir A, Zhang H G, Ogoke O, Amine J C, Lu J and Wu G 2016 Silicon-based anodes for lithium-ion batteries: effectiveness of materials synthesis and electrode preparation *Nano Energy* **27** 359–76

[187] Su X, Wu Q L, Li J C, Xiao X C, Lott A, Lu W Q, Sheldon B W and Wu J 2014 Silicon-based nanomaterials for lithium-ion batteries: a review *Adv. Energy Mater.* **4** 1300882

[188] Liang B, Liu Y P and Xu Y H 2014 Silicon-based materials as high capacity anodes for next generation lithium ion batteries *J. Power Sources* **267** 469–90

[189] Ogata K, Salager E, Kerr C J, Fraser A E, Ducati C, Morris A J, Hofmann S and Grey C P 2014 Revealing lithium-silicide phase transformations in nano-structured silicon-based lithium ion batteries via *in situ* nmr spectroscopy *Nat. Commun.* **5** 3217

[190] Myung S T, Maglia F, Park K J, Yoon C S, Lamp P, Kim S J and Sun Y K 2017 Nickel-rich layered cathode materials for automotive lithium-ion batteries: achievements and perspectives *ACS Energy Lett.* **2** 196–223

[191] Manthiram A, Song B H and Li W D 2017 A perspective on nickel-rich layered oxide cathodes for lithium-ion batteries *Energy Stor. Mater.* **6** 125–39

[192] Liu W, Oh P, Liu X, Lee M J, Cho W, Chae S, Kim Y and Cho J 2015 Nickel-rich layered lithium transition-metal oxide for high-energy lithium-ion batteries *Angew. Chem. Int. Ed.* **54** 4440–57

[193] Zheng J M, Kan W H and Manthiram A 2015 Role of mn content on the electrochemical properties of nickel-rich layered $LiNi_{0.8\text{-}x}Co_{0.1}Mn_{0.1+x}O_2$ $(0.0 \leq x \leq 0.08)$ cathodes for lithium-ion batteries *ACS Appl. Mater. Interfaces* **7** 6926–34

[194] Zhang J N *et al* 2019 Trace doping of multiple elements enables stable battery cycling of $LiCoO_2$ at 4.6 V *Nat. Energy* **4** 594–603

[195] Wu F X and Yushin G 2017 Conversion cathodes for rechargeable lithium and lithium-ion batteries *Energy Environ. Sci.* **10** 435–59

[196] Chiang H H and Kuo C L 2017 A comparative first-principles study of the structural and electronic properties of the liquid Li-Si and Li-Ge alloys *J. Chem. Phys.* **146** 064502

[197] Chiang H H, Lu J M and Kuo C L 2016 First-principles study of the structural and dynamic properties of the liquid and amorphous Li-Si alloys *J. Chem. Phys.* **144** 034502

[198] Nitta N and Yushin G 2014 High-capacity anode materials for lithium-ion batteries: choice of elements and structures for active particles *Part. Part. Syst. Charact.* **31** 317–36

[199] Yamada Y, Usui K, Chiang C H, Kikuchi K, Furukawa K and Yamada A 2014 General observation of lithium intercalation into graphite in ethylene-carbonate-free superconcentrated electrolytes *ACS Appl. Mater. Interfaces* **6** 10892–9

[200] Lago N, Garcia-Calvo O, del Amo J M L, Rojo T and Armand M 2015 All-solid-state lithium-ion batteries with grafted ceramic nanoparticles dispersed in solid polymer electrolytes *ChemSusChem* **8** 3039–43

[201] Inaguma Y and Nakashima M 2013 A rechargeable lithium-air battery using a lithium ion-conducting lanthanum lithium titanate ceramics as an electrolyte separator *J. Power Sources* **228** 250–5

[202] Fergus J W 2010 Ceramic and polymeric solid electrolytes for lithium-ion batteries *J. Power Sources* **195** 4554–69

[203] Abe T, Sagane F, Ohtsuka M, Iriyama Y and Ogumi Z 2005 Lithium-ion transfer at the interface between lithium-ion conductive ceramic electrolyte and liquid electrolyte – a key to enhancing the rate capability of lithium-ion batteries *J. Electrochem. Soc.* **152** A2151–4

[204] Tsai Y J and Kuo C L 2020 Effect of structural disorders on the Li storage capacity of graphene nanomaterials: a first-principles study *ACS Appl. Mater. Interfaces* **12** 22917–29

[205] Tien N T, Thao P T B, Phuc V T and Ahuja R 2020 Influence of edge termination on the electronic and transport properties of sawtooth penta-graphene nanoribbons *J. Phys. Chem. Solids* **146** 109528

[206] On V V, Thanh L N and Tien N T 2020 The electronic properties and electron transport of sawtooth penta-graphene nanoribbon under uniaxial strain: ab-initio study *Philos. Mag.* **100** 1834–48

[207] Tien N T, Thao P T B, Phuc V T and Ahuja R 2019 Electronic and transport features of sawtooth penta-graphene nanoribbons via substitutional doping *Physica* E **114** 113572

[208] Tien N T, Phuc V T and Ahuja R 2018 Tuning electronic transport properties of zigzag graphene nanoribbons with silicon doping and phosphorus passivation *AIP Adv.* **8** 085123

[209] Lin K H and Kuo C L 2017 Lithiation mechanisms and lithium storage capacity of reduced graphene oxide nanoribbons: a first-principles study *J. Mater. Chem.* A **5** 4912–22

[210] Tran N T T, Nguyen D K, Glukhova O E and Lin M F 2017 Coverage-dependent essential properties of halogenated graphene: a DFT study *Sci. Rep.* **7** 17858

[211] Chung H C, Chang C P, Lin C Y and Lin M F 2016 Electronic and optical properties of graphene nanoribbons in external fields *Phys. Chem. Chem. Phys.* **18** 7573

[212] Tien N T, Thao D N, Thao P T B and Quang D N 2016 Key scattering mechanisms limiting the lateral transport in a modulation-doped polar heterojunction *J. Appl. Phys.* **119** 214304

[213] Chung H C, Lin Y T, Lin S Y, Ho C H, Chang C P and Lin M F 2016 Magnetoelectronic and optical properties of nonuniform graphene nanoribbons *Carbon* **109** 883

[214] Tran N T T, Lin S Y, Glukhova O E and Lin M F 2016 pi-Bonding-dominated energy gaps in graphene oxide *RSC Adv.* **6** 24458–63

[215] Li L J, Hu L H, Lin C T and Wu F Y 2014 Electrochemical graphene and electrode composite materials and lithium ion battery including the same Taiwan patent no. I441947

[216] Chung H C, Yang P H, Li T S and Lin M F 2014 Effects of transverse electric fields on Landau subbands in bilayer zigzag graphene nanoribbons *Philos. Mag.* **94** 1859

[217] Hu L H, Wu F Y, Lin C T, Khlobystov A N and Li L J 2013 Graphene-modified $LiFePO_4$ cathode for lithium ion battery beyond theoretical capacity *Nat. Commun.* **4** 1687

[218] Chung H C, Su W P and Lin M F 2013 Electric-field-induced destruction of quasi-Landau levels in bilayer graphene nanoribbons *Phys. Chem. Chem. Phys.* **15** 868

[219] Chung H C, Lee M H, Chang C P and Lin M F 2011 Exploration of edge-dependent optical selection rules for graphene nanoribbons *Opt. Express* **19** 23350–63

[220] Chung H C, Lee M H, Chang C P, Huang Y C and Lin M F 2011 Effects of transverse electric fields on quasi-Landau levels in zigzag graphene nanoribbons *J. Phys. Soc. Jpn.* **80** 044602

[221] Chung H C, Huang Y C, Lee M H, Chang C C and Lin M F 2010 Quasi-Landau levels in bilayer zigzag graphene nanoribbons *Physica* E **42** 711

[222] Novoselov K S, Jiang D, Schedin F, Booth T J, Khotkevich V V, Morozov S V and Geim A K 2005 Two-dimensional atomic crystals Proc. Natl. Acad. Sci. 102 10451

[223] Novoselov K S, Geim A K, Morozov S V, Jiang D, Zhang Y, Dubonos S V, Grigorieva I V and Firsov A A 2004 Electric field effect in atomically thin carbon films *Science* **306** 666

[224] Tran N T T, Gumbs G, Nguyen D K and Lin M F 2020 Fundamental properties of metal-adsorbed silicene: a DFT study *ACS Omega* **5** 13760–9

[225] Chung H C, Chiu C W and Lin M F 2019 Spin-polarized magneto-electronic properties in buckled monolayer GaAs *Sci. Rep.* **9** 2332

[226] Bhimanapati G R *et al* 2015 Recent advances in two-dimensional materials beyond graphene *ACS Nano* **9** 11509

[227] Liu H, Neal A T, Zhu Z, Luo Z, Xu X F, Tomanek D and Ye P D 2014 Phosphorene: an unexplored 2D semiconductor with a high hole mobility *ACS Nano* **8** 4033

[228] Matthes L, Gori P, Pulci O and Bechstedt F 2013 Universal infrared absorbance of two-dimensional honeycomb group-IV crystals *Phys. Rev.* B **87** 035438

[229] Roy P and Srivastava S K 2015 Nanostructured anode materials for lithium ion batteries *J. Mater. Chem.* A **3** 2454–84

Chapter 17

Concluding remarks

Thi Dieu Hien Nguyen, Shih-Yang Lin, Hsien-Ching Chung, Wei-Bang Li, Ngoc Thanh Thuy Tran, Nguyen Thi Han, Hsin-Yi Liu, Hai Duong Pham and Ming-Fa Lin

The current book is mainly focused on the theoretical framework from first-principles simulations, being suitable for any condensed-matter systems. The developed strategy is clearly illustrated by 3D multi-component compounds of cathode/electrolyte/anode materials lithium-sodium-/potassium-/magnesium-/aluminum-/iron-based batteries (details in chapters 4–14). There exist diverse physical/chemical/material phenomena due to significant multi-orbital mixings in chemical bonds and specific atom- and orbital-induced spin configurations. The critical mechanisms, quasiparticle properties in charges (spins), are thoroughly examined from the optimal crystal structures (the orbital-dependent net magnetic moments), the atom-dominated valence and conduction states (the spin-split energy spectra), the spatial charge density distributions (the real-space spin arrangements), the atom- and orbital-decomposed density of states (the spin-decomposed ones), and the specific relations between the energy-dependent orbital hybridizations and the optical absorption peaks. Very interestingly, the many-body excitonic effects have been successfully included to clarify the whole physical/chemical/material viewpoints. Highly potential applications, e.g., energy storages, are thoroughly discussed. Part of theoretical predictions agree with the high-resolution measurements, while systematic verifications are required for most of them. Certain important issues, which are strongly associated with the theoretical and experimental progress of up-to-date batteries, will be proposed in the next chapter. For example, how to present a complete theory in fully comprehending the ion transport with the core systems becomes the most urgent research work.

By the delicate VASP calculations and qualitative analyses, the main features of quasiparticle properties are well presented and explained by the concise physical/chemical/material pictures. Furthermore, the development of theoretical framework

doi:10.1088/978-0-7503-4685-6ch17

is clarified from the essential properties of the core materials in ion-based batteries. The calculated results cover the optimal geometric, electronic, magnetic and optical properties, such as, the concentration-dependent lattice symmetries [1, 2], the atomic positions and bond-length fluctuations within a moiré superlattice [1, 3, 4], a lot of valence and conduction energy subbands along the high-symmetry points of the greatly reduced first Brillouin zone [1, 3, 4], the orbital-hybridization-induced charge density distributions [1, 3, 4], the orbital-generated magnetic moments [1], the spin-up and spin-down spatial arrangements (ferromagnetism, anti-ferromagnetism and non-magnetism), the orbital- and spin-projected van Hove singularities [1, 3, 4], the real- and imaginary-part transverse dielectric functions [5, 6], their reflectance spectra [5, 6], the optical absorption coefficients [5, 6], and the energy loss functions [7, 8]. The direct combinations of the above-mentioned physical quantities are responsible for the featured quasiparticle behaviors, i.e., the significant multi-/single-orbital mixings in chemical bonds, as well as the orbital-created spin configurations, are well characterized within a moiré superlattice, the different energy-spectrum ranges, and the specific absorption frequencies. The successful development of the theoretical framework would be generalized to more broad ranges, e.g., the co-dominance of charges and spins in transition-metal- and rare-earth-metal-related 2D emergent materials [9, 10] in terms of electronic, magnetic, optical and transport properties. The strong cooperation/competition between orbital hybridizations and spin configurations are expected to present diverse phenomena. On the other hand, the phenomenological models [11, 12] are almost absent in exploring the ternary lithium oxides, especially for the ion transports (discussed later in chapter 18). However, the tight-binding model might be useful for studying the x-ray diffraction patterns. Interestingly, how to develop a unified theory for solving the average current density through cathode/electrolyte/anode material becomes the most important issue. In addition, first-principles simulations should be very efficient to fully comprehend the rich and unique optical properties of carbon-sp^2-bonding systems. The excitonic effects, arising from bulk graphites [13], layered graphenes [14, 15], carbon nanotubes [16, 17], graphene nanoribbons [18, 19], and carbon fullerenes [20], are worthy of systematic studies. This is under current investigation.

High-resolution experimental measurements on cathode/electrolyte/anode materials can confirm the essential properties and test the developed theories, thus, promoting highly potential applications of ion-based batteries. When these samples are successfully synthesized under the various chemical and physical environments, their crystal structures are well identified from the x-ray diffraction patters, e.g., certain **G**-dependent peaks in pristine graphite/alkali-atom graphite intercalation compounds [21]/[22], and $Li_4Ti_5O_{12}$, Li_2GeO_3/$LiFe_2O_3$ [23]/[24]/[25]. The featured peak structures, which rely on the θ dependence, number, form, intensity, are mainly determined by a whole charge density distribution within normal/moiré superlattice. How to estimate (evaluate) them from the measured discrete peaks (the phenomenological model; discussed in chapter 18) will become one of most important issues in near-future science research. However, the intermediate crystal structures, being revealed during the charging and discharging processes, are very difficult to detect using this spectroscopy. Perhaps time-resolved x-ray experiments [26, 27] could be developed to resolve some of the excited states. As for scanning tunneling microscopy (STM) and scanning electronic microscopy (SEM), their surface morphologies are, respectively, available in examining the stacking configuration of layered graphites [28–30] and the uniform or non-uniform atom densities of multi-component lithium oxides [31]. Angle-resolved photoemission spectroscopy (ARPES) observations have verified the semi-metallic behaviors of valence and conduction band overlaps in Bernal graphite [32], and the n-type dopings in stage-*n* alkali-atom graphite intercalation compounds [33], where the latter clearly show metal-like electric conductivities. These examinations seem to be useless for electronic energy spectra arising from moiré superlattices. For example, the unusual band structures of lithium oxide compounds, which exist along the high-symmetry points of the greatly reduced first Brillouin zone, cannot be well tested through ARPES to date. The low-energy quasiparticle properties are further directly reflected in optical vertical excitations by observing the reflection/absorption/transmission spectra [6, 34]; therefore, low-lying energy bands, Fermi level and free carrier density are achieved using such measurements. In addition, the low-frequency [11, 12], π [11, 12] and π+σ plasmon modes [11, 12] in layered graphite systems could be verified from the electron energy loss spectra (EELS) measurements. Specifically, the theoretical predictions, being conducted on significant absorption structures of lithium oxides [5, 35], would be very difficult to verify from optical experiments. The main reason is that there are too many prominent peak structures due to a unique band structure [5]. But for the meta-stable excitons [4, 35, 36], their main features could be efficiently clarified from the high-resolution photoluminescence spectra [35, 36] such as threshold energy, number, and intensity of excitonic peaks. Up-to-date results only show their contributions to the first absorption peaks, and delicate observations and analyses are required in current optical measurements. Very interestingly, the current density of ion transport in a whole battery has been confirmed to be very stable and sufficiently high for commercial purposes [37, 38]. How to thoroughly realize the critical mechanisms is under current systematic investigation [37].

In chapter 4, the quaternary Li_xFePO_4 compounds of $x = 1.0, 0.75, 0.5$ and 025 clearly illustrate the diverse phenomena of quasiparticle change and spin distributions. The lithium concentrations make it easy to modulate the essential properties [39, 40]. The x-induced crystal structures are successfully examined from the x-ray diffraction peaks [40]. The highly non-uniform chemical environments within moiré superlattices are responsible for the main features of the geometric, electronic, and magnetic properties. The featured behaviors cover the atom- and position-dependent chemical bonds (in terms of the observable fluctuations of bond lengths and x-created crystal distortion; [41]), the rich and unique electronic properties (many valence and conduction energy subbands, their large asymmetry about E_F, non-monotonous energy dispersion relations, partial flat bands near E_F different atom dominances within certain energy ranges, van Hove singularities, highly anisotropy, spin splitting or degeneracy, a finite/vanishing bad gap), the whole orbital hybridizations in the spatial charge distributions, the emergences of the atom-, orbital- and spin-projected van Hove singularities, the atom and orbital-decomposed magnetic moments, and the spin-up and spin-down spatial arrangements. The distinct physical/chemical quantities are thoroughly utilized to analyze the critical mechanisms for the diversified quasiparticle properties; that is, the active orbital hybridizations and spin configurations are examined and identified from the current delicate analyses. Li–, Fe– and P–O bonds, respectively, 2s-[2s, $2p_x$, $2p_y$, $2p_z$], [4s, $3d_{x^2-y^2}$, $3d_{z^2}$, $3d_{xy}$, $3d_{yz}$, $3d_{xz}$]-[2s, $2p_x$, $2p_y$, $2p_z$], and [3s, $3p_x$, $3p_y$, $3p_z$]-[2s, $2p_x$, $2p_y$, $2p_z$]. Furthermore, 3d orbitals dominate the spin configurations and almost fully suppress other s and p ones. Their spin-up valence states determine the FM moments, while the spin-down configuration across the Fermi level determines the metallic behavior. Interestingly, the lithium concentration makes it easy to modulate crystal symmetry and thus other essential properties. The $x = 1$ case presents antiferromagnetic (AFM) and ferromagnetic (FM) spin phases with the almost identical ground state energy and geometry, but rather distinct electronic energy spectra. $x = 0.75$, 0.5 and 0.25 belong to FM metals. Band structures of semiconductors and metals could be divided into four and five energy ranges, respectively.

Although Vienna *ab initio* simulation package (VASP) simulations have provided very useful information on certain physical/chemical/material behaviors, their combinations with the analytic methods are required to fully understand the other ones, e.g., the theoretical predictions on the discrete x-ray diffraction peaks [2] and electrical conductivities of Li^+-ion transports. The complicated quasiparticle interactions, with the single- and many-body forms due to charges and spins, are expected to be quite difficult in building the suitable Hamiltonian matrix. Whether the phenomenological models are available in near-future investigation will become an emergent issue.

$AlCl_4^-$-ion and $AlCl_4$-molecule graphite intercalation compounds exhibit diversified quasiparticle phenomena, mainly owing to large intercalants and distinct charge cases [42]. Based on delicate VASP simulations and theoretical analyses, all the active chemical bonds are covered in the intralayer/interlayer C–C orbital hybridizations, interlayer C-intercalant interactions, and intra- and intercalant ones. They have been thoroughly examined from stage-1 crystal symmetries, C-, Al-, and Cl-dominated energy spectra at specific energy ranges, charge distributions before and after intercalations/de-intercalations, and atom- and orbital-projected van Hove singularities, as well as their emergences. Very interestingly, the greatly diversified fundamental properties are revealed as follows: periodical arrangements along the longitudinal and transverse directions simultaneously, absence of stage-1 and stage-2 cases, moiré superlattices, generation of many valence and conduction energy subbands, changes of initial high-symmetry valleys, strengthened asymmetry of hole and electron spectra about $E_F = 0$, the obvious reduce or enhancement of band overlaps (diversified free carrier denisities), almost isotropic/highly anisotropic features near/away from $E_F = 0$, diverse energy dispersions at the distinct critical points, frequent band crossings and anti-crossings, undefined behaviors of the whole π and σ valence bands, carbon-, aluminum- and chloride-dominated band structures within the different energy ranges. The specific orbital hybridizations in significant C–C/C–Cl/Al–Cl/Cl–Cl bonds are [2s, $2p_x$, $2p_y$]-[2s, $2p_x$, $2p_y$] and $2p_z$–$2p_z$/[2s, $2p_x$, $2p_y$, $2p_z$]-[3s, $3p_x$, $3p_y$, $3p_z$]/[3s, $3p_x$, $3p_y$, $3p_z$]-[3s, $3p_x$, $3p_y$, $3p_z$]/[3s, $3p_x$, $3p_y$, $3p_z$]-[3s, $3p_x$, $3p_y$, $3p_z$].

However, the obvious evidence is absent for C–Al and Al–Al bonds. In addition, the stage-n systems, with $n \geqslant 2$ under the highest intercalant concentrations, are under current investigation. There are certain significant differences between molecule-related and alkali-atom graphite intercalation compounds in terms of featured properties, e.g., the different n- or p-type dopings [43], single-particle and collective excitations of π electrons, and optical initial absorption structures [3, 44]. By the delicate analyses on the VASP simulations, the intralayer/interlayer C–C orbital hybridizations, the interlayer C-intercalant interactions, and the intra- and intercalant ones can be reached thoroughly. This is capable of providing very useful information about the suitable hopping integrals and site energies of active orbitals within the phenomenological tight-binding model [45]. It should be notice that the recent experimental measurements only observe the stage-3 and stage-4 cases, but not the stage-1 and stage-2 ones [46, 47]. The method of molecular dynamics could be utilized to investigate the existence of stage-1 and stage-2 configurations. As for the complicated intercalation and de-intercalation during the charging and discharging processes of the high-performance batteries, the critical mechanism on the large-ion transports are very important research topics and worthy of thorough studies [47], especially for the theoretical framework developments .

In chapter 6, the calculated results of $FeCl_3$ graphite intercalation compounds show detailed features. The stage-1 $FeCl_3$-graphite intercalation compounds present diversified electronic properties due to the intercalant. The first-principles calculations on VASP are utilized to analyze the essential properties, such as the geometric structures, spatial charge distributions, charge variations, band structures and density of states. The density of states displays full information for an explanation of the hybridizations with the special structures van Hove singularities on them. The van Hove singularities in graphite-related systems are very important and can provide full information for examining the intercalation effects. The orbital-decomposed density of states for C atoms shows that the π bondings are orthogonal to the sp^2 bondings and the C–C bondings are retained in the intralayer C atoms. The Fe atoms and Cl atoms form the Fe–Cl bondings at some unique energy range, presenting the multi-orbital hybridizations of [4s, $3d_{xy}$, $3d_{yz}$, $3d_{xz}$, $3d_{x^2-y^2}$, $3d_{z^2}$]-[$3p_x$, $3p_y$, $3p_z$]. For C–Cl and Cl–Cl bonds, the unique van Hove singularities exhibit their coupling interactions, revealing multi-orbital hybridizations of [$2p_x$, $2p_y$, $2p_z$]-[$3p_x$, $3p_y$, $3p_z$] and [3s, $3p_x$, $3p_y$, $3p_z$]-[3s, $3p_x$, $3p_y$, $3p_z$], respectively. The Fe–Cl bondings arise from multi-orbital hybridizations of [4s, $3d_{xy}$, $3d_{yz}$, $3d_{xz}$, $3d_{x^2-y^2}$, $3d_{z^2}$]-[$3p_x$, $3p_y$, $3p_z$]. Due to the band structures and density of states, the multi-orbital interactions between intercalants and graphene layers dominate in the low-lying energy range. The charge transfers per atom (electrons/atom) for C, Fe, Cl are -0.02 *e*/atom, -0.28 *e*/atom and $+0.46$ *e*/atom, respectively. Thus, the C atoms in graphene layers present as positive ones after intercalation, i.e., the graphite system exhibits p-type doping features in agreement with previous study [48]. Furthermore, the graphene layer possesses free holes after intercalation and its conductance is significantly enhanced [49]. Interestingly, the stage-1 $FeCl_3$ graphite intercalation compounds perform at higher capacity than pure graphite but lower capacity than the stage-2 cases in previous experiment [50]. It is worth constructing the complete theoretical framework of stage-*n* $FeCl_3$ graphite intercalation compounds [50–53] by VASP simulations to further understand the complicated mechanics.

In chapter 7, the 3D binary compound V_2O_5 is successfully accomplished based on the numerical calculation. The orbital hybridization theory in the chemical bonds can clearly identify the essential properties of the V_2O_5 system. V_2O_5 has the indirect gap $E_g^i = 2.308$ eV semiconductor. This compound belongs to a moiré superlattice with an anisotropic and highly non-uniform environment. This can lead to the ease of ion transport back and forth at the cathode of rechargeable batteries. The specific orbital hybridizations in significant V–O bonds are [4*s*, $3d_{x^2-y^2}$, $3d_{xy}$, $3d_{yz}$, $3d_{xz}$, $3d_{z^2}$] -[2s, $2p_x$, $2p_y$, $2p_z$]. The connection between the geometry, electronic, and optical properties is thoroughly covered in this chapter under the band-edge state and orbital transitions. Furthermore, the single and many-particle interactions are systematically provided under the dielectric properties, which show the screening ability of electrons in the electromagnetic fields. Many prominent peaks and shoulder absorptions are achieved, associated with the inner transition and the orbital-channel creation of vanadium and oxygen atoms. The numerical analysis has been confirmed by the experimental measurements, which are worth investigating for accuracy by further developed techniques.

In chapter 8, first-principles calculation was used to investigate the geometric, electronic, and optical properties of the Li_2SiO_3 compound. Under this state-of-the-art analysis, a concise physical/chemical picture, and the crucial characteristics of the electrolyte in Li^+ related materials are achieved, through which the diversified essential properties of cathode/electrolyte/anode materials in ion-related materials in general could be fully understood. The Li_2SiO_3 compound exhibits special properties, a large moiré cell with extremely non-uniform physical/chemical environments, wide-indirect insulator gap, a significant spatial charge density distribution generated by the orbital hybridizations, and atom-/orbital-projected density of states. Consequently, the distinct orbital hybridization related to the strong optical response could be identified. The outstanding optical properties involve the un-isotropic optical behaviors under three electronic dipole directions, the robust excitonic states with a largely reduced optical gap (E_g^O= 6.95 eV) much smaller than the band gap (E_g^i = 8.4 eV), strongly modified single particle interband transitions, the reflectance/high transparency of electromagnetic waves for $\omega < E_g^o$, a high absorption coefficient, twenty prominent single-particle excitation peaks, and many pronounced peaks in energy loss function (ELF) at the plasmon resonance frequency ω_p, 17–19 eV. The main optical features of the Li_2SiO_3 compound could be verified by experimental optical spectroscopy measurements. The current theoretical framework could be developed for further studies of other anode/cathode/electrolyte materials for lithium-ion batteries (LIBs).

The search map for predicting the electron affinity (EA) and ionization energy (IE) values for LIB applications has been established by means of artificial neural networks based on Matlab and Python codes. A large-scale of more than 20 000 molecules database obtained from the Materials Project was trained and tested, and then from the validation results to estimate the expected values by the self-developed code. Further work is needed to clarify some unrevealed mechanisms of molecule decomposition and improve the properties of electrolyte additives. Moreover, applying artificial neural networks to predict modified molecules lead to a new approach to create brand-new functional molecules not only for electrolyte additives but also for electrolyte molecules. This could serve as a first step toward a better design of electrolytes and electrolyte additives for LIBs and be useful for experimental fabrications.

First-principles calculations are used to investigate the essential properties on stage-1 Li- and Li^+-graphite-intercalation compounds (GICs) under various concentrations, covering geometric, electronic, and charge bonding scheme. The LiC_x/ Li^+C_x with $x = 6$, 12, 18, 24, 32 and 36 are thoroughly studied and compared with the stage-n cases under the same concentrations. The main results of Li/Li^+ GICs include the increase/decrease of interlayer geometric structures, variation of Li/Li^+–C spatial charge distributions, and energy shift of band structures. The Li-GICs has its energy bands shift toward lower energy and so does the density of states. However, the Li^+-GIC has its electronic structure remaining almost the same, i.e., semimetal. In addition, Li^+–C interactions are found to be the critical factor to dominate the electronic properties. Importantly, the Li–C interaction belongs to 2s-2p_z orbital–orbital interaction, while the orbital hybridizations are relatively weak in Li^+ and C systems.

Sodium-, potassium- and magnesium-graphite intercalation compounds are verified to be outstanding candidates for developments of quasiparticle framework [22, 54–56], progress of experimental measurements [33, 57], as well as the potential applications of electronic and optical devices [57]. The critical multi-/single-orbital hybridizations are achieved from highly active environments, atom-dominated band structure within the specific energy ranges, spatial charge density distributions and their variations due to intercalation, and merged van Hove singularities of atom- and orbital-projected density of states. The active forms cover 2p_z–2p_z and [2s, 2p_x, 2p_y]-[2s, 2p_x, 2p_y] in C–C bonds, 2p_z–3s/2p_z–4s/2p_z–3s in C–Na/C–K/C–Mg bonds, and 3s-3s/4s-4s/3s-3s in Na–Na/K–K/Mg–Mg bonds. Such intralayer and interlayer atomic interactions, being assisted by the intercalant concentrations and distributions, are responsible for the diversified low-energy electronic properties, such as the enhanced asymmetry of hole and electron energy spectra relative to the Fermi level, an observable blue shift of E_F, the free conduction electron density, the modification about the isotropic and linear Dirac-cone structure near the K or Γ valley, the well-characterized, but enriched π and σ subbands, and few valence subbands associated with the intercalants. Since the various chemical bonds possess concise orbital hybridizations, the tight-binding model, with reliable on-site Coulomb potentials and neighboring orbital-dependent hopping integrals [58, 59], could be established from well-fitting through VASP. This is very useful in exploring the Coulomb excitations [58, 59] and decay rates [58, 59]. Furthermore, its generalization form, the generalized tight-binding model [60], can thoroughly investigate the magnetic quantization phenomena, e.g., magneto-electronic properties [58], magneto-optical selection rules [58], and quantum transports [58]. Only part of the theoretical prediction on crystal structures has been verified by x-ray diffraction patterns [26, 27, 61]. Delicate analyses of theoretical and experimental research are required for the further examinations of the meta-stable configurations. Their strong cooperation is very suitable for the confirmations of the occupied energy subbands below the Fermi level and the spatial charge density distributions. It should also be noticed that direct verifications of van Hove singularities by STS measurements are forbidden because of the intrinsic limits. However, the featured density of states can induce the prominent structures in Coulomb- [6] and electromagnetic-field excitation spectra [6]. Up-to-date experiments clearly show that graphite and graphite intercalation compounds display a lot of commercial applications, e.g., batteries [33, 62], carbon fibers [22, 63], fighters [22, 63], and vehicles [22, 63]. Basic science research would be very useful in fully understanding the rich and unique transport properties of ion-based batteries, such as the similarities and significant differences among Li-, Na-, K- and Mg-ion transports.

1D armchair silicon and carbon nanotubes, with a unique cylindrical symmetry, are very suitable for studying the basic and applied sciences. The geometric structure and electronic properties of armchair nanotubes are investigated by the first-principles method. Both nanotubes exhibit diverse geometries and electronic properties, as a result of the significant chemical bonding of four outer orbitals and the spin-orbital couplings. The band structure is dependent on the chirality and radius, charge density and orbital-projected density of state, which are useful in exploring the orbital-dependent essential properties. The former possesses buckled and planar honeycomb lattices of 3.54–3.785 Å; the ground state energy E_0 and buckling height Δh decrease with an increase of the radius, and E_0 of the carbon nanotubes decreases with an increase of the radius as well. The bond lengths b_1 and b_2 of silicon and carbon nanotubes increase when the diameter of the tube becomes smaller. The results of the band structure from silicon and carbon nanotubes are quite different. Compared to carbon nanotubes, silicon nanotubes exhibit direct and indirect band gaps due to their asymmetry structure.

Armchair silicon nanotubes (aSiNTs) from (2,2) to (11,11) possess semiconducting behavior, whereas armchair carbon nanotubes (aCNTs) are from (3,3) to (12,12) behave like gapless metals. The energy dispersion of aSiNTs and aCNTs are reflected in projected denisties of states (PDOSs) as many anti-symmetric and symmetric peaks. There are five kinds of low-lying peak structures characteristic of the main features of the essential properties. The predicted geometric structure, energy bands and PDOSs are reliably obtained by the experimental methods of STM, ARPES and STS, respectively. ARPES and scanning tunneling spectroscopy (STS) are particularly useful to determine the spin resolution. 1D armchair carbon and silicon nanotubes, with their unique cylindrical symmetry, are very suitable for studying the basic and applied sciences. Chirality- and radius-dependent semiconductors and metals are available in functionalized nano devices, such as electronic and optoelectronic devices. The intrinsic 1D band structures, with the decoupled angular momenta, dominate the quantized electrical and optical properties, e.g., the geometry-dependent resistance, capacitance, inductance, and optical conductance [64–66]. At present, field-effect transistors (FETs) based on semiconducting carbon nanotubes have been widely developed, mainly owing to the advantage of high carrier mobility under low scatterings. The first FETs composed of single- or multi-walled carbon nanotubes were reported in 1998 [67, 68]; they possess superior electrical conductance at room temperature using the gate-voltage modulation. A very high carrier mobility (> 105 cm^2 Vs^{-1}) is identified to be related to the diffusive transport of holes under the electron–phonon scattering [65, 66]. The nano-scaled radii allow the gate's ability to control the potential of the conducting channel in the ultimate thin FETs, while suppressing short-channel effects [69, 70]. Moreover, the semiconducting carbon nanotubes have highly potential applications in optoelectronic devices, including electroluminescent light emitters [71, 72], super capacitors [73] and photodetectors [74–76]. The band-edge states in 1D nanotubes, with high PDOSs, can create very prominent optical excitations [64]. The excited electrons and holes on a cylindrical surface are further driven towards each other by using an appropriate bias between the source and drain of the FET. The electron–hole recombination will emit strong electroluminescence. This highly efficient photoemission process has been extensively utilized in light-emitting diodes (LEDs) [77–79]. The application of photodetector is based on the electric current generated by the resonant excitations. On the other hand, metallic carbon nanotubes could be selected as candidates in future interconnect materials of integrated electronic devices [80, 81].

Chapter 13 provides the theoretical framework on studying the featured properties of the lithium titanate ($Li_4Ti_5O_{12}$) compound. Covering the whole book chapter, the multi-orbital hybridization in Li–O and Ti–O chemical bonds is linking all the electronic properties such as the atom-dominated valence and conduction bands, the spatial charge density, the atom- and orbital-decomposed density of states [3]. The optical properties also study in detail through the interband transition of the specific band-edge states, and the orbital-decomposed density of states in Li, Ti, O [2, 82]. Apparently, Li atoms have a minor contribution in the band spectrum and atom-, orbital-density of states, however, their existence in the compound is extremely significant. The single- and many-particle effects are examined through prominent peak/shoulder absorptions in the complex dielectric functions. The current orbital chemical bonds include $2s$-($2p_x$, $2p_y$, $2p_z$), ($4s$, $3d_{x^2-y^2}$, $3d_{xy}$, $3d_{yz}$, $3d_{zx}$, $3d_{z^2}$)-($2p_x$, $2p_y$, $2p_z$), which are corresponding Li–O and Ti–O, respectively [1]. The critical mechanism is worthy of systematic comprehension in terms of physical, chemical, material properties. This can be extended into the other related system in anodes, cathodes, and electrolytes of Li^+-ion based batteries. Besides, other Li–Ti–O compounds such as $LiTi_2O_4$ [82, 83], $Li_2Ti_2O_4$ [82, 84], and $Li_7Ti_5O_{12}$[23, 82] can become our further research work; they present diverse geometric symmetries [23, 82–85], specific electronic [23, 82–84], optical [82], and superconducting properties [23, 82–85]. Due to large variations of chemical bond lengths in $Li_4Ti_5O_{12}$, the relatively easy modulations on chemical bonds might be supportive in explaining the open issue, which is related to the structural transformation between different metastable configurations during the charging or discharging processes of Li^+-ion based batteries.

Transition metal dichalcogenides (TMDs) have paved the way for the fundamental and practical development of pseudo-capacitors [86] that put them among the electroactive materials that are promising candidates for such applications. As mentioned, MoS_2 [87–90] and beyond-MoS_2 such as WS_2 [91, 92], WTe_2 [93] present the essential properties promising for electrochemical energy storage applications. These materials exhibit significant characteristics such as tunable band gap properties, photo-electrochemistry, supercapacitors. With expectation that there will be a dramatic increase in the number of applications in electrochemistry of these materials, beyond-MoS_2 compounds are attracting the concern from material research society. ZrX_2 and HfX_2 are also the classes of beyond-MoS_2 which are expected to be implemented in a part of batteries. They are layered conversion-type TMDs ranged from semiconductors to semimetals which are studied to bring attention to electrochemical energy storage. They all show that the electronic properties depend on the geometry, chemical bonding, and orbital hybridization. It should be noted that the crystalline structure and electrochemical properties can be turned by exchanging transition metal or chalcogen atoms. In the other words, constructing and investigating a new class of TMDs are expected to enhance the electrochemistry applications.

In chapter 15, the designed test procedures according to UL 1974 and the charge/discharge profile dataset of the $LiFePO_4$ repurposed batteries are provided. Researchers/engineers can use the characteristic curves in estimating the repurposed of batteries under UL 1974. Furthermore, the profile dataset can be applied in the model-based engineering of repurposed batteries, e.g., fitting the variables of an empirical model or validating the results of a theoretical model.

In chapter 16, relying on the battery management system and energy management system, current potential applications of Li-ion battery systems in various size scales work efficiently. Li-ion batteries power almost every smartphone, laptop, and tablet sold today around the world. Their role will be even more important in the future, as electric vehicles are still an emerging market, including not only electric cars, but also electric motorcycles, buses, and trucks. Batteries are also under-exploited in power supply systems, especially in combination with PVs and wind power, where they can massively reduce carbon emission. Since Li-ion batteries were first commercially introduced in 1991, they have become an integral part of modern technology and, consequently, of our modern way of life. The future is bright for Li-ion energy storage.

References

[1] Lin S-Y, Liu H-Y, Nguyen D K, Tran N T T, Pham H D, Chang S-L, Lin C-Y and Lin M-F 2020 *Silicene-Based Layered Materials: Essential Properties* (Bristol: IOP Publishing)

[2] Tran N T T L, Lin S Y, Lin C Y and Lin M F 2017 *Geometric and Electronic Properties of Graphene-Related Systems: Chemical Bonding Schemes* (Boca Raton, FL: CRC Press)

[3] Nguyen T D H, Pham H D, Lin S Y and Lin M F 2020 Featured properties of Li+-based battery anode: $Li_4Ti_5O_{12}$ *RSC Adv.* **10** 14071–9

[4] Khuong Dien V, Thi Han N, Nguyen T D H, Huynh T M D, Pham H D and Lin M-F 2020 Geometric and electronic properties of Li_2GeO_3 *Front. Mater.* **7** 288

[5] Dien V K, Pham H D, Tran N T T, Han N T, Huynh T M D, Nguyen T D H and Fa-Lin M 2020 Orbital-hybridization-created optical excitations in Li_2GeO_3 arXiv:2009.02160

[6] Fox M 2001 *Optical Properties of Solids* (Oxford: Oxford University Press)

[7] Nguyen-Truong H T 2014 Energy-loss function including damping and prediction of plasmon lifetime *J. Electron. Spectrosc. Relat. Phenom.* **193** 79–85

[8] Sun Y, Xu H, Da B, Mao S-f and Ding Z-j 2016 Calculations of energy-loss function for 26 materials *Chin. J. Chem. Phys.* **29** 663–70

[9] Schulz S, Nechaev I A, Güttler M, Poelchen G, Generalov A, Danzenbächer S, Chikina A, Seiro S, Kliemt K and Vyazovskaya A Y 2019 Emerging 2D-ferromagnetism and strong spin–orbit coupling at the surface of valence-fluctuating $EuIr_2Si_2$ *npj Quantum Mater.* **4** 1–8

[10] Dierolf V, Ferguson I and Zavada J M 2016 *Rare Earth and Transition Metal Doping of Semiconductor Materials: Synthesis, Magnetic Properties and Room Temperature Spintronics* (Cambridge: Woodhead Publishing)

[11] Lin C-Y, Wu J-Y, Chiu C-W and Lin M-F 2019 *Coulomb Excitations and Decays in Graphene-Related Systems* (Boca Raton, FL: CRC Press)

[12] Ho J-H, Lu C, Hwang C, Chang C and Lin M-F 2006 Coulomb excitations in AA-and AB-stacked bilayer graphites *Phys. Rev.* B **74** 085406

[13] Trevisanutto P E, Holzmann M, Côté M and Olevano V 2009 High-energy excitonic effects in graphite and graphene arXiv:0909.1682

[14] Mak K F, Shan J and Heinz T F 2011 Seeing many-body effects in single-and few-layer graphene: observation of two-dimensional saddle-point excitons *Phys. Rev. Lett.* **106** 046401

[15] Yang L, Deslippe J, Park C-H, Cohen M L and Louie S G 2009 Excitonic effects on the optical response of graphene and bilayer graphene *Phys. Rev. Lett.* **103** 186802

[16] Ando T 2010 Environment effects on excitons in semiconducting carbon nanotubes *J. Phys. Soc. Jpn.* **79** 024706

[17] Pedersen T G 2004 Exciton effects in carbon nanotubes *Carbon* **42** 1007–10

[18] Yang L, Cohen M L and Louie S G 2007 Excitonic effects in the optical spectra of graphene nanoribbons *Nano Lett.* **7** 3112–5

[19] Prezzi D, Varsano D, Ruini A, Marini A and Molinari E 2008 Optical properties of graphene nanoribbons: the role of many-body effects *Phys. Rev.* B **77** 041404

[20] Golden M S, Knupfer M, Fink J, Armbruster J, Cummins T, Romberg H, Roth M, Sing M, Schmidt M and Sohmen E 1995 The electronic structure of fullerenes and fullerene compounds from high-energy spectroscopy *J. Phys. Condens. Matter* **7** 8219

[21] Augustsson A, Herstedt M, Guo J-H, Edström K, Zhuang G, Ross P Jr, Rubensson J-E and Nordgren J 2004 Solid electrolyte interphase on graphite Li-ion battery anodes studied by soft x-ray spectroscopy *Phys. Chem. Chem. Phys.* **6** 4185–9

[22] Li Y, Lu Y, Adelhelm P, Titirici M-M and Hu Y-S 2019 Intercalation chemistry of graphite: alkali metal ions and beyond *Chem. Soc. Rev.* **48** 4655–87

[23] Tanaka S, Kitta M, Tamura T, Maeda Y, Akita T and Kohyama M 2014 Atomic and electronic structures of $Li_4Ti_5O_{12}/Li_7Ti_5O_{12}$ (001) interfaces by first-principles calculations *J. Mater. Sci.* **49** 4032–7

[24] Rahman M M, Sultana I, Yang T, Chen Z, Sharma N, Glushenkov A M and Chen Y 2016 Lithium germanate (Li_2GeO_3): a high-performance anode material for lithium-ion batteries *Angew. Chem.* **128** 16293–7

[25] Narendhran S and Vinoth V 2019 Biosynthesis and characterization of cassia fistula mediated lithium coated iron oxide nanoparticle ($LiFe_2O_3$) and evaluation of its antibacterial activity *National Conf.* p 33

[26] Baltes H, Yacoby Y, Pindak R, Clarke R, Pfeiffer L and Berman L 1997 Measurement of the x-ray diffraction phase in a 2D crystal *Phys. Rev. Lett.* **79** 1285–8

[27] Li J and Sun J L 2017 Application of x-ray diffraction and electron crystallography for solving complex structure problems *ACC. Chem. Res.* **50** 2737–45

[28] Pong W-T and Durkan C 2005 A review and outlook for an anomaly of scanning tunnelling microscopy (STM): superlattices on graphite *J. Phys. D: Appl. Phys.* **38** R329

[29] Sun H-L, Shen Q-T, Jia J-F, Zhang Q-Z and Xue Q-K 2003 Scanning tunneling microscopy study of superlattice domain boundaries on graphite surface *Surf. Sci.* **542** 94–100

[30] Stöberl U, Wurstbauer U, Wegscheider W, Weiss D and Eroms J 2008 Morphology and flexibility of graphene and few-layer graphene on various substrates *Appl. Phys. Lett.* **93** 051906

[31] Scrosati B, Hassoun J and Sun Y-K 2011 Lithium-ion batteries. A look into the future *Energy Environ. Sci.* **4** 3287–95

[32] Mahatha S, Menon K S and Balasubramanian T 2011 Unoccupied electronic structure of graphite probed by angle-resolved photoemission spectroscopy *Phys. Rev.* B **84** 113106

[33] Gunasekara N, Takahashi T, Maeda F, Sagawa T and Suematsu H 1988 Angle-resolved ultraviolet photoemission study of first stage alkali-metal graphite intercalation compounds *Z. Phys. B Condens. Matter* **70** 349–55

[34] Matthes L, Pulci O and Bechstedt F 2014 Optical properties of two-dimensional honeycomb crystals graphene, silicene, germanene, and tinene from first principles *New J. Phys.* **16** 105007

[35] Trukhin A N, Rogulis U and Spingis M 1997 Self-trapped exciton in Li_2GeO_3 *J. Lumin.* **72** 890–2

[36] Trukhin A 1997 Self-trapped excitons in SiO_2, GeO_2, Li_2GeO_3, $AIPO_4$ and $GaPO_4$ *Mater. Sci. Forum* (College Park, MD: Trans Tech Publ) 531–6 pp

[37] van Scrosati W A and Scrosati B 2002 *Advances in Lithium-Ion Batteries* (Cham: Springer)

[38] Pistoia G 2013 *Lithium-Ion Batteries: Advances and Applications* (London: Newnes)

[39] Malik R, Zhou F and Ceder G 2011 Kinetics of non-equilibrium lithium incorporation in $LiFePO_4$ *Nat. Mater.* **10** 587–90

[40] Andersson A S, Kalska B, Häggström L and Thomas J O 2000 Lithium extraction/insertion in $LiFePO_4$: an x-ray diffraction and Mössbauer spectroscopy study *Solid State Ionics* **130** 41–52

[41] Wang L, He X, Sun W, Wang J, Li Y and Fan S 2012 Crystal orientation tuning of $LiFePO_4$ nanoplates for high rate lithium battery cathode materials *Nano Lett.* **12** 5632–6

[42] Novko D, Zhang Q and Kaghazchi P 2019 Nonadiabatic effects in Raman Spectra of $AlCl_4$-graphite based batteries *Phys. Rev. Appl.* **12** 024016

[43] Csányi G, Littlewood P, Nevidomskyy A H, Pickard C J and Simons B 2005 The role of the interlayer state in the electronic structure of superconducting graphite intercalated compounds *Nat. Phys.* **1** 42–5

[44] Tassin P, Koschny T, Kafesaki M and Soukoulis C M 2012 A comparison of graphene, superconductors and metals as conductors for metamaterials and plasmonics *Nat. Photonics* **6** 259

[45] Konschuh S, Gmitra M and Fabian J 2010 Tight-binding theory of the spin–orbit coupling in graphene *Phys. Rev.* B **82** 245412

[46] Wang D-Y, Huang S-K, Liao H-J, Chen Y-M, Wang S-W, Kao Y-T, An J-Y, Lee Y-C, Chuang C-H and Huang Y-C 2019 Insights into dynamic molecular intercalation mechanism for AlC battery by operando synchrotron x-ray techniques *Carbon* **146** 528–34

[47] Childress A S, Parajuli P, Zhu J, Podila R and Rao A M 2017 A Raman spectroscopic study of graphene cathodes in high-performance aluminum ion batteries *Nano Energy* **39** 69–76
[48] Smith D and Eklund P 1983 Intercalated graphite Proc. Mater. Res. Soc. Symp. p 99
[49] Liu W, Kang J and Banerjee K 2016 Characterization of $FeCl_3$ intercalation doped CVD few-layer graphene *IEEE Electron Device Lett.* **37** 1246–9
[50] Wang L, Zhu Y, Guo C, Zhu X, Liang J and Qian Y 2014 Ferric chloride-graphite intercalation compounds as anode materials for Li-ion batteries *ChemSusChem.* **7** 87–91
[51] Qi X, Qu J, Zhang H-B, Yang D, Yu Y, Chi C and Yu Z-Z 2015 $FeCl_3$ intercalated few-layer graphene for high lithium-ion storage performance *J. Mater. Chem.* A **3** 15498–504
[52] Wang Z and Inagaki M 1991 Intercalation of $FeCl_3$ into graphite films in molten salts *Synth. Met.* **44** 165–76
[53] Zhan D, Sun L, Ni Z H, Liu L, Fan X F, Wang Y, Yu T, Lam Y M, Huang W and Shen Z X 2010 $FeCl_3$-based few-layer graphene intercalation compounds: single linear dispersion electronic band structure and strong charge transfer doping *Adv. Funct. Mater.* **20** 3504–9
[54] Okamoto Y 2014 Density functional theory calculations of alkali metal (Li, Na, and K) graphite intercalation compounds *J. Phys. Chem.* C **118** 16–9
[55] Lin W-B, Tran N T T and Lin S-Y 2019 Diverse fundamental properties in stage-n graphite alkali-intercalation compounds: anode materials of Li^+-based batteries arXiv:2001.02042
[56] Li W-B, Lin S-Y, Tran N T T, Lin M-F and Lin K-I 2020 Essential geometric and electronic properties in stage-n graphite alkali-metal-intercalation compounds *RSC Adv.* **10** 23573–81
[57] Assouik J, Hajji L, Boukir A, Chaouqi M and Lagrange P 2017 Heavy alkali metal-arsenic alloy-based graphite intercalation compounds: Investigation of their synthesis and of their physical properties *C.R. Chim.* **20** 116–24
[58] Chen S, Wu J, Lin C and Lin M 2016 Theory of magnetoelectric properties of 2d systems *New J. Phys.* **18** 103024
[59] Lin C-Y, Ho C-H, Wu J-Y, Do T-N, Shih P-H, Lin S-Y and Lin M-F 2019 *Diverse Quantization Phenomena in Layered Materials* (Boca Raton, FL: CRC Press)
[60] Foulkes W M C and Haydock R 1989 Tight-binding models and density-functional theory *Phys. Rev.* B **39** 12520
[61] Speakman S A 2013 Introduction to x-ray powder diffraction data analysis *Center for Materials Science and Engineering at MIT*
[62] Kim D-M, Jung S C, Ha S, Kim Y, Park Y, Ryu J H, Han Y-K and Lee K T 2018 Cointercalation of Mg^{2+} ions into graphite for magnesium-ion batteries *Chem. Mater.* **30** 3199–203
[63] Xu J, Dou Y, Wei Z, Ma J, Deng Y, Li Y, Liu H and Dou S 2017 Recent progress in graphite intercalation compounds for rechargeable metal (Li, Na, K, Al)-ion batteries *Adv. Sci.* **4** 1700146
[64] Lin M-F and Shung K W-K 1994 Plasmons and optical properties of carbon nanotubes *Phys. Rev.* B **50** 17744
[65] Pennington G and Goldsman N 2003 Semiclassical transport and phonon scattering of electrons in semiconducting carbon nanotubes *Phys. Rev.* B **68** 045426
[66] Dürkop T, Getty S, Cobas E and Fuhrer M 2004 Extraordinary mobility in semiconducting carbon nanotubes *Nano Lett.* **4** 35–9
[67] Tans S J, Verschueren A R and Dekker C 1998 Room-temperature transistor based on a single carbon nanotube *Nature* **393** 49–52

[68] Martel R, Schmidt T, Shea H, Hertel T and Avouris P 1998 Single-and multi-wall carbon nanotube field-effect transistors *Appl. Phys. Lett.* **73** 2447–9

[69] Shlafman M, Tabachnik T, Shtempluk O, Razin A, Kochetkov V and Yaish Y 2016 Self aligned hysteresis free carbon nanotube field-effect transistors *Appl. Phys. Lett.* **108** 163104

[70] Myodo M, Inaba M, Ohara K, Kato R, Kobayashi M, Hirano Y, Suzuki K and Kawarada H 2015 Large-current-controllable carbon nanotube field-effect transistor in electrolyte solution *Appl. Phys. Lett.* **106** 213503

[71] Niimi Y, Matsui T, Kambara H, Tagami K, Tsukada M and Fukuyama H 2006 Scanning tunneling microscopy and spectroscopy of the electronic local density of states of graphite surfaces near monoatomic step edges *Phys. Rev.* B **73** 085421

[72] Misewich J, Martel R, Avouris P, Tsang J, Heinze S and Tersoff J 2003 Electrically induced optical emission from a carbon nanotube FET *Science* **300** 783–6

[73] Yuksel R, Sarioba Z, Cirpan A, Hiralal P and Unalan H E 2014 Transparent and flexible supercapacitors with single walled carbon nanotube thin film electrodes *ACS Appl. Mater. Interfaces* **6** 15434–9

[74] Liang S, Ma Z, Wu G, Wei N, Huang L, Huang H, Liu H, Wang S and Peng L-M 2016 Microcavity-integrated carbon nanotube photodetectors *ACS Nano* **10** 6963–71

[75] Zhang T-F, Li Z-P, Wang J-Z, Kong W-Y, Wu G-A, Zheng Y-Z, Zhao Y-W, Yao E-X, Zhuang N-X and Luo L-B 2016 Broadband photodetector based on carbon nanotube thin film/single layer graphene Schottky junction *Sci. Rep.* **6** 38569

[76] He X, Léonard F and Kono J 2015 Uncooled carbon nanotube photodetectors *Adv. Opt. Mater.* **3** 989–1011

[77] Mueller T, Kinoshita M, Steiner M, Perebeinos V, Bol A A, Farmer D B and Avouris P 2010 Efficient narrow-band light emission from a single carbon nanotube p–n diode *Nat. Nanotechnol.* **5** 27–31

[78] Pyatkov F, Fütterling V, Khasminskaya S, Flavel B S, Hennrich F, Kappes M M, Krupke R and Pernice W H 2016 Cavity-enhanced light emission from electrically driven carbon nanotubes *Nat. Photonics* **10** 420–7

[79] Wang S, Zeng Q, Yang L, Zhang Z, Wang Z, Pei T, Ding L, Liang X, Gao M and Li Y 2011 High-performance carbon nanotube light-emitting diodes with asymmetric contacts *Nano Lett.* **11** 23–9

[80] Mittal J and Lin K 2017 Carbon nanotube-based interconnections *J. Mater. Sci.* **52** 643–62

[81] Singh K and Raj B 2015 Temperature-dependent modeling and performance evaluation of multi-walled CNT and single-walled CNT as global interconnects *J. Electron. Mater.* **44** 4825–35

[82] Liu Y, Lian J, Sun Z, Zhao M, Shi Y and Song H 2017 The first-principles study for the novel optical properties of $LiTi_2O_4$, $Li_4Ti_5O_{12}$, $Li_2Ti_2O_4$ and $Li_7Ti_5O_{12}$ *Chem. Phys. Lett.* **677** 114–9

[83] Feng C, Li L, Guo Z, Shi D, Zeng R and Zhu X 2009 Synthesis and properties of Li–Ti–O spinel ($LiTi_2O_4$) *J. Alloys Compd.* **478** 767–70

[84] Liu Y, Lian J, Yang X, Zhao M, Shi Y, Song H and Dai K 2017 First-principles calculation for point defects in $Li_2Ti_2O_4$ *Mater. Res. Express* **4** 106502

[85] Zhao M, Lian J, Jia Y, Jin K, Xu L, Hu Z, Yang X and Kang S 2016 Investigation of the optical properties of $LiTi_2O_4$ and $Li_4Ti_5O_{12}$ spinel films by spectroscopic ellipsometry *Opt. Mater. Express* **6** 3366–74

[86] Seman R, Azam M A and Ani M H 2018 Graphene/transition metal dichalcogenides hybrid supercapacitor electrode: status, challenges, and perspectives *Nanotechnology* **29** 502001
[87] Barik G and Pal S 2019 Defect induced performance enhancement of monolayer MoS_2 for Li- and Na-ion batteries *J. Phys. Chem.* C **123** 21852–65
[88] Liu H, Su D W, Zhou R F, Sun B, Wang G X and Qiao S Z 2012 Highly ordered mesoporous MoS_2 with expanded spacing of the (002) crystal plane for ultrafast lithium ion storage *Adv. Energy Mater.* **2** 970–5
[89] Soares D M, Mukherjee S and Singh G 2020 TMDs beyond MoS_2 for electrochemical energy storage *Chem.-A Eur. J.* **26** 6320–41
[90] Parida S, Mishra A, Chen J, Wang J, Dobley A, Carter C B and Dongare A M 2020 Vertically stacked 2H-1T dual-phase MoS_2 microstructures during lithium intercalation: a first principles study *J. Am. Ceram. Soc.* **103** 6603–14
[91] Srinivaas M, Wu C Y, Duh J G and Wu J M 2019 Highly rich 1T metallic phase of few-layered WS_2 nanoflowers for enhanced storage of lithium-ion batteries *ACS Sustain. Chem. Eng.* **7** 10363–70
[92] Liu H, Su D W, Wang G X and Qiao S Z 2012 An ordered mesoporous WS_2 anode material with superior electrochemical performance for lithium ion batteries *J. Mater. Chem.* **22** 17437–40
[93] Soares D M and Singh G 2020 Superior electrochemical performance of layered WTe(2)as potassium-ion battery electrode *Nanotechnology* **31** 455406

IOP Publishing

First-Principles Calculations for Cathode, Electrolyte and Anode Battery Materials

Thi Dieu Hien Nguyen, Shih-Yang Lin, Hsien-Ching Chung, Ngoc Thanh Thuy Tran and Ming-Fa Lin

Chapter 18

Open issues and potential applications

Thi Dieu Hien Nguyen, Shih-Yang Lin, Hsien-Ching Chung, Wei-Bang Li, Ngoc Thanh Thuy Tran, Nguyen Thi Han, Hsin-Yi Liu, Hai Duong Pham and Ming-Fa Lin

The rich and unique essential properties in anode, electrolyte, and anode materials of ion-based batteries have been successfully explored by the Vienna *ab initio* simulation package (VASP) simulations, as thoroughly explored in chapters 4–13. Most/part of mainstream materials possess moiré superlattices/high-symmetry lattices, with a lot of/few atoms in a primitive unit cell. From the viewpoint of phenomenological methods, the unusual tight-binding models are worth further development, since they can provide very useful information about the critical roles of multi-/single-orbital hybridizations, as well as their significant effects on the various physical/chemical/material properties. For example, the ternary lithium oxide of $Li_4Ge_8O_{12}$ could be chosen for a model study. Li, Ge, and O atoms, respectively, present 2s, [4s, $4p_x$, $4p_y$, $4p_z$] and [2s, $2p_x$, $2p_y$, $2p_z$] orbitals in creating a crystal structure. The mutual mixings in chemical bonds (the nearest-neighbor hopping integrals due to the various orbitals in figure 18.1; [1]) and the ionization energies of different orbitals need to be regarded as intrinsic interactions simultaneously. The uniform/non-uniform Li–O and Ge–O bonds correspond to the ideal/real case [1, 2], in which the length-dependent hopping integrals could be estimated through the Harrison rule [3] . All the atoms and available orbitals will determine a 216×216 Hamiltonian matrix, in which the diagonal and non-diagonal elements correspond to the on-site energies and the orbital-mixing strengths. After diagonalization of this large Hermitian matrix, the predicted energy spectrum, which is conducted along the high-symmetry points of the largely reduced first Brillouin zone, is compared with that of the VASP result. The numerical modulation of the suitable parameters is very difficult during the delicate calculation analysis process, and it is very sensitive to various materials. The reliable tight-binding model could be further utilized to investigate the optical excitation in the absence of excitonic

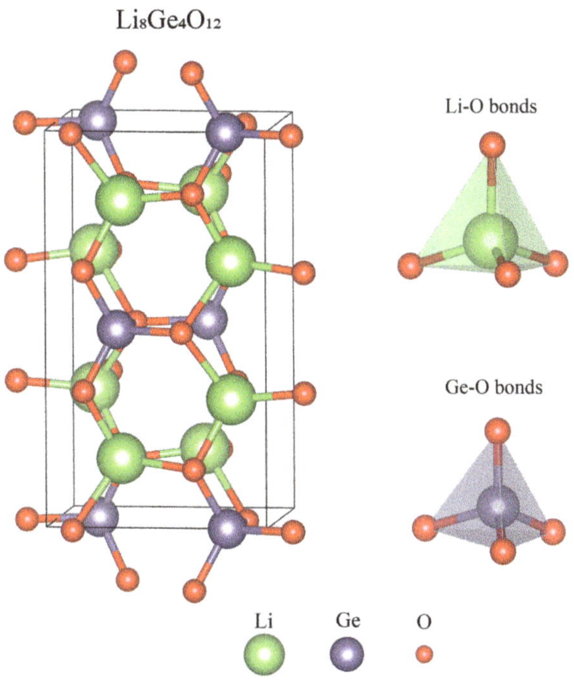

Figure 18.1. The position- and orbital-dependent hopping integrals in the tight-binding model for $Li_8Ge_4O_{12}$: 2s-[2s, 2p$_x$, 2p$_y$, 2p$_z$] and [4s, 4p$_x$, 4p$_y$, 4p$_z$]-[2s, 2p$_x$, 2p$_y$, 2p$_z$] multi-orbital hybridizations of Li–O and Ge–O bonds, respectively.

effects. In addition, the stable excitons and their contributions to optical properties might be tough research topics for ternary lithium oxides, or the other different ones.

According to the elastic scatterings of an incident photon beam [4, 5], the theoretical calculations on the x-ray diffraction peaks could be done for the various anode/electrolyte/anode materials under the well-established tight-binding models. Such simulations are able to provide the reliable wave-vector-dependent energy spectra and wave functions, being approximately consistent with first-principles evaluations [1, 6]. Most importantly, each unusual charge density distribution in a real crystal structure is clearly achieved from the total summation for all the products between the orbital-decomposed probabilities of the occupied valence states and the orbital-created charge densities (the hydrogen-like results; [7]). That is to say, the carrier densities, which mainly come from the active atomic orbitals, are clearly attributed to the position-dependent chemical bonds in a primitive unit cell. And then, the scattering form factor is determined by the specific atomic positions and their contributions to the localized charge distributions. The predicted results should cover the discrete reciprocal lattice vectors (**G**s) in accounting for the x-ray measurements [4, 8]. Furthermore, they present the different features for the various peak structures, such as the spectral intensities, forms (symmetric or asymmetric structure with a broadening width) and 2θ dependences [4, 8]. For example, the outstanding systems, being suitable for model studies, include AA- [9], AB- [9] and ABC-stacked [9] bulk graphites, alkali-intercalation graphite compounds [9], $Li_4Ti_5O_{12}$ [6, 10], $LiGe_2O_3$ [1, 2], and $LiFeO_2$ [11]. On the other side, VASP simulations about the x-ray patterns are never found in previous publication works. Such simulations are very reliable for the atom- and orbital-decomposed charge density distributions. The direct summation with the position-/atom-/orbital-driven scattering contributions might be responsible for the diversified x-ray measurements [4, 8] and thus clarify the critical mechanisms. How to link the phenomenological and simulation methods and the x-ray patterns will become very interesting topics in near-future science research.

Time-resolved x-ray diffraction measurements, which are urgently required in further experiments, are very useful in understanding certain metastable/stable crystal structures of core materials in ion-based batteries. Each set of 2θ-dependent peak structures needs to be finished within a sufficiently short time. Of course, it is almost mission impossible to thoroughly comprehend the whole chemical reactions. The charging or discharging processes have generated the Li^+ ion transport inside the cathode, electrolyte, and anode materials, or vice versa, such as, the steady ion current flowing among LiFeO, LiGeO and graphite. Under the former case, the lithium ions gradually escape from the cathode (reduction of their concentration), continuously flow into/out from electrolyte (almost simultaneous increase and decrease), and finally reach the anode until saturated configuration (slow increment). Obviously, the Li^+-ion concentrations change quickly, leading to the creation of many intermediate configurations. It is very interesting to fully explore the non-well-behaved crystal structures. The various defects, which mainly arise from vacancies, ion/atom dislocations, distortions of chemical bonds, and non-uniform concentrations, frequently come to exist during the dynamic reactions. How to characterize and classify their main features should be the study focus of highly precise x-ray measurements [4, 5, 8]. On the theoretical side, the VASP simulations [12] are available in thoroughly investigating the significant differences (the diversified phenomena; [12]) among the intermediate crystal materials, e.g., the loss of lattice symmetries in the presence of various defects [13], the dramatic transformations in band structures, spatial charge densities and van Hove singularities [13, 14], and the total changes of optical excitation spectra [13, 14].

The non-perfect crystal structures, which are generated by the chemical reactions [15], are able to provide mainstream science research [15] and the necessary conditions in highly potential battery applications [15]. The various defects mainly arise from vacancies (figure 18.2(a); [16]), guest-atom substitutions (figure 18.2(b); [17]), extra impurities [16, 17], chemical-bond distortions [16, 17], and mixed crystal structures (twin- or multi-boundary inside the same material; [16, 17]); therefore, there exist position-dependent orbital hybridizations (hopping integrals; [3, 18]) Furthermore, they will frequently come to exist in the core materials of batteries. Very interestingly, the charging and discharging processes, directly reflecting the natural transformation of electric powers and chemical energies [19, 20], will repeat many circles during specific purposes [19, 20]. The solid-state cathode [19, 20], electrolyte [19, 20] and anode materials [19, 20] have been clearly examined to

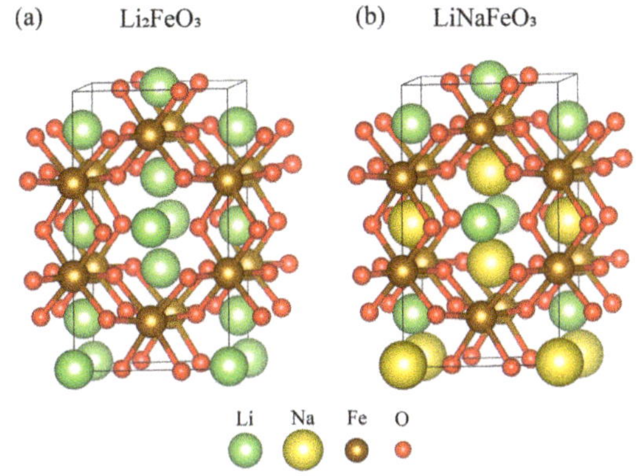

Figure 18.2. Defect-enriched Li_xFeO_3 crystal structures: (a) vacancies and (b) guest-atom substitutions (e.g., Na).

present the maximum component variations during ion transport within the inside circuit. For example, the Li_xFeO_3/Li_xFePO_4 cathode might exhibit the largest reduction of lithium ions, $x = 0.25$ and 0.25, based on theoretical predictions and experimental x-ray estimations [21]. Certain defect-enriched atomic configurations could be achieved from the VASP simulations, only if their high ground state energies agree with the highly on- uniform/distorted chemical environments (similar to those observed in the experimental examinations; [22, 23]). Such meta-stable crystal structures should have larger moiré superlattices. The enhanced unit cell and the drastic changes of chemical bonds (the highly anisotropic multi-orbital hybridizations; [24]), will play critical roles in greatly diversifying the geometric [24], electronic [24], optical [24] and magnetic properties [24]. Such interesting topics are worthy systematic investigations in the current status, since they are strongly related to the significant mechanism of the high-performance ion transport (discussed in the twelfth paragraph).

A specific boundary, which is successfully synthesized between two neighboring core materials in an ion-based battery [1, 6], deserves thorough examinations from both theoretical [1, 6] and experimental viewpoints [2, 25]. The ternary lithium oxides might be outstanding samples in fully illustrating the boundary-diversified physical/chemical/materials phenomena [1, 6, 11], such as a composite cathode/electrolyte/anode system corresponding to $LiFeO_3$/$LiGe_2O_3$/$Li_4Ti_5O_{12}$. Using the first-principles method [1, 6], it would be very interesting to achieve an optimal heterojunction structure with rich and complicated chemical bonds/spin configurations. How to take the suitable widths of the neighboring materials needs to be delicately tested by high-resolution simulation processes. This dominates an unusual moiré superlattice (a dramatic transformation of the first Brillouin zone; [1, 6]), leading to totally different fundamental properties. The low-energy electronic properties, e.g., band gap, band-edge states, energy dispersions, and spin degeneracy, are expected to be thoroughly modified by the orbital mixings and/or the spin configurations [24]. The critical roles of quasiparticle charges and spins could be clearly examined from the various physical/chemical properties [7, 9, 26]; that is,

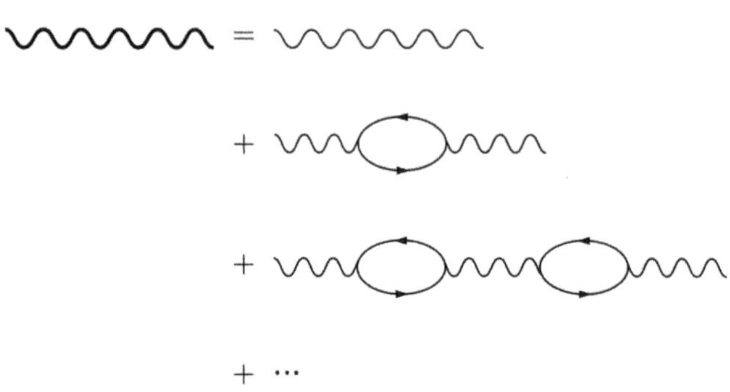

Figure 18.3. The random-phase approximation, being due to the Coulomb-field perturbation under the transferred momentum and energy [**q** ω], accounts for the charge screening ability using the ladder form of the Feynman diagrams.

these works can present concise pictures in fully understanding the boundary-enriched behaviors. It should be noticed that the unusual heterojunctions are also revealed in the hybridized compounds: a few-layer 2D subsystem on the various substrates. For example, layered silicene/germanege/tinene/plumbene on the different metal surfaces, which are successfully synthesized by the MBE growth method [27], are another outstanding sample in exploring the various heterojunction phenomena. The distinct layer numbers of two subsystems will dominate the main features of fundamental properties, e.g., the VASP predictions on monolayer/bilayer silicene/Ag(111) about the featured crystal structures near a specific heterojunction, as well as the orbital-mixing- and spin-enrich energy spectra, charge density distributions and van Hove singularities (figures 18.5(a) and (b); details in [7]). In short, the unusual heterojunctions [7], being induced by the close combination of distinct crystal structures, might appear in a few-layer material on the specific substrate [7], two infinite distinct subsystems [7], and twin/multi boundaries inside the same material [7, 9]. As for the diversified physical/chemical/material phenomena [7, 9], the theoretical and/or experimental frameworks [7, 9] are worthy of current development, as done for layered graphene systems [9], AA-, AB- and ABC-stacked bulk graphites [9], graphene nanoribbons [26], carbon nanotubes [26], and silicene-related emergent materials [7].

The scattering matrix elements, which are generated by the Coulomb- or electromagnetic-field perturbations, will play critical roles in determining the response abilities and thus generating the diversified physical phenomena. Obviously, they are well characterized by the initial- and final-state wave functions and the perturbed Hamiltonian. Under the tight-binding model, the orbital-dependent scattering amplitudes are required in the delicate calculations [3]. For example, the Coulomb matrix element mainly arises from the average of a specific plane wave associated with the transferred momentum (**q**; [3]) during the electron–electron interaction. As to graphene-related systems [3], the C-2p_z-induced screening behaviors, being taken under RPA (Feynman diagrams in figure 18.3), are very successful in providing the π-electronic Coulomb excitations, e.g., the stacking-enriched (momentum, frequency)-phase diagrams in trilayer AAA-, ABA-, ABC- and

AAB-stacked graphene systems (details in [9]). When the other three σ orbitals are included in model calculations simultaneously, their single-particle and plasmon modes are expected to greatly widen the windows of excitation spectra [3] and thus fully comprehend the composite screening behaviors of π and σ charge carriers [3]. In addition, such theoretical works seem to be thoroughly absent up to now. The above-mentioned calculations among the different orbitals are also required in the optical excitonic effects, where the carrier screening will come to exist in any the scattering events [3]. The latter should be another issue in the theoretical predictions (figure 18.4).

Up-to-date theoretical works show that the numerical VASP simulations [28, 29], but not the phenomenological methods, are frequently used to fully explore the excitonic effects on the optical excitations of emergent semiconductors, as those done for large-gap insulators in this book [29]. A lot of many-body scattering processes come to exist because of the attractive Coulomb potentials. In general, similar Feynman diagrams, with continuous orders, are taken into account and summarized together, leading to the so-called Bethe–Salpeter equation of two-particle scatterings [28, 29]. The calculated optical properties, the frequency-dependent dielectric function, absorption coefficient, reflectance, and energy loss spectrum, are sufficient in identifying the close relations between the main features and the specific/significant orbital hybridizations in chemical bonds. The diversified optical properties are expected to be revealed in many mainstream materials, such as those of 1D silicon [30], germanium [31, 32] and carbon nanotubes/nanoribbons [33, 34].

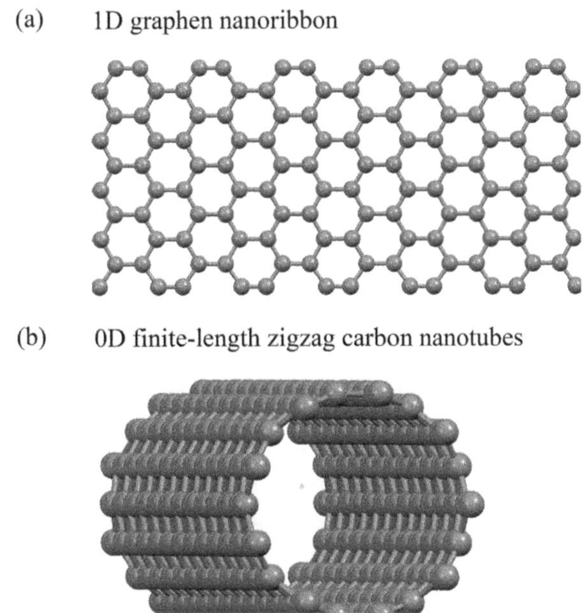

Figure 18.4. Geometric structures for (a) 1D zigzag graphene nanoribbons and 0D finite-length zigzag carbon nanotubes.

According to first-principles study (details in chapter 12), the first systems only present semiconducting behaviors, even for very small radii, i.e., armchair and zigzag silicon nanotubes belong to middle-gap semiconductors. Their direct- or indirect-gap features will strongly affect the threshold absorption structure and the excitonic peaks. It is very interesting in determining the orbital-hybridization-induced prominent channels from the imaginary-part dielectric function, as well as the radius-, chirality- and polarization-related optical selection rules (the width-, edge- and polarization-dependent ones in silicon nanoribbons; [35]). There exist significant similarities and differences between 1D nanotube and nanoribbons. The diversified optical properties can be modulated by the dimensionality [29, 36], layer number [31, 36], stacking configuration [37], buckling [31, 32], and external gate voltage [31, 32], since such critical factors dominate the intrinsic charge and spin interactions [7, 9, 37]. On the experimental side, a lot of highly precise optical measurements are conducted on the single- or multi-wall carbon nanotubes [33, 34], showing strong evidence of geometry-dominated excitonic effects. However, no examinations are done for the other group-IV/group-V nanomaterials. They are urgently required in further tests [38, 39]

The co-dominated roles of quasiparticle charge and spins [1, 40] on the rich and unique phenomena will become the mainstream science research, especially for those associated with the latter. A lot of condensed-matter materials, have been successfully synthesized in experimental laboratories, and are outstanding candidates in verifying spin-diversified physical/chemical/material phenomena. For example, very popular condensed-matter systems, which show the unusual magnetic spin configurations, might cover 1D zigzag graphene nanoribbons/finite-length zigzag carbon nanotubes (figure 18.4(a)/18.4(b); [26]/[41]), few-layer graphene/silicene/germanene with chemical modifications due to the transition- and rare-earth-metal atoms ([9]/[42]/[17]), layered silicene/germanene on distinct substrates [43]/[44], and $LiFeO_3$ and $LiFePO_4$ compounds (details in chapter 4). According to the up-to-now theoretical predictions [45], three types of magnetic configurations, ferro-magnetic [45], anti-ferromagnetic [45] and nonmagnetic spin distributions [45], are attributed to the open zigzag edges [26], the specific heterojunctions [43], and five/seven kinds of *d*/*f* orbitals. They are directly reflected in band structures, spatial spin density distributions and van Hove singularities under the spin-split or spin-degenerate phenomena. However, the spin-enriched optical excitations are never explored in the previous works, e.g., the spin-induced threshold frequency and the creation of spin-related exciton bound states. By the available vertical transitions, the prominent absorption structures might be determined by the spin-dependent electronic states in valence and conduction subbands. The featured optical spectra could be utilized to identify the strong relations between the orbital hybridizations of chemical bonds and the atom- and orbital-decomposed spin configurations. That is, after the direct combination with the near-future optical measurements, this can clarify the critical roles of quasiparticle charge and spins.

Gate voltages, which create uniform electric fields [35, 46] and thus on-site Coulomb potential energies [35, 46], have been successfully applied in condensed-matter systems, with obvious quantum-confinement structures [46], both theoretically [46] and experimentally [46]. For example, their external modulations are, respectively, available through the perpendicular modulation of few-layer graphene systems under various stacking configurations (figure 18.5(a); [3]), the transverse direction (the finite-width one) of graphene nanoribbons [47], and the azimuthal-angle variations of coaxial carbon nanotubes (figure 18.5(b); [48]). Systematic studies, which cover their diversified effects on the single- and many-particle properties, are clear in the recently published books [3, 9, 26]. Similar studies could be generalized to emergent materials. For example, the electric-field-enriched lattice symmetries [3], electronic properties [3], Coulomb excitations [3], magnetic spin configurations [3], and optical absorption spectra [3] could be thoroughly explored for silicon/germanium nanotubes [33, 35]/[49], layered silicene/germanene/tinene/plumbene structures [14]/[31, 42]/[50]/. Specifically, unique behaviors, dramatic changes of the spatial charge/spin density distributions, semiconductor-metal transitions [40], rich (momentum, frequency)-phase diagrams (drastic variations of electron–hole regions and plasmon modes), edge-related magnetisms [40], meta-stable/non-stable excitons [40], and extra absorption peaks [40], are the current focus of study. From a theoretical viewpoint, the direct combination of electric and magnetic fields is easy to greatly diversify all the physical phenomena. However, experimental examinations might find it very difficult to verify the predicted results, e.g., the calculated magneto-plasmon modes in graphene systems [3] under the absence of high-resolution measurements up to now.

Very interestingly, a uniform/modulated magnetic field can create the diverse quantization phenomena, as clearly revealed in 2D layered mainstream materials [7, 9]. There exist systematic investigations on few-layer group-IV and group-V systems [7]. Most importantly, the generalized tight-binding model has been successfully developed to present/identify/examine diversified magneto-electronic

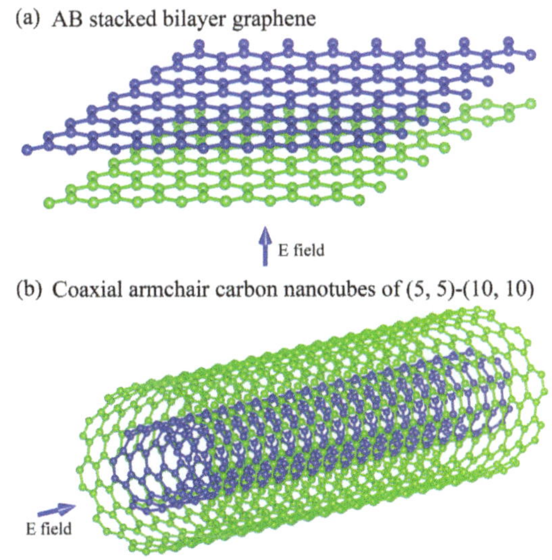

Figure 18.5. Under a uniform perpendicular electric field for (a) AB-stacked bilayer graphene and (b) coaxial armchair carbon nanotubes of (5, 5)-(10, 10) with a specific C_5 symmetry.

properties, such as well characterizations for the various Landau levels with featured energy spectra and oscillatory wave functions [3]. It is suitable for a lot of unusual Hamiltonians, being associated with the single-/multi-orbital hybridizations [6, 9], spin-orbital couplings [3], layer-dependent charge and spin interactions [3, 9, 36], deformed/normal crystal structures, moiré superlattices due to vector potentials [1, 6] and guest-atom dopings [17], and gate voltages [9]. As for core materials of Li^+-related batteries, stage-n alkali-ion/alkali-atom graphite intercalation compounds are outstanding candidates in clearly illustrating the diverse quantization behaviors in terms of electronic, optical and transport properties. Their Hamiltonians should, at least, cover the C-$2p_z$-induced hopping integrals on the same and neighboring graphitic layers, their orbital hybridizations with the outmost s-orbital of alkali atoms, and alkali-atom/alkali-ion interactions. After the application of a uniform perpendicular magnetic field (figure 18.6(a) and figure 18.6(b); [3]), the rich Landau subbands, with the significant k_z-dispersion relations, will come to exist frequently under any magnetic-field strengths. Their magneto-wave functions are well defined using the available A and B sublattices [3, 9]. The main features of the essential properties can be thoroughly clarified through the critical mechanisms (the concise physical pictures; [9]), such as the crossing and anti-crossing magneto-electronic energy spectra during the variation of field strength [46, 51], the normal and extra magneto-optical selection rules [46, 51], and the existence/absence of quantum Hall conductivities [46, 51]. Obviously, the magnetic quantization phenomena cannot be investigated by the VASP simulations because of too large Hamiltonian matrices [51]. On the other side, the ternary lithium oxides might not exhibit meaningful magneto-related properties, since their pristine crystal structures

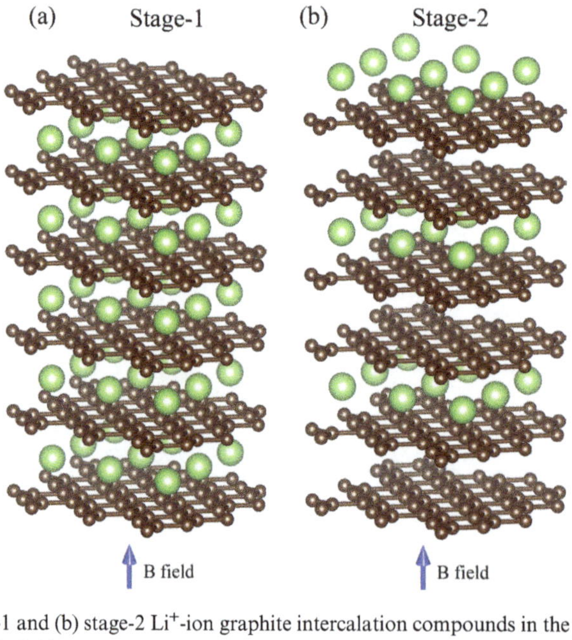

Figure 18.6. (a) Stage-1 and (b) stage-2 Li^+-ion graphite intercalation compounds in the presence of a uniform perpendicular magnetic field.

belong to moiré superlattices with many significant orbital hybridizations [6] and they have large indirect/direct band gaps. The strongly combined effects due to the highly-nonuniform chemical environment and the vector-potential-enhanced unit cell have resulted in very large and complicated magneto-Hamiltonian matrices.

How to thoroughly comprehend plenty of intermediate crystal structures and electronic/ionic states during the charging and discharging processes of ion-based batteries must be the most important research topics. The VASP/physical simulations and the measured x-ray patterns are not sufficient in exploring their mean features, since they are only focused on few stable/metal-stable configurations in the presence of higher lattice symmetries (standard moiré superlattices without a giant number of atoms in a greatly enhanced unit cell; [6]). The near-future studying focus should clarify the dramatic transformations between any two neighboring intermediate crystal structures, in which they are attributed to be driven by the irreversible chemical reactions. Specific numerical software [52–54], capable of dealing with the continuous chemical reactions according to the conservation of the total charge and energy, could be utilized to present rapid variations of core-material crystals, e.g., $Li_4Ti_5O_{12}$, $LiFePO_4$ and V_2O_5 [6, 22, 24, 55–58]. Both chemical and physical pictures/viewpoints/calculations should be combined/unified/taken simultaneously, since the whole intermediate configurations are hardly solved in the absence of any experimental evidence. While certain meta-stable crystals, with larger moiré superlattices, are identified in theoretical predictions and/or experimental tests, their featured fundamental properties could be achieved by the VASP simulations. Configuration-created diverse phenomena would come to exist among the highly non-uniform chemical environments associated with the various defects.

Very interestingly, any materials might present outstanding or very poor electronic/ionic transport properties, obviously leading to the diversified applications in various devices. In general, a typical condensed-matter system is set up between two heat reservoirs with the different bias voltages and temperatures (figure 18.7). The external probes, which mainly come from a weak, but steady electric field and/or thermal excitations [59], could induce the observable electric and/or thermal current densities, respectively [59]. The theoretical predictions have

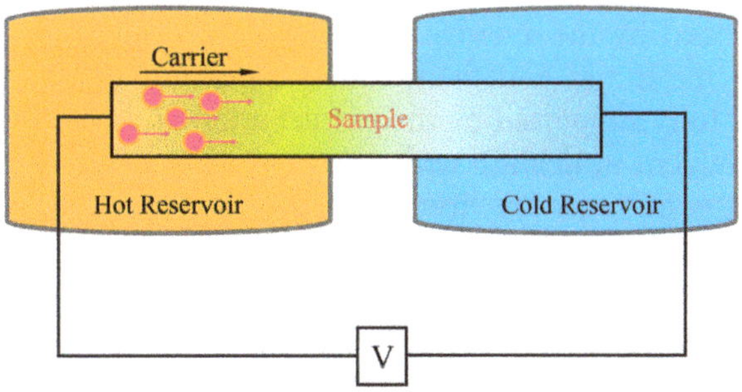

Figure 18.7. The electrical/thermal current densities of free electrons or holes of a typical sample between two reservoirs with distinct bias voltages and temperatures.

been successful under the different cases, e.g., the Boltzmann equation [60], static Kubo formula [60], and Landauer–Buttiker ballistic transport [60]. The linear relations of the measured result versus perturbation are thoroughly examined by the high-resolution transport measurements [4, 8], e.g., the full temperature-dependences of the electrical and thermal conductivities in graphite/graphite intercalation compounds [61, 62], layered graphene systems [9], single- and multi-wall carbon nanotubes [26], graphene nanoribbons [26], and carbon fullerenes. As for a lot of theoretical and experimental studies, the diversified transport phenomena are clearly shown to come from the electron-/hole-impurity and carrier-phonon scatterings [3], in which the Coulomb screenings of the whole valence and conduction electrons play the critical role during the transformations of the initial and final electronic states (discussed earlier; [3]). The former and the latter, respectively, belong to the elastic [3] and inelastic processes [3]. A charged impurity is able to generate a Coulomb-field perturbation, where its effective Coulomb potential is determined by all the carrier screenings. And then, the Fermi-golden rule/the Kubo formula is available in evaluating the residue resistivity at very low temperature [63]. However, it is very difficult to achieve the electron–phonon scattering rates for the various materials with diverse phonon properties. The quasiparticle energy spectra and wave functions of atomic coherent oscillations need to be taken into account for evaluating the wave-vector-dependent scattering probability. Furthermore, their scattering orders and amplitudes [3] act as critical barriers of analytic calculations and numerical simulations. In addition, the main features of quantized phonons, which survive in ternary lithium oxides (cathode/electrolyte/anode), are almost absent to date. The first-principles method could be utilized to examine the collective atomic oscillations within a moiré superlattice with many atoms [1, 6] .

It is well known that transport properties of lithium-/aluminum-/iron-ion-based batteries [20, 64–67]/[67–70]/[71] will dominate the researches of basic sciences [20, 64–71] and thus potential applications [20, 64–71]. The charge flows are, respectively, driven by ions and electrons inside and outside the core components. Specifically, the transport performance of the former is strongly dependent on the cathode, electrolyte, and anode materials. A completely theoretical framework, being able to successfully deal with the ion conductivities through three main components (the observable current density; [72]), has been absent up to now. According to the basic rules of quantum mechanics [60], the measured current density is well defined by the average value of the time-dependent current density operator between the initial and final states. Generally speaking, the Heisenberg picture [60], being fully supported by the perturbation viewpoints of Kubo formulas [63], is more convenient in dealing with the many-body scattering processes. Most importantly, the ion currents are driven by electric powers and chemical energies during the charging and discharging cases, respectively [20, 65, 67]. They are closely related to the intermediate crystal structures [20, 65, 67], in which the critical roles, the spatial charge density distributions, are greatly diversified by the rapid variations of host-ion concentrations and arrangement configurations. The dominating scattering mechanisms, which are responsible for the resistive forces [9, 40], might originate from the giant charge distortions due to the various defects (discussed earlier; [73, 74]). That is, the Coulomb scatterings between charged ions and the highly non-uniform chemical environments are deduced to play the dominant mechanisms [1, 6]. Such mechanisms lead to the quasi-steady Bose–Einstein ion distributions [60, 75, 76] from the recovery of non-equivalent perturbation distribution. Also, it can be noticed that a lot of identical ions belong to the same bosons [75], since each atomic configuration is able to generate the active orbitals under an even number in terms of defect-induced charge distributions [13, 74]. How to develop an efficiently theoretical framework in evaluating the ion transports of batteries should be very interesting issues within near-future studies. For example, the VASP simulations [77, 78] might be available for the specific charge density distributions of a few intermediate crystals (the tight-binding model cannot get them up to now), while such results cannot link with the transport theories.

Both chemical adsorptions and substitutions are frequently utilized to clearly observe the diversities of quasiparticle properties [73, 79], greatly enhance the specific functionalities [17, 73], and largely widen the potential applications [17, 79]. For example, the oxidization of carbon-germanium-/silicon-/tin-/lead-based emergent materials [80]/[81]/[82]/[83]/[84] frequently come to exist as the 3D [1, 6], 2D [17], 1D [48, 85, 86] and 0D [87] crystal structures for the various allotropes. Furthermore, the multi-component compounds (binary/ternary/quaternary/pentanary forms in [81]/[88–90]/[91]/[92]) could be further generated by the active participations of extra elements/molecules/ions, such as, the well-known lithium germanium oxides with the distinct Li^+-ion concentrations and distributions during the charging and discharging processes [20, 65, 93]. Such semiconductor compounds might present unusual physical/chemical/material properties and thus play critical roles in the applied sciences [2, 94]. As for the dimensionality-, morphology- and component-enriched semiconductor oxide compounds [81, 95], the close relations among the crystal symmetries, other essential properties and specific functionalities could be well established from the current quasiparticle framework by delicate first-principles simulations [81]. All the theoretical predictions need to be thoroughly examined by high-resolution experimental measurements [96–98]. The combined mechanisms, active orbital hybridizations within the distinct energy ranges and significant spin-split configurations, must be clearly identified from the highly non-uniform chemical environments [99–101], the rich electronic properties [100, 102], as well as the featured optical transitions [103]. Also, a series of studies could be conducted on the mainsteam topics of hydrogenation [104–106], halogenation [101], and alkalization [107, 108].

As for lithium-ion-based batteries, silicone [109], polymerized siloxanes or polysiloxanes, has successfully served as a super-high-performance anode in replacing the critical role of graphite [110, 111]. Such emergent material is capable of greatly enhancing the Li^+-ion intercalations and de-intercalations through the largely reduced reaction time [112]. The charging times of Li^+-ion batteries are highly shortened in the commercial applications, e.g., those related to electric vehicles [113–115]. This well-known condensed-matter system belongs to an intermediate polymer compound which corresponds to a giant moiré superlattice with a straight inorganic Si–O bone tightly bound to organic benzene-like functional groups. Its chemical formula is $[-R_2SiO-]_n$, where Rs denote the benzene-based organic molecules, and they possess strong covalent bondings with silicon atoms. Obviously, the position-dependent chemical bonds are revealed to illustrate highly non-uniform chemical environments, being fully supported by the atom-dependent multi-orbital hybridizations. These crystal structures, with many significant atoms and orbitals, are outstanding candidates in examining, unifying and extending the theoretical development of quasiparticle framework (detailed in books [7, 9]). Three parts of chemical bondings, which survive in quasi-benzene molecules, group-silicon heterojunctions, and silicon–oxygen bonds, play a critical role in the diversified crystal symmetries. They are expected to dominate very complicated essential properties under first-principles simulations, e.g., the unusual energy dispersions [7], atom- and orbital-decomposed van Hove singularities [109], spatial charge density distributions [116], as well as single-particle and many-body optical excitations spectra [117]. The well-characterized orbital mixings in different chemical bonds will be a near-future challenge.

Lithium-, phosphorous- and sulfur-based materials can generate various crystal symmetries, diversify the essential properties and play critical roles in plenty of commercial applications, e.g., high-efficiency lithium sulfide batteries [118]. According to the combinations of core elements, such three-dimensional condensed-matter systems could be classified into single-element crystals (Li, P and S lattices; [119–121]), binary compounds [Li_2S [120], Li_5P [122], P_4S_x of $1 \leqslant x \leqslant 10$; [123, 124], ternary materials of LiPS [125], and quaternary ones of LiPSO [125]. Very interestingly, these three elements are capable of building the different 3D allotropes with stable/quasi-stable crystal structures. Each compound is expected to be a moiré superlattice in the presence of position-dependent chemical bonds [126], clearly illustrating a highly non-uniform chemical environment [125–127]. Furthermore, their special functionalities are easily modulated through the sensitive variations of their relative weights and distribution configurations; as a result, the up-to-date various compounds show many successful applications [128]. The different composites are outstanding candidates in fully exploring the diversified quasiparticle behaviors and testing the development of the theoretical framework in the current and near-future books (a next book proposal). It is very interesting for systemically comprehending the close relations among the multi-element-enriched crystal structures, the significant chemical bondings/spin configurations [7, 9], and the unusual fundamental properties (details in books [7, 9, 26]). The diverse orbital hybridizations, which survive in various chemical bonds of the same system/different compounds, need to be further identified from the spatial charge density distributions [9], compared with those of isolated atoms [9]. Such delicate identifications are thoroughly supported by the orbital-emerged van Hove singularities and the atom-dominated band structures [9]. The above-mentioned quasiparticle viewpoints are almost absent in previously published works [7, 9], while they are thoroughly examined under the current investigations. These could be generalized to phosphorene-related 2D and 1D systems (buckled planes, nanotubes, and nanoribbons; [129–131]) for understanding the dimension-enriched physical/chemical/material phenomena [129–131].

The intercalations/de-intercalations of group-I (Li/Na/K/Rb/Cs; [132–134]), group-II (Be/Mg/Ca; [108, 135, 136]) and group-III (B/Al [108, 134]) into the spacings of graphitic layers frequently come to exist during the charging/discharging process of ion-based batteries. Such an intercalant might appear as the neutral or ionic cases. The crystal symmetries, electronic properties and optical transitions of these graphite intercalation compounds could be thoroughly investigated by first-principles simulations [12], but not the phenomenological models [18, 137]. The critical factors cover the different elements, atomic configurations, as well as guest-atom concentrations and distributions. Apparently, they will determine the optimal geometries, corresponding to the most stable crystal structures. To reach all the active orbital hybridization on the same and neighboring layers, the chemical environment, the atom-dominated band structure, the spatial charge density in a unit cell, and the orbital-merged van Hove singularities within a whole energy range, are combined together under a quasiparticle framework [9]. All the intrinsic interactions, which survive on the honeycomb lattices, the intercalant layers, and the neighboring layers, will be determined in detail. As a result of their concise chemical pictures [9, 133], the phenomenological models could be further developed by getting the reliable parameters, e.g., the on-site Coulomb potentials [3, 46] and orbital-dependent hopping integrals [18, 137]. These graphite intercalation compounds are expected to exhibit unusual quasiparticle properties, such as excitonic effects of carrier-doped metallic systems [138], magneto-optical selection rules of Landau subbands [139, 140], 3D quantum Hall effects [141, 142], and various mageto-plasmon modes [48, 143].

Solar cells are becoming important in everyday life, mainly owing to the creation of Earth friendly environments. In general, in the mainstream they possess unusual perovskite crystal structures, being very suitable for basic [144] and applied science research [145]. Their chemical formula is ABX_3, where A and B are cations and X is a halogen anion (chloride, bromine and iodine; [146, 147]), such as CH_3NH_3PbX/$CH_3NH_3SnX_3$ [144, 148], $H_2NCHNH_2PbX_3$/$H_2NCHNH_2SnX_3$ [149], as well as $CsPbX_3$/$CsSnX_3$ [150, 151]/[152]. That is to say, they could be regarded as the heavy group-IV- and halide-based multi-component compounds, being surrounded by organic–inorganic or inorganic absorbers. Up-to-date experimental works show that the best energy-conversion efficiencies of the former and the latter are about 26 % [153] and 29 % [153], respectively. Very interestingly, the two cations possess prominent chemical bondings with active halogen anions. That is, the ionic bonds appear between large-molecule positive and halogen negative ions. This should be totally different from the weak chemical bondings (the interlayer atomic interactions of the saturated ions and carbon-honeycomb lattices) under significant intercalations/de-intercalations of the large-molecule negative ions within the graphitic layers, as clearly illustrated in the chemical reactions of aluminum-/iron-ion-based batteries due to $AlCl_4^-$ and $Al_2Cl_7^-$/$FeCl_3^-$ [154, 155]/[156]. In addition, there also exist the intercalation layers of guest molecules. Most importantly, the complicated orbital hybridizations are induced by three kinds of ions, especially for the wide charge density distributions near the ion-form functional groups before and after the active participations. The concise physics and chemistry pictures could be clearly presented under a unified quasiparticle framework, becoming one of the study focuses on the current investigations [154, 155]. The diverse crystal symmetries and electronic properties will be responsible for the featured optical transitions. The optical threshold gap [65], special absorption structures [103], and excitonic effects, need to be thoroughly explored from the transverse dielectric function [157], energy loss function [158], reflectance [158], absorption coefficient [1, 157] and transmission spectrum [157] by the VASP calculations. These optical properties are worthy of systematic investigations, being able to reach the critical mechanisms for rich quasiparticle phenomena [7, 9, 26].

Hydrogenations of mainstream materials are able to generate geometry stability (e.g., H-passivation of open edges [100, 105], unusual phenomena [100, 105, 159], modified functionalities [105, 159], and specific applications [104]). Most chemical environments in condensed-matter are very friendly for hydrogen atoms/molecules. For example, monolayer graphene and silicene/germanene/stanene/plumbene have planar and buckled honeycomb structures, respectively. Furthermore, they possess active π and $\pi+\sigma$ orbitals perpendicular to the (x, y)-plane. Hydrogen chemisorption is capable of creating very strong bonding with the group-IV elements. The hydrogenated materials are predicted/expected to be totally different from the pristine systems in terms of quasiparticle properties, such as, crystal symmetries [99, 160], band structures [100, 105], magnetic moments [105, 159, 160], charge/spin densities [100, 105, 159], van Hove singularities [100, 105], optical absorption structures [99], and exciton bound states [99, 159]. It should be noticed that these emergent materials might play dominating roles in hydrogen atom storage [106]. Up to now, systematic investigations on hydrogenated group-IV/group-V systems are absent in the previously theoretical predictions [99] and experimental syntheses/ measurements [161]; that is, the concise physical and chemical pictures are not proposed to link the relevant quantities. This will be finished under the quasiparticle framework of theoretical development [7, 9, 26], especially for the geometric, electronic, and optical properties. Quasiparticle phenomena, which are greatly diversified by the crystal symmetries, strong s-sp^3 covalent bondings and significant spin-orbital couplings, are thoroughly examined from first-principles simulations.

In summary, various condensed-matter systems, which are generated by the chemical [1, 17, 40] and physical [7, 17] modifications, provide diverse crystal symmetries in basic and applied sciences. The quasiparticle framework is further developed through the phenomenological models [3, 18, 46] and first-principles simulations [12, 162], being based on the featured materials. The close relations among the different physical/chemical/material properties and the significant orbital hybridizations/spin configurations will be established for the mainstream materials under systematic investigation, as clearly illustrated in the previously published books [7, 9, 163]. There exist mainstream topics which are worthy of systematic investigation, e.g. the diverse quasiparticle properties in silicene-/germanene-/ tinene-/plumbene-related 2D systems [7]/[42]/[31],[32]/[50], and their optical excitations after/before hydrogenation in the presence of excitonic effects [36]. Similar studies are required for the other quasiparticles. Very interestingly, a grand/ comparable theoretical framework could be developed for significant coupling/

separation among electrons, phonons, and photons. For example, delicate VASP simulations are able to provide quasiparticle energy spectra of plasmons [12], phonons [12], polarons [164], photons [12] and polaritons [165]. This is under our current study.

References

[1] Khuong Dien V, Thi Han N, Nguyen T D H, Huynh T M D, Pham H D and Lin M-F 2020 Geometric and electronic properties of Li_2GeO_3 *Front. Mater.* **7** 288

[2] Rahman M M, Sultana I, Yang T, Chen Z, Sharma N, Glushenkov A M and Chen Y 2016 Lithium germanate (Li_2GeO_3): a high-performance anode material for lithium-ion batteries *Angew. Chem.* **128** 16293–7

[3] Lin C-Y, Wu J-Y, Chiu C-W and Lin M-F 2019 *Coulomb Excitations and Decays in Graphene-Related Systems* (Boca Raton, FL: CRC Press)

[4] Li J and Sun J L 2017 Application of X-ray diffraction and electron crystallography for solving complex structure problems *Acc. Chem. Res.* **50** 2737–45

[5] Speakman S A 2013 Introduction to X-ray powder diffraction data analysis *Center for Materials Science and Engineering at MIT*

[6] Nguyen T D H, Pham H D, Lin S Y and Lin M F 2020 Featured properties of Li^+-based battery anode: $Li_4Ti_5O_{12}$ *RSC Adv.* **10** 14071–9

[7] Lin S-Y, Liu H-Y, Nguyen D K, Tran N T T, Pham H D, Chang S-L, Lin C-Y and Lin M-F 2020 *Silicene-Based Layered Materials: Essential Properties* (Bristol: IOP Publishing)

[8] Baltes H, Yacoby Y, Pindak R, Clarke R, Pfeiffer L and Berman L 1997 Measurement of the x-ray diffraction phase in a 2D crystal *Phys. Rev. Lett.* **79** 1285–8

[9] Tran N T T L, Lin S Y, Lin C Y and Lin M F 2017 *Geometric and Electronic Properties of Graphene-Related Systems: Chemical Bonding Schemes* (Boca Raton, FL: CRC Press)

[10] Sandhya C P, John B and Gouri C 2014 Lithium titanate as anode material for lithium-ion cells: a review *Ionics* **20** 601–20

[11] Buyukyazi M and Mathur S 2015 3D nanoarchitectures of a-$LiFeO_2$ and alpha-$LiFeO_2$/C nanofibers for high power lithium-ion batteries *Nano Energy* **13** 28–35

[12] Hafner J 2008 Ab-initio simulations of materials using VASP: density-functional theory and beyond *J. Comput. Chem.* **29** 2044–78

[13] Raebiger H 2010 Theory of defect complexes in insulators *Phys. Rev.* B **82**

[14] Gao J F, Zhang J F, Liu H S, Zhang Q F and Zhao J J 2013 Structures, mobilities, electronic and magnetic properties of point defects in silicene *Nanoscale* **5** 9785–92

[15] Thakur T S, Dubey R and Desiraju G R 2015 Crystal structure and prediction *Annu. Rev. Phys. Chem.* **66** 21–42

[16] Ji L, Toprakci O, Alcoutlabi M, Yao Y, Li Y, Zhang S, Guo B, Lin Z and Zhang X 2012 α-Fe_2O_3 nanoparticle-loaded carbon nanofibers as stable and high-capacity anodes for rechargeable lithium-ion batteries *ACS Appl. Mater. Interfaces* **4** 2672–9

[17] Pham H D, Nguyen T D H, Vo K D, Huynh T M D and Lin M-F 2020 Rich essential properties of boron, carbon, and nitrogen substituted germanenes *Appl. Phys. Express* **13** 085502

[18] Foulkes W M C and Haydock R 1989 Tight-binding models and density-functional theory *Phys. Rev.* **B 39** 12520

[19] Mekonnen Y, Sundararajan A and Sarwat A I 2016 A review of cathode and anode materials for lithium-ion batteries *SoutheastCon 2016* (Piscataway, NJ: IEEE) 1–6

[20] Nitta N, Wu F, Lee J T and Yushin G 2015 Li-ion battery materials: present and future *Mater. Today* **18** 252–64
[21] Ohmer N, Fenk B, Samuelis D, Chen C-C, Maier J, Weigand M, Goering E and Schütz G 2015 Phase evolution in single-crystalline $LiFePO_4$ followed by *in situ* scanning x-ray microscopy of a micrometre-sized battery *Nat. Commun.* **6** 1–7
[22] Nishimura S-i, Kobayashi G, Ohoyama K, Kanno R, Yashima M and Yamada A 2008 Experimental visualization of lithium diffusion in Li_xFePO_4 *Nat. Mater.* **7** 707–11
[23] Fongy C, Gaillot A-C, Jouanneau S, Guyomard D and Lestriez B 2010 Ionic vs electronic power limitations and analysis of the fraction of wired grains in $LiFePO_4$ composite electrodes *J. Electrochem. Soc.* **157** A885
[24] Zaghib K, Mauger A, Goodenough J B, Gendron F and Julien C M 2007 Electronic, optical, and magnetic properties of $LiFePO_4$: small magnetic polaron effects *Chem. Mater.* **19** 3740–7
[25] Purwamargapratala Y, Sujatno A, Sabayu Y L and Kartini E 2019 Synthesis of $Li_4Ti_5O_{12}$ (LTO) by sol-gel method for lithium ion battery anode *IOP Conf. Series: Mater. Sci. Eng.* **553** 012062
[26] Lin S Y T, Tran N T T, Chang S L, Su W P and Lin M F 2018 *Structure- and Adatom-Enriched Essential Properties of Graphene Nanoribbons* (Boca Raton, FL: CRC Press)
[27] Grazianetti C and Molle A 2019 Engineering epitaxial silicene on functional substrates for nanotechnology *Research* **2019** 8494606
[28] Wang X M, Meng W W and Yan Y F 2017 Electronic band structures and excitonic properties of delafossites: a GW-BSE study *J. Appl. Phys.* **122** 085104
[29] Dien V K, Pham H D, Tran N T T, Han N T, Huynh T M D, Nguyen T D H and Fa-Lin M 2020 Orbital-hybridization-created optical excitations in $Li_8Ge_4O_{12}$ arXiv:2009.02160
[30] Green M A 2013 Improved value for the silicon free exciton binding energy *AIP Adv.* **3** 112104
[31] Matthes L, Pulci O and Bechstedt F 2014 Optical properties of two-dimensional honeycomb crystals graphene, silicene, germanene, and tinene from first principles *New J. Phys.* **16** 105007
[32] Matthes L, Pulci O and Bechstedt F 2013 Massive Dirac quasiparticles in the optical absorbance of graphene, silicene, germanene, and tinene *J. Phys.-Condens. Matter* **25** 395305
[33] Spataru C D, Ismail-Beigi S, Capaz R B and Louie S G 2007 *Carbon Nanotubes* (Berlin: Springer) pp 195–227
[34] Tries A, Osella S, Zhang P, Xu F, Ramanan C, Kläui M, Mai Y, Beljonne D and Wang H I 2020 Experimental observation of strong exciton effects in graphene nanoribbons *Nano Lett.* **20** 2993–3002
[35] Lu D-b, Song Y-l and Huang X-y 2019 Electric and optical properties modulations of armchair silicene nanoribbons by transverse electric fields *Curr. Appl Phys.* **19** 31–6
[36] Wei W, Dai Y, Huang B B and Jacob T 2013 Many-body effects in silicene, silicane, germanene and germanane *Phys. Chem. Chem. Phys.* **15** 8789–94
[37] Chen Z F and Wang X Q 2011 Stacking-dependent optical spectra and many-electron effects in bilayer graphene *Phys. Rev.* B **83** 081405
[38] Fan J and Chu P K 2010 Group IV nanoparticles: synthesis, properties, and biological applications *Small* **6** 2080–98

[39] Zhou X, Zhang Q, Gan L, Li H, Xiong J and Zhai T 2016 Booming development of group IV–VI semiconductors: fresh blood of 2D family *Adv. Sci.* **3** 1600177
[40] Nguyen T D H, Pham H D, Lin S-Y and Lin M-F 2020 Featured properties of Li^+-based battery anode: $Li_4Ti_5O_{12}$ *RSC Adv.* **10** 14071–9
[41] Kasti N A 2007 Zigzag carbon nanotubes—molecular/structural mechanics and the finite element method *Int. J. Solids Struct.* **44** 6914–29
[42] Roome N J and Carey J D 2014 Beyond graphene: stable elemental monolayers of silicene and germanene *ACS Appl. Mater. Interfaces* **6** 7743–50
[43] Lin C-L, Arafune R, Kawahara K, Tsukahara N, Minamitani E, Kim Y, Takagi N and Kawai M 2012 Structure of silicene grown on Ag (111) *Appl. Phys. Express* **5** 045802
[44] Lin C-H, Huang A, Pai W W, Chen W-C, Chen T-Y, Chang T-R, Yukawa R, Cheng C-M, Mou C-Y and Matsuda I 2018 Single-layer dual germanene phases on Ag (111) *Phys. Rev. Mater.* **2** 024003
[45] Jiles D 2015 *Introduction to Magnetism and Magnetic Materials* (Boca Raton, FL: CRC Press)
[46] Lin C-Y, Ho C-H, Wu J-Y, Do T-N, Shih P-H, Lin S-Y and Lin M-F 2019 *Diverse Quantization Phenomena in Layered Materials* (Boca Raton, FL: CRC Press)
[47] Chung H-C, Chang C-P, Lin C-Y and Lin M-F 2016 Electronic and optical properties of graphene nanoribbons in external fields *Phys. Chem. Chem. Phys.* **18** 7573–616
[48] Lin M-F and Shung K W-K 1994 Plasmons and optical properties of carbon nanotubes *Phys. Rev.* B **50** 17744
[49] Timerkaev B, Skvortsov S, Feltsinger V and Shakirov B 2019 Electric arc synthesis of germanium nanotubes *J. Physics: Conf. Ser.* **328** 012081
[50] Das D K, Sarkar J and Singh S K 2018 Effect of sample size, temperature and strain velocity on mechanical properties of plumbene by tensile loading along longitudinal direction: a molecular dynamics study *Comput. Mater. Sci.* **151** 196–203
[51] Chen S C, Wu J Y, Lin C Y and Lin M F 2017 *Theory of Magnetoelectric Properties of 2D Systems* (Bristol: IOP Publishing Ltd)
[52] Higham D J 2008 Modeling and simulating chemical reactions *SIAM Rev.* **50** 347–68
[53] Naukkarinen J and Sainio T 2018 Supporting student learning of chemical reaction engineering using a socially scaffolded virtual laboratory concept *Educ. Chem. Eng.* **22** 61–8
[54] Koretsky M and Kelly C 2010 Enhancement of student learning in experimental design using virtual laboratories *2010 Annual Conf. & Exposition* pp 15.506. 1–15 9
[55] Yamanaka T, Abe T, Nishio K and Ogumi Z 2018 Diffusion of Li-deficient phases in large $LiFePO_4$ single crystals during chemical delithiation *J. Mater. Chem.* A **6** 11005–11
[56] Yuan L-X, Wang Z-H, Zhang W-X, Hu X-L, Chen J-T, Huang Y-H and Goodenough J B 2011 Development and challenges of $LiFePO_4$ cathode material for lithium-ion batteries *Energy Environ. Sci.* **4** 269–84
[57] Fiermans L, Clauws P, Lambrecht W, Vandenbroucke L and Vennik J 1980 Single crystal V_2O_5 and lower oxides. A survey of their electronic, optical, structural, and surface properties *Phys. Status Solidi (A)* **59** 485–504
[58] Wang H g, Ma D I, Huang Y and Zhang X b 2012 Electrospun V_2O_5 nanostructures with controllable morphology as high-performance cathode materials for lithium-ion batteries *Chem. Eur. J.* **18** 8987–93

[59] Potok R, Rau I, Shtrikman H, Oreg Y and Goldhaber-Gordon D 2007 Observation of the two-channel Kondo effect *Nature* **446** 167–71
[60] Kittel C 1996 *Introduction to Solid State Physics* (New York: Wiley)
[61] Issi J-P, Heremans J and Dresselhaus M 1983 Electronic and lattice contributions to the thermal conductivity of graphite intercalation compounds *Phys. Rev.* B **27** 1333
[62] Qian X, Gu X, Dresselhaus M S and Yang R 2016 Anisotropic tuning of graphite thermal conductivity by lithium intercalation *J. Phys. Chem. Lett.* **7** 4744–50
[63] Nakayama T and Shima H 1998 Computing the Kubo formula for large systems *Phys. Rev.* E **58** 3984
[64] Tarascon J-M and Armand M 2011 *Materials for Sustainable Energy: A Collection of Peer-Reviewed Research and Review articles from Nature Publishing Group* (Singapore: World Scientific) pp 171–9
[65] Goodenough J B and Park K-S 2013 The Li-ion rechargeable battery: a perspective *J. Am. Chem. Soc.* **135** 1167–76
[66] Goodenough J B and Kim Y 2010 Challenges for rechargeable Li batteries *Chem. Mater.* **22** 587–603
[67] Li H, Wang Z, Chen L and Huang X 2009 Research on advanced materials for Li-ion batteries *Adv. Mater.* **21** 4593–607
[68] Zhang Y, Liu S Q, Ji Y J, Ma J M and Yu H J 2018 Emerging nonaqueous aluminum-ion batteries: challenges, status, and perspectives *Adv. Mater.* **30** 1706310
[69] Zhang K Q, Kirlikovali K O, Suh J M, Choi J W, Jang H W, Varma R S, Farha O K and Shokouhimehr M 2020 Recent advances in rechargeable aluminum-ion batteries and considerations for their future progress *ACS Appl. Energy Mater.* **3** 6019–35
[70] Ambroz F, Macdonald T J and Nann T 2017 Trends in aluminium-based intercalation batteries *Adv. Energy Mater.* **7** 1602093
[71] Saroja A, Samantaray S S and Sundara R 2019 A room temperature multivalent rechargeable iron ion battery with an ether based electrolyte: a new type of post-lithium ion battery *Chem. Commun.* **55** 10416–9
[72] Elbashar Y H 2021 A rare earth doped glass for laser medium application: a review *Nonlinear Optics Quantum Optics-Concepts in Modern Optics* **53** 61–98
[73] Shiota K and Kawai T 2017 Li atom adsorption on graphene with various defects for large-capacity Li ion batteries: first-principles calculations *Jpn. J. Appl. Phys.* **56** 06GE11
[74] Yang L, Chen H-S, Song W-L and Fang D 2018 Effect of defects on diffusion behaviors of lithium-ion battery electrodes: *in situ* optical observation and simulation *ACS Appl. Mater. Interfaces* **10** 43623–30
[75] Lincoln D C, Mary A L and William P R 2000 Dynamics of the Bose-Einstein condensate: quasi-one-dimension and beyond *J. Phys. B: Atom. Mol. Opt. Phys.* **33** 3983
[76] Wächtler F and Santos L 2016 Quantum filaments in dipolar Bose-Einstein condensates *Phys. Rev.* A **93** 061603
[77] Ohba N, Miwa K, Aoki M, Noritake T, Towata S-i, Nakamori Y, Orimo S-i and Züttel A 2006 First-principles study on the stability of intermediate compounds of $LiBH_4$ *Phys. Rev.* B **74** 075110
[78] Yonezawa S, Kim J-H and Takashima M 2015 *Advanced Fluoride-Based Materials for Energy Conversion* (Amsterdam: Elsevier) pp 33–50
[79] Lin S-Y, Lin Y-T, Tran N T T, Su W-P and Lin M-F 2017 Feature-rich electronic properties of aluminum-adsorbed graphenes *Carbon* **120** 209–18

[80] Compton O C and Nguyen S T 2010 Graphene oxide, highly reduced graphene oxide, and graphene: versatile building blocks for carbon-based materials *Small* **6** 711–23
[81] Liu Q-J, Liu Z-T, Feng L-P and Tian H 2010 First-principles study of structural, elastic, electronic and optical properties of rutile GeO_2 and α-quartz GeO_2 *Solid State Sci.* **12** 1748–55
[82] Trukhin A 1997 Self-Trapped Excitons in SiO_2, GeO_2, Li_2GeO_3, $AIPO_4$ and $GaPO_4$ *Materials Science Forum* (Bäch: TransTech Publ.) pp 531–6
[83] Dai Z, Gole J, Stout J and Wang Z 2002 Tin oxide nanowires, nanoribbons, and nanotubes *J. Phys. Chem.* B **106** 1274–9
[84] Venkataraj S, Geurts J, Weis H, Kappertz O, Njoroge W K, Jayavel R and Wuttig M 2001 Structural and optical properties of thin lead oxide films produced by reactive direct current magnetron sputtering *J. Vacuum Sci. Technol. A: Vacuum, Surfaces, Films* **19** 2870–8
[85] Ho Y, Chang C, Shyu F, Chen R, Chen S and Lin M-F 2004 Electronic and optical properties of double-walled armchair carbon nanotubes *Carbon* **42** 3159–67
[86] Shyu F L and Lin M F 2002 Electronic and optical properties of narrow-gap carbon nanotubes *J. Phys. Soc. Jpn.* **71** 1820–3
[87] Omacrsawa E 2012 *Perspectives of Fullerene Nanotechnology* (Cham: Springer)
[88] Liu Y, Lian J, Yang X, Zhao M, Shi Y, Song H and Dai K 2017 First-principles calculation for point defects in $Li_2Ti_2O_4$ *Mater. Res. Express* **4** 106502
[89] Tanaka S, Kitta M, Tamura T, Maeda Y, Akita T and Kohyama M 2014 Atomic and electronic structures of $Li_4Ti_5O_{12}/Li_7Ti_5O_{12}$ (001) interfaces by first-principles calculations *J. Mater. Sci.* **49** 4032–7
[90] Liu Y, Lian J, Sun Z, Zhao M, Shi Y and Song H 2017 The first-principles study for the novel optical properties of $LiTi_2O_4$, $Li_4Ti_5O_{12}$, $Li_2Ti_2O_4$ and $Li_7Ti_5O_{12}$ *Chem. Phys. Lett.* **677** 114–9
[91] Zhou F, Kang K, Maxisch T, Ceder G and Morgan D 2004 The electronic structure and band gap of $LiFePO_4$ and $LiMnPO_4$ *Solid State Commun.* **132** 181–6
[92] Malik R, Zhou F and Ceder G 2011 Kinetics of non-equilibrium lithium incorporation in $LiFePO_4$ *Nat. Mater.* **10** 587–90
[93] Scrosati W A v S B 2002 *Advances in Lithium-Ion Batteries* (Cham: Springer)
[94] Suslick K S 2001 Encyclopedia of physical science and technology II *Sonoluminescence and Sonochemistry* 3rd edn. (New York: Elsevier Science Ltd) pp 1–20
[95] Sevik C and Bulutay C 2007 Theoretical study of the insulating oxides and nitrides: SiO_2, GeO_2, Al_2O_3, Si_3N_4, and Ge_3N_4 *J. Mater. Sci.* **42** 6555–65
[96] Kordyuk A 2014 ARPES experiment in fermiology of quasi-2D metals *Low Temp. Phys.* **40** 286–96
[97] Damascelli A 2004 Probing the electronic structure of complex systems by ARPES *Phys. Scr.* **2004** 61
[98] Bussolotti F, Chi D, Goh K J, Huang Y L and Wee A T 2020 *2D Semiconductor Materials and Devices* (Amsterdam: Elsevier) pp 199–220
[99] Pontes R B, Mançano R R, da Silva R, Cótica L F, Miwa R H and Padilha J E 2018 Electronic and optical properties of hydrogenated group-IV multilayer materials *Phys. Chem. Chem. Phys.* **20** 8112–8
[100] Lin S-Y, Chang S-L, Tran N T T, Yang P-H and Lin M-F 2015 H–Si bonding-induced unusual electronic properties of silicene: a method to identify hydrogen concentration *Phys. Chem. Chem. Phys.* **17** 26443–50

[101] Tran N T T, Nguyen D K, Glukhova O E and Lin M-F 2017 Coverage-dependent essential properties of halogenated graphene: a DFT study *Sci. Rep.* **7** 1–13
[102] Crespo E A, Ruda M, de Debiaggi S R, Bringa E M, Braschi F U and Bertolino G 2012 Hydrogen absorption in Pd nanoparticles of different shapes *Int. J. Hydrogen Energy* **37** 14831–7
[103] Fox M 2001 *Optical Properties of Solids* (Oxford: Oxford University Press)
[104] Sherif S A, Goswami D Y, Stefanakos E L and Steinfeld A 2014 *Handbook of Hydrogen Energy* (Boca Raton, FL: CRC Press)
[105] Huang H-C, Lin S-Y, Wu C-L and Lin M-F 2016 Configuration-and concentration-dependent electronic properties of hydrogenated graphene *Carbon* **103** 84–93
[106] Tozzini V and Pellegrini V 2013 Prospects for hydrogen storage in graphene *Phys. Chem. Chem. Phys.* **15** 80–9
[107] Li W-B, Lin S-Y, Tran N T T, Lin M-F and Lin K-I 2020 Essential geometric and electronic properties in stage-n graphite alkali-metal-intercalation compounds *RSC Adv.* **10** 23573–81
[108] Li Y, Lu Y, Adelhelm P, Titirici M-M and Hu Y-S 2019 Intercalation chemistry of graphite: alkali metal ions and beyond *Chem. Soc. Rev.* **48** 4655–87
[109] Lai J and Hodges J 1999 Effects of processing parameters on physical properties of the silicone maxillofacial prosthetic materials *Dental Mater.* **15** 450–5
[110] Belenkov E and Ali-Pasha V 2011 3D-graphite structure *Crystallogr. Rep.* **56** 101–6
[111] Lin C-Y, Chen R-B, Ho Y-H and Lin M-F 2017 *Electronic and Optical Properties of Graphite-Related Systems* (Boca Raton, FL: CRC Press)
[112] Khatun F, Gafur M, Ali M, Islam M and Sarker M 2014 Impact of lithium composition on structural, electronic and optical properties of lithium cobaltite prepared by solid-state reaction *J. Sci. Res.* **6** 217–31
[113] Higashimoto K, Homma H, Uemura Y, Kawai H, Saibara S and Hironaka K 2011 Automotive lithium-ion batteries *Hitachi Rev.* **60** 17–21
[114] Lee S S, Kim T H, Hu S J, Cai W W and Abell J A 2010 Joining technologies for automotive lithium-ion battery manufacturing: a review *Int. Manufacturing Science and Engineering Conf.* pp 541–9
[115] Barré A, Deguilhem B, Grolleau S, Gérard M, Suard F and Riu D 2013 A review on lithium-ion battery ageing mechanisms and estimations for automotive applications *J. Power Sources* **241** 680–9
[116] Pilania G, Slenes K and Ramprasad R 2013 First principles study of the interface between silicone and undoped/doped $BaTiO_3$ *J. Appl. Phys.* **113** 064316
[117] Mahan G D 2013 *Many-Particle Physics* (Cham: Springer)
[118] Li G, Chen Z and Lu J 2018 Lithium-sulfur batteries for commercial applications *Chem.* **4** 3–7
[119] Kudu Ö U, Famprikis T, Fleutot B, Braida M-D, Le Mercier T, Islam M S and Masquelier C 2018 A review of structural properties and synthesis methods of solid electrolyte materials in the Li_2S– P_2S_5 binary system *J. Power Sources* **407** 31–43
[120] Zhang L, Sun D, Feng J, Cairns E J and Guo J 2017 Revealing the electrochemical charging mechanism of nanosized Li_2S by *in situ* and operando x-ray absorption spectroscopy *Nano Lett.* **17** 5084–91
[121] Ertaş E, Öztürk T and Ösken I 2018 Purification of phosphorus decasulfide P_4S_{10} *Google Patents*

[122] Zhao Z, Liu L, Yu T, Yang G and Bergara A 2017 Pressure-induced stable Li_5P for high-performance lithium-ion batteries *J. Phys. Chem.* C **121** 21199–205

[123] Tahri Y, Chermette H and Hollinger G 1991 Electronic structures of phosphorus oxide P_4O_{10} and phosphorus sulfides P_4S_{10} and P_4S_7 *J. Electron. Spectrosc. Relat. Phenom.* **56** 51–69

[124] Jason M E 1997 Transfer of sulfur from arsenic and antimony sulfides to phosphorus sulfides. Rational syntheses of several less-common P_4S_n species *Inorg. Chem.* **36** 2641–6

[125] Lepley N, Holzwarth N and Du Y A 2013 Structures, Li^+ mobilities, and interfacial properties of solid electrolytes Li_3PS_4 and Li_3PO_4 from first principles *Phys. Rev.* B **88** 104103

[126] Liu Z, Hubble D, Balbuena P B and Mukherjee P P 2015 Adsorption of insoluble polysulfides Li_2S_x ($x = 1, 2$) on Li_2S surfaces *Phys. Chem. Chem. Phys.* **17** 9032–9

[127] Ozturk T, Ertas E and Mert O 2010 A Berzelius reagent, phosphorus decasulfide (P_4S_{10}), in organic syntheses *Chem. Rev.* **110** 3419–78

[128] Yang L, Li Q, Wang Y, Chen Y, Guo X, Wu Z, Chen G, Zhong B, Xiang W and Zhong Y 2020 A review of cathode materials in lithium-sulfur batteries *Ionics* **26** 5299–318

[129] Peng X, Copple A and Wei Q 2014 Edge effects on the electronic properties of phosphorene nanoribbons *J. Appl. Phys.* **116** 144301

[130] Wu Q, Shen L, Yang M, Cai Y, Huang Z and Feng Y P 2015 Electronic and transport properties of phosphorene nanoribbons *Phys. Rev.* B **92** 035436

[131] Guo H, Lu N, Dai J, Wu X and Zeng X C 2014 Phosphorene nanoribbons, phosphorus nanotubes, and van der Waals multilayers *J. Phys. Chem.* C **118** 14051–9

[132] Okamoto Y 2014 Density functional theory calculations of alkali metal (Li, Na, and K) graphite intercalation compounds *J. Phys. Chem.* C **118** 16–9

[133] Lin W-B, Tran N T T and Lin S-Y 2019 Diverse fundamental properties in stage-n graphite alkali-intercalation compounds: anode materials of Li^+-based batteries arXiv:2001.02042

[134] Xu J, Dou Y, Wei Z, Ma J, Deng Y, Li Y, Liu H and Dou S 2017 Recent progress in graphite intercalation compounds for rechargeable metal (Li, Na, K, Al)-ion batteries *Adv. Sci.* **4** 1700146

[135] Kaneko T and Saito R 2017 First-principles study on interlayer state in alkali and alkaline earth metal atoms intercalated bilayer graphene *Surf. Sci.* **665** 1–9

[136] Kim D-M, Jung S C, Ha S, Kim Y, Park Y, Ryu J H, Han Y-K and Lee K T 2018 Cointercalation of Mg^{2+} ions into graphite for magnesium-ion batteries *Chem. Mater.* **30** 3199–203

[137] Konschuh S, Gmitra M and Fabian J 2010 Tight-binding theory of the spin-orbit coupling in graphene *Phys. Rev.* B **82** 245412

[138] Kira M and Koch S W 2006 Many-body correlations and excitonic effects in semiconductor spectroscopy *Prog. Quantum Electron.* **30** 155–296

[139] Giesbers A, Ponomarenko L, Novoselov K, Geim A, Katsnelson M, Maan J and Zeitler U 2009 Gap opening in the zeroth Landau level of graphene *Phys. Rev.* B **80** 201403

[140] Yuan S, Roldán R and Katsnelson M I 2011 Landau level spectrum of ABA-and ABC-stacked trilayer graphene *Phys. Rev.* B **84** 125455

[141] Barlas Y, Yang K and MacDonald A 2012 Quantum Hall effects in graphene-based two-dimensional electron systems *Nanotechnology* **23** 052001

[142] Jiang Z, Zhang Y, Tan Y-W, Stormer H and Kim P 2007 Quantum Hall effect in graphene *Solid State Commun.* **143** 14–9

[143] Shyu F-L and Lin M-F 2000 Plasmons and optical properties of semimetal graphite *J. Phys. Soc. Jpn.* **69** 3781–4

[144] Peng L, Hu Y and Xu L 2016 Theoretical and experimental research on the bulk photovoltaic effect in hybrid organic–inorganic perovskites $CH_3NH_3PbI_2$ X (X = Cl, Br, I) *Sci. Adv. Mater.* **8** 2223–30

[145] Yamada Y, Nakamura T, Endo M, Wakamiya A and Kanemitsu Y 2014 Photocarrier recombination dynamics in perovskite $CH_3NH_3PbI_3$ for solar cell applications *J. Am. Chem. Soc.* **136** 11610–3

[146] Qian J, Xu B and Tian W 2016 A comprehensive theoretical study of halide perovskites ABX_3 *Org. Electron.* **37** 61–73

[147] Li C, Lu X, Ding W, Feng L, Gao Y and Guo Z 2008 Formability of ABX_3 (X = F, Cl, Br, I) halide perovskites *Acta Crystallogr., Sect. B: Struct. Sci.* **64** 702–7

[148] Liu H-F, Zhang S-C, Liu L-J, Zhang Z-W and Wang J-L 2020 Density functional theory analysis of electronic and optical properties of orthorhombic perovskite $CH_3NH_3SnX_3$ (X = Br, I) *Chem. Phys. Lett.* **740** 137062

[149] Rhodes C J 2016 Recent developments in solar energy applications based on perovskites: a current commentary *Sci. Prog.* **99** 335–45

[150] Fu Y, Zhu H, Stoumpos C C, Ding Q, Wang J, Kanatzidis M G, Zhu X and Jin S 2016 Broad wavelength tunable robust lasing from single-crystal nanowires of cesium lead halide perovskites ($CsPbX_3$, X = Cl, Br, I) *ACS Nano* **10** 7963–72

[151] Nedelcu G, Protesescu L, Yakunin S, Bodnarchuk M I, Grotevent M J and Kovalenko M V 2015 Fast anion-exchange in highly luminescent nanocrystals of cesium lead halide perovskites ($CsPbX_3$, X = Cl, Br, I) *Nano Lett.* **15** 5635–40

[152] Kang C, Rao H, Fang Y, Zeng J, Pan Z and Zhong X 2021 Antioxidative stannous oxalate derived lead-free stable $CsSnX_3$ (X = Cl, Br, and I) perovskite nanocrystals *Angew. Chem.* **133** 670–5

[153] Kim S, Eom T, Ha Y-S, Hong K-H and Kim H 2020 Thermodynamics of multicomponent perovskites: a guide to highly efficient and stable solar cell materials *Chem. Mater.* **32** 4265–72

[154] Bhauriyal P, Mahata A and Pathak B 2017 The staging mechanism of $AlCl_4$ intercalation in a graphite electrode for an aluminium-ion battery *Phys. Chem. Chem. Phys.* **19** 7980–9

[155] Wu M, Xu B, Chen L and Ouyang C 2016 Geometry and fast diffusion of $AlCl_4$ cluster intercalated in graphite *Electrochim. Acta* **195** 158–65

[156] Wei-Bang L, Shih-Yang L, Ming-Shuei T, Ming-Fa L and Kuang-I L 2020 Geometric and electronic properties on stage-1 $FeCl_3$-graphite intercalation compounds arXiv:2001.02042

[157] Dien V K, Pham H D, Tran N T T, Han N T, Huynh T M D, Nguyen T D H and Fa-Lin M 2021 Orbital-hybridization-created optical excitations in Li_2GeO_3 *Sci. Rep.* **11** 1–10

[158] Jackson J D 2007 *Classical Electrodynamics* (New York: Wiley)

[159] Boukhvalov D, Katsnelson M and Lichtenstein A 2008 Hydrogen on graphene: electronic structure, total energy, structural distortions and magnetism from first-principles calculations *Phys. Rev.* B **77** 035427

[160] Pumera M and Wong C H A 2013 Graphane and hydrogenated graphene *Chem. Soc. Rev.* **42** 5987–95

[161] Crisostomo C P, Yao L-Z, Huang Z-Q, Hsu C-H, Chuang F-C, Lin H, Albao M A and Bansil A 2015 Robust large gap two-dimensional topological insulators in hydrogenated III–V buckled honeycombs *Nano Lett.* **15** 6568–74

[162] Lozhnikov V E, Mamonov A V, Borzilov V O, Mamonova M V, Prudnikov P V, Sorokin A A and Baksheev G G 2019 Estimating the performance of Ab Initio calculation by VASP on openpower high performance system *CEUR Workshop Proc.* pp 24–9

[163] Lin S-Y, Tran N T T, Chang S-L, Su W-P and Lin M-F 2018 *Structure- and Adatom-Enriched Essential Properties of Graphene Nanoribbons* (Boca Raton, FL: CRC Press)
[164] Kweon K E, Hwang G S, Kim J, Kim S and Kim S 2015 Electron small polarons and their transport in bismuth vanadate: a first principles study *Phys. Chem. Chem. Phys.* **17** 256–60
[165] Alvertis A M, Pandya R, Quarti C, Legrand L, Barisien T, Monserrat B, Musser A J, Rao A, Chin A W and Beljonne D 2020 First principles modeling of exciton-polaritons in polydiacetylene chains *J. Chem. Phys.* **153** 084103

IOP Publishing

Thi Dieu Hien Nguyen, Shih-Yang Lin, Hsien-Ching Chung, Ngoc Thanh Thuy Tran and Ming-Fa Lin

Chapter 19

Battery-related problems

Shih-Yang Lin, Thi Dieu Hien Nguyen, Hsien-Ching Chung, Wei-Bang Li, Ngoc Thanh Thuy Tran, Nguyen Thi Han, Hsin-Yi Liu, Hai Duong Pham and Ming-Fa Lin

Problem 19.1. Use the sp^3 tight-binding to investigate electronic properties of monolayer graphene (figure 19.1; [1, 2]) the perpendicular π and σ bondings. (a) Establish a 4×4 Hermitian Hamiltonian matrix in the presence of different four-orbital site energies, (b) plot band structure along the high-symmetry points, (c) solve valence- and conduction-state wave functions in terms of [A, B]-sublattice tight-binding functions, and (d) sum all the spatial probability distributions for the whole occupied valence states using the product of amplitude square and hydrogen-like-orbital charge densities. (e) And then, the calculated result in (d) is compared with the first-principles evaluation. (f) The similar simulations could be generalized to few-layer graphene systems, e.g., AAA-, ABA-, ABC- and AAB-stacked trilayer graphenes.

Problem 19.2. Similar Hamiltonian matrices could be done for three kinds of graphite crystals with AA, AB, and ABC stackings (figure 19.2: [3–5]). (a) They, respectively, have the dimensions of 4, 8 and 12. The diversified electronic energy spectra, wave functions and total charge densities are clearly presented under the exact Hamiltonian diagonalization [6]. (b) Clearly illustrate the featured phenomena, such as the rich and unique stacking dependences in low-lying energy dispersions (the existence or absence of Dirac-cone structures), critical points, spectrum asymmetry about valence and conduction bands, free valence hole and conduction electron densities at different high-symmetry points, π and σ band widths, and 3D total charge distributions (summing all the occupied states and the

doi:10.1088/978-0-7503-4685-6ch19

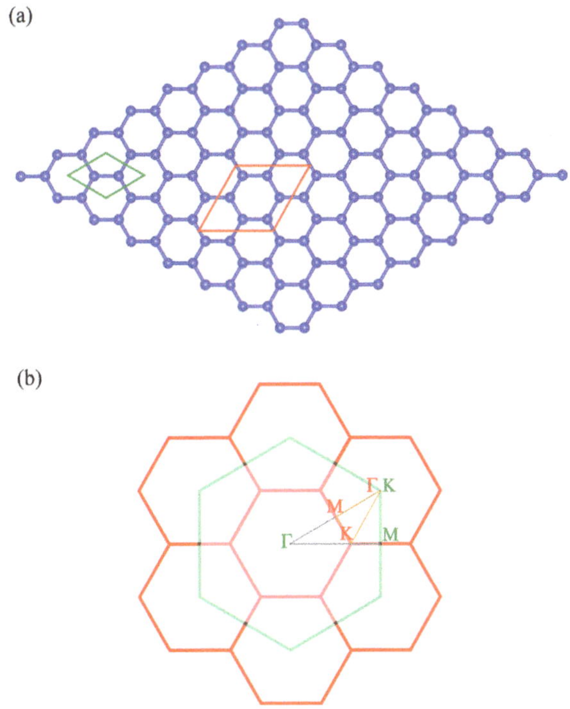

Figure 19.1. (a) A honeycomb lattice and (b) the first Brillouin zone for (a) monolayer graphene.

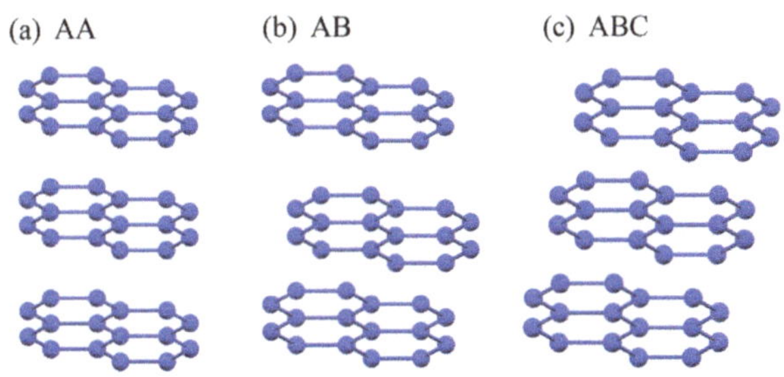

Figure 19.2. Similar plots as figure 19.1, but shown for (a) simple hexagonal, (b) Bernal, and (c) rhombohedral graphites.

orbital-related contributions). Detailed comparisons with the VASP simulations [7] are required for the accurate evaluations.

Problem 19.3. Predict the x-ray diffraction patterns for AA-, AB- and ABC-stacked graphite crystals by the sp^3 tight-binding model. (a) Utilize charge densities due to the position-dependent carbon atoms in a unit cell, calculate their scattering form factor, sum all the contributions, and then obtain the **G**-determined spectral intensities. (b) Investigate whether the theoretical predictions are consistent with the experimental measurements [8]. A similar generalization can be conducted on stage-1 alkali-graphite compounds [9]: LiC_6 and KC_8 with AA and AB stackings, respectively. For example, (c) discuss the *n*-type doping effects about the low-lying energy bands, the spatial charge density distributions, and the x-ray diffraction peaks [10].

Problem 19.4. Utilize the linear Kobo formula to explore the unusual optical properties of the simple hexagonal, Bernal and rhombohedral layered graphite crystals [11, 12]. The vertical optical transitions between the occupied and unoccupied states, with the same wave vector, come from the electric-dipole perturbation of a normal incident electromagnetic wave. Furthermore, they are successfully evaluated within the gradient approximation (the equivalence for an average momentum value and a Hamiltonian matrix-element derivative versus wave vector). (a) Establish the complex dielectric function for three kinds of bulk graphites [3, 4, 13]. Fully explore the main features of (b) the real and imaginary parts of frequency-dependent dielectric functions, (c) reflectance, (d) absorption coefficient, and (e) energy loss function (their definitions in chapter 2 section 2.1.1; [14]). (g) Identify the close relations of the frequency-specified prominent absorption structures and the intralayer/interlayer π- and σ-orbital hybridizations. (h) Summarize the significant effects due to the stacking configurations in terms of chemical bondings. (i) Generalize all the tight-binding model calculations to stage-1 LiC_6 and KC_8, clearly illustrating the *n*-type doping effects on the electron–hole excitations and plasmon modes in the different optical properties [6].

Problem 19.5. Explore the charge and spin features of quasiparticle in a buckled monolayer stanene under the tight-binding model, in which the intrinsic Hamiltonian has the significant sp^3 multi-orbital hybridizations and spin-orbital couplings (geometric and interacting details in [15]; similarly for plumbene/germanene/silicene). (a) By solving a 16×16 Hermitian matrix, thoroughly investigate the orbital and spin dependences for the main features of energy spectra and wave functions. (b) Plot the whole spatial carrier density closely related to the active four orbitals of [5s, $5p_x$, $5p_y$, $5p_z$]. Require a detailed comparison with the VASP simulations (problem 19.8; [7]). (c) Moreover, utilize the featured electronic properties to fully explore the optical absorption structures.

Problem 19.6. Explore the unusual sp^3 chemical bonding on an armchair carbon nanotube with the well-behaved armchair and zigzag structures along the azimuthal and tubular directions, respectively (figure 19.3(a) and (b); [16–18]). Consider the significant misorientation of π bondings and mixing of π and σ orbitals (figure 19.3(c)), and (a) built an 8×8 Hamiltonian under a specific periodical boundary condition. (b) Show band structure and determine a suitable quantum number (angular momentum; [18]) for each valence/conduction energy subband by using the spatial probability distributions. (c) Evaluate the total charge density on a cylindrical surface by summing the whole contributions due to [2s, $2p_x$, $2p_y$, $2p_z$] orbitals of the occupied valence states. (d) Clearly illustrate the radius dependence of multi-orbital hybridizations

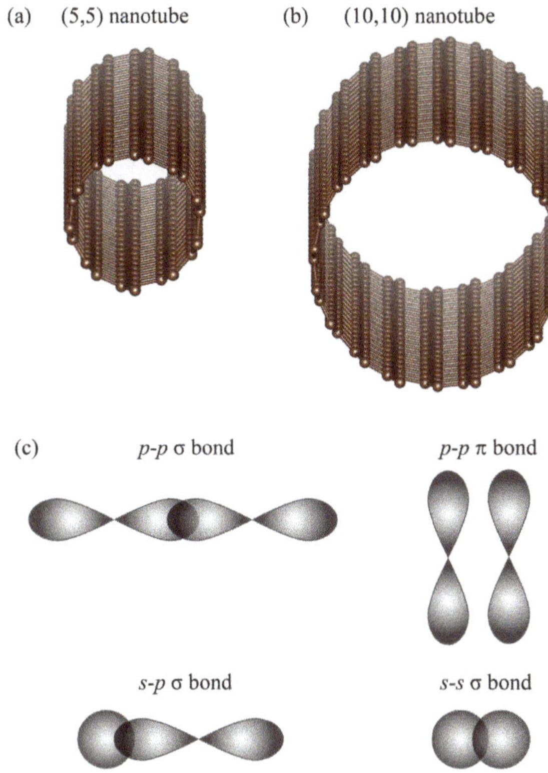

Figure 19.3. Geometric structures of (a) (5, 5) and (b) (10, 10) armchair carbon nanotubes, with (c) four type multi-orbital hybridizations: ppπ, ppσ, spσ and ssσ.

and compare with VAST calculations [19], e.g., the carrier distributions of the (5, 50, (10, 10), (15, 15) and (20, 20) armchair carbon nanotubes. Moreover, the man features of electronic properties are very useful in identifying the prominent absorption peaks, as well as the decomposed orbital hybridizations.

Problem 19.7. Investigate the close relations between lattice symmetries and armchair graphene nanoribbon widths. Each nanoscale graphene nanoribbon has two hydrogen-passivated boundaries [20, 21] and finite-size quantum confinement along the transverse direction (figures 19.4(a) and (b); [22]). The number of dimer lines (N_A) well defines the width of an armchair system. Only utilize the π-bonding tight-binding mode to illustrate the low-energy essential properties. (a) Build a $2N_A$x$2N_A$ Hamiltonian matrix in the presence/absence of bond-length fluctuations [23], in which the optimal geometries are obtained from the VASP calculations, e.g., those of N_A = 10 and N_A = 20 systems. Furthermore, the strengths of hopping integrals are inversely proportional to the square of C–C bond length according to the Harrison rule [24]. (b) Explore the radius dependence for three types of armchair graphene nanoribbons: N_A = 3I, 3I + 1 and 3I + 2 [25]. (c) Evaluate the single-particle optical absorption structures through the gradient approximation,

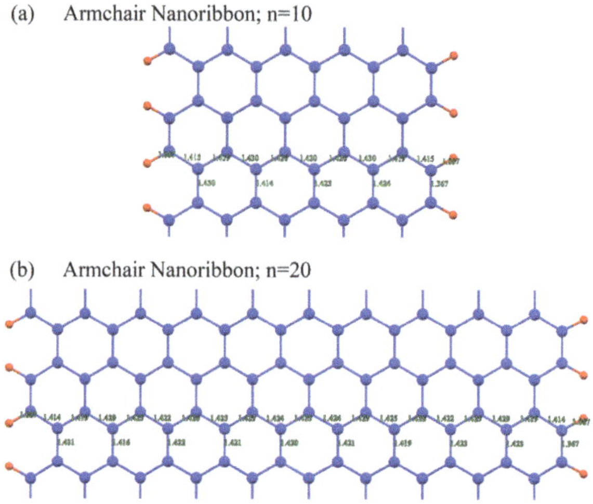

Figure 19.4. Optimal geometric symmetries for (a) $N_A = 10$ and (b) $N_A = 20$ armchair graphene nanoribbons with hydrogen passivation on two boundaries using VASP simulations. They clearly show the significant fluctuations of C–C bond lengths.

determine the width-dependent optical selection rules, and illustrate their physical pictures [20, 21]. (d) Comprehend the effects due to the highly non-uniform chemical bonds by a detailed comparison. Also, comprehend certain important similarities and differences between 1D graphene nanoribbons and carbon nanotubes [26, 27].

Problem 19.8. Use the first-principles method [7] to fully understand the diversified fundamental properties for monolayer graphene/silicene/germanene/tinene/plumbene/antimonene/bismuthene (figures 19.5(a) and (b); [15, 28–33]). (a) Calculate the spatial charge density distributions, the rich electronic energy, and orbital-decomposed van Hove singularities. (b) Account for their main features through the multi-single-orbital hybridizations and the spin–orbital interactions. (c) Evaluate the optical excitation properties, the frequency-dependent dielectric function, reflectance, absorption coefficient and energy loss function, in the presence/absence of excitonic effects. (d) Discuss whether π- and σ-electronic energy subbands, excitation channels, and plasmon modes could be distinguished from the different systems with the various bucklings. (e) Estimate the stable crystal structures of 1D nanotubes and nanoribbons for the various elements, such as certain important similarities and differences. Moreover, external gate voltages, being easily achieved by the experimental equipment, can strongly modulate the featured absorption structures, e.g., the frequency, number, intensity, form, and selection rule of prominent absorption structures.

Problem 19.9. Investigate the diversified essential properties in bilayer composite systems of silicene-, germanene-, tinene- and plumbene-graphene under the similar case by the VASP simulations (figures 19.6(a) and (b); [34–37]). (a) Calculate the

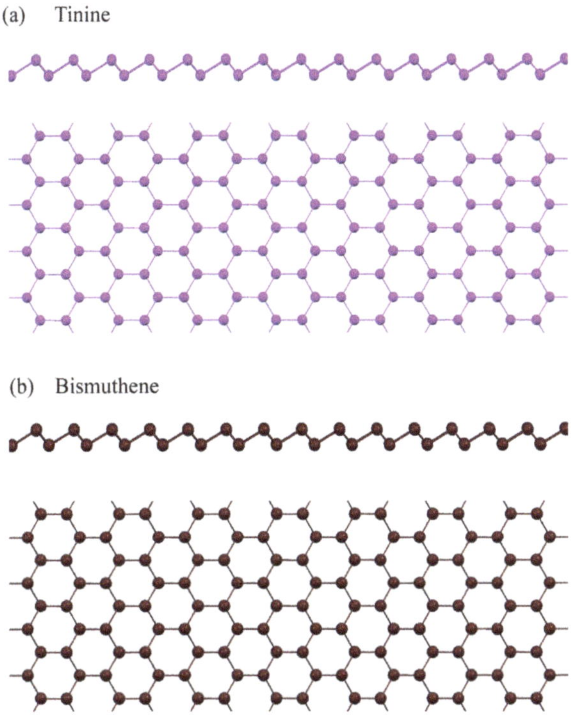

Figure 19.5. Optimal crystal structures of (a) tinene and (b) bismuthene.

optimal geometries with/without buckled structures through both top and side view, band structures along the high-symmetry points in the first Brillouin zone, spatial charge density distributions, and orbital-decomposed van Hove singularities. (b) Emphasize the diverse behaviors among them from the viewpoints about the significant mechanisms: moiré superlattices, weak, but rich van Der Waals interactions, single- or multi-orbital hybridizations, and spin-orbital couplings. (c) Discuss how to detect the theoretical predictions by STM [38–42], ARPES [39, 42–44] and STS measurements ([45, 46]; details in chapter 3). (d) Whether such semi-metallic or narrow-gap systems could exhibit strongly excitonic effects is worth systematic investigation in terms of dielectric function, absorption coefficient, reflectance and energy loss spectrum [47].

Problem 19.10. Identify the rich and unique optical properties of layered graphene systems using the first-principles method, as done by the phenomenological model (details in [48, 49]). In general, the well-stacked honeycomb lattices present AAA [50], ABA [50, 51], ABC [51, 52] and AAB configurations [53]. The main features of vertically electronic excitations are expected to be greatly diversified by the geometric symmetries and external electric/magnetic fields [49]. Very interestingly, the VASP calculations are available in thoroughly understanding the diverse excitation phenomena in the presence/absence of many-body effects. It is well known that all

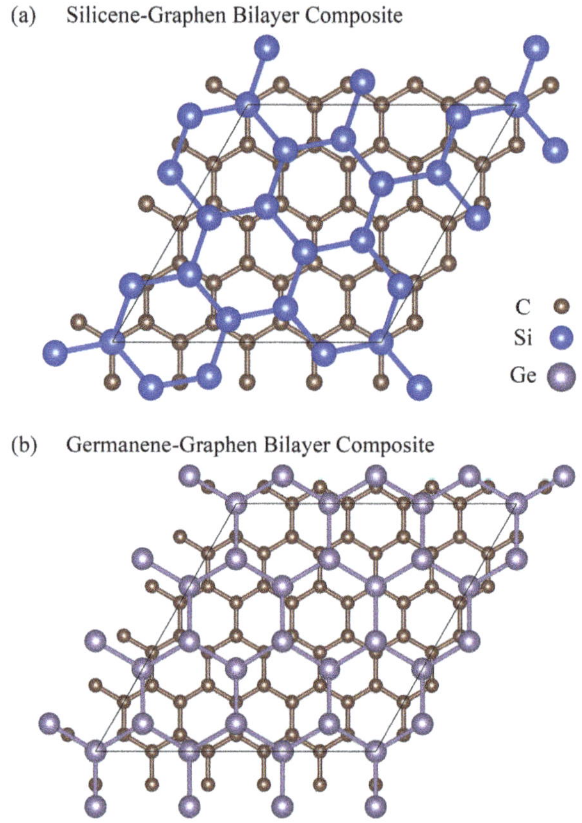

Figure 19.6. Bilayer composites: (a) silicene- and (b) germanene-graphene.

the high-symmetry graphene systems belong to the gapless/semi-metallic materials except under certain cases (e.g., the strong distortion and substrate effects; [54–56]). (a) Clearly illustrate whether the bound exciton states could be formed under the pristine case, or sufficiently high gate voltage. (b) Furthermore, establish the close relations between the featured absorption structures and the spatial charge distributions (unusual orbital hybridizations). (c) Similar works could be generalized to few-layer silicene [57] and germanene systems [58], with rather strong interlayer atomic interactions due to the largely enhanced buckling. Their optical properties would be totally different from one another.

Problem 19.11. Thoroughly explore spin-dominated optical excitations under VASP examinations for the transition- and rare-earth-metal chemical adsorptions of monolayer graphene (figures 19.7(a) and (b); [59, 60]). Both active *d* and *f* orbitals (the outer ones), respectively, possess five and seven kinds of density distribution forms. In addition to isolated atoms, their condensed-matter subsystems are frequently identified to display the spin-dependent magnetic configurations by reflecting in the unusual quasiparticle behaviors, such as, the spin-split valence

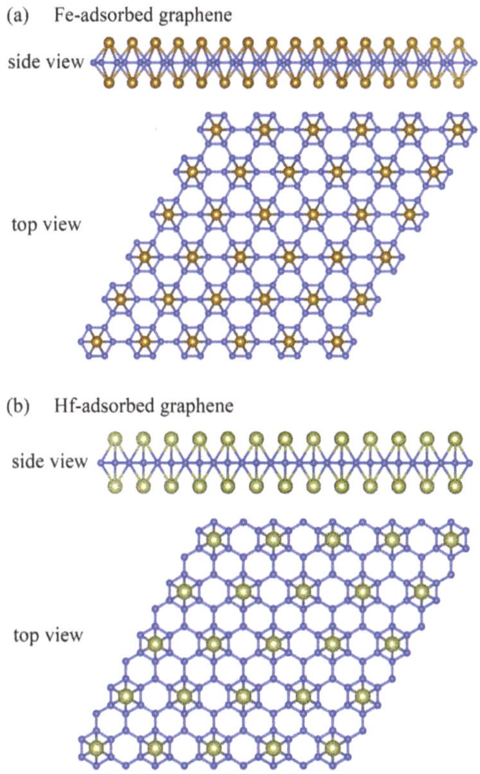

Figure 19.7. Saturated chemisorptions of (a) transition- and (b) rare-earth-metal adatoms on monolayer carbon-honeycomb lattice under the double-side case, such as, Fe and Hf decorations.

and conduction subbands across the Fermi level, the net magnetic moments, the spatial spin density distributions, and the spin-projected van Hove singularities [60]. (a) Only consider the high-concentration adatom chemisorptions, with a small unit cell, to show the optimal lattice symmetries. (b) The featured quasiparticle properties, the geometric, electronic, magnetic and optical properties, are presented and unified together to achieve clear classification of significant orbital hybridizations and atom- and orbital-induced spin configurations. For example, the non-negligible role of spin-up and spin-down states would strongly modify the main features of prominent absorption structures. (c) The strong competition or cooperation of two critical mechanisms need to be discussed in detail. This viewpoint is very suitable in developing a complete theoretical framework with the quasiparticle charges and spins simultaneously [23].

Problem 19.12. Thoroughly explore the rich and unique orbital hybridizations due to the unusual chemical bondings and spin configurations associated with the various heterojunctions [61] [65]. Non-graphene group-IV few-layer systems have successfully grown different substrates by molecular beam epitaxy, such as silicene/Ag(111) (figures 19.8(a) and (b); [28, 62, 63]) and germanene/Al(111) [64]. Such emergent materials are outstanding candidates for verifying the diversified quasi-particle properties, especially for the charge- and spin-enriched fundamental properties. (a) VASP simulations could be used to obtain optimal crystal structures with the specific lattice symmetries on the (x,y)-plane, highly non-uniform chemical bonds within the 2D, boundary and substrate, spin-split/spin-degenerate band structures, atom- and orbital-decomposed magnetic moment/vanishing one, spatial charge density distributions, spin-up and spin down arrangements in a whole crystal, and atom-, orbital- and spin-projected van Hove singularities. (b) Clearly clarify the strongly cooperative/competitive relations among the critical mechanisms of the unique geometries, multi-/single-orbital mixings and magnetic spin configurations. They are predicted to play critical roles in the diverse phenomena in terms of geometric, electronic, and magnetic properties.

Problem 19.13. Comprehend the dominating roles of the anti-ferromagnetic spin configurations in 1D zigzag graphene nanoribbons (details in [23]). Such emergent materials are successfully synthesized in experimental laboratories [20], in which the open boundaries/their density of states are clearly examined by STM/STS measurements [65–67]. The first-principles calculations have predicted the antiferromagnetic (ferromagnetic) spin arrangements cross the ribbon center (the ferromagnetic ones along the open edge structure). The 1D quantum-confinement creates unique spin density distribution, while this magnetic configuration cannot generate the net

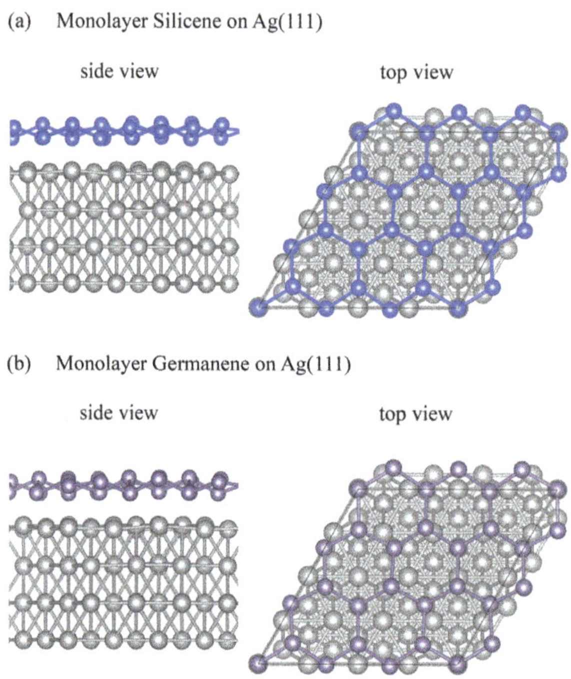

Figure 19.8. The optimal crystal structures of (a) monlayer silicene/(b) germanene on Ag(111)/Cu(111).

magnetic moment and spin-spit band structures and van Hove singularities. That is, direct examinations by STS and ARPES [43, 44] would become mission impossible. However, the unusual anti-ferromagnetism strongly affects the low-lying band gap and energy spacing between two partial flat bands near the specific zone boundary. (a) The vertical optical excitations due to them, being further conducted by VASP simulations, are expected to be greatly enriched by the widths, chiralities and polarization directions of graphene nanoribbons. (b) Furthermore, the critical roles of spin magnetisms could be easily modulated by the alkali-adatom chemisorptions, as done for their geometric and electronic properties (details in [68]). The spin-induced magnetisms are dramatically changed from the chemical modifications; therefore, the various magnetisms of doped graphene nanoribbon compounds are suitable candidates for experimental verifications.

Problem 19.14. The VASP method is available in exploring the diversified quasi-particle properties of 3D LiS_2 (figure 19.9(a); [69, 70]) and P_4S_{10} binary compounds (figure 19.9(b); [71, 72]), in which these two materials are closely related to lithium sulfide batteries [73]. The multi-component compounds, corresponding to the lithium-, phosphorus- and sulfur-based compositions, have been successfully synthesized in experimental laboratories [74]. (a) Theoretical studies could be initiated from the single-element crystals of lithium, phosphorus, and sulfur, e.g., their featured crystal symmetries, band structures, wave functions, charge density distributions, van Hove singularities, as well as optical absorption spectra. (b) Explore important differences between two binary 3D systems from the viewpoint of orbital hybridizations. (c) Evaluate whether the reliable tight-binding model could be established from well-fitting with the VASP calculations. (d) Many allotropes of P_4S_x [$x \leqslant 10$; [75]] show various crystal symmetries. (e) Furthermore, which structure has the lowest ground state energy. (f) Explore how to identify the physical and chemical pictures in determining the most stable and quasi-stable configurations. (g) Apparently, their electronic and optical properties will exhibit the diverse quasi-particle phenomena.

Problem 19.15. Tin/lead and chloride/bromine/iodine binary compounds form the backbone of commercial solar cells [76, 77]. (a) First-principles simulations can fully explore how many allotropes will be revealed in those real synthesized. The most stable and quasi-stable crystal structures are optimized and then identified from the delicate calculations and analyses. (b) Clearly illustrate the main features of geometric symmetries for well defining the various crystals, e.g., the position-dependent chemical bonds, and the observable fluctuations of bond lengths. (c) Present the featured band structures to examine the atom dominances within the specific energy ranges. (d) Check whether there exist well-defined valence subbands with certain chemical bondings, as revealed as π and σ bondings in graphene-related systems [78]. (e) Clearly illustrate the multi-orbital hybridizations/the orbital-related spin configurations through the spatial charge/spin densities before and after the material formation, being fully assisted by the featured band structure and the

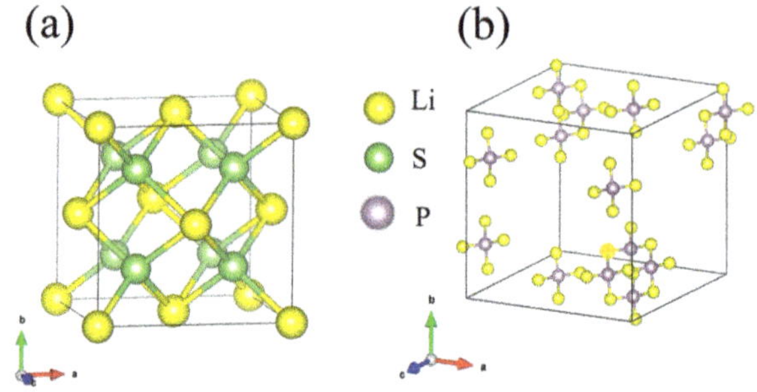

Figure 19.9. The crystal structures of (a) Li_2S and P_4S_{10}.

merged van Hove singularities. (f) Calculate optical properties, the real- and imaginary-part dielectric function, electron energy loss spectrum, reflectance, and refractive index. These quantities provide very useful information about the bare and screened response functions [79]. Most importantly, (g) directly link all the relevant physical quantities under a unified quasiparticle framework, i.e., charge and spin interactions will dominate all the fundamental properties [23, 78].

Problem 19.16. Boron-atom intercalation between graphitic layers can present p-type doing, as revealed in graphene-related chemisorption and substitution. This unusual behavior is identified to be related to the specific orbital hybridizations in B–C bonds, according to the VASP simulations [80, 81]. The π and σ are almost well defined in binary boron–carbon compounds, indicating their orthogonal relations. (a) Plot the primitive unit cells under the cases of chemical modifications, in which boron positions are obtained from the VASP calculations [80, 81]. (b) Establish the reliable parameters in the tight-binding model, e.g., the orbital-dependent on-site Coulomb potentials and hopping integrals (three kinds of σ bondings and one kind of π-orbital hybridization; [80, 81]). (c) And then, evaluate and discuss the main features of band structures, wave functions, and van Hove singularities. (d) Emphasize the sensitive dependences of adatom/guest-atom concentrations and configurations. (e) Finally, the Fermi golden rule, being under the gradient approximation for the electric dipole moment [82], is available in investigating the rich optical absorption spectra, especially for the direct relations between the prominent structures and the specific orbital hybridizations. (f) Examine the agreement of phenomenological models and first-principles simulations.

Problem 19.17. Electronic properties and optical excitations for graphene [83, 84], graphane [85, 86] and graphone [84, 87]. The tight-binding models, respectively, with sp^3, s-sp^3 and s-sp^3 bondings, might be suitable in qualitatively understanding the essential properties. The hydrogen adatoms are situated at the top sites, where the optimal height is determined from the VASP simulations [88]. (a) Build the tight-binding Hamiltonian matrix with the dimension suitable for the number of atoms and active orbitals in a primitive unit cell. This problem is only focused on the double- and single-side saturated chemisorptions. Both electronic energy spectra and wave functions could be obtained from the exact diagonalization. (b) Discuss the important differences among three materials, being fully supported by the unusual chemical bondings. (c) This quasi-particle viewpoint is also responsible for the diversified van Hove singularities. (d) The chemical bonding becomes complex, and so do the optical excitation structures. (e) Similar investigations could be done for the other hydrogenated group-IV systems. Such works are very helpful in understanding element-enriched quasiparticle behaviors.

References

[1] Novoselov K S, Geim A K, Morozov S V, Jiang D, Zhang Y, Dubonos S V, Grigorieva I V and Firsov A A 2004 Electric field effect in atomically thin carbon films *Science* **306** 666–9

[2] Geim A K and Novoselov K S 2010 *Nanoscience and Technology: A Collection of Reviews from Nature Journals* (Singapore: World Scientific) pp 11–9

[3] Bernal J D 1924 The structure of graphite *Proc. R. Soc. Lond.* A **106** 749–73

[4] Lee J-K, Lee S-C, Ahn J-P, Kim S-C, Wilson J I and John P 2008 The growth of AA graphite on (111) diamond *J. Chem. Phys.* **129** 234709

[5] Ho J-H, Lu C, Hwang C, Chang C and Lin M-F 2006 Coulomb excitations in AA-and AB-stacked bilayer graphites *Phys. Rev.* B **74** 085406

[6] Lin C-Y, Chen R-B, Ho Y-H and Lin M-F 2017 *Electronic and Optical Properties of Graphite-Related Systems* (Boca Raton, FL: CRC Press)

[7] Kresse G and Furthmüller J 1996 Efficient iterative schemes for *ab initio* total-energy calculations using a plane-wave basis set *Phys. Rev.* B **54** 11169

[8] Li Z, Lu C, Xia Z, Zhou Y and Luo Z 2007 X-ray diffraction patterns of graphite and turbostratic carbon *Carbon* **45** 1686–95

[9] Aurbach D and Ein-Eli Y 1995 The study of Li-graphite intercalation processes in several electrolyte systems using *in situ* x-ray diffraction *J. Electrochem. Soc.* **142** 1746

[10] Okojie R S, Holzheu T, Huang X and Dudley M 2003 X-ray diffraction measurement of doping induced lattice mismatch in n-type 4H-SiC epilayers grown on p-type substrates *Appl. Phys. Lett.* **83** 1971–3

[11] Norimatsu W and Kusunoki M 2010 Selective formation of ABC-stacked graphene layers on SiC (0001) *Phys. Rev.* B **81** 161410

[12] Bacon R 1960 Growth, structure, and properties of graphite whiskers *J. Appl. Phys.* **31** 283–90

[13] Kopnin N, Ijäs M, Harju A and Heikkilä T 2013 High-temperature surface superconductivity in rhombohedral graphite *Phys. Rev.* B **87** 140503

[14] Hightower A, Ahn C, Fultz B and Rez P 2000 Electron energy-loss spectrometry on lithiated graphite *Appl. Phys. Lett.* **77** 238–40

[15] Zhu F-f, Chen W-j, Xu Y, Gao C-l, Guan D-d, Liu C-h, Qian D, Zhang S-C and Jia J-f 2015 Epitaxial growth of two-dimensional stanene *Nat. Mater.* **14** 1020–5

[16] Zhou Z, Steigerwald M, Hybertsen M, Brus L and Friesner R A 2004 Electronic structure of tubular aromatic molecules derived from the metallic (5, 5) armchair single wall carbon nanotube *J. Am. Chem. Soc.* **126** 3597–607

[17] Liu Y, Dobrinsky A and Yakobson B I 2010 Graphene edge from armchair to zigzag: the origins of nanotube chirality? *Phys. Rev. Lett.* **105** 235502

[18] Ho Y, Ho G, Chen S, Ho J-H and Lin M-F 2007 Low-frequency excitation spectra in double-walled armchair carbon nanotubes *Phys. Rev.* B **76** 115422

[19] Rangel T, Caliste D, Genovese L and Torrent M 2016 A wavelet-based projector augmented-wave (PAW) method: reaching frozen-core all-electron precision with a systematic, adaptive and localized wavelet basis set *Comput. Phys. Commun.* **208** 1–8
[20] Chen Y-C, De Oteyza D G, Pedramrazi Z, Chen C, Fischer F R and Crommie M F 2013 Tuning the band gap of graphene nanoribbons synthesized from molecular precursors *ACS Nano* **7** 6123–8
[21] da Rocha C G, Clayborne P A, Koskinen P and Häkkinen H 2014 Optical and electronic properties of graphene nanoribbons upon adsorption of ligand-protected aluminum clusters *Phys. Chem. Chem. Phys.* **16** 3558–65
[22] Chen Z, Lin Y-M, Rooks M J and Avouris P 2007 Graphene nano-ribbon electronics *Physica* E **40** 228–32
[23] Lin S Y T, Tran N T T, Chang S L, Su W P and Lin M F 2018 *Structure- and Adatom-Enriched Essential Properties of Graphene Nanoribbons* (Boca Raton, FL: CRC Press)
[24] Harrison W A 2012 *Electronic Structure and the Properties of Solids: The Physics of the Chemical Bond* (North Chelmsford, MA: Courier Corporation)
[25] Li X, Wang X, Zhang L, Lee S and Dai H 2008 Chemically derived, ultrasmooth graphene nanoribbon semiconductors *Science* **319** 1229–32
[26] Iijima S 1991 Helical microtubules of graphitic carbon *Nature* **354** 56–8
[27] Sharma R and Iqbal Z 2004 *In situ* observations of carbon nanotube formation using environmental transmission electron microscopy *Appl. Phys. Lett.* **84** 990–2
[28] Fukaya Y, Mochizuki I, Maekawa M, Wada K, Hyodo T, Matsuda I and Kawasuso A 2013 Structure of silicene on a Ag (111) surface studied by reflection high-energy positron diffraction *Phys. Rev.* B **88** 205413
[29] Li L, Lu S z, Pan J, Qin Z, Wang Y q, Wang Y, Cao G y, Du S and Gao H J 2014 Buckled germanene formation on Pt (111) *Adv. Mater.* **26** 4820–4
[30] Lin X, Lu J and Zhu H 2017 The stability and electronic properties of a new allotrope of silicene and silicon nanotube *Superlattices Microstruct.* **101** 480–7
[31] Lei T, Liu C, Zhao J-L, Li J-M, Li Y-P, Wang J-O, Wu R, Qian H-J, Wang H-Q and Ibrahim K 2016 Electronic structure of antimonene grown on Sb_2Te_3 (111) and Bi_2Te_3 substrates *J. Appl. Phys.* **119** 015302
[32] Ares P, Palacios J J, Abellán G, Gómez-Herrero J and Zamora F 2018 Recent progress on antimonene: a new bidimensional material *Adv. Mater.* **30** 1703771
[33] Reis F, Li G, Dudy L, Bauernfeind M, Glass S, Hanke W, Thomale R, Schäfer J and Claessen R 2017 Bismuthene on a SiC substrate: a candidate for a high-temperature quantum spin Hall material *Science* **357** 287–90
[34] Berdiyorov G, Neek-Amal M, Peeters F and Van Duin A C 2014 Stabilized silicene within bilayer graphene: a proposal based on molecular dynamics and density-functional tight-binding calculations *Phys. Rev.* B **89** 024107
[35] Pflugradt P, Matthes L and Bechstedt F 2014 Unexpected symmetry and AA stacking of bilayer silicene on Ag (111) *Phys. Rev.* B **89** 205428
[36] Yaokawa R, Ohsuna T, Morishita T, Hayasaka Y, Spencer M J and Nakano H 2016 Monolayer-to-bilayer transformation of silicenes and their structural analysis *Nat. Commun.* **7** 1–6
[37] Chen X, Meng R, Jiang J, Liang Q, Yang Q, Tan C, Sun X, Zhang S and Ren T 2016 Electronic structure and optical properties of graphene/stanene heterobilayer *Phys. Chem. Chem. Phys.* **18** 16302–9

[38] Liu M 2017 *Controlled Synthesis and Scanning Tunneling Microscopy Study of Graphene and Graphene-Based Heterostructures* (Berlin: Springer)

[39] Crepaldi A, Pons S, Frantzeskakis E, Calleja F, Etzkorn M, Seitsonen A, Kern K, Brune H and Grioni M 2013 Combined ARPES and STM study of Pb/Au (111) Moiré structure: one overlayer, two symmetries *Phys. Rev.* B **87** 115138

[40] Reetz M T, Helbig W, Quaiser S A, Stimming U, Breuer N and Vogel R 1995 Visualization of surfactants on nanostructured palladium clusters by a combination of STM and high-resolution TEM *Science* **267** 367–9

[41] Osváth Z, Vértesy G, Tapasztó L, Wéber F, Horváth Z, Gyulai J and Biró L 2005 Atomically resolved STM images of carbon nanotube defects produced by Ar^+ irradiation *Phys. Rev.* B **72** 045429

[42] Hwang E and Sarma S D 2008 Quasiparticle spectral function in doped graphene: electron-electron interaction effects in ARPES *Phys. Rev.* B **77** 081412

[43] Bussolotti F, Chi D, Goh K J, Huang Y L and Wee A T 2020 *2D Semiconductor Materials and Devices* (Amsterdam: Elsevier) pp 199–220

[44] Damascelli A 2004 Probing the electronic structure of complex systems by ARPES *Phys. Scr.* **2004** 61

[45] Matsui T, Kambara H, Niimi Y, Tagami K, Tsukada M and Fukuyama H 2005 STS observations of Landau levels at graphite surfaces *Phys. Rev. Lett.* **94** 226403

[46] Zhang Y, Brar V W, Wang F, Girit C, Yayon Y, Panlasigui M, Zettl A and Crommie M F 2008 Giant phonon-induced conductance in scanning tunnelling spectroscopy of gate-tunable graphene *Nat. Phys.* **4** 627–30

[47] Lu J, Loh K P, Huang H, Chen W and Wee A T 2009 Plasmon dispersion on epitaxial graphene studied using high-resolution electron energy-loss spectroscopy *Phys. Rev.* B **80** 113410

[48] Lin C-Y, Do T-N, Huang Y-K and Li M-F 2017 *Optical Properties of Graphene in Magnetic and Electric Fields* (Bristol: IOP Publishing)

[49] Lin C-Y, Ho C-H, Wu J-Y, Do T-N, Shih P-H, Lin S-Y and Lin M-F 2019 *Diverse Quantization Phenomena in Layered Materials* (Boca Raton, FL: CRC Press)

[50] Bao C, Yao W, Wang E, Chen C, Avila J, Asensio M C and Zhou S 2017 Stacking-dependent electronic structure of trilayer graphene resolved by nanospot angle-resolved photoemission spectroscopy *Nano Lett.* **17** 1564–8

[51] Lui C H, Li Z, Mak K F, Cappelluti E and Heinz T F 2011 Observation of an electrically tunable band gap in trilayer graphene *Nat. Phys.* **7** 944–7

[52] Cong C, Yu T, Sato K, Shang J, Saito R, Dresselhaus G F and Dresselhaus M S 2011 Raman characterization of ABA-and ABC-stacked trilayer graphene *ACS Nano* **5** 8760–8

[53] Lin C-Y, Huang B-L, Ho C-H, Gumbs G and Lin M-F 2018 Geometry-diversified Coulomb excitations in trilayer AAB stacking graphene *Phys. Rev.* B **98** 195442

[54] Ishigami M, Chen J, Cullen W, Fuhrer M and Williams E 2007 Atomic structure of graphene on SiO_2 *Nano Lett.* **7** 1643–8

[55] Zhou S Y, Gweon G-H, Fedorov A, First P, De Heer W, Lee D-H, Guinea F, Neto A C and Lanzara A 2007 Substrate-induced bandgap opening in epitaxial graphene *Nat. Mater.* **6** 770–5

[56] Fratini S and Guinea F 2008 Substrate-limited electron dynamics in graphene *Phys. Rev.* B **77** 195415

[57] Lin S-Y, Liu H-Y, Nguyen D K, Tran N T T, Pham H D, Chang S-L, Lin C-Y and Lin M-F 2020 *Silicene-Based Layered Materials: Essential Properties* (Bristol: IOP Publishing)

[58] Dávila M E and Le Lay G 2016 Few layer epitaxial germanene: a novel two-dimensional Dirac material *Sci. Rep.* **6** 1–9

[59] Sevinçli H, Topsakal M, Durgun E and Ciraci S 2008 Electronic and magnetic properties of 3d transition-metal atom adsorbed graphene and graphene nanoribbons *Phys. Rev.* B **77** 195434

[60] Pi K, McCreary K, Bao W, Han W, Chiang Y, Li Y, Tsai S-W, Lau C and Kawakami R 2009 Electronic doping and scattering by transition metals on graphene *Phys. Rev.* B **80** 075406

[61] Young A F and Kim P 2009 Quantum interference and Klein tunnelling in graphene heterojunctions *Nat. Phys.* **5** 222–6

[62] Arafune R, Lin C-L, Kawahara K, Tsukahara N, Minamitani E, Kim Y, Takagi N and Kawai M 2013 Structural transition of silicene on Ag (111) *Surf. Sci.* **608** 297–300

[63] Du Y, Zhuang J, Wang J, Li Z, Liu H, Zhao J, Xu X, Feng H, Chen L and Wu K 2016 Quasi-freestanding epitaxial silicene on Ag (111) by oxygen intercalation *Sci. Adv.* **2** e1600067

[64] Derivaz M, Dentel D, Stephan R, Hanf M-C, Mehdaoui A, Sonnet P and Pirri C 2015 Continuous germanene layer on Al (111) *Nano Lett.* **15** 2510–6

[65] Söde H, Talirz L, Gröning O, Pignedoli C A, Berger R, Feng X, Müllen K, Fasel R and Ruffieux P 2015 Electronic band dispersion of graphene nanoribbons via Fourier-transformed scanning tunneling spectroscopy *Phys. Rev.* B **91** 045429

[66] Tapasztó L, Dobrik G, Lambin P and Biro L P 2008 Tailoring the atomic structure of graphene nanoribbons by scanning tunnelling microscope lithography *Nat. Nanotechnol.* **3** 397–401

[67] Ridene M, Girard J, Travers L, David C and Ouerghi A 2012 STM/STS investigation of edge structure in epitaxial graphene *Surf. Sci.* **606** 1289–92

[68] Kahn L and Ying S 1975 Alkali-metal chemisorption *Solid State Commun.* **16** 799–801

[69] Luo G, Zhao J and Wang B 2012 First-principles study of transition metal doped Li_2S as cathode materials in lithium batteries *J. Renew. Sustain. Energy* **4** 063128

[70] Eithiraj R, Jaiganesh G, Kalpana G and Rajagopalan M 2007 First-principles study of electronic structure and ground-state properties of alkali-metal sulfides–Li_2S, Na_2S, K_2S and Rb_2S *Physica Status Solidi (B)* **244** 1337–46

[71] Ertaş E, Öztürk T and Ösken I 2018 Purification of phosphorus decasulfide P_4S_{10} *Google Patents*

[72] Tahri Y, Chermette H and Hollinger G 1991 Electronic structures of phosphorus oxide P_4O_{10} and phosphorus sulfides P_4S_{10} and P_4S_7 *J. Electron. Spectrosc. Relat. Phenom.* **56** 51–69

[73] Kang W, Deng N, Ju J, Li Q, Wu D, Ma X, Li L, Naebe M and Cheng B 2016 A review of recent developments in rechargeable lithium–sulfur batteries *Nanoscale* **8** 16541–88

[74] Kudu Ö U, Famprikis T, Fleutot B, Braida M-D, Le Mercier T, Islam M S and Masquelier C 2018 A review of structural properties and synthesis methods of solid electrolyte materials in the Li_2S– P_2S_5 binary system *J. Power Sources* **407** 31–43

[75] Nizamov I y S, Al'metkina L A, Garifzyanova G G, Sergeenko G n G and Batyeva E S 1995 reactions of phosphorus sulfides (P_4S_5, P_4S_7 and P_4S_{10}) and 2, 4-bis (alkylthio)-2, 4-dithioxo-1, 3, 2λ5, 4λ5-dithiadiphosphetanes with dialkyldisulfides and thioacetals in the presence of iodine *Phosphorus, Sulfur, Silicon Relat. Elem.* **102** 71–81

[76] Prasanna R, Leijtens T, Dunfield S P, Raiford J A, Wolf E J, Swifter S A, Werner J, Eperon G E, de Paula C and Palmstrom A F 2019 Design of low bandgap tin–lead halide perovskite solar cells to achieve thermal, atmospheric and operational stability *Nat. Energy* **4** 939–47

[77] Kumari N, Patel S R and Gohel J V 2018 Enhanced stability and efficiency of Sn containing perovskite solar cell with $SnCl_2$ and SnI_2 precursors *J. Mater. Sci., Mater. Electron.* **29** 18144–50

[78] Tran N T T, Lin S-Y, Lin C-Y and Lin M-F 2017 *Geometric and Electronic Properties of Graphene-Related Systems: Chemical Bonding Schemes* (Boca Raton, FL: CRC Press)

[79] Fox M 2001 *Optical Properties of Solids* (Oxford: Oxford University Press)

[80] Xu A, Shi L, Zeng L and Zhao T 2019 First-principle investigations of nitrogen-, boron-, phosphorus-doped graphite electrodes for vanadium redox flow batteries *Electrochim. Acta* **300** 389–95

[81] Suarez-Martinez I, El-Barbary A, Savini G and Heggie M 2007 First-principles simulations of boron diffusion in graphite *Phys. Rev. Lett.* **98** 015501

[82] Jackson J D 1999 *Classical Electrodynamics* (College Park, MD: American Association of Physics Teachers)

[83] Lin C-Y, Wu J-Y, Chiu C-W and Lin M-F 2019 *Coulomb Excitations and Decays in Graphene-related Systems* (Boca Raton, FL: CRC Press)

[84] Tian X-Q, Kiani M, Wei Y-D, Feng N, Gong Z-R, Wang X-R, Du Y and Yakobson B I 2021 Electronic and magnetic diversity of graphone/graphene superlattices *Chem. Mater.* **33** 2090–8

[85] Cudazzo P, Tokatly I V and Rubio A 2011 Dielectric screening in two-dimensional insulators: implications for excitonic and impurity states in graphane *Phys. Rev.* B **84** 085406

[86] Şahin H, Ataca C and Ciraci S 2009 Magnetization of graphane by dehydrogenation *Appl. Phys. Lett.* **95** 222510

[87] Şaşıoğlu E, Hadipour H, Friedrich C, Blügel S and Mertig I 2017 Strength of effective Coulomb interactions and origin of ferromagnetism in hydrogenated graphene *Phys. Rev.* B **95** 060408

[88] Mapasha R E, Ukpong A M and Chetty N 2012 *Ab initio* studies of hydrogen adatoms on bilayer graphene *Phys. Rev.* B **85** 205402

www.ingramcontent.com/pod-product-compliance
Ingram Content Group UK Ltd.
Pitfield, Milton Keynes, MK11 3LW, UK
UKHW051908100926
8556IPUK00005B/803